Staffing Organizations
Contemporary Practice and Theory

Staffing Organizations
Contemporary Practice and Theory

Third Edition

Robert E. Ployhart
University of South Carolina

Benjamin Schneider
*Valtera
and University of Maryland*

Neal Schmitt
Michigan State University

CRC Press
Taylor & Francis Group
Boca Raton London New York

CRC Press is an imprint of the
Taylor & Francis Group, an **informa** business

Senior Acquisitions Editor: Anne Duffy
Editorial Assistant: Rebecca Larsen
Cover Design: Tomai Maridou
Full-Service Compositor: TechBooks

This book was typeset in 10.5/13 pt. Palatino Roman, Bold, and Italic.
The heads were typeset in Palatino, Palatino Bold, and Palatino Bold Italic.

Copyright © 2006 by Lawrence Erlbaum Associates, Inc.
All rights reserved. No part of this book may be reproduced in
any form, by photostat, microform, retrieval system, or any
other means, without prior written permission of the publisher.

Lawrence Erlbaum Associates, Inc., Publishers
10 Industrial Avenue
Mahwah, New Jersey 07430
www.erlbaum.com

Library of Congress Cataloging-in-Publication Data
Ployhart, Robert E., 1970-
 Staffing organizations : contemporary practice and theory /
Robert E. Ployhart, Benjamin Schneider, Neal Schmitt.– 3rd ed.
   p. cm.
 Previous edition: Glenview, Ill.: Scott, Foresman, 1986; lists Benjamin
Schneider as first author.
 Includes bibliographical references and index.
 ISBN 0-8058-5579-3 (casebound : alk. paper)—ISBN 0-8058-5580-7
(pbk.   )
 1. Employees—Recruiting. I. Schneider, Benjamin, 1938- II. Schmitt,
Neal. III. Title.
 HF5549.5.R44.P56 2006
 658.3'11—dc22
                                                       2005023426

To Lynn and Matt
For Brenda, still
To Kara

Contents

Series Foreword	xi
Preface	xv

Chapter 1 Introduction   1
    Aims of the Chapter   1
    Staffing Organizations Defined   1
    A Focus on the Changing Nature of Work   2
    A Focus on Individual Differences   9
    A Focus on the Work and Work Environment   16
    A Focus on How Staffing Contributes to
        Organizational Effectiveness   23
    A Focus on the Influence of
        Federal Employment Legislation and Court Cases   25
    Summary   28

Chapter 2 Measurement Concepts and Data
        Analytic Tools   33
    Background   33
    Aims of the Chapter   35
    Nature of Measurement   37
    Manifest and Latent Variables   42
    Summarizing Measurements of Individual Differences   43
    Covariation   49
    The Pearson Product-Moment Correlation
        Coefficient (r)   54
    Predictions Based on Correlation   61

Predictions Using Multiple Predictors	66
Reliability	71
Validity	77
Moderated Regression	78
Structural Equation Modeling	81
Hierarchical Linear Modeling	87
Item Response Theory	90
Summary	97

Chapter 3 Job Analysis — **100**

Aims of the Chapter	100
Job Analysis is the Foundation of Staffing	104
The Larger Environment	106
Historical Perspectives on Job Analysis	113
Modern Perspectives on Jobs and Job Specifications	119
Current Challenges for Job Analysis	131
Summary	147

Chapter 4 Performance and Criterion Development — **151**

Aims of the Chapter	151
The Importance of Organizational Context and Goals	153
Criteria Defined	156
Models of the Job Performance Domain	160
Desirable Aspects of Criteria	165
Objective Criterion Measures	170
Judgmental Criterion Measures	178
A Comparison of Objective and Judgmental Criteria	194
Criterion Measurement Considerations	195
Current Challenges for Performance and Criterion Development	205
Summary	224

Chapter 5 Recruitment: Retention and Attraction — **237**

Aims of the Chapter	237
Internal Recruitment	239
External Recruitment	261
Current Challenges for Internal and External Recruitment	288
Summary	293

Chapter 6 Validation Strategies and Utility — **300**

Aims of the Chapter	300
Approaches to Validation	302

CONTENTS

Validity Based on Test Content	313
Validity Based on the Theoretical Meaningfulness of Tests	319
Validity Based on the Consequences of Testing	339
Test Utility	344
Summary	359

Chapter 7   Hiring Procedures: An Overview   **365**

Aims of the Chapter	365
Predictors: A General Definition	366
A Model of the Predictor-Response Process	367
The Structure and Function of the Predictor Domain	370
Evaluating the Adequacy of Predictors	379
Use of Predictors in Staffing Practice	385
Current Issues in Predictor Development	389
Summary	397

Chapter 8   Hiring Procedures I: Cognitive Ability and Certification Exams   **402**

Aims of the Chapter	402
Paper-and-Pencil Tests of General Ability	402
Racial Subgroup Differences and Cognitive Ability	416
Licensure and Certification Exams	426
Summary	429

Chapter 9   Hiring Procedures II: Personality Tests, Affective Measures, Interests, and Biodata   **434**

Aims of the Chapter	434
Personality and Personality Tests	436
Interests and Interest Testing	469
Biographical Data	476
Summary	484

Chapter 10   Hiring Procedures III: Interviews, Performance-Based Tests, Simulations, and Physical Ability   **491**

Aims of the Chapter	491
Interviews	492
Performance-Based and Simulation Testing	510
Physical Ability Testing	542
Summary	550

Chapter 11 The Practice of Staffing: Legal, Professional,
and Ethical Concerns **559**
Aims of the Chapter 559
A Tripartite View of Staffing Practice 560
I. Legal Issues 562
II. Professional Standards 576
III. Ethical Considerations 581
Current Issues in Staffing Practice 582
Summary 590

Chapter 12 Staffing Organizations: Review
and Implications **592**
Aims of the Chapter 592
Effective Staffing: Emphasizing Some Key Points 595
Recommendations for Practical Staffing Problems 596
Speculation on the Future Innovations in Staffing 610
Implications for the Job Seeker 614
Summary 619

Author Index **623**

Subject Index **637**

Series Foreword

Series Editors

Jeanette N. Cleveland
The Pennsylvania State University

Edwin A. Fleishman
George Mason University

There is a compelling need for innovative approaches to the solution of many pressing problems involving human relationships in today's society. Such approaches are more likely to be successful when they are based on sound research and applications. This Series in Applied Psychology offers publications that emphasize state-of-the-art research and its application to important issues of human behavior in a variety of social settings. The objective is to bridge both academic and applied interests.

We are pleased to welcome the third edition of the classic *Staffing Organizations* into our Series. Since the first edition in 1976 by Schneider and the second edition in 1986 by Schneider and Schmitt, both scholars and practitioners in applied work psychology have waited for a revision and update of this classic book. We know of few serious students who do not have earlier editions of *Staffing Organizations* on their bookshelves. The current edition by Ployhart, Schneider, and Schmitt will be required reading for undergraduate and graduate students, academics, and practitioners in organizations. The authors are recognized scholars and experts in the field of staffing. Two of the authors (Schneider & Schmitt) have received the Distinguished Scientific Contributions awards from the Society for Industrial and Organizational Psychology (SIOP) and were past presidents of SIOP, and Fellows of Division 14 of the American Psychological Association and American Psychological Society. Ployhart has received multiple dissertation awards. Furthermore, each author has applied experience in small- and large-scale projects sponsored by U.S. and local governments, by enforcement agencies, military branches, and many sectors of private industry. The book presents a meticulously balanced content coverage including major conceptual or theoretical approaches to staffing, critical empirical findings, and examples of real life organizational staffing challenges, based on their combined and extensive consulting experiences.

Between the first and second editions, *Staffing Organizations* grew from 8 to 11 chapters. From the second to third edition, the number of chapters has remained deceptively constant (from 11 to 12 chapters). However, an incredible number of important developments have emerged that affect staffing effectiveness. A sample of these issues include the Civil Rights Act of 1991, the Americans with Disabilities Act, recent globalization/cross-cultural issues, advances in technology, new models of personality, competency modeling, test score banding, research on contextual performance, groups and teams, and large scale projects such as Project A and O*NET. The new edition is up to date with these and other developments and provides a cutting edge view into the future.

Other innovations include discussions of a number of "foundation issues" such as the history of staffing, comparison of traditional and current measurement methods, as well as a modern discussion of job analysis. The first three chapters are followed by a comprehensive discussion of job performance, internal and external recruiting mechanisms, meta-analysis, and utility analysis of staffing issues. Because there has been an explosion in new types of predictors, four chapters are devoted to them, including cognitive ability testing, personality testing, biographical predictors and finally, on performance based testing. The book concludes with two chapters that review key U.S. laws that affect staffing, and discusses the tradeoffs that practitioners frequently make among legal issues, ethical issues, and professional guidelines and the authors conclude with their predictions of emerging staffing innovations.

In summary, the book's discussion of modern staffing has several unique characteristics. First, there is a focus on the recent challenges provided by the technological and global changes, and legal developments in the United States. The authors also recognize how staffing contributes to various aspects of organizational performance. Finally, the authors identify a number of ways that small business can incorporate and benefit from formal staffing procedures. They also provide recommendations for practice, and for needed future research development.

The primary audiences for the book include students and their instructors, as well as consultants and practicing managers. They can rely on this book as an authoritative resource on the most up-to-date practices based on sound research concerning staffing within organizations.

Preface

In this third edition of *Staffing Organizations: Contemporary Practice and Theory*, we consider how organizations of all sizes can use effective staffing procedures as a source of competitive advantage. Our approach is both practical and academic. It is practical because we show how to develop and administer effective staffing procedures, including conducting a job analysis, defining and measuring performance, identifying predictors of performance that are legally defensible, and using this information to make sound hiring decisions. All three of the authors are active practitioners, and recommendations based on this experience are woven into the fabric of this book. Yet, this edition is also academic because we are personally also grounded in a scientific, theoretical perspective that informs the practice of staffing. We review a wealth of cutting-edge theory and research in diverse areas of importance to the practice of staffing, and we identify several areas where more research is necessary.

The bottom line is that if organizations correctly do the things we discuss in this book, their employees will be more productive, produce higher-quality work, and provide better service, resulting in greater satisfaction, less absenteeism, and lower turnover. In times when every source of competitive advantage is important, this can be the difference between organizational success and failure.

CHANGES SINCE THE SECOND EDITION

Since the last edition of this book (published in 1986), there has been an incredible number of important developments. When we started the revision, we came up with a noncomprehensive list of some of the major ones:

- The Five Factor Model of personality
- Situational judgment tests
- Organizational citizenship/ contextual performance
- Counterproductive work behaviors
- Explosion in teams/groups
- Attraction-selection-attrition model
- Fit
- Multilevel issues
- Globalization/cross-cultural issues
- New revisions of the *Standards* and *Principles*
- Civil Rights Act of 1991
- Americans with Disabilities Act of 1990

- Competency modeling
- Internet and growth of personal computers
- Macro/strategic human resources (HR) management
- Applicant reactions
- Project A
- Campbell's performance model
- Shifts toward knowledge/ service work
- Increasing recognition of subgroup differences and strategies for reducing adverse impact
- The unitarian validity model
- Test score banding
- O*NET
- Hierarchical Linear Modeling

This is no small list of developments! As another way to illustrate how much has changed since 1986, consider this—there was no electronic copy of the second edition when we started the revision. We had to scan in the older book for the small amount of materials we kept from the prior edition, and then convert the scanned images to text.

UNIQUE FEATURES OF THE THIRD EDITION

This new edition is up to date with respect to these and other developments, and offers a cutting-edge review with a look toward the future.

PREFACE xvii

In addition, we have implemented several new and unique features into the third edition:

- *Focus on modern challenges.* These include technological advancements, legal developments, and global issues. In each chapter, we illustrate how staffing has adapted, or is trying to adapt, to these challenges.
- *Adopting a multilevel perspective.* Although the focus of this book is primarily at the individual level, we make greater reference to the growing research on macro staffing and HR strategy to show how staffing practices contribute to organizational performance. Likewise, as research on groups and teams has exploded, we show how staffing may be implemented in such situations, as well as the unique features of staffing for teams or interpersonal/service positions.
- *Suggestions for small business.* We identify ways small businesses can benefit from formal staffing procedures, and show how this might be done.
- *Providing straightforward recommendations.* In each chapter, we summarize the research by providing straightforward recommendations for practice. These recommendations decipher several streams of research to provide a more accessible set of suggestions for practitioners.
- *An emphasis on future research directions.* We also describe current theoretical challenges that require further research. As such, each chapter is written to provide many research directions for students and staffing scholars.
- *An emphasis on personal experience.* We have been actively involved in innumerable projects of both small and large scales. We have worked with all levels of government (including federal, state, county, and city); enforcement agencies such as the FBI, police, and state troopers; civil service agencies (e.g., firefighters); military units in multiple countries; and all variations of private industry, from small businesses to large multinational corporations. This practice and the research that emerged has resulted in numerous awards for us. Schneider, for example, has been awarded the Society for Industrial and Organizational Psychology (SIOP) year 2000 Distinguished Scientific Contributions Award, he has won the Scholarly Contributions Award for the Best Paper from both the

Organizational Behavior and Human Resources Management Divisions of the Academy of Management, and he is a Fellow of SIOP, the Academy of Management, the American Psychological Association, and the American Psychological Society. Schmitt has been awarded the SIOP Distinguished Scientific Contributions Award in 1999, SIOP's Distinguished Service Contributions Award in 1998, the Heneman Career Achievement Award in 2000 from the Human Resources Division of the Academy Management, SIOP's Applied Research Award in 2004, and is a Fellow of SIOP, the American Psychological Association, and the American Psychological Society. Ployhart has won awards for his research on situational judgment testing, customer service, and multilevel staffing (S. Rain Wallace Dissertation Award from SIOP, Best Dissertation Award and Best Student Paper Award from the Human Resources Division of the Academy of Management). In each chapter, we pass along our observations and lessons learned from many staffing projects and research efforts through the use of "side stories" illustrating applications of important concepts.

OVERVIEW OF CONTENT

Chapter 1 provides a historical overview of staffing and emphasizes five points of focus for the book: the changing nature of work, individual differences, work and the work environment, how staffing contributes to organizational effectiveness, and the influence of federal employment legislation and case law.

Chapter 2 is an overview of traditional and current statistical and measurement methods. The classic concepts of correlation, multiple regression, and reliability are presented, and the current edition also includes discussion of structural equation modeling, item response theory, and even newer approaches such as hierarchical linear modeling.

Chapter 3 is an entirely revamped review of job analysis methodologies. In particular, we emphasize the use of O*NET and competency models, and review a number of current challenges facing job analysis.

A lot has changed since 1986 in our theories and models of job performance, and consequently, chapter 4 has been dramatically revised. This chapter discusses models of job performance, distinctions between objective and judgmental criteria, criterion measurement considerations (e.g., typical/maximum, longitudinal), and current challenges.

PREFACE   xix

Chapter 5 provides an up-to-date, comprehensive review of internal and external recruiting mechanisms, as well as new material on applicant reactions and the role of the Internet/technology in recruitment processes.

Chapter 6 covers validity, meta-analysis, and utility, which comprise perhaps some of the most important topics in staffing. We have substantially updated this chapter to reflect current professional and legal thinking on validity and the demonstration of validity evidence.

Chapter 7 is an entirely new chapter that provides an overview of hiring procedures. Since 1986, there has been an explosion of different predictors. It is also apparent that we need to pay attention to more than just the validity of predictors; we must consider demographic differences, applicant reactions, and utility. Thus, chapter 7 tries to bring clarity to the predictor space by providing an organizing framework for comparing and contrasting possible predictors—always with a focus on usability and validity.

Chapter 8 examines the important, and frequently controversial, topic of cognitive ability testing. Given our recognition of the facts that cognitive ability tests are among the best predictors of job performance available but demonstrate important racial subgroup differences, we now devote half the chapter to this potential dilemma. In particular, we draw on our experience and research to consider ways of reducing subgroup differences in cognitive ability tests, while still retaining the strength they reveal as important predictors of job performance.

Chapter 9 tackles the difficult task of describing a huge volume of research on personality testing, affect, interest measures, and biographical information. It is on these topics that research has expanded dramatically during the 1990s, and we summarize this information and consider several issues not frequently discussed with the use of personality tests in practice.

Likewise, chapter 10 considers the many developments in performance-based testing, including interviews, simulations, assessment centers, work samples, and physical abilities tests. There have been numerous changes in these areas, and we provide a current review of research and best practices in the use of these performance-based predictors.

Chapter 11 is a new chapter that serves two purposes. First, it reviews the key U.S. federal and case law that affects staffing practice, including the new decisions examining the University of Michigan's use of diversity considerations in student admission decisions. Second, it presents a tripartite model of the staffing practitioner, recognizing the frequent trade-offs that require consideration between the legal system, professional guidelines, and ethical issues.

Finally, chapter 12 reviews and reinforces key concepts raised throughout the book. It also provides a series of practical recommendations for handling specific staffing problems, such as those facing small businesses or organizations having difficulty with recruiting. We also provide some predictions concerning where we believe the next major innovations in staffing research will occur, and conclude with recommendations for the job seeker who will encounter the procedures we describe.

Readers familiar with the first and second editions may be surprised at how the book has grown. This largely reflects the expansion of knowledge and techniques over the past 18 years. And since the number of authors has been tripled since the first edition...

INTENDED AUDIENCE

This book is intended for students and practitioners. It is hoped that student readers of this book will come away with the impression that we have powerful models for identifying and selecting talent for work organizations, and that they will learn some ways of making this happen. At a minimum, they will be aware of some of the opportunities to make good personnel staffing decisions and some of the pitfalls of trying the seductive "quick fix." We present many real world examples in the text so the novice to the topic will be able to see how some of these procedures look and work. The examples, combined with a comprehensive treatment of topics, and what we hope is a straightforward writing style, make the material available to senior undergraduate students, as well as graduate students in business administration, psychology, human resources, labor and industrial relations, and MBA programs.

Our hope is that practitioners in personnel, human resources management, labor relations, and so forth will also use the book. Although the procedures and options we describe may seem idealized, our examples should make them real enough to use. The extensive coverage we give to a wide variety of issues in staffing, from recruitment to cost–benefit analysis and from job analysis to interviewing, should help create new ideas for effective staffing strategies and tactics. Perhaps of most interest will be chapters 7 through 10, which are devoted to the various kinds of assessment techniques that now exist for gathering data on job applicants. Chapter 11 may also be of interest because it discusses the challenges of being a staffing practitioner and balancing off seemingly competing demands associated with legal issues, professional standards, and ethics. However, each chapter contains material that is as up to date as we could find so the book will be a valuable desk reference and a useful source of various staffing procedures.

SPECIAL THANKS

Writing this revision took place during many changes in our own lives. Ployhart and Schneider both moved, Schmitt took over the role of department chair (we still think he must have had a "senior moment" when he said yes to that one), Schneider "retired" (if writing a book can be called retirement), Ployhart had a baby son (reading the chapters aloud helped him fall asleep... and the baby, too), and we each took on more projects than we should have. However, through it all, we were and continue to be friends. Of course, we did not do this book by ourselves. Our wives Lynn, Brenda, and Kara all allowed us the chance to write. We could not have done it without them.

On a more technical level, a number of our colleagues helped us by carefully reviewing the new edition. Useful critiques and suggestions were obtained from Timothy A. Judge, University of Florida, Ronald S. Landis, Tulane University, and Manuel London, SUNY Stonybrook. We also appreciate the assistance and guidance of series editors Jeanette N. Cleveland, The Pennsylvania State University, and Edwin A. Fleishman, George Mason University.

Finally, the real workers... numerous graduate research assistants assisted in this revision. Deirdre Lozzi had the thankless task of scanning in and formatting every page of the second edition. Justin Lebiecki, Mike Camburn, and Vesela Vassileva had the joy of doing countless literature reviews, obtaining obscure articles and books, and proofreading draft chapters. From Michigan State University, Marcy Shafer did an incredible job of checking references and providing careful edits throughout the book. From Erlbaum, Anne Duffy and Rebecca Larsen have been absolutely wonderful to work with (we are not sure they would say the same about us). As you can see, this has been a team effort and we sincerely thank all the people we named for their help.

—Rob, Ben, & Neal

Ben, Rob, and Neal at the annual conference for the Society for Industrial and Organizational Psychology, April 2005.

Staffing Organizations
Contemporary Practice and Theory

1

Introduction

AIMS OF THE CHAPTER

In this first chapter of *Staffing Organizations: Contemporary Practice and Theory*, we set a historical and conceptual foundation against which you can view contemporary issues and practices in staffing organizations. To provide this background information, we first define "staffing," and then provide beginning thoughts on the relationship between "staffing" and "organizations." This relationship is an intimate one because it reveals our simultaneous interest in the people who enter and live in work organizations and the organizations in which they live and work. In the chapter, we provide a brief history of, and orientation to, the major foci of this book: (a) the changing nature of work, (b) the role of individual differences in staffing, (c) the role of work and work context, (d) how staffing contributes to organizational effectiveness, and (e) how federal employment legislation and court cases influence staffing. However, let us first define what we mean by the term "staffing organization."

STAFFING ORGANIZATIONS DEFINED

Staffing may be defined as the processes involved in finding, assessing, placing, and evaluating individuals at work. These processes operate through recruiting, selecting, appraising, and promoting individuals. *Organizations* add breadth and scope to these processes and indicate how important the characteristics of the job and the organization are to the entire staffing process. The nature of the organization not only influences the staffing process itself, but also the kinds of people who will be recruited, selected, and promoted—and these, in turn, have a large effect on what the organization is. This is so because the attributes and behavior of people in an organization have a lot to do with what we think of an organization. So, *staffing organizations* in our minds represents a reciprocal process, whereby the staffing process is influenced by the organization in which that process exists and the organization is influenced by the outcomes of the staffing process. The best staffing programs take great care to specify (a) the kinds of people who will be effective and satisfied with the job and the organization into which they are hired, and (b) the kinds of people required by the organization today and in the future to promote long-term organizational effectiveness. In brief, then, the goals of staffing an organization are to improve organizational functioning and effectiveness by attracting, selecting, and retaining people who will facilitate the accomplishment of organizational goals and meet their own individual goals.

A FOCUS ON THE CHANGING NATURE OF WORK

Figure 1.1 presents a working framework as a convenient way to think about the many factors that influence an individual's behavior at work and, in the aggregate, influence organizational effectiveness. This framework is also helpful in illustrating the many changes taking place in the modern workforce; changes that are summarized in Table 1.1 and have a fundamental impact on the science and practice of staffing.

Individual Attributes

Individual attributes refer to the characteristics people bring with them to work, such as knowledge, skills, abilities, personality, values, interests, and experiences. They also refer to the knowledge, skills, and abilities people learn after entering the organization, through both formal and informal training and interactions with coworkers and the system at large. These individual attributes are psychological in nature; therefore, research from other disciplines of psychology, such as cognitive, differential, personality, and social, is helpful in identifying and understanding these attributes. This is also the reason why organizational staffing is most commonly studied by applied psychologists.

A FOCUS ON THE CHANGING NATURE OF WORK

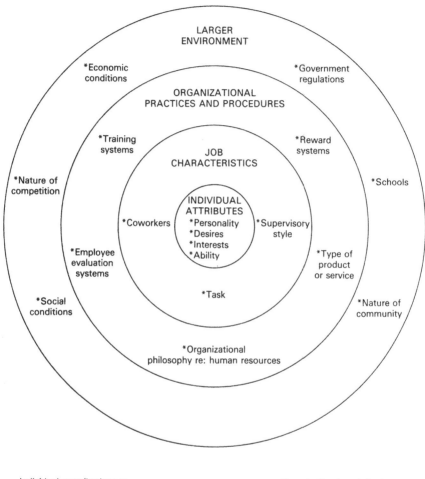

Individual contributions to attainment of organizational goals; satisfactoriness.

Organizational contributions to attainment of individual goals; satisfaction.

FIG. 1.1. Some variables of impact on employee satisfactoriness and satisfaction in the work setting.

Most staffing systems attempt to identify the individual characteristics that will increase the probability that people hired will be able to do the work they are hired for and will enjoy the work at least long enough to repay the organization's investment in hiring and training them. A lot of evidence presented later indicates that (a) people who bring job-relevant skills and abilities with them are likely to perform better on the job, and (b) when their personality fits the rewards the job

1. INTRODUCTION

TABLE 1.1

Modern Challenges Affecting Staffing

Challenge
I. Individual Attributes 1. Aging of the workforce 2. Increasing demographic diversity 3. Large educational and skills shortages 4. Group differences in ability, skills, and experience
II. Job Characteristics 5. Technology and the Internet 6. Shift from manufacturing to service-based and knowledge-based work 7. Increased movement toward team-based work
III. Organizational Practices and Procedures 8. Increase in contract, part-time, or project-based work 9. Outsourcing and shifting work to other countries with cheaper labor costs 10. Flatter organizations push autonomy to lower levels 11. Increasing mergers and acquisitions
IV. The Larger Environment 12. Increased immigration 13. Globalization and global competition 14. Ever-changing and increasing employment laws and regulations

has to offer and the way the larger organization functions, they will be more satisfied, more committed, and will stay longer. We call it *satisfactoriness* when people meet the performance standards set for them at work. When people's own personality and interests are fulfilled, then we refer to *satisfaction* (Lofquist & Dawis, 1969).

There are several difficult staffing challenges facing modern employers. Among these are the increasing age of the U.S. workforce as Baby Boomers (those born after World War II) grow older and begin to retire. This wave of retirement is already starting to create challenges such as labor shortages and not enough skilled labor to replace the retiring employees. Although there was a large amount of immigration during the 1990s, much of this labor lacks basic skills and training (e.g., proficiency with English). This means the number of qualified employees needed to replace retiring workers and/or fill newly created positions is not comparable to the number of such vacancies.

Job Characteristics

Jobs include the tasks at which people work, the people with whom they work, and the people for whom they work. Jobs not only include the formal tasks people do that are assigned to them by their boss, but also the informal tasks they perform as part of a work group or team. These informal tasks are called organizational citizenship behaviors (OCBs; Organ, 1988), and they turn out to be important for organizational effectiveness. For example, repairing a computer may be the technical part of a computer technician's job, but being service oriented, friendly, and willing to work over one's lunch break to fix computers quickly are examples of OCBs. Interestingly, OCBs are more probable when people feel fairly treated by the organization and their boss and when they experience job satisfaction (Organ & McFall, 2004).

The characteristics of the formal tasks at which people work have been changing a lot recently, especially due to increased use of information technology. Information technology is itself changing rapidly, so the speed of the change in tasks at work appears to be escalating. However, this is true mostly for older workers who did not grow up in a world of information technology and for whom the mouse and the iMac are recent invaders from the world of science fiction. All generations go through rapid changes in the tasks they perform at work, with each generation believing the changes it experiences are the most dramatic ever. Still, it is clear that in today's work, continually changing tasks and responsibilities will require more adaptive skills from the workforce and a willingness to learn these new tasks.

Interestingly, at the same time jobs require greater technical adaptability, they also require greater social and cultural adaptability. This has occurred for a variety of reasons, including an increased use of teamwork instead of individual work, the continuing globalization of industry and use of multinational collaboration, and the new economy's emphasis on service and knowledge production. Therefore, today's workforce must also learn how to work with individuals from cultures and backgrounds quite different from their own. This means many job tasks are not only technical, but also social in nature, with interpersonal skills and personality being important determinants for effective performance on such tasks.

Changes in the tasks at work, however, do not change the way we think about the necessity to have people in those jobs with the skills and knowledge required to perform them, the personality and motivation necessary to exert effort in pursuit of job and organizational goals, or the importance of the nature of the relationships people have with others at work, including their bosses.

1. INTRODUCTION

Organizational Practices and Procedures

The goals of organizations (e.g., to be the low-cost producer versus the innovator), the industries in which organizations function (e.g., service versus manufacturing), and the nature of the reward and benefit plans of organizations (e.g., incentive system and health benefits versus straight salary and no benefits) play important roles in the people attracted to, selected by, and retained by organizations (Schneider, 1987). In short, why the organization exists and the world in which it exists have important implications for the kinds of people whom you are likely to find employed there. For example, different kinds of people work at your local YMCA than work at your local stock brokerage dealership. Furthermore, at the local YWCA, the accountants will be different than the physical trainers. In brief, not only does the nature of the job have an effect on the people who staff them, but the nature of the organization in which the job exists also has an effect on the kinds of people employed there (Schaubroeck, Ganster, & Jones, 1998).

The primary concern with organizational practices and procedures in this book is in understanding the importance of organizational goals on the kinds of people hired by an organization. We argue that the best staffing systems are those designed around a careful specification of strategic organizational goals. The logic goes this way: If an organization is able to be specific about its strategic goals, knows where it wants to go, and has defined a set of jobs linked to the accomplishment of its goals, it is likely that the goals will be accomplished. The chore for the staffing specialist becomes one of defining in clear terms what the goals are that specific people must meet, and then finding people able and willing to meet those goals—either immediately or with appropriate training. Further, as jobs change the people hired to meet organizational goals, as well as immediate job goals, can be shifted to new jobs as they emerge. As we show, many organizations have done a poor job of specifying their goals, and this makes finding people who are able and willing to meet them difficult. That is, if the standards against which people will be evaluated once they are on the job are vague, it is going to be difficult to predict who will be able to do the job well.

The relationship between workers and the organizations in which they work has also been undergoing some changes more recently, and these changes have implications for staffing as well. Perhaps the most important example concerns the fact that some companies hire contract workers, as well as permanent employees, with contract workers

A FOCUS ON THE CHANGING NATURE OF WORK 7

actually being paid by a third party and brought into the organization on a temporary basis. Contract workers can range from a chief financial officer to a janitor and everything in between. When the formal contract between workers and the organization in which they work changes, there are associated with those formal changes a host of psychological changes—changes in what we call the *psychological contract* (Rousseau, 1995). In short, people behave in ways that for them fulfill their psychological contract, and these differences in the relationships between workers and the organizations in which they work can have important implications for staffing processes.

The Larger Environment

The issues listed here, such as the nature of competition and federal regulations, have played an increasingly important role in the way organizations staff themselves. Thus, an organization requiring highly skilled people in an environment filled with competitors for those same people will behave differently from one that does not face these circumstances. As noted earlier, an important challenge facing modern organizations is a severe shortage of qualified labor in many sectors of the economy. This has led to severe organizational competition for those who are qualified, not to mention other strategies to make up for labor shortages (including technological advancements, mergers and acquisitions, outsourcing, and moving jobs overseas). However, perhaps the most obvious external force on current staffing procedures in the United States has been U.S. government legislation concerning equal employment opportunity (EEO). Both here in chapter 1 and throughout the book, this influence on staffing is presented in considerable detail (see especially chapter 11).

One feature of Fig. 1.1 that is not obvious is the reciprocal nature of these four critical elements: larger environment; organizational practices; job characteristics; and individual attributes. By reciprocal, we mean the four features interact with each other and influence each other. So, for example, the kinds of people who are attracted to, selected by, and remain in a particular job will influence the kinds of pay and incentive systems of the organization and even the nature of the larger environment in which the organization functions. That is, different kinds of people seem to start up different kinds of organizations, and those organizations end up competing in different kinds of environments—people who start halfway houses are not the same as those who start banks (Schein, 1992; Schneider, 1987). We believe it is clear that organizations are differentially effective to the extent that the four elements in Fig. 1.1 fit or match, or are aligned with, each other. In other words, the kinds of people in an organization should fit their jobs, the technology should fit the industry, and so forth (Thompson, 1967). It becomes clear in the remainder of the book that "fit" and "match" are central constructs in staffing organizations.

Although the four elements shown in Fig. 1.1 are likely to be reciprocal, we can draw from the theory and research on organizational levels of analysis to more precisely define how they are interrelated. Levels of analysis research is concerned with understanding how processes at one level (e.g., individual) influence and are influenced by processes at another level (e.g., group, organization). Figure 1.1 shows four such levels, and there are three major implications of levels of analysis research for linking the elements in Fig. 1.1 together. First, the principle of *bond strength* suggests that adjacent levels will be more strongly interconnected than levels farther apart (Simon, 1973). This means the individual attributes and job characteristic levels are more closely intertwined than the individual attribute and larger environment levels. Second, different levels operate on different time scales, such that higher levels (i.e., the larger environment, organizational practices and procedures) generally produce effects more slowly than lower levels on individuals. Third, there is an asymmetry between the reciprocal relationships across the levels, such that higher levels have stronger effects (and operate more quickly) on lower levels than lower levels have on higher levels (Kozlowski & Klein, 2000). That is, it is usually (but not always; Goodman, 2000) easier for an organization to influence the behavior of its employees than it is for a single employee to influence the organization (an exception would be when an individual holds a position with organizational-level consequences, such as a chief executive officer, or when the behavior of one person, such as the rogue investment broker at Baring's Bank, causes financial ruin). We raise these concepts now because such multilevel issues are considered throughout this book.

Figure 1.1 helps provide a context around which staffing is researched and practiced, but we should pinpoint the emphasis of this book: We are concerned with fitting the attributes of people to jobs in legal ways that permit the organization to be effective in accomplishing its goals and for the people themselves to achieve their goals. To make this happen, we must carefully consider the role that staffing plays in job and organizational effectiveness (this chapter); have considerable information about how to assess and measure people, job, and organizational attributes (chapter 2) and the jobs to be staffed (chapter 3); the standards against which the effectiveness of the staffing process will be evaluated (chapter 4); how to find sufficient staff both externally and internally (chapter 5); the various ways of evaluating or choosing staffing procedures (chapter 6); and the variety of procedures available to help make valid staffing decisions (e.g., ability and personality tests, interviews; chapters 7–10). In addition, considerations for these chapters must take into account the legal and political issues surrounding staffing (chapter 11), and by way of summary, we present a series of recommendations of best practices that can serve as standards against which the world of practice functions (chapter 12).

Prior to exploring these issues, we want to give you a flavor of the historical roots of the procedures and conclusions we present. This is not a long history; it encompasses only about 100 years in terms of concepts and methods similar to those used today. As we discover, however, some of the concepts on which those methods rest are quite a bit older.

A FOCUS ON INDIVIDUAL DIFFERENCES

The fact that people differ from each other in important ways has been known and recognized at least as long as the existence of the written word. It is clear, furthermore, that early peoples not only recognized these differences, but also made use of them—not everyone performed the same functions in the hunt, not everyone was a chief, and the medicine man was often selected after some competition.

Aristotle suggested the use of measures of physical prowess for the selection of soldiers. Plato, speaking of soldier selection for the Republic, not only described the physical characteristics required, but also the kind of "personality" necessary—obviously implying that not everyone possessed the personality required:

> Then it will be our duty to select, if we can, natures which are fitted for the task of guarding the city? *It will.*
> And the selection will be no easy matter, I said; but we must be brave and do our best. *We must.*
> Whereas, I said, they [those selected] ought to be dangerous to their enemies, and gentle to their friends; if not they will destroy themselves without waiting for their enemies to destroy them. *True, he said.*

> What is to be done then? I said; how shall we find a gentle nature which has also a great spirit, for the one is the contradiction of the others? (Plato, 1952, pp. 319–320)

The beauty of this quote is how relevant it is to issues confronted today. For example, in Afghanistan and Iraq, how can soldiers who serve there be both vicious killers and peacekeepers? Is this not the contradiction Plato mentions? Such careful consideration of the potentially competing requirements in jobs is something staffing specialists must identify and then deal with because many jobs have such seemingly competing requirements.

Reading further in Plato, we find a discussion of differences in personal attributes among women:

> One woman has a gift of healing, another not; one is a musician, and another has no music in her nature? *Very true.*
> And one woman has a turn for gymnastic and military exercises, and another is unwarlike and hates gymnastics? (Plato, 1952, p. 359)

Those same kinds of differences are the subject of this book. How do we identify people with knowledge, skill, ability, and the personality to perform well at a set of tasks we call a job? Even more difficult, how do we do this before we have ever seen the person perform on the job?

Given the early and continuing interest in identifying how people differ from each other, it is surprising that the refinement of psychological methods of measuring the range and kind of individual differences did not take place in the Western world until the late 1800s. We specify psychological here because the first recorded scientific study of individual differences was accomplished by early astronomers, who depended on human observation of the heavens for their calculations. They found that observers did not agree with each other on issues such as the speed with which heavenly bodies moved across the sky. Astronomers thus had to "calibrate" different observers; to do this they had to "measure," in some degree, how people differed from one another (see Boring, 1950). We specify Western world here because beginning around 700 in China, formal written examination processes were instituted as ways to enhance the quality and variety of people entering the civil service. The fact that these procedures enhanced the potential to counter the influence of the landed gentry was also useful: "Although civil service examinations that went beyond the less formalized selection process used in the Han empire had been instituted in the sixth century by the Sui dynasty (581–618) and reinstituted under the Tang by Emperor Kao-tsu (r. 618–26) and Tai-tsung (r. 627–650) in the seventh century, it was not until Empress Wu (r. 690–705) that rulers in China discovered that officials selected by open examination served as a useful countervailing force to the power of entrenched aristocrats in capital politics" (Elman, 1991, p. 9).

Darwin

In the West, the belief that individuals differ from each other in important ways that determine or are related to differences in work behavior underlies the science of staffing. This philosophy can be traced historically to the work of Charles Darwin (Jenkins & Paterson, 1961), whose theory of natural selection showed that those organisms best suited to an environment are the most likely to survive and prosper (a notion termed *functional utility*). Darwin's work provides a rationale for the scientific study of staffing, and subsequent research in statistics and testing provided the procedures for relating the measured capabilities and personality of individuals to their performances in organizations (e.g., schools, business and industry, the military). In fact, Darwin's theory proposed that survival and thriving was attributable to the fact that organisms that displayed behaviors and evolved attributes that were maximally functional for the situation were those most likely to survive and thrive. This emphasis on the functional usefulness of organism attributes and behavior is what has dominated the thoughts of people interested in individual differences relevant for work organizations. However, the application of the study of individual differences did not occur overnight.

In England itself, Sir Francis Galton, following on Darwin's theory, began looking into the heritability of competencies and ways to statistically document relationships between the attributes of parents and the attributes of their children. In France, Binet and Simon creatively revealed how the achievements of French schoolchildren could be predicted by knowing how well they performed on what they called "little experiments"—what we call test items today. In the United States, a young psychologist, James McKeen Cattell, returned from Germany after studying in Wilhelm Wundt's laboratory (some say Wundt's laboratory was the first psychology laboratory) and applied what he had learned about individual differences to the development of what he called "mental tests." This occurred over a period of about 25 years in the late 1800s and early 1900s, but we are getting ahead of ourselves.

The Correlation Coefficient

For the science of staffing, a most important breakthrough in procedures was due to the Englishman, Sir Francis Galton. Galton sought a way to determine whether his belief that the mental capabilities of parents were co-related with the mental capabilities of their children was supportable with data. He collected considerable evidence to document his hypothesis but was having trouble summarizing his conclusions in a single numerical index showing the relationship between parents' and children's intelligence. A colleague, Karl Pearson, solved Galton's problem in 1896, by developing what is now called the Pearson product-moment correlation coefficient—correlation for short.

The statistical concepts underlying correlation are presented in some detail in chapter 2. Here, it is important to stress the usefulness of the correlation because in one number we can summarize how well two kinds of data collected on a group of people are related. For example, if one knew that 50 people differed in their arithmetic ability as measured by a paper-and-pencil test, one might wonder whether these assessed differences in ability were related to differences in their performance as bank tellers. To answer this question, their arithmetic ability test scores could be statistically correlated with an assessment of their success as a bank teller.

The correlation coefficient is the statistical procedure used to indicate the degree of relationship between any two measured variables when those data are collected on a number of people (or groups or organizations). With this statistical tool, then, we are not only in a position to be able to relate individual differences to each other (e.g., differences in arithmetic ability to differences in performance as a bank teller), but also to relate group differences to each other (e.g., differences in the level of group interaction to turnover rates in the groups) and organizational differences to each other (e.g., differences in the degree to which organizations focus rewards on service quality to differences in the customer satisfaction in those organizations). The point cannot be overemphasized that this correlational procedure, this ability to establish statistical relationships, is at the very foundation of the science of staffing organizations and of all research where a statistical index of the relationship between two variables is useful.

Mental Tests

Sir Francis Galton was not overly interested in practical matters; his concern was with documenting the extent and nature of individual differences and their interrelationships. In contrast, Alfred Binet and Theophile Simon, two French experimental psychologists, approached the problem of identifying children of subnormal intelligence—children who would not benefit from the typical French school system. Thus, although Binet and Simon were not the first authors of psychological tests, they were the first to demonstrate the practical usefulness of testing procedures—that is, they were the first to validate their mental test. By validate, we mean that they showed that assessments made of children's ability to respond correctly to mental problems at one point in time were predictive of how well they actually behaved in school at a later point in time. They wrote in 1905 about validity:

> The use of tests is today very common, and there are even contemporary authors who have made a specialty of organizing new tests according to theoretical views, but who have made no efforts to patiently try them out in schools. Theirs is an amusing occupation, comparable to a person's making a colonizing expedition into Algeria, advancing always upon the map, without taking off his dressing gown. We place but slight confidence in the tests invented by these authors and we have borrowed nothing from them. All the tests which we propose have been repeatedly tried, and have been retained from among many which after trial have been discarded. We can certify that those which are here presented have proved themselves valuable. (Binet & Simon, 1948, pp. 415–416)

The process of "trying" tests and continuing to use only those that prove valuable is known as *validating* a test. It is the process of validation that distinguishes the science of staffing from armchair guesswork in the prediction of behavior. This process of validation and the development of mental tests are two of the most important technological contributions of psychology to staffing.

The Binet-Simon scales, as noted previously, were not the first tests, although they became (as the Stanford-Binet Intelligence Test) the most frequently used, individually administered measure of childhood general scholastic aptitude. In England, Galton had developed a series of tests of physiological capabilities (hearing, hand strength, and so forth), and in 1890 in the United States, Cattell (1948) proposed that a series of 10 tests could be used "in discovering the constancy of mental processes, their interdependence, and their variation under different circumstances" (p. 347). Cattell also suggested that these tests would be "perhaps, useful in regard to training, mode of life or indication of disease."

The 10 tests Cattell (1948) proposed were as follows:

1. Dynamometer pressure
2. Rate of movement
3. Sensation areas
4. Pressure causing pain
5. Least noticeable difference in weight
6. Reaction time for sound
7. Time for naming colors
8. Bisection of a 50-cm line
9. Judgment of 10 seconds time
10. Number of letters remembered on once hearing (p. 347)

Despite the fact that Cattell coined the term "mental tests," it is clear from this list that in 1890, testing was not as we know it today. Compare Cattell's mental tests with the Binet-Simon indicators of intelligence. A 7-year-old should

1. Indicate omissions in a drawing
2. Give the number of fingers
3. Copy a written sentence
4. Copy a triangle and a diamond
5. Repeat five figures
6. Describe a picture
7. Count 13 single sous (coins)
8. Name four pieces of money (Binet & Simon, 1948, p. 420)

The Binet-Simon scales had a different purpose from Cattell's tests. Cattell was primarily interested in understanding the range of individual differences, and he suggested that his measures would "perhaps be useful" in accomplishing this purpose. Binet and Simon, in contrast, made the functional usefulness of their measures a criterion for the success of their work. They were interested in the constancy of mental processes and the range of individual differences primarily as these processes and differences were useful to society. They believed the ability to identify subnormal intelligence (and, conversely, identify normals who had been heretofore perhaps deemed subnormal) offered proof that psychology was "in a fair way to become a science of great social utility" (p. 412).

This functional orientation to the assessment of people's attributes proved quite effective. During World War I, large-scale assessments of soldiers were attempted—the armies of England, France, Germany, and especially the United States were staffed through the use of group-administered mental tests. The fairly successful use of tests of individual differences in ability during World War I was carried over into civilian pursuits after the war. Perhaps the primary reason for this was the development of tests that could be administered to large groups of people rather than to one person at a time, as was required with tests such as the Binet-Simon or Stanford-Binet procedures. Hull's book *Aptitude Testing* (1928) provided a useful summary of the range and kind of individual differences and of the consistent statistical validity of tests in staffing all kinds of organizations (to staff schools with pupils, as well as to staff manufacturing and service organizations with semiskilled, skilled, and clerical employees). Viteles (1932) summarized the early knowledge related to the effectiveness of tests in government, business, and industry in the United States, and Welch and Miles (1935) published a practical guide in England that summarized validated procedures for selecting weavers, salesmen, dressmakers, machine workers, cigarette packers, tram drivers, and engineers, among others.

Not only psychologists were writing about the success of studying individual differences as a means of identifying the more from the less capable. By 1923, *Personnel Management* by Scott and Clothier was in its first of five editions. These authors were already oriented toward matching the skill requirements of tasks to the ability of workers. In their book, they stated their view of the essential principles of individual differences as follows:

> First, one individual differs from another in those personal aptitudes, those special abilities with which he is equipped and which he is able to contribute to the work of his company in exchange for his salary. Second, individuals differ in interest and motive and respond best to varying stimuli. Third, the same individual changes from day to day and from year to year in ability (both degree and kind) and in interest. Fourth, different kinds of work require different kinds of personal ability in the persons who are to perform them. Fifth, granting equal ability, different kinds of work are done best by persons who, temperamentally, are particularly interested in them. Sixth, the work in each position in a company changes as time goes on; duties are added and taken

away. Sometimes the change is negligible, sometimes it is great. In the measure in which it takes place, a similar change is apt to take place in the abilities and interests the work requires of the worker. Seventh, environment—working conditions, supervision, relations with the employer and with fellow employees, opportunity and so forth—exercises a tremendous influence on personal efficiency and consequently on group production. (p. 12)

It is interesting how similar the opening is to what Plato had to say about individual differences so many years ago. Points six and seven—the impact of kinds of work and the working environment on people—did not receive much early attention from those interested in staffing organizations; individuals and their own personal characteristics were viewed as the sole factor in performance. To quote the early guru of industrial psychology, Morris Viteles (1932): "Industrial Psychology [which was mostly personnel selection at the time] is based on the study of *individual differences*—of human variability—the importance of which as an objective of scientific psychology seems to have first been definitely recognized and stressed by Sir Francis Galton" (p. 29, emphasis in original).

World War II was an important proving ground for the social utility of the methods and personnel selection applied to understanding and assessing individual differences. Literally millions of men had to be selectively assigned to fill all kinds of jobs, and *new* jobs as well—pilots, submarine personnel- and thousands of officers had to be commissioned. Psychologists were assigned to interview, test, appraise, and place personnel. These psychologists were successful and, especially in the U.S. Army Air Corps, they developed large-scale programs that were effective in identifying men who would be more likely to succeed as pilots. In addition, in all the branches of the military, success in identifying those who would perform effectively as officers and leaders was achieved. By the end of the war, a clear set of procedures was available for placing people in those jobs for which they had the highest probability of success (Thorndike, 1949).

A FOCUS ON THE WORK AND WORK ENVIRONMENT

Personnel psychology, with its emphasis on individual differences in ability, reached its zenith during World War II. Both sides in this war found the application of systematic procedures for staffing the war machines effective, but not as effective as they had hoped. Beginning

A FOCUS ON THE WORK AND WORK ENVIRONMENT 17

just prior to and after the end of World War II, there were important changes in the way work and work organizations were studied and understood. These changes concerned conceptualizing and studying the role of the work situation (especially social issues at work), the work itself, and the roles of leadership and the larger organization as important contributors to individual and organizational performance and individual job satisfaction.

Perhaps most important was the growing realization that a lot more than the individuals themselves had an impact on their behavior and attitudes—and the behavior of the groups and organizations to which people belonged. The introduction of sociological thinking about work and the workplace revealed that more than individual attributes were related to individual performance—that, as we noted in explicating Fig. 1.1, the work, and especially the work group, and the larger organization and environment, also mattered. In our present view of things, it is perfectly reasonable to study both individual attributes and the nature of the work team in understanding performance and satisfaction at work, but in the late 1930s and 1940s the focus was on the individual. It was a group of Harvard sociologists led by Elton Mayo who showed that groups have consequences for performance and satisfaction. The work they did, subsequently referred to as "The Hawthorne Studies" (cf. Roethlisberger & Dickson, 1939), took a long time to have its impact, but it is fair to say that today the dominant paradigm for understanding behavior at work is to take into account the individual attributes of workers, the work they do, and the work situation in which they behave.

Early in the long-term project at Hawthorne, the goal was to study the effects of environmental characteristics (e.g., lighting, temperature) on work performance. The researchers discovered that by increasing the light available to workers, productivity increased steadily. What was surprising was that as the light available to workers was subsequently decreased, productivity continued to go up! This seemingly strange paradox was far more interesting than the effects of light on performance because something other than light intensity was at work. To make a long and fruitful research effort short, what the researchers concluded was that paying attention to people can influence their behavior and performance. This notion of the potential for the social situation to influence behavior became the focus of the research program and resulted in some quite dramatic findings regarding the ways in which groups and group pressure might influence individual behavior. For example, in the bank wiring room project, newly hired workers who worked very productively when they first arrived at the job were punished for such productivity ("rate busting") both verbally and physically to bring their production more in line with the old-timers. What really was surprising to the researchers was that production output was being restricted, despite the fact that there was in place at the time a group incentive system that was hypothetically designed to boost productivity to the group's full potential (see Roethlisberger & Dickson, 1939, for a full description of this fascinating study).

Our point here is that more than the individual attributes of workers must be considered when understanding performance and productivity of workers. For example, the least productive worker in the bank wiring room was the most intelligent one there, this despite the fact that we have mountains of evidence to show that intelligence is positively related to performance (cf. Hunter, 1986)—usually but not always because situations can enhance or depress the display of intelligence.

Scientists, being people, respond to the *Zeitgeist,* or climate of the times. Occasionally, scientists anticipate the climate of the times and are prepared to facilitate change in a societally useful way. It is clear that during and just after World War II psychologists began making in-depth studies of the impact of situations on people's behavior. Studies of leadership, management style, organizational reward systems, power, and the effects of formal and informal groups, were accomplished both in social psychology laboratories and in field settings.

Although Darwin suggested early on that survival depends on the fitness of the organism, of equal importance to him was his concept of the environment as a sculptor, molding the behavior (and psychology) of the organism. The molding process was believed to encourage behavior that was appropriate to the ever-changing environment; only those organisms so deviant from the environment in form and function, or so rigid in form and function that they could not adapt, failed to survive. Although Darwin had stressed understanding the role of the environment as a determinant of the behavior of organisms, early psychologists concentrated on the assessment of which organisms could adapt rather than on understanding the environments to which they had to adapt.

After World War II, however, there was a veritable explosion of interest and research that addressed the impact of work and the work environment on work motivation, in particular, and work behavior, in general. In fact, between about 1955 and 1965, a number of books appeared about the role of the work environment in human motivation and in the general quality of life that have had a lasting effect on the study of behavior at work. Perhaps the father of this surge in both scholarly and practical theorizing was Kurt Lewin.

Lewin

Before and during World War II, the work of Kurt Lewin showed the importance of the effect of the social environment on people's behavior. In one set of experiments, he showed how young boys' behavior depended on whether their teacher was democratically or autocratically oriented. In another study, during the war, he was able to obtain the cooperation of housewives in serving kinds of meat they had previously found distasteful (e.g., beef lung) by having groups of women discuss the problem and come to a consensus about the benefits of these meats (cf. Marrow, 1970). Perhaps Lewin's major contribution to our future thinking was his idea that human behavior is a function of the person and the environment in natural interaction with each other. As he put it (Lewin, 1935): "From a certain total constellation—comprising a situation and an individual—there results certain behavior, i.e., $(E_I, P_a) - B_A$, or in general $B = f(P, E)$" (p. 73, italics in original). The "constellation" to which Lewin refers includes people, their personal attributes and situations, and their attributes; the term $B = f(P, E)$ means that behavior (B) of individuals is a function of their attributes (P) and the environment (E) in which they are located.

Lewin's notion that behavior is a function of both person and environment had a tremendous impact on the study of humans at work. This impact extends into theories of motivation and management, which stress the concept of person and situation alike. His influence would also seem to extend to issues and approaches to staffing organizations, but this has not been the case as much as one might expect. A major purpose of our approach in this book is to make the connection between individuals and their work and work environments more explicit than has been done previously from the staffing vantage point.

For now, we want to provide the reader with some flavor of the more motivational and organizational theories about behavior that emerged in the early 1960s because they provided for the development of modern industrial and organizational (I/O) psychology—the home of contemporary research and thinking about staffing organizations.

Vroom

V. H. Vroom's (1964) theory of motivation illustrates Lewin's concepts. Vroom's statements regarding motivation take the following form: (a) people put out effort when they think that they work in a place where effort results in rewards they value; and (b) people put out effort to perform when they think that they can put out enough effort to reach the levels of performance they see are required to get the rewards they value. This sounds complicated but it is not. This view of human motivation is one that conceptualizes a system of values attached to particular rewards, establishes an environmental relationship between effort and reward, and determines the extent to which workers can be expected to perform at the level necessary to obtain rewards. Note that individuals may differ in the rewards they desire so two people in the same environment may not be equally "motivated." Further, individuals may differ in the degree to which they think they are able to perform at the levels they see are required to obtain rewards; again, they will not be equally "motivated."

The implications of this theory for understanding human behavior at work are great. For example, Vroom suggests employee effort is a function of individual desires and organizational feedback (i.e., the relationship the organization creates between worker effort and worker rewards). This truly involves an interaction between individual and organizational characteristics. A second implication of this theory is that people are thinking, rational, decision makers; they are organisms that make all kinds of choices—about where to work and how much effort they should put into their work. The kinds of decisions are based on some assessment of the extent to which where they work and how much effort they put out will be instrumental in obtaining desired rewards. In the Hawthorne studies, for example, we saw that workers might restrict their production to gain social approval (a valued reward) rather than raise their production to obtain money.

Vroom's (1964) work tied together much of the early work in psychology concerning studies of learning and motivation, as well as studies of behavior in the work setting. His theory seemed to offer a relatively parsimonious (straightforward) way to summarize findings in occupational choice, job satisfaction, and motivation. Although the optimism surrounding Vroom's formulation has not been fully realized, his formulation was important because it directed the efforts of I/O psychologists to worker motivation and the role of the work situation in facilitating effort. Other theorists came along to push this mode of thinking, including use of his concepts for understanding leader effectiveness (e.g., House & Mitchell, 1975), and today there are literally hundreds of studies that can be attributed to Vroom's influence (see Van Eerde & Thierry, 1997, for an excellent review of the research done in this paradigm).

McGregor

Vroom's (1964) theory was focused on individuals and individual rationality (i.e., thinking, decision-making people). An alternative view on worker motivation was promoted by McGregor (1960), following the theories of Maslow (1954). McGregor's portraits of worker motivation are more inclusive and less individual because they emphasize workers, in general (Schneider, 1985). These theories, and others of the same time period (e.g., Argyris, 1957; Herzberg, Mausner, & Snyderman, 1959), emphasized feelings as the major goal of human behavior. For example, feelings of security, or self-esteem or self-actualization, were pictured as goals toward which workers moved. Since the 1980s, these portrayals of human motivation have been very influential in the design of new forms of work organizations and the renewal of old ones as organizations attempt to improve the general quality of life at work (cf. Burke, 2002).

McGregor (1960) proposed that depending on a manager's views of what motivates people, different strategies for motivating workers may be adopted. If the management of an organization takes the view that people want to experience self-actualization, then jobs are designed in ways that permit people to display their abilities and take responsibility—the jobs have challenge and variety and permit people to work autonomously, and supervision is designed to be supportive and participative rather than exclusively directive. McGregor called this a Theory Y philosophy about what motivates people. In contrast, if managers take the view that employees work only for money, must be coerced into putting forth effort at work, and must be closely supervised and monitored or they would be goofing off, then the result would be close and directive supervision and a focus on money as a reward. McGregor called this a Theory X philosophy about what motivates workers. The important point here is that, depending on one's philosophy or theory regarding the motivational bases of behavior, one adopts certain guidelines and procedures vis-à-vis employees that are consistent with those prior conceptions.

McGregor's writing has had a continuing impact on theories of organizational design and organizational effectiveness, despite the fact that there is little quantitative data to support it as he originally wrote it. In contrast, if one accepts his theory as indicating that a manager's expectations about what motivates employees and a manager's expectations about what is possible in the way of effort and performance on the part of subordinates, the evidence is quite clear that this is true. This is true because managers seem to behave toward subordinates in accord with their expectations of them—and this can produce both high performance and low performance as a self-fulfilling prophecy depending on the expectations (cf. Eden, 1990).

Of course, motivation is not all about deep internal feelings because there are key concepts about motivation that are much more of the thinking or cognitive sort. Lord and Hanges (1987), for example, have proposed what has come to be called a "control theory" of motivation. This theory suggests that people set goals and that they continuously monitor how their performance is helping them obtain those goals. Monitoring their progress through obtaining this kind of feedback on progress serves to adjust the behaviors displayed such that the probability of goal attainment is increased. You can see this kind of model of motivation is very adaptive due to the important role of feedback into the system.

Summary

Our point is not to summarize motivation theory but to show the complexity of the ideas surrounding attempts to understand behavior in the workplace. What we learn from this brief exploration of Lewin, Vroom, McGregor, and Lord and Hanges is that there is more than individual differences in abilities and personality that lie behind individual behavior in organizations. At a minimum, it becomes clear that behavior is jointly determined by attributes of individuals and the situations in which they work. More specifically, it becomes clear that we must think about behavior in the workplace as at least partially being under the control of the rewards available to workers and their perceptions of what they must do to attain valued rewards, and that the expectations management has about workers can determine the way management behaves toward workers and the way they design jobs and the work environment for workers (Eden, 1990). This leads inexorably to the conclusion that the attributes of managers—especially the philosophies they have about what motivates employees—should be a focus of the staffing efforts in organizations. We return to this topic of managerial and leadership selection later.

A FOCUS ON HOW STAFFING CONTRIBUTES TO ORGANIZATIONAL EFFECTIVENESS

So far, we have discussed the role of individual differences, and the effects of work and the work environment, on individual behavior. These are among the most central topics in staffing, but perhaps the main practical purpose for staffing is to improve organizational performance. There is a growing body of research that shows organizations that develop and implement staffing practices in accordance with the types of guidelines we offer in this book have superior outcomes of many kinds—revenues, customer satisfaction, lower employee turnover, fewer accidents, and so forth. Thus, at the organizational level of analysis, there is a relationship between the kinds of staffing practices organizations implement and organizational effectiveness, although this relationship may be affected by the type of industry the organization operates in (e.g., Terpstra & Rozell, 1993). This reiterates a point raised earlier, that there needs to be a match between people, jobs, organizations, and environments.

This begs the question of how staffing contributes to organizational effectiveness. We hire at the individual level but want to produce results for the organizational level. It is ultimately the skill and ability composition of the organization's workforce—its human capital—that will contribute to such organizational differences. Essentially, we argue that when organizations select certain types of people, they build homogeneity in employee skill and ability attributes within the organization (Schneider, 1987), and these attributes in the aggregate contribute to organizational level differences. Thus, it is the cumulative and long-term consequences of hiring certain types of individuals that contributes to organizational effectiveness—or ineffectiveness.

To fully understand how hiring individuals contributes to organizational consequences requires an understanding of levels of analysis and the many means through which individual attributes form and aggregate to higher levels of analysis. Earlier, we discussed this topic briefly

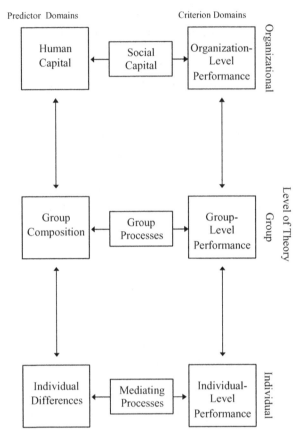

FIG. 1.2. How staffing contributes to organizational consequences. Note. Adapted from Ployhart, 2004; Ployhart & Schneider, 2002.

and provided more detail on how it relates to staffing in other places (Ployhart, 2004; Ployhart & Schneider, 2005; Schneider, 1987; Schneider, Smith, & Sipe, 2000). For our present purposes, we summarize the main implications in Fig. 1.2.

Figure 1.2 shows three levels of theory: individual, group, and organizational. When organizations hire, they hire individuals and hence staffing occurs primarily at the individual level. By only hiring people with specific attributes (e.g., personality, cognitive ability), it creates homogeneity in terms of those attributes. Over time, these attributes become homogeneous within groups (group composition, or the proportion of different attributes within the group) and ultimately within organizations (human capital, or the proportion of different attributes within the organization). It is the nature of the strength of human capital that contributes to differences in organizations and also effectiveness. Therefore, through a process of aggregation from lower to higher levels, we see how hiring individuals contributes to organizational effectiveness.

However, the relationships between these attributes and effectiveness are not necessarily direct but rather mediated. At the individual level, individual differences in ability and personality influence performance through their effects on knowledge and motivation. At the group level, group composition influences group coordination and cohesion processes, which in turn influence individual and group performance. At the organizational level, human capital influences social capital, which represents the types of relationships and relationship processes between members of different units and organizations, and determines organizational effectiveness. In short, the evidence is now clear that organizations with more formal hiring and staffing practices turn out to be competitively more effective than those who do not (cf. *Academy of Management Journal*, 1996).

A FOCUS ON THE INFLUENCE OF FEDERAL EMPLOYMENT LEGISLATION AND COURT CASES

Throughout this book, we see how the U.S. federal government has taken an active role in legislating the use of tests as a basis for employment decisions. Sometimes it is easy to forget how important the rest of the environment is for business and for the practices that go on in business, and one of those forces in the environment is the federal government. From a staffing point of view, the most important event was the passage of the 1964 Civil Rights Act during the Johnson administration.

The act had several parts (called titles). For staffing purposes, the most important title was Title VII because it dealt with employment. Thus, it dealt with all forms of employment issues—pay, promotion, working conditions—and selection. Title VII made it illegal to use age, race, sex, national origin, or religion as a basis for making a hiring decision. One outcome of this legislation was a close examination of procedures that were being used for making these kinds of decisions.

Because written tests constituted a prevalent source of information used in making hiring decisions, some employees who believed they had been denied employment because of poor performance on a test brought suit against that company, arguing that the test had wrongfully discriminated against them. By wrongfully discriminated, they meant that the test was used to discriminate against them because they were Black, or female, or older. This was a serious charge because paper-and-pencil tests, as well as all measures of people used to collect information as a basis for making employee decisions, are designed to discriminate more capable from less capable persons. So, what some claimants were arguing was that these tests were discriminating against them because of their sex or race and not because they were more or less capable. Although we review case law concerning what has transpired since the mid- to late 1960s in chapter 11, here it is important to summarize a few key points concerning the influence of legislation on personnel selection:

1. From the standpoint of the staffing process, the most important outcome of numerous court cases and tremendous effort on the part of staffing practitioners and researchers has been a focus on job relevance in the design of measures used as a basis for making selection decisions. By job relevance, we mean the degree to which the selection procedure is shown to be relevant to the specific job for which people are being considered—an ability to do mathematical calculations involving calculus and geometry is not relevant to a job as a bank teller, although an ability to do addition and subtraction will be. In brief, due to enforcement of Title VII, any procedure used for making hiring decisions is currently more likely to accurately reflect the demands of jobs than was perhaps true prior to federal Equal Employment Opportunity (EEO) pressures.

2. Thorough review of what tests do in providing information to decision makers has revealed that tests designed according to the principles we cover here are not wrongfully discriminating. This is not to say that tests do not make discriminations among people, for that is precisely what they are designed to do. However, they do not make wrongful (i.e., unwarrantedly biased) discriminations. In the language of the prestigious National Academy of Sciences (NAS; 1982) report on ability testing:

> The Committee has seen no evidence of alternatives to testing that are equally informative, equally adequate technically, and also economically

A FOCUS ON THE INFLUENCE OF FEDERAL LEGISLATION 27

and politically viable,... and little evidence that well-constructed and competently administered tests are more valid predictors for a population subgroup than for another: individuals with higher scores tend to perform better on the job regardless of group identity. (p. 144)

What the NAS was saying, then, is that the presumed bias of tests against different subgroups based on sex, race, and so on had not been demonstrated to exist. Further, the tests were really quite useful.

3. Another interesting outcome of the legal battles was a definition of a test. Typically, we think of a test as an examination of some kind responded to with paper and pencil, usually on some form of answer sheet. We think a test is like the multiple-choice exam we took in some college course or like the SAT or ACT. In fact, I/O psychologists and the "Uniform Guidelines on Employee Selection Procedures" (1978) have defined the word "test" in much broader terms, and the courts have adopted this definition. In brief, a test is defined as any form of collecting information on individuals when that information is used as a basis for making an employment decision. So, interviews are tests, as are application blanks, training programs that some people may fail, performance appraisals used as a basis for making promotions (which, obviously, are selection decisions), and any other kind of information used for making employment decisions.

We try in the rest of the book to continually remind the reader of this definition of a test because many of the most important technical principles underlying the evaluation of selection procedures require us to include all sources of information used in making decisions in our definition of a test.

4. The finding by the NAS referred to previously made it clear that an attack on tests was not going to be viable as a way to redress past discriminations based on race or gender in the employment world. There is no doubt there have been such discriminations, but that does not mean carefully designed and validated selection procedures are doing something wrongful, unscientific, or invalid.

One way in which the courts have tried to ensure tests do not wrongfully discriminate against various subgroups in an organization's workforce is by means of applying the *four fifths rule*. An organization is "violating" the four fifths rule if the rate at which it employs minority individuals is less than four fifths the rate at which it employs members of the majority group. When this is true, the organization must show its hiring procedures are based on people's ability to perform important job tasks (job relevance), not on the basis of their racial, ethnic, or gender status. If the organization can provide this evidence, it is usually taken as "proof" that no unwarranted racial, ethnic, or gender discrimination was present.

We cover a number of the court cases that resulted in these conclusions in chapter 11. Here, it is important to note the cases brought to the courts may have been a mistake in that they attacked paper-and-pencil tests when the tests themselves were not usually at fault; it was the people using the tests, not the tests themselves, that were artificially biased. In contrast, one reason why the legislation was necessary was because there were poor-quality selection practices being used from companies buying off-the-shelf procedures with little demonstrable validity, blindly applying them as a basis for making hiring decisions. This model fails to meet the Binet and Simon standard: "All the tests which we propose have been repeatedly tried, and have been retained from among many which after trial have been discarded" (1948, p. 416)

Although many may disagree, federal intervention may have improved the quality of the selection procedures employed and, thus, improved organizational functioning. In this vein, most companies failed to recognize the pain they suffered when suits were brought against them probably resulted in them eventually implementing a far more effective selection program than they had previously. For example, as we note in the previous section, organizations that adopt sound staffing practices tend to show greater levels of organizational effectiveness. However, while legal costs have been tremendous, as chapter 6 shows, an efficient selection system can have rapid dollar benefits for an organization. The moral here is that the pressure to be accurate in selection decisions should have long-term positive consequences for organizations.

SUMMARY

Considerable background information is presented in this chapter, and some of it may seem somewhat contradictory. For example, first we introduce the idea that many individual, job, organizational, and environmental factors influence people's behavior in the work setting. At the end of that discussion, however, was the idea that all those factors were likely in reciprocal relationship, with people interacting with each

SUMMARY

other to determine many of those factors at the same time those factors were, in turn, influencing them. We note how these reciprocal relationships are asymmetric, operate on different time scales, and differ in the strength of their interrelationships. Following this discussion, we present a brief history of ideas relevant to staffing organizations. This discussion emphasizes both the role of individual differences in understanding individual behavior and the role of situational factors beyond individual control in influencing the behavior of people.

A subtle but important distinction is introduced—the distinction between predicting and understanding the behavior of individuals and predicting and understanding the behavior of people in the aggregate in the form of organizations. This is an important distinction, and one that is introduced and reintroduced throughout the remainder of this book. Staffing organizations in its most simple form still refers to the idea that staffing practitioners are always concerned with making a choice from a pool of applicants and the challenge is to make the best choice—to choose the best individual. At the same time, however, staffing researchers need to maintain an organizational perspective and be sensitive to the cumulative effects of decisions on the functioning of the organization, in the aggregate. To do this, staffing specialists need to remember that pay, leadership, job design, and management philosophy will all influence organizational performance at least as much as selection decisions will—and that managers chosen for higher-level positions help determine these situational variables employees encounter. We know that other factors influence the behavior of managers in organizations besides their own personal attributes—like federal legislation, the behavior of markets, and so forth—so the people responsible for staffing organizations cannot bear the total burden of responsibility for the success and failure of the organization. However, they must be sensitive to the cumulative effects of what they do because, in the aggregate, the result of a comprehensive staffing program will help determine many of the organizational factors that are reflected in organizational effectiveness.

It is clear that most of this book emphasizes the making of choices among applicants for a job; that is, it focuses on the individual and not on the cumulative effects or overall organizational performance. We do not yet, unfortunately, know much about the cumulative effects of staffing on organizational performance (Ployhart & Schneider, 2002; Schneider et al., 2000), or about how staffing practices might be differentially important depending on the needs of the organization, but work such as that developed by Schmidt and Hunter (1977, 1981, 1998) suggests that the effects can be positive indeed. We proceed under that assumption and point out where and how that assumption can be tested.

REFERENCES

Academy of Management Journal. (1996). Special research forum on human resource management and organizational performance, *39*, 4.
Argyris, C. (1957).*Personality and organization.* New York: Harper.
Binet, A., & Simon, T. (1948). The development of the Binet-Simon scale. In W. Dennis (Ed.), *Readings in the history of psychology* (pp. 412–424). New York: Appleton-Century-Crofts.
Boring, E. G. (1950). *History of experimental psychology* (2nd ed.). New York: Appleton-Century-Crofts.
Burke, W. W. (2002). *Organization change: Theories and practice.* Thousand Oaks, CA: Sage.
Cattell, J. M. (1948). Mental tests and measurements. In W. Dennis (Ed.), *Readings in the history of psychology* (pp. 347–354). New York: Appleton-Century-Crofts.
Eden, D. (1990). *Pygmalion in management: Productivity as a self-fulfilling prophecy.* Lexington, MA: Lexington Books.
Elman, B. A. (1991). Political, social, and cultural reproduction via civil service examinations in late imperial China. *The Journal of Asian Studies, 50,* 7–28.
Gerhart, B. (1999). Human resource management and firm performance: Challenges in making causal inferences. In P. Wright et al. (Eds.), *Research in personnel and human resources management* (pp. 31–74). Oxford, UK: Elsevier.
Goodman, P. S. (2000). *Missing organizational linkages: Tools for cross-level research.* Thousand Oaks, CA: Sage.
Herzberg, F., Mausner, B., & Snyderman, B. (1959). *The motivation to work* (2nd ed.). New York: Wiley.
House, R. J., & Mitchell, T. R. (1975). Path-goal theory of leadership. *Journal of Contemporary Business, 3,* 81–97.
Hull, C. L. (1928). *Aptitude testing.* Tarrytown, NY: World.
Hunter, J. E. (1986). Cognitive ability, cognitive aptitudes, job knowledge, and job performance. *Journal of Vocational Behavior, 29,* 340–362.
Jenkins, J. J., & Paterson, D. G. (Eds.). (1961). *Studies in individual differences: The search for intelligence.* New York: Appleton-Century-Crofts.
Kozlowski, S. W. J., & Klein, K. J. (2000). A multilevel approach to theory and research in organizations: Contextual, temporal, and emergent processes. In K. J. Klein & S. W. J. Kozlowski (Eds.), *Multilevel theory, research, and methods in organizations: Foundations, extensions, and new directions* (pp. 3–90). San Francisco: Jossey-Bass.
Lewin, K. (1935). Environment forces in child behavior and development. In K. Lewin (Ed.), *A dynamic theory of personality* (pp. 66–113). New York: McGraw-Hill.
Lofquist, L. H., & Dawis, R. V. (1969). *Adjustment to work.* New York: Appleton-Century-Crofts.
Lord, R. G., & Hanges, P. J. (1987). A control system model of organizational motivation: Theoretical development and applied implications. *Behavioral Science, 32,* 161–178.
Marrow, A. J. (1970). *The practical theorist.* New York: Basic Books.
Maslow, A. H. (1954). *Motivation and personality.* New York: Harper.

REFERENCES

McGregor, D. M. (1960). *The human side of enterprise.* New York: McGraw-Hill.
National Academy of Sciences. (1982). *Ability testing: Uses, consequences, and controversies* (Vol. 1). Washington, DC: National Academy Press.
Organ, D. W. (1988). *Organizational citizenship behavior: The good soldier syndrome.* Lexington, MA: Lexington Books.
Organ, D. W., & McFall, J. B. (2004). Personality and citizenship behavior in organizations. In B. Schneider & D. B. Smith (Eds.), *Personality and organizations* (pp. 291–314). Mahwah, NJ: Erlbaum.
Plato. (1952). *The dialogues of Plato* (B. Jowett, Trans.). Chicago: Encyclopedia Britannica.
Ployhart, R. E. (2004). Organizational staffing: A multilevel review, synthesis, and model. In J. Martocchio (Ed.), *Research in personnel and human resource management* (Vol. 23, pp. 121–176). Oxford, UK: Elsevier.
Ployhart, R. E., & Schneider, B. (2002). A multilevel perspective on personnel selection: Implications for selection system design, assessment, and construct validation. In F. J. Dansereau & F. Yammarino (Eds.), *Research in multi-level issues. Volume 1: The many faces of multi-level issues* (pp. 95–140). Oxford, UK: Elsevier Science.
Ployhart, R. E., & Schneider, B. (2005). Multilevel selection and prediction: Theories, methods, and models. In A. Evers, O. Smit-Voskuyl, & N. R. Anderson (Eds.), *Handbook of personnel selection* (pp. 495–516). Chichester/London: Wiley.
Roethlisberger, F. J., & Dickson, W. J. (1939). *Management and the worker.* Boston: Harvard.
Rousseau, D. M. (1995). *Psychological contracts in organizations: Understanding written and unwritten agreements.* Thousand Oaks, CA: Sage.
Schaubroeck, J., Ganster, D. C., & Jones, J. R. (1998). Organization and occupation influences in the attraction-selection-attrition process. *Journal of Applied Psychology, 83,* 869–891.
Schein, E. A. (1992). *Organizational culture and leadership* (2nd ed.). San Francisco: Jossey-Bass.
Schmidt, F. L., & Hunter, J. E. (1977). Development of a general solution to the problem of validity generalization. *Journal of Applied Psychology, 62,* 529–540.
Schmidt, F. L., & Hunter, J. E. (1981). The future of criterion-related validity. *Personnel Psychology, 33,* 41–60.
Schmidt, F. L., & Hunter, J. E. (1998). The validity and utility of selection methods in personnel psychology: Practical and theoretical implications of 85 years of research findings. *Psychological Bulletin, 124,* 262–274.
Schneider, B. (1985). Organizational behavior. *Annual Review of Psychology, 36,* 573–611.
Schneider, B. (1987). The people make the place. *Personnel Psychology, 40,* 437–453.
Schneider, B., Smith, D. B., & Sipe, W. P. (2000). Personnel selection psychology: Multilevel considerations. In K. J. Klein & S. W. J. Kozlowski (Eds.), *Multilevel theory, research, and methods in organizations* (pp. 91–120). San Francisco: Jossey-Bass.
Scott, W. D., & Clothier, R. C. (1923). *Personnel management.* New York: McGraw-Hill.
Shaw, J. D., Delery, J. E., Jenkins, G. D., & Gupta, N. (1998). An organization-level analysis of voluntary and involuntary turnover. *Academy of Management Journal, 41,* 511–525.
Simon, H. A. (1973). The organization of complex systems. In H. H. Pattee (Ed.), *Hierarchy theory* (pp. 1–27). New York: Braziller.
Terpstra, D. E., & Rozell, E. J. (1993). The relationship of staffing practices to organizational level measures of performance. *Personnel Psychology, 46,* 27–48.
Thompson, J. D. (1967). *Organizations in action.* New York: McGraw-Hill.
Thorndike, R. L. (1949). *Personnel selection.* New York: Wiley.
Uniform guidelines on employee selection procedures. (1978). *Federal Register, 43,* 38290–38315.

Van Eerde, W., & Thierry, H. (1997). Vroom's expectancy models and work-related criteria: A meta-analysis. *Journal of Applied Psychology, 81* , 575–586.
Viteles, M. S. (1932). *Industrial psychology*. New York: Norton.
Vroom, V. H. (1964). *Work and motivation*. New York: Wiley.
Welch, H. J., & Miles, G. H. (1935). *Industrial psychology in practice*. London: Sir Isaac Pitman & Sons.

2

Measurement Concepts and Data Analytic Tools

BACKGROUND

Perhaps the single most important contribution American psychology has made to the Western world is the pragmatic approach to understanding and predicting behavior. As in other sciences, Americans have not been consistently superior in developing psychological theories, but they have excelled in applying and evaluating theories and procedures. A major contribution of American psychologists to organizational functioning and effectiveness has been the development and application of measurement procedures for evaluating the validity and usefulness of staffing programs. It has been this concern for the careful evaluation of staffing procedures—to actually test whether they work—that has characterized the field and made it useful to business and industry.

Many companies maintain significant internal staffs of selection specialists who have been trained in the development and use of these measurement procedures. For example, AT&T has had such people, and they have made major contributions to practice through the careful evaluations they have done of their procedures; we read more about their work in subsequent chapters. In addition, consulting firms that specialize in the development and evaluation of staffing procedures are numerous and are increasingly used by companies to try to hire the best talent for their firms. Finally, these measurement concepts and data analytic tools have proved critical in the legal arena because court cases rest heavily on evaluations of staffing procedures, and those evaluations are based on the kinds of measurement and statistical tools we discuss in this chapter.

This chapter, then, is specifically concerned with some of the concepts and methods useful in evaluating the procedures used to staff organizations. Specifically, the chapter presents some basic methods of measurement (measures of central tendency, variation, and covariation) and measurement/evaluation concepts (reliability and validity). We also introduce the basics of regression analyses, including multiple regression and the use of regression to test moderator and mediator hypotheses. Multiple regression is a technique that permits us to combine predictors in ways that maximize the predictability of outcomes of interest. The issues of moderator and mediator variables refer to the idea that direct prediction of outcomes of interest may not always be possible—that prediction involves various contingencies. For example, suppose you wanted to predict someone's job performance but you knew that the kind of leadership they work under will affect how well you can make that prediction. In the language of regression, you have just said that leadership moderates the relationship between the selection procedure and performance. In contrast, suppose you wanted to predict someone's job performance based on a test you administered at hiring, but you knew the test actually predicted who got chosen for training and it was training performance that predicted job performance. In the language of regression, you have just said that training performance mediates the relationship between test performance and job performance.

The sophistication of the analytic tools available to the staffing researcher has increased dramatically since the mid-1980s. We introduce the reader to some of these modern techniques, including structural equation modeling (SEM), item response theory (IRT), and hierarchical linear modeling (HLM). SEM lets us test configurations of predictors in the way they relate to each other and to outcomes of interest. IRT facilitates the evaluation of items that comprise tests so the eventual tests are as efficient and reliable as possible. HLM is a technique that helps us evaluate the degree to which variables at different levels of analysis enable us to predict the outcome of interest (recall Fig. 1.1 in chapter 1). For example, returning to the leadership issue raised earlier when we discussed moderators, it is also reasonable to hypothesize that people as a group perform more effectively when they work under democratic leaders than when they work under autocratic leaders. If you knew this to be true, then you would not only want to have test performance data on people to predict their performance, but also data on the kind of leaders they work for—with leadership being a variable at another level of analysis because it is not an individual attribute or characteristic of the people whose performance you are trying to predict. These topics will prove useful in later chapters as we explore ways to determine the nature of jobs (chapter 3), assess performance (chapter 4), evaluate recruiting effectiveness (chapter 5), evaluate staffing effectiveness and dollar value (chapter 6), and evaluate the relative effectiveness of different staffing procedures and methods (chapters 7–10). Being able to fully appreciate these later topics requires a solid understanding of measurement and statistics, which comprise the basis of this chapter.

AIMS OF THE CHAPTER

Personnel selection is rooted in the psychology of individual differences. The basis of this area of psychology is obviously the fact that people differ along a wide variety of dimensions. We all have eyes, but we vary considerably in our capacity to see the world around us.

Obviously, if a measurement is documenting a real individual difference, then another measurement of the same phenomenon should yield a similar result or, as social scientists say, the measurement should be *reliable*. Further, if the score resulting from the measurement has some practical consequence for the person measured, or for society, we say it is *valid*. Different phenomena are measured with varying degrees of reliability and validity. We also know that some human attributes that psychologists have tried to measure have more important consequences in the workplace than others. For example, cognitive ability tests are generally more valid than personality tests for predicting productivity or task performance.

In this chapter, we intend to describe

1. The nature of measurement in psychology, in general, and staffing, in particular
2. Methods of summarizing our measurements

3. Ways of quantifying or operationalizing the relationships among measures, including ways of demonstrating the reliability and validity of our staffing procedures
4. Use of data and analysis methods to test theories about the ways variables are likely related to each other.

Box 2.1 provides a summary of key statistical notation used throughout this chapter.

Box 2.1 Glossary of Key Statistical Symbols

Term or Symbol	Definition
X	The predictor test or measure used to make staffing decisions (e.g., interviews, cognitive ability tests, personality tests, application blanks). In experimental designs, it would be called the independent variable.
Y	The criterion or what we are trying to predict (e.g., job performance, absenteeism, turnover, accidents, sales dollars). In experimental designs, it would be called the dependent variable.
N or n	Usually refers to sample size (i.e., the number of people being studied).
i	Subscript usually refers to an individual-level observation in a sample (e.g., a given person).
j	Subscript usually refers to a group or unit-level observation (e.g., a given group or organization).
Σ	Summation. For example, "ΣX" would indicate one should add all the X scores.
Mean	The average of a sample. Frequently denoted by $\bar{X}$ or M.
Var or σ^2	Variance.
SD or σ	Standard deviation.
Z	Usually refers to a z score
T or t	Capital T refers to a T score. Lowercase t refers to a true score or t test.

r or R	Correlation or relationship between variables; ranges from -1 to $+1$. Lowercase r is the correlation between a criterion and predictor; capital R is the multiple correlation between a criterion and two or more predictors. Sometimes the correlation uses subscripts, such that r_{xy} indicates the correlation between the variable X and the variable Y. With more than one predictor, the symbols may be r_{Y1} and r_{Y2}, indicating the correlation between Y and the first (X_1) and second (X_2) predictors, respectively. When referring to reliability, r_{xx} is the reliability of X.
r^2 or R^2	Variance explained in a criterion by a single predictor (r^2) or multiple predictors (R^2).
A and b	Regression coefficients. A represents the intercept, b represents the slope (how much of a change in Y occurs with a one unit change in the predictor X).
e	Error.
SD_{est}	Standard error of estimate.
CI	Confidence interval around an effect size.
k	Number of predictors in a regression equation.
ρ_c	Estimate of cross-validity.
ρ	Estimate of reliability or population correlation.
α	Coefficient alpha, an estimate of internal consistency reliability.
θ	Latent true score in item response theory.

NATURE OF MEASUREMENT

Measurement has been defined as the assignment of numbers to objects or events (Stevens, 1946). Sometimes measurement more precisely involves assigning numbers to properties of things or to events. In other words, the thing itself is not measured but some sign or indicant of it is. Persons who continually try to beat others at every pursuit they undertake are said to possess a great deal of competitiveness, and their behavior in interactions with others is a *sign* of competitiveness. Much of the measurement related to staffing issues concerns these signs rather than direct measurement. For example, the tests we design to assess ability and personality are not direct measurements of those abilities or personality, but rather they are windows through which we get signs with regard to what the true ability or personality may be. We have developed a number of measurement tools to help us evaluate the extent to which these signs are valid.

The assignment of numbers to objects or events was a phenomenal intellectual accomplishment. Prior to the development of measurement, no scientific progress was possible. It was impossible because of the wide variety of nonspecific (loose) ways of describing nature. Even with the advent of language, including written language that had some precision, specification of the properties of objects, events, or their indicants was extremely cumbersome until the advent of measurement.

Because we assign numbers to properties of things, it is necessary to know the features of numbers because all numbers are not created equal. So, some numbers we assign to objects or events provide more information than others. From lowest to highest amount of information, we can say that there are three features to numbers:

1. Numbers are ordered. This means that when we assign numbers to properties of things, we can order the things, for example, rank them according to the numbers. If it makes sense only to order things after assigning numbers to them, we say that our scale is *ordinal*.
2. Differences between numbers are ordered. This means that when numbers are assigned to properties of things, not only can the things be ordered, but also the size of the differences between things is known and can be ordered. This type of measurement is called *interval* measurement because the size of the differences or intervals among ordered objects is known and can also be ordered.
3. The number series has a unique origin. When the assignment of numbers to properties of things allows the assignment of zero (e.g., zero weight or height), we say that we have *ratio* scales of measurement. Not only can the size of the difference between numbers be ordered, but also numbers can be presented as ratios of each other. If we can reasonably multiply or divide numbers to obtain ratios like "half as tall" or "twice as heavy," we have ratio measurement.

NATURE OF MEASUREMENT 39

In addition to ordinal, interval, and ratio forms of measurement (also called *scales of measurement*), when we simply assign a number to something as a way to identify it, we say we have a *nominal* scale. Nominal scales include numbers on football jerseys or baseball uniforms. The numbers do not signify any ordering of uniforms or players, hence no higher level of measurement.

Gender and race are measured at a nominal level. That is, we can reliably classify individuals as male or female, or Black or White, but we do not typically order gender subgroups or racial subgroups. The color of hair of a group of individuals can be ordered from light to dark, so we would say that the lightness or darkness of one's hair color is an ordinal variable. The order requirement means that if Person A's hair is darker than B's hair, and B's hair in turn is darker than C's hair, then A's hair must be darker than C's. If we assigned the number 10 to A's hair color, 6 to B's, and 2 to C's, then our scale would be an interval scale only if we can establish that the differences between A, B, and C or any other difference of four on our scale was equivalent. Finally, if the property we are measuring satisfies the order and interval rules, and also has a meaningful zero point, we say we are measuring that property with a ratio scale. Height, weight, distance, and temperature measured on the Kelvin scale are examples of variables in which zero is meaningful. The zero point makes it possible to make ratio statements in comparing individuals. If Person A weighs 200 lb and B 100 lb, then we can say that A weighs twice as much as B. However, if 200 and 100 were scores on an intelligence test, it would not be correct to say A was twice as intelligent as B because zero intelligence has no operational meaning (even though it might seem some people have zero intelligence!).

The level of measurement with which we index variables has practical consequences when we start summarizing our measurements or performing arithmetic operations on the numbers we use to represent the objects of interest. For example, we can transform nominal data in any way that maintains the identity implied by the original assignment of numbers. So, for example, we could multiply all the baseball uniform numbers by two, and each player would still have a unique identity or number (this would, however, mess up the sports announcer). Numbers assigned to objects using an ordinal scale may be added, subtracted, multiplied, divided, squared, or transformed in any way that preserves the original order of the objects. In contrast, transformations of interval data must preserve both the order and relative size of the scale intervals; that is, multiplying, dividing, adding, or subtracting by a constant is fine, but taking the square root or squaring would change the size of the intervals between or among the numbers, so those transformations are not permissible. Finally, the ratio properties of a scale would be preserved only by multiplication or division by a constant; addition or subtraction of a constant would change the zero point.

There are two important practical consequences of the scales of measurement. The first consequence is that the higher the scale of measurement, the finer the distinctions we can make. As one moves from the nominal, ordinal, interval, and finally ratio, scales of measurement, one can make better discriminations. Consider a supervisor evaluating an employee's performance. A nominal scale would only ask if the performance was acceptable or unacceptable. An ordinal scale would ask the supervisor to rank order the employees. An interval scale might ask the supervisor to rate each employee on a 7-point scale. The second consequence is that with higher scales of measurement, more powerful statistical and measurement tools can be used. Thus, by allowing finer distinctions and more powerful statistical methods, higher scales of measurement provide more information about the concept we are trying to measure.

Most psychological variables are measured close to an interval scale. So we almost invariably use statistics such as the mean, standard deviation, and Pearson product-moment correlation (explained later), which are appropriate when we have interval data. There are two reasons for using these statistics even with data whose interval nature may be somewhat questionable. The first argument involves the observation that variables with known interval and/or ratio properties such as height and weight are normally distributed. Typically, the measures we use in staffing programs, especially those we use as predictors, yield a normal distribution of measurements (see Fig. 2.1 for a normal distribution on four different tests), so we make the inference they have interval properties. We do not claim ratio properties, however, because we do not have knowledge of a zero point. Because our psychological variables are assumed to be normally distributed and our measures yield scores that are normally distributed, we believe our scales possess interval properties (Magnusson, 1966). In fact, psychological measures can be constructed in such a manner that they yield a normal distribution, or we can transform the data in a way that yields a normal distribution. Of course, both methods rely on the implicit assumption that individual differences on psychological variables are indeed distributed normally, but because so many of the measures of natural

NATURE OF MEASUREMENT

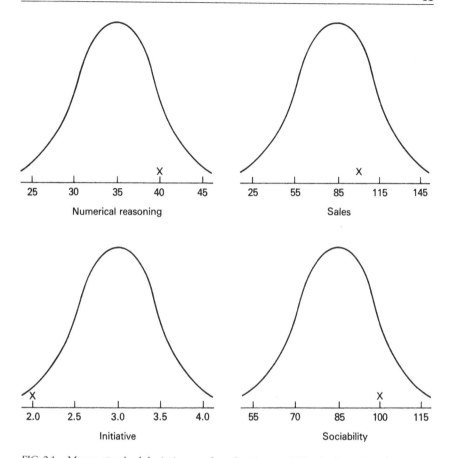

FIG. 2.1. Means, standard deviations, and applicant scores (X in the figures) on four tests.

characteristics (tree size, fish speed, amount of rain) take this form, it is perfectly reasonable to make the same assumption for our measures. Some scholars (e.g., Baker, Hardyck, & Petrinovich, 1966) have even shown that the kinds of statistics we use in staffing, when computed on distributions of noninterval data, yield essentially the same decisions as did statistics based on distributions of interval data. It is important to recognize that the nature of measurement and the implications for the social and behavioral sciences continue to be debated (e.g., Stine, 1989).

In summary, we have explained that a number does not always mean the same thing; what it means depends on the kind of measurement from which it was derived. The number 100, then, can mean nothing more than a football jersey, or it can mean something as numerically useful as the temperature at which water boils. Most personnel selection measurement is neither nominal nor ratio in form; it is usually somewhere between ordinal and interval, and we treat it as interval. For practical purposes, this means we can rank order people from high to low on various measures of individual differences and that we can say something about how much better or how much more superior one person is compared with other people. The use of interval scales of measurement also permits us to summarize many persons' scores using various descriptive statistics. In the remainder of this chapter, we mention the importance of scales of measurement when we describe individual differences or use information about individual differences on one variable to predict the behavior measured on another variable.

MANIFEST AND LATENT VARIABLES

At the beginning of the previous section, we note a distinction between the things we actually observe and count (what we called signs) and the property we think (hope) we are measuring. This is a critically important distinction in psychological measurement, and it must be considered further. To elaborate on the competitiveness example in that section, suppose we are interested in measuring the competitiveness of a group of students. We might count the number of times they express disappointment at not winning a sporting event in which they or their school participated. Or, we might ask them directly to rate how competitive they think they would be in a number of different situations. Or, their physiological responses could be measured (e.g., their galvanic skin response could be electronically recorded) while engaging in a competitive word game. All these would be *manifest* or *observed* variables—signs—because they are variables we can see and measure. In contrast, the *construct* or *latent* variable we think we are measuring using these different methods is competitiveness. Note we cannot directly observe the latent construct of students' competitiveness, but instead must infer it from a set of manifest (observed) variables (i.e., the number of expressions of disappointment, their self-reports, and the level of physiological arousal during a competitive event). If competitiveness is a meaningful latent explanation of these varying responses/behaviors, then these different manifest observations should be related to each other. We would also expect that some combination of these different measures of the latent construct of competitiveness would relate to other behavior or outcomes that make sense—such as

MEASUREMENTS OF INDIVIDUAL DIFFERENCES

attendance at sporting events, participation in athletics or other games or contests, ratings by others of their competitiveness, and so forth.

Obviously, we ultimately are interested in latent psychological constructs, but because we cannot directly observe the latent construct, we must make inferences of the construct based on manifest measures—signs. It is important to understand this distinction because it is at the basis of many of the data analytic techniques we discuss later.

SUMMARIZING MEASUREMENTS OF INDIVIDUAL DIFFERENCES

When we gather a set of measurements, we usually report, in addition to the raw (actually obtained) scores, some summary indices of the total set of scores. These summary indices include measures of central tendency and variability. At times, we are also interested in the relationship of scores, say some test scores, to other variables of interest—for example, the relationship between scores on job attitude measures and scores on job performance measures. In the latter case, our summary measure of the relationship is called a *correlation*.

Central Tendency Measures

When we receive a score on an exam, we have little or no idea what that score means until we have some framework for understanding it. So a common question is, "What was the class average or *mean*?" Formally,

$$\text{Mean} = \sum X/N$$

where $\sum$ = the sum of all Xs
 X = each raw score
 N = the number of persons, objects, items, and so on in the set

In other words, this formula adds all scores and then divides this sum by the total number of observations in the dataset. This measure of central tendency allows us to make some sense of our score; that is, we are either at above or below the mean, and we can know by how much. Even more important, the mean is our best guess or estimate about any one person's score from the sample (unless we have additional information about the person, as discussed later). The mean is appropriate only when data with interval properties have been collected. If we have only

2. MEASUREMENT AND DATA ANALYTIC TOOLS

TABLE 2.1
Measures of Central Tendency and Variability

1. 13	Mean = Sum of scores/Number of scores = 400/20 = 20
2. 15	Median = Middle score in the ranked distribution = 20
3. 17	Mode = Most frequently occurring score = 20
4. 17	Range = 13–27
5. 18	Standard Deviation = $\sqrt{\frac{\sum(X_i - M)^2}{N}} = \sqrt{\frac{214}{20}} = 3.27$
6. 18	where X_i is an individual score, M is the mean, N is the number of
7. 18	scores, and $\sum$ indicates the sum of $(X_i - M)$ values.
8. 19	Very rarely are the mean, median, and mode of a score distribution
9. 20	precisely equal. We made them equal to provide a simple example.
10. 20	They will be equal in a normal distribution of scores, as is depicted
11. 20	at the top of Fig. 2.2.
12. 20	
13. 21	
14. 21	
15. 22	
16. 22	
17. 23	
18. 24	
19. 25	
20. 27	
Total 400	

ordinal data, we correctly report the median that is the middle score in a distribution of scores. We also report the median when the mean is not representative of the scores in the distribution. For example, the median of income statistics is commonly reported as the measure of central tendency because a few millionaires in a group of people who typically have salaries around $50,000 will seriously distort the mean. The mode is the appropriate index of central tendency when one has only nominal data. In Table 2.1, we present a distribution of scores and the computed mean, median, and mode. It should be noted that the mean, median, and mode are identical when the distribution of scores is perfectly normally distributed (see Fig. 2.1). Figure 2.1 also shows many other ways of representing central tendency and a distribution (e.g., z scores), which are discussed shortly.

Variability Measures

In receiving feedback on a course exam, we are first interested in our score. Our next concern is the average score or mean of the group or class. Then, we are likely to ask what the highest and/or lowest scores

MEASUREMENTS OF INDIVIDUAL DIFFERENCES 45

were. In other words, we are asking what the variability of the score distribution is. We know intuitively that a score 2 points above the mean is more impressive when the mean is 8 and the range of scores is 6 to 10 than when the range of scores is 0 to 16. A simple measure of variability—the *range* or difference between the highest and lowest score—is appropriate when we have ordinal data. An arithmetically more complex measure of variability, the standard deviation, is used as the measure of variability when we have interval data. The standard deviation is commonly used in staffing research because it represents the average deviation of each person's score from the mean and therefore estimates the spread of the distribution.

The *standard deviation* (*SD*) is computed for the set of data presented in Table 2.1 and is particularly useful because of its relationship to the normal distribution or normal curve pictured at the top of Fig. 2.2. The frequency of people obtaining each score in the distribution is indicated

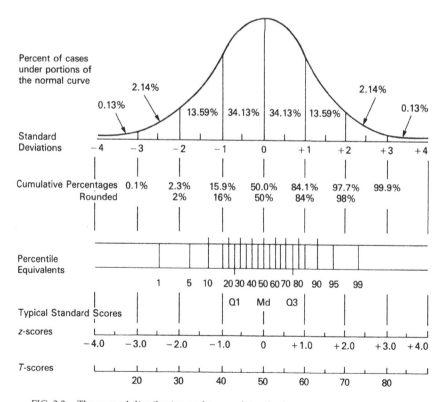

FIG. 2.2. The normal distribution and types of standard scores.
Source: Reprinted from the Psychological Corporation, Test Service Bulletin number 48, January 1955.

under the line in Fig. 2.2. The resulting bell-shaped curve (when the distribution is normal) reflects a set of people (objects, things) that have been counted, measured, or assessed on some trait or attribute and then ordered by the attribute according to the frequency or number of people who obtained each score. Because attributes of people typically distribute themselves in a curve of this shape it is also called a normal distribution. If this distribution is approximately normal, the percentage of cases falling in any portion of the normal curve is known and presented in tabular form in statistics books. For example, we know that a person whose score is 1 SD above the mean lies above approximately 84% of the people represented by the distribution. Most cases (approximately 96%) are found within plus or minus 2 SDs of the mean. In Table 2.1, this would mean most scores lie between 26.54 and 13.46 [20.00 + or − 2(3.27)]. Some scores will obviously fall outside these figures, but the probability of this occurring is rather low. When behavioral scientists are interested in determining whether they have produced a real change, whether two variables are really related, or whether the conclusions from a set of observations can be explained by chance alone, they use a form of the standard deviation called the standard error to form confidence intervals around a particular score beyond which it is unlikely to observe scores. This procedure is described in detail later.

Standard Deviations and Standard Scores

When a person applies for a job, we may administer a battery of tests. Because tests are made by different groups of researchers, the possible range of scores on any test is decided by each test maker. Suppose our job applicant scores

- 40 on a numerical reasoning test
- 100 on a sales inventory
- 2.0 on a test of initiative
- 100 on a sociability scale.

We want to know two things: (a) how well, relative to others, did the applicant perform on each of the tests; and (b) is the applicant better at sales than at numerical reasoning? How well have others done on these tests; that is, what is the average performance? Say average scores are

Numerical reasoning = 35
Sales inventory = 85

Initiative = 3
Sociability = 85.

Such information demonstrates our knowledge that a score or number does not have inherent meaning. A number only has meaning in some context.

Unscrupulous test publishers capitalize on people's assumptions about the context of numbers. A score of 100 is always better than a score of 80, right? Wrong. A score of 100 is always the same as another score of 100, right? Wrong. Given two scores of 100 on two different tests, and given that 85 is the average performance on those two tests, then the two scores are the same, right? Wrong. Test users must know how to interpret scores; this requires that they know the scores' contexts. Of course, most test developers would not create a test and use a particular scale (e.g., 1–5, 0–100) simply to deceive the public. Instead, different tests have different score ranges because they are more convenient for the purpose of the test.

Average performance on a test, as we learned from the classroom example cited previously, demonstrates only one piece of contextual information. The mean gives us some idea about where our applicant stands, but not a very good picture. The mean alone provides no information about the relative spread of the distributions on tests. But now, referring back to the tests our applicant took, we find that

SD numerical reasoning = 5
SD sales inventory = 30
SD initiative = 0.5
SD sociability = 15.

Figure 2.1 shows distributions for the four tests listed previously, along with the respective mean, standard deviation, and our hypothetical applicant's scores (denoted by an X). Note that our applicant achieved the same numerical score on the sales inventory and the sociability scale. Also note that the average performance on these two tests is the same in the general population. However, the standard deviations are different. An examination of the standard deviation indicates that on sociability our applicant is 1 SD above the mean ($X = 100$, $SD = 15$, Mean = 85). On the sales inventory, however, our applicant is 0.5 SD above the mean ($X = 100$, $SD = 30$, Mean = 85).

The concept of standard deviation indeed implies a standard. The standard deviation tells something about the degree of variation around the mean. This permits us to locate how far above or below the mean any score lies. Indeed, given the mean, the standard deviation, and a particular score, we can tell from the distribution precisely how far above or below the mean the score is—"how far" indicating percentage of people when the distribution of scores is normal. We can take such percentages and convert them into *percentiles*, which represent the percentage of people who fall below a given score.

Figure 2.2 presents these percentiles schematically. Note in Fig. 2.2 that the standard deviations are presented as $-1\,SD$, $+1\,SD$, $+2\,SD$, and so forth. Note also the mean of the distribution is zero. Given the mean and standard deviation of a distribution of scores, we may convert the distribution to a common score unit called a *standard score*. Standard scores are useful because they allow us to compare the scores people obtain on different measures when those measures are reported in different raw score form (as in Fig. 2.1). Although a number of potential standard scores exist, the two most frequently used are z scores (Mean = 0, $SD = 1$) and T scores (Mean = 50, $SD = 10$). The formula for these scores is

$$Z = (X_i - \text{Mean})/SD \text{ and } T = 10z + 50 \qquad (1)$$

Notice that in the z score formula, we basically express any score in units that are standard units. That is, a person's score has meaning if we know how others typically did (the mean) and the spread of the scores (the standard deviation). Incidentally, T scores are sometimes preferred to z scores because z scores range from -5 to $+5$, so at the extremes they can be negative.

Let us apply the previous formulas to our applicant's test scores:

Test	X	Mean	SD	z	T	Percentile
Numerical reasoning	40	35	5	1.00	60	84
Sales inventory	100	85	30	0.50	55	70
Initiative	2	3	0.5	−2.00	30	2
Sociability	100	85	15	1.00	60	84

$$Z = (X_i - \text{Mean})/SD = (40 - 35)/5$$
$$= 5/5 = 1.00 \text{ (Numerical reasoning)}$$
$$= (100 - 85)/30 = 15/30 = 0.50 \text{ (Sales inventory)}$$
$$= (2 - 3)/.5 = -1/0.5 = -2.00 \text{ (Initiative)}$$
$$= (100 - 85)/15 = 15/15 = 1.00 \text{ (Sociability)}$$

Now we can see that on numerical reasoning and sociability our applicant is 1 SD above the mean ($z = 1.00$, $T = 60$, Percentile = 84). Thus, her strongest traits relative to others are numerical reasoning and sociability. Her weakest trait is initiative ($z = -2.00$, $T = 30$, Percentile = 2). The percentiles are derived from Fig. 2.2.

Transforming scores to standard scores is a linear transformation that preserves the original distribution of scores. If the raw score distribution is not normal, then our scores are not distributed symmetrically around the mean as in the top figure in Fig. 2.2. Standardized scores will also not be normally distributed but will retain their original distribution. Standardized scores from distributions with very different shapes are not directly comparable; hence, we occasionally apply a more complex nonlinear transformation called *normalization*. For most score distributions, this is probably not necessary.

It may be helpful to reiterate a few key concepts before moving into more detailed information. First, there is an important distinction between latent and manifest variables, such that latent variables are the psychological constructs we cannot directly see or observe, but must do so indirectly through manifest variables. Second, manifest variables can be assessed using nominal, ordinal, interval, and (less common) ratio scales. The higher the scale of measurement, the more information it provides. Third, latent constructs and manifest measures are most often normally distributed. When manifest variables are normally distributed, the distribution can be summarized by the mean and standard deviation. The measurement scale used to describe the distribution can be transformed into a number of different kinds of equivalent scores, such as z scores, T scores, and percentiles.

COVARIATION

What we really want to be able to accomplish in most staffing contexts is to make predictions about applicants. That is, given test scores such as those shown previously, and given the scores of a particular person's performance on tests, we want to make some statements about the person in regard to a certain criterion or criteria of effectiveness. Are different scores on these tests related to different levels of effectiveness? Do test scores covary with criterion scores? To deal with these questions requires extending the concepts of means and standard deviations to the issue of covariation.

2. MEASUREMENT AND DATA ANALYTIC TOOLS

When we refer to making predictions based on tests (including interviews and all forms of assessments), the tests are usually called "predictors" and the "thing" we are trying to predict (usually job performance, but also turnover, accidents, and absenteeism) is referred to as the *criterion* (or criteria, if there are multiple variables). For those with a background in experimental design, another way to think about such predictions is that the predictors are like the independent variables and the criteria are like the dependent variables. It is convention that predictors are denoted by X and criteria are denoted by Y.

Two Scores on the Predictor

Suppose for a moment that you administered some predictor (e.g., test, interview) to a group of applicants. Suppose further that for some reason the applicants scored either X_1 or X_2 on the predictor such that you have only two scores. You then follow up on those who scored X_1 and those who scored X_2 to see how effective they are on the job. You find those who scored X_1 on the predictor tend to score lower on the criterion than those who scored X_2. You note, however, that for both those who scored X_1 and those who scored X_2 on the predictor, there are distributions of criterion scores. That is, not all who scored X_1 on the predictor were equally effective on the job. The situation is portrayed in Fig. 2.3. Obviously, the criterion performance of a person scoring X_1 would be best characterized by the mean of Y_1, the average performance of all those who scored X_1. We would hope, of course, that the deviation around the mean of Y_1 was small, because as the size of the standard deviation increases relative to the mean, our estimate of the mean is more likely to be in error as a measure of central tendency.

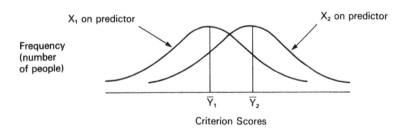

FIG. 2.3. Distribution of criterion scores for those scoring X_1 on predictor and those scoring X_2 on predictor.

Multiple Scores on the Predictor

Deviation around the mean of criterion scores becomes clearer when there are more than two possible scores on the predictor. When there are many predictor scores (as is often the case), there is theoretically a corresponding distribution of criterion scores for each predictor score. In Fig. 2.4, the hypothetical distribution of criterion scores for 8 of 40 predictor scores is presented; another set of analogous distributions is presented in Fig. 2.5. For any predictor score in Fig. 2.4, the range of criterion scores is lower than for the same predictor score in Fig. 2.5.

Figure 2.4 represents a more accurate prediction than Fig. 2.5 because Fig. 2.4 shows less deviation from the means of criterion distributions

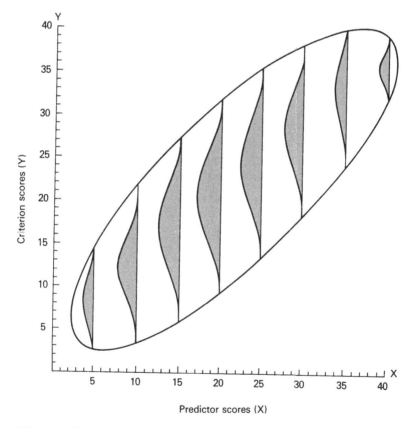

FIG. 2.4. Different criterion score distributions for selected predictor scores.
Note: It is a convention to label the horizontal (X) axis as the predictor and the vertical axis (Y) as the criterion.

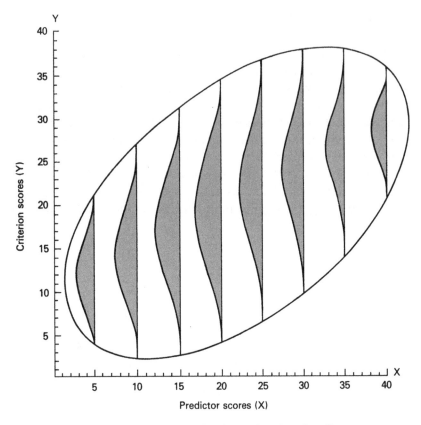

FIG. 2.5. Different criterion score distribution for selected predictor scores.

for each predictor score than does Fig. 2.5. Although the same overall range of criterion and predictor scores (from 1 to 40) is observed in both figures, the deviation from the mean of each distribution of criterion scores is smaller in Fig. 2.4; the mean of each criterion distribution in Fig. 2.4 would be a better estimate of criterion performance for its predictor score than would be true in Fig. 2.5.

In both figures, the best estimate of criterion behavior for any predictor score is the mean of the criterion distribution for that predictor score. Once again, as when we discussed standard deviations, if one were putting money on a particular criterion score, one would bet more on the situation pictured in Fig. 2.4 because the standard deviation of the criterion is smaller for every predictor score.

Now think about the following: If the best estimate of criterion behavior for any predictor score is the mean of the criterion distribution for that predictor score, the reverse is also true. Not only is there a

COVARIATION

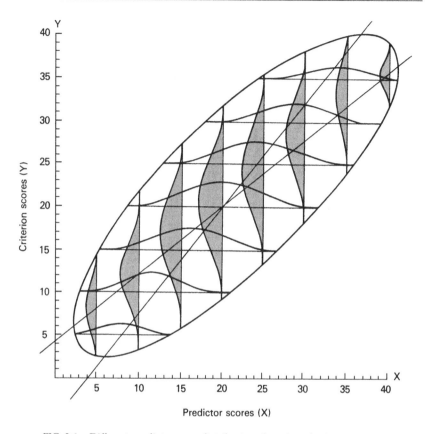

FIG. 2.6. Different predictor score distributions for selected criterion scores.

distribution of criterion scores for each predictor score, but there is a distribution of predictor scores for each criterion score. Figures 2.6 and 2.7 show this schematically. Figures 2.4 and 2.5 are reproduced in Figs. 2.6 and 2.7, respectively, and overlaid with a graph of the predictor score distributions for each criterion score. In addition, a line passes through the means of the distributions, from X (predictor) to Y (criterion). This is because the best estimate of a distribution of scores is the mean.

Let us take stock. For each predictor score, there is a distribution of criterion scores for all people with that predictor score. This indicates that for a person's score of, say, 15 on a test, we may not know exactly what the criterion score will be, but our best estimate is the mean of the criterion scores for all those who also have a score of 15 on the predictor. The probability of being right by guessing the mean is a function of the standard deviation of the distribution from which the mean comes. That is, if the standard deviation of criterion scores is very large, then

2. MEASUREMENT AND DATA ANALYTIC TOOLS

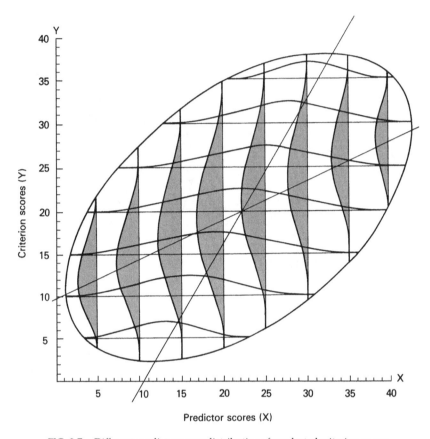

FIG. 2.7. Different predictor score distributions for selected criterion scores.

the mean of those criterion scores is not a very good estimate. The same would hold true in estimating the mean of the predictor scores for all those scoring, say, five on the criterion; predictions from Y to X are more accurate when the distribution of possible predictor scores has a small standard deviation.

To summarize all this information about the degree of relationship between the predictor and criterion, we use a statistic called the correlation coefficient. This statistic and its usefulness are described in the next several sections of the chapter.

THE PEARSON PRODUCT-MOMENT CORRELATION COEFFICIENT (r)

Like the mean and the standard deviation, the correlation coefficient (represented by r) is a convention. Staffing researchers have used r as

THE PEARSON PRODUCT-MOMENT CORRELATION

their index of relationship because of (a) its parsimony—the result is only one number, and (b) it meets the same criterion as the mean and standard deviation—it is an index of relationship that is a mean.[1]

The technicalities of the arguments underlying the use of r need not concern us here. What is important, though, because of the extensive use of correlation coefficients in staffing work, is to begin to grasp the concepts of (a) the range of r, (b) the wonder of r, (c) the limitations of r, (d) the uses of r in reliability and validity analyses, and (e) alternatives to r.

The Range of r

The correlation coefficient may range in size from +1.00 through zero to −1.00. In staffing work, we rarely observe correlations above 0.90, and then generally only in relation to reliability (i.e., consistency or reproducibility). Typical observed validity coefficients between predictors and criteria range between 0.10 and 0.40.

A correlation of 1.00, positive or negative, indicates that information about an individual regarding X also yields perfect information regarding Y; it provides an "if...then" situation. For instance, in Figs. 2.6 and 2.7 this circumstance would be indicated by one straight line, rather than by an ellipse. You need not worry about this problem as far as staffing procedures are concerned because it never happens!

The "positiveness" or "negativeness" of a correlation refers to the covariance of scores in X and Y. No "goodness" or "badness" is implied. If scores on a predictor go up as scores on the criterion go up, we speak of positive relationships. For example, we know that higher scores on mechanical aptitude tests are related to higher automotive mechanic performance—a case of a positive correlation. We also know that high job satisfaction is related to low turnover and absenteeism—a case of a negative correlation. Both cases are "good" if we are interested in predicting job performance or turnover and absenteeism.

---

[1] It is a mean in the sense that it is the mean of the cross products of standard scores:

$$r_{xy} = \sum Z_x Z_y / N$$

The spread of the distributions of both the predictors and criterion scores affects the size of the correlations, which makes sense because the correlation is between X and Y. Note also that the correlation is the average sum of the z scores of X times the z scores of Y. Because z scores are being multiplied, it is clear that the units of measurement for test and predictor need not be the same; the raw score formula for calculating the correlation coefficient converts them both to z scores.

The correlation coefficient, r, being a summary statistic, represents a picture of how two variables covary. You should be able to picture what an r of a particular level portrays. For example, go back and reexamine Figs. 2.6 and 2.7. Figure 2.6 shows the relationship between a relatively accurate predictor and criterion behavior. Figure 2.7 indicates a weaker relationship between predictor and criterion. Because the distribution of criterion scores for each predictor score has greater variability in Fig. 2.7, an ellipse enclosing the set of distributions in Fig. 2.7 is more circular or wider than an ellipse enclosing the set of distributions in Fig. 2.6. The less spread there is—that is, the more the ellipse is cigar shaped rather than circular—the stronger the relationship between the two variables.

Figure 2.8 presents three bivariate scatterplots for correlations, positive and negative, of different sizes. A bivariate scatterplot is a pictorial presentation of the scores of a group of people (organizations, products) on two variables (bivariate). In scatterplots, predictors are always represented on the horizontal (X) axis, and the criterion is always represented on the vertical (Y) axis. The two scores for each person are plotted on paper as one point. This is called a scatterplot because the points are usually pretty well scattered!

Figure 2.8a presents test scores and criterion data for 10 people. Usually, data would be collected on a much larger number of people before calculating a correlation, but the data in Fig. 2.8 are just for illustrative purposes. Each person was scored on the predictor variable X and on the criterion variable Y. For each person, then, there are two scores, X and Y, which are plotted from the data presented for Fig. 2.8a. The correlation between X and Y is +0.76 in Fig. 2.8a. Notice how the scatterplots (and hence correlations) change in Figs. 2.8b (where the correlation is weaker but still positive) and 2.8c (where the correlation is strong and negative).

The Wonder of r

It is difficult to convey the wonder of r because it is a truly creative statistic. Think of it: In one number, the magnitude of the relationship between two sets of data can be summarized, regardless of the scale units on which one or both sets of data are measured and regardless of the number of people (objects, things) that have been assessed.

Because of its simplicity (one number), statistical and mathematical purity, and the fact that it subsumes the mean and standard deviation, r

THE PEARSON PRODUCT-MOMENT CORRELATION 57

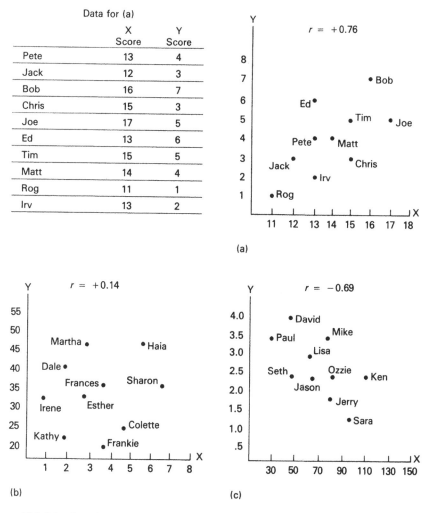

FIG. 2.8. Correlation coefficients of different sizes plotted as bivariate scatterplots. Any relationship between these names and actual people is purely *intentional*. Source: The shape of the figures was suggested by J. P. Guilford, *Fundamental Statistics in Psychology and Education*, 2nd ed. (New York: McGraw-Hill, 1950), pp. 156–57. Used with permission.

is the major statistic of interest in all psychological measurement, including staffing. Both reliability and validity are assessed using r. In addition, advanced concepts in measurement such as factor analysis (a procedure useful in grouping or clustering concepts that are related to each other), multiple regression (a procedure for combining measures to help understand a phenomenon of interest), structural equation modeling (the simultaneous solving of a set of multiple regression equations), and hierarchical linear modeling (a regression-based method that accounts for clustered data) are also based on r and are described shortly.

The Limitations of r

The correlation coefficient is technically called the Pearson product-moment coefficient of correlation. It is named after Karl Pearson, the British statistician, who solved the problem of presenting the extent of a linear relationship between two variables in one number. Linear relationship means that the trend of relationship between two variables is a straight, rather than a curved line. As with any summary, r may not be an adequate summary of the actual state of affairs; if the relationship between two variables is nonlinear, then the correlation ratio or *eta*, which summarizes the extent of nonlinear relationship between variables, is a more appropriate statistic. Although the primary limitation of r is that it is not always the appropriate index to use to show the relationship between two variables, the overwhelming proportion of situations encountered in staffing is adequately described by r.

For our purposes, it is important to note that (a) eta may be a more appropriate index of relationship than r when one or both variables to be correlated deviate markedly from the normal bell-shaped distribution; (b) eta varies from -1.00 to $+1.00$ and is interpreted in the same way as r; and (c) eta is a summary index that can be applied to any bivariate distribution—if r and eta yield markedly different results, eta is the more appropriate statistic.

There are other instances where the Pearson correlation coefficient is not appropriate, but we want to represent the relationship between two variables. In particular, we need to use a different estimate of the relationships when one or both variables are dichotomous. When one variable is continuous and another is dichotomous, we use something called a point-biserial correlation. When both variables are dichotomous, we can use a phi coefficient or a biserial correlation. The phi coefficient assumes both variables are truly dichotomous, whereas the biserial correlation assumes they are truly continuous at the latent level but dichotomized at the manifest level.

Both a wonder and a limitation of r is that it yields only one number; this makes people think it easy to interpret. It is obvious that we had to present some background information so you could understand the

THE PEARSON PRODUCT-MOMENT CORRELATION 59

concepts underlying r because r is quite mathematically sophisticated. Simple explanations of r are difficult to make and would be quite misleading. Some things to remember are as follows:

1. An r of 0.50 does not mean 50% of this or half of that (likewise for any other values of r).
2. Under some circumstances (in addition to those described in discussing eta), r cannot be legitimately calculated.
3. In validity analyses, r is difficult to describe to the statistically unsophisticated, and alternative procedures are required to explain relationships.

Alternative Ways of Describing the Correlation

Although the correlation is a simple, yet powerful, statistical description of the relationship between two variables, it can be difficult to describe the full meaning and consequences of the correlation to those not familiar with statistics. Therefore, pictorial representations of correlations called expectancy charts are good substitutions for r when explaining correlations to unsophisticated people. The principle underlying expectancy charts is as follows: With any criterion, it is possible to say that below a certain point, performance on that criterion is unacceptable. With reference to Fig. 2.9a, we can see that when such a performance criterion is set, the proportion of people who meet or exceed the criterion given any predictor score increases as one moves from left to right on the predictor axis. Translating the bivariate scatterplot to an expectancy chart is a simple matter. The percent of persons above the point on the criterion that separates successful and unsuccessful employees is calculated and then graphed as in Fig. 2.9b. In this graph, the likelihood of success increases as test scores increase, indicating a significant degree of correlation. The point on the job performance criterion at which we decide people are successful or unsuccessful must, of course, be set at the appropriate level, but this should not be a problem if we remember that this is a standard against which we judge potential improvements in the workforce selected with the procedures we have evaluated. Figure 2.9b, incidentally, can be contrasted with a simple "$r = 0.50$" to see how really remarkable the Pearson r is.

Another alternative way to describe the correlation is to refer to its *effect size*, or the magnitude of the relationship between two variables.

60   2. MEASUREMENT AND DATA ANALYTIC TOOLS

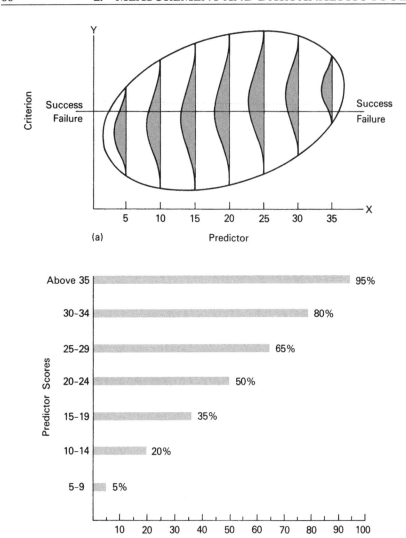

FIG. 2.9. Expectancy chart showing percent who succeed for each of seven different predictor score ranges. *Source:* An excellent discussion of the preparation and use of expectancy charts is provided by Lawshe and Balma (1966).

We know that large correlations indicate a stronger relationship and, in personnel staffing, indicate that a selection test is highly related to some performance criterion. For example, if a cognitive ability test correlates 0.50 with performance and an interview correlates 0.30, one would say that the cognitive ability test has a larger effect size than the interview.

When interpreting the correlation, one must always be cognizant of the effect size of the correlation because different predictor measures show different effect sizes for different criteria. Thus, we care not only that there is a relationship, but also how strong this relationship is (i.e., the effect size). It is the effect size that ultimately comprises test validity and utility. In chapter 7, we introduce the concept of small, medium, and large effect sizes for correlations (validities) in staffing contexts so one can compare the relative effect sizes across different predictor measures.

PREDICTIONS BASED ON CORRELATION

The reader should now see the relationship between correlation and probability statements about "success" using an expectancy chart. The important point to remember is that, the higher the calculated correlation, the more certain the probability statement. The higher the correlation is, the more cigar-shaped the ellipse encircling the bivariate scatterplot, that is, the less deviation there is of points in the scatterplot from a line passed through the means of distributions (see Figs. 2.6 and 2.7). Using the correlation coefficient between job performance and predictor scores and the knowledge of a person's score on the predictor, we can also make exact predictions of a person's job performance. These exact predictions, though, may not be perfectly accurate (see, for example, Fig. 2.10). To make these predictions, we use a technique called *regression analysis*, which gives us an equation by which we make predictions.

When the predictor and criterion are both in standardized form (i.e., z scores), then the prediction of Y from X is simply $Z_p = r_{xy} Z_x$, where Z_p represents predicted job performance in standard score form and Z_x is the standardized test score. That is, the standardized criterion is equal to some weight (the correlation, r_{xy}) times the standardized predictor. If a person's standard score on a test was 0.50 and the correlation between X and Y was 0.30, the best prediction one could make regarding the person's standard score on Y would be 0.15, or slightly above the mean of the group because all standard scores in z score form have averages of 0.00 and standard deviations equal to 1.00.

If we want to make predictions in raw score terms, as would usually be the case, the prediction formula becomes more complicated:

$$y = r_{xy}(SD_y/SD_x)(X) + A_{yx} \qquad (2)$$

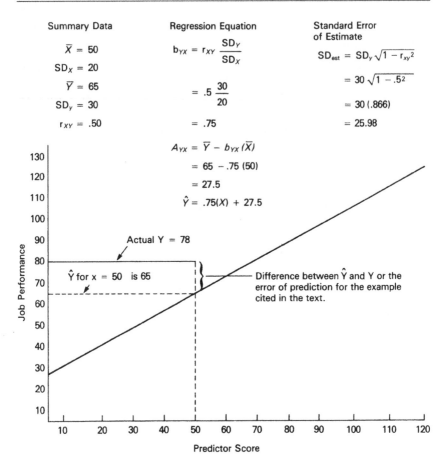

FIG. 2.10. Demonstration of calculation of a regression equation, the plot of a regression line and the standard error of estimate.

SD_y and SD_x are the standard deviations of X and Y and A_{yx} is a constant that is used to make adjustments for the difference between the means of X and Y. The r_{xy} (SD_y/SD_x) term in the regression equation is frequently referred to as the regression coefficient (b_{yx}) and represents the slope of the line that best fits the data in the scattergram. It represents the amount of increase in Y that results from a one unit increase in X. The constant is calculated by subtracting from the mean of Y the product of the mean of X and the regression coefficient. This equation describes the line of best fit in the sense that it minimizes the sum of the squared distances from the regression line to each data point in the scatterplot. An illustration of a regression line and its components is presented in Fig. 2.10. Consider a person whose score on the test is 50. The test has a correlation (validity) of 0.50, a mean of 50, and a standard deviation of 20; the standard deviation of Y is 30 and Y's mean is 65. The regression coefficient is 0.50 (30/20) or 0.75. The constant for the equation is 65 − 0.75(50), or 27.5. These calculations yield the following regression equation:

$$Y = 0.75(X) + 27.5 \quad (3)$$

The best prediction concerning the job performance of a person with a score of 50 is 0.75(50) + 27.5, or 65.0. However, if this person's actual job performance score equaled 78, we would be wrong by 13 (78 − 65).

The scatter in a scatterplot represents these errors in prediction. That is, the more scatter there is in the criterion scores for each predictor score, the more errors in prediction we would make. The nice feature of the regression line is that it is the one line that will minimize the total amount of errors we will make. In other words, using the regression line as our basis for prediction, the sum of squared errors will be smaller than using any other basis for predicting a criterion score for a particular predictor score. This is true even when we have human experts make predictions. Box 2.2 discusses the conditions where even the best human judges can make predictions no better than regression equations.

If we were to examine all our mistakes in prediction and compute the standard deviation of those mistakes, they would be equal to the standard error of estimate (SD_{est}); computationally:

$$SD_{est} = SD_y \left(\sqrt{1 - r_{xy}^2} \right) \quad (4)$$

From this formula, we see that if r_{xy} is equal to 1.00, the standard error of estimate is 0.00. This is another way of saying two thing: First, we are making no errors of prediction, and second, the scatter plot is reduced to a straight line. Alternately, if r_{xy} is 0.00, then the standard error of estimate is equal to the standard deviation of Y.

Because the worst case in terms of making errors of prediction is their standard deviation would be equal to SD_y, psychometricians frequently use the percent reduction in the size of the SD_{est} relative to SD_y as a measure of the usefulness of a test. In this case, the regression coefficient reduces to zero, and the constant is equal to the mean of Y.

> **Box 2.2 Optimal Prediction: Human Experts Versus Linear Regression**
>
> More than 50 years ago, Paul Meehl (1954) published a book that created quite a stir by pitting the accuracy of predictions based on regression against the accuracy of predictions based on human experts. Strikingly, he demonstrated that at worst regression tied with human experts' predictions, but most of the time regression outperformed human experts and was more accurate. Since then, there have been numerous demonstrations of this effect, with the general conclusion being that predictions made by human experts are inferior to those based on the linear regression model (Hastie & Dawes, 2001). In fact, as long as the sign of the regression weights remains the same, using randomly generated regression weights will outpredict human experts.
>
> Of course, one should never rely on blindly "mechanical" predictions (i.e., predictions based on regression or statistical methods without human judgment), and research suggests that human experts are good at identifying the types of information that are important to consider when making predictions (e.g., which selection tests should be used). There may also be instances where new information must be considered that could not have been used in the regression analysis. Thus, humans' expert judgment serves as an important and necessary means of identifying the important information to be used in the prediction task, but these experts should leave the actual predictions to the linear regression model.
>
> The implications of this for staffing are clear—we need qualified experts to help identify the types of knowledge, skills, abilities, and other characteristics necessary for successful job performance. However, the actual combination of these characteristics should be determined by the linear regression model because no expert can provide more accurate predictions.

(The reader can confirm this is true by using the formula for the regression coefficient and the constant presented previously.) So, for all values of the predictor variable, we would predict the mean of Y and the deviations ($Y_{predicted} - Y$) would be the same as those we calculate when we find the standard deviation of Y (because predicted Y is the mean; see Table 2.1).

PREDICTIONS BASED ON CORRELATION 65

As r_{xy} increases, of course, the standard error of estimate decreases, indicating better prediction and more confidence in the results we obtain. Using the standard error of estimate, we can place *confidence intervals* around our predictions of any given person's score. Confidence intervals tell us the likely range of effect sizes one might expect in the population, given the existing sample size and thus the standard error of estimate. Because we know the correlation will be affected by sample size and sample specific variance (i.e., unique characteristics of a particular sample), we know the correlation may differ to a certain extent if we had calculated it with a different sample. Thus, confidence intervals estimate how much the correlation might differ. For example, in the case cited in Fig. 2.10, the standard error of estimate is equal to $30\sqrt{1 - 0.5^2}$ or almost 26. So, we are 68% confident that the person whose test score is 50 has a Y score somewhere between 65 ± 26 (recall that 68% of the cases fall within 1 SD of the mean in the normal curve).

Clearly, prediction of single cases may not be as accurate as we would like, given such a large standard error of estimate. In chapter 6, we discuss the standard error of estimate as a utility measure and show how it is an unnecessarily pessimistic view of the total worth of a selection procedure. It is useful, however, to remember the size of the standard error of estimate when we are trying to make decisions in individual cases or when we encounter a single case in which use of a valid selection instrument resulted in an error. Obviously, errors are possible and inevitable but, as shown in chapter 6, the procedures we have developed are still quite useful, and it makes sense to use them even in the presence of imperfect prediction.

We should also discuss the degree of confidence we have in the validity coefficient itself; that is, whether the estimated relationship between X and Y is really nonzero. The degree of confidence we have is related to the size of the sample on which the correlation is calculated, as well as the size of the validity coefficient itself. If our correlation coefficient is based on two cases, it will always be equal to 1.00 because the line we would draw between the two points on a scatterplot would always allow perfect prediction for these two cases. As the sample size (N) increases, we will no longer be able to draw such a perfectly fitting line, but the question remains as to whether our estimate of r_{xy} depends on the particular sample we have drawn. Obviously, as N increases, we have more confidence that our results are not just flukes, but that there is a relationship of some magnitude between X and Y. The degree to which our estimate of r_{xy} varies from sample to sample can be estimated from the standard error associated with r_{xy} (the standard error of the correlation), which is equal to $1/\sqrt{N-1}$. If our estimate of r_{xy} is more than two standard errors (1.96 precisely) above zero, then we say we are 95% confident that our validity is above zero when the sample size is large and we hypothesize that r is zero. Stated differently, we can be 95% confident that the population correlation lies above zero.

We can see from the standard error of the correlation that we can expect much variability in our estimates of correlation, even when they come from relatively large samples. For example, r_{xy} estimated from samples equal to 101 has a standard error of 0.10. If the true population coefficient is 0.30, we can expect about 32% of the coefficients based on samples of 100 to be less than 0.21 or greater than 0.39.[2] We return to this point in chapter 6 when we discuss criterion-related validity and the feasibility of conducting a meaningful criterion-related study. It is important to remember this point in our later discussion of validity generalization because this sample variability is in large part responsible for the mistaken notion that the validity of tests depends on the circumstances in which they are used or that test validities are situationally specific.

PREDICTIONS USING MULTIPLE PREDICTORS

When we introduced the regression equation and issues concerning the prediction of criterion performance, one question that may have occurred to you is what happens when we want to use two or more predictors? It would seem that if we can predict a criterion with one predictor, then using two or more predictors should permit even better prediction of the criterion of interest. This is a good question and staffing researchers have grappled with the problem and arrived at a number of conclusions about it.

First, it is important to try to weight predictors when using two or more predictors. To weight predictors means to derive the relative importance of two or more predictors in helping predict a criterion. The weight a predictor will receive depends on

- How strongly it correlates with the criterion
- How strongly it correlates with the other predictor(s)

---

[2] For values of r greater than 0.00, the standard error is $(1-r^2)/\sqrt{N-1}$; hence, as r increases, the standard error decreases. We use $1/\sqrt{N-1}$ because we are testing the hypothesis that r equals zero.

PREDICTIONS USING MULTIPLE PREDICTORS

Next, it is important to estimate the correlation of the combinations of predictors with the criterion. Whenever we combine predictors to correlate with a criterion, we speak of a multiple correlation or multiple regression situation, called a "multiple r" and symbolized by R.

Calculating Weights

The multiple regression formula for two predictors is

$$\text{Predicted } Y = b_1 X_1 + b_2 X_2 + A_{Y12} \qquad (5)$$

where b_1 and b_2 are the derived weights for predictors 1 and 2 and A_{Y12} is the constant in the two predictor regression equation. The idea, then, is to calculate b_1 and b_2 so we know how much to weight each predictor (X_1 and X_2) to minimize errors. This calculation works out so the multiple R is as high as possible. In other words, the weights that are calculated are *optimal weights*; no other weights will yield as high a multiple R (see Box 2.2). As noted previously, the weights will depend on how much each predictor correlates with the criterion and how much it correlates with the other predictor(s).

Various possible multiple predictor–criterion relationships are illustrated in Fig. 2.11. In Situation 1, we have two valid predictors that are moderately intercorrelated (represented by the overlap in the circles or Venn diagrams). Each predictor is somewhat related to the criterion and somewhat related to a portion of the criterion to which the other predictor is unrelated (areas A and C). There is also a portion where all three overlap (area B). If our goal is the best possible prediction

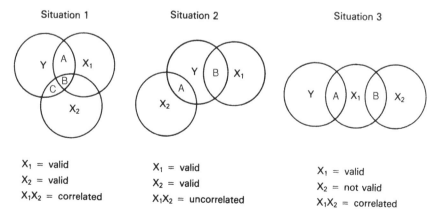

FIG. 2.11. Predictor-criterion relationships in the two-predictor case.

of the criterion, then we want to minimize overlap in predictors (area B) while maximizing the unique overlap of each predictor with the criterion (validity; areas A and C).

In Situation 2, we have pictured such a case: Both predictors are valid, and they overlap with different portions of Y (areas A and B). You can see that in both Situation 1 and Situation 2, the predictors overlap (correlate) with the criterion to the same extent. However, because in Situation 2 the predictors do not overlap (correlate) with each other, when the two predictors are used together they explain more of the criterion variance in Situation 2 than they do in Situation 1. The rule to follow, then, is that when decisions are made about the composition of a test battery, they should be made in light of the intercorrelations among the various predictors and in terms of their independent validity with the criterion. High predictor intercorrelations reduce the effectiveness of a test battery because each test provides little unique prediction of the criterion.

Situation 3 is more difficult to interpret. In Situation 3, X_2 will add to predictability (area A) when used in combination with X_1, even though it is not valid by itself. This occurs because X_2 accounts for some of the variance in X_1 (area B), which is not related to Y. Or, if we examine the formula for R [given in Eq. (9)], we see that the term in the numerator (which is subtracted) will be zero because r_{y2} is zero. The denominator, however, will be substantially lower because r_{12} is high. This combination of predictor–criterion relationships is called a *suppressor* effect and results in a negative regression coefficient for X_2.

Just for your information, we present the formulae for calculation of the weights b_1 and b_2 for the two-predictor case. We present the two-predictor case not only because of this complexity, but also because the method of dealing with additional predictors is identical to that which allows an optimal combination of two predictors. Equations for the regression coefficients, or the optimal predictive weights, in the two-predictor case read as follows:

$$b_1 = \left(\frac{r_{y1} - r_{y2} r_{12}}{\sqrt{(1 - r_{y2}^2)(1 - r_{12}^2)}} \right) \left(\frac{SD_y \sqrt{1 - r_{y2}^2}}{SD_1 \sqrt{1 - r_{12}^2}} \right) \quad (6)$$

$$b_2 = \left(\frac{r_{y2} - r_{y1} r_{12}}{\sqrt{(1 - r_{y1}^2)(1 - r_{12}^2)}} \right) \left(\frac{SD_y \sqrt{1 - r_{y1}^2}}{SD_2 \sqrt{1 - r_{12}^2}} \right) \quad (7)$$

PREDICTIONS USING MULTIPLE PREDICTORS

where r_{y1} and r_{y2} are, respectively, the validities (correlations) of the first and second predictors; r_{12} is the correlation of the two predictors with each other; and SD_y, SD_1, and SD_2 are the standard deviations of the criterion and the two predictors. The equation for the constant in the two-predictor regression is as follows:

$$A_{y12} = \text{mean of } Y - b_1(\text{mean of } X_1) - b_2(\text{mean of } X_2) \qquad (8)$$

where A_{y12} is the constant.

The formula for multiple R is

$$R_{y12} = \sqrt{\frac{r_{y1}^2 + r_{y2}^2 - 2(r_{y1})(r_{y2})(r_{12})}{1 - r_{12}^2}} \qquad (9)$$

It is easy to see that the complexity of these equations is due to the intercorrelation of the two predictors, r_{12}, and the fact that all three variables overlap. In other words, X_1 and X_2 may predict the same portion of the variability in job performance. If they do not overlap (i.e., $r_{12} = 0.00$), then the previous equations become simple.

We go through all these steps in evaluating two (or more) predictors of a criterion of interest so we can obtain an estimate of the likely prediction of that criterion when we use the same predictors but on a new sample. That is, because the R we obtain in one sample is the best possible description of the data only for that sample, we then should be concerned about what will happen in a new sample. Figuring out how large R will be in a new sample is called cross-validation. That is, because the original R estimate includes idiosyncrasies (error) associated with the particular sample on which it is calculated, we need some way to judge how R is going to behave in a new set of data.

Cross-Validation of Multiple Regression

There are two procedures to use to estimate cross-validated R (how well our prediction equation works in a new sample). The more traditional proposal (Mosier, 1951), called double cross-validation, involves the following steps:

1. Randomly split the validation sample into two equal groups.
2. Compute R and regression weights for the predictors in both samples.

3. Use the regression equation computed in each sample to predict job performance values in the other sample.
4. Estimate cross-validated R by correlating these predicted values with actual values of Y.
5. Average the two resulting Rs to obtain an estimate of the R one can expect in subsequent samples.

The double cross-validation procedure is not used much any more because it has been shown that estimating regression weights based on using all the data available is more effective than splitting the sample for double cross-validation (Schmitt, Coyle, & Rauschenberger, 1977).

If one has not preselected predictors (based on earlier evidence) prior to estimating R, a much more desirable method of estimating cross-validated R is to use the formula provided by Burket (1964):

$$\rho_c = (NR^2 - k)/(R(N - k)) \qquad (10)$$

where R is the sample multiple R, ρ_c is the estimated population cross-validity, N is the number of people in the sample, and k is the number of predictors in the regression equation. The population cross-validity is the value of R we would expect if we cross-validated a prediction equation in an infinite number of new samples. Consider the following example. We have $R = 0.40$ in a sample of 80 with 12 predictors. Applying the formula provided by Burket, we get the following:

$$\rho_c = ((80 \times 0.4 \times 0.4) - 12)/(0.4(80 - 12)) = 0.03 \qquad (11)$$

Clearly, our sample R of 0.40 is much larger than the R we would expect if we applied the original regression equation to predict values of Y. The role of the N/k ratio in this equation is obvious and critical. If this ratio is close to one (the sample size is close to the number of predictors), then a great deal of reduction in the multiple R will occur. The larger this ratio is, the less shrinkage will occur. The lesson for validation research is obvious: Large sample sizes are important—and, obviously, one must choose one's predictors carefully to keep (a) the number of them low and (b) their validity with the criterion as high as possible.

RELIABILITY

There are three very practical uses of correlation coefficients: (a) reliability estimates, (b) examination of the underlying dimensions that may account for the correlations among our measures, and (c) estimates of the relationships between job performance and predictor data. In the following section, we discuss various ways of estimating reliability and determining the underlying structure of our measures. We define validity, but defer discussion of the concept of validity and various validation designs for presentation in chapter 6.

Reliability

Reliability of measurement is necessary if the measure is to be useful in making any type of decision, including selection decisions. *Reliability* can be generally defined as the reproducibility of test scores. For example, if you take a test today and retake the same test tomorrow, you should get about the same score. However, what if when you retook the test you received a completely different score? Would you place any faith in the results of either test score? This is why reliability is so important for interpreting test scores and making decisions; lacking reliability, one can place little faith in test scores. It is for this reason that reliability is often considered to be the upper limit for validity.

Reliability can be assessed in several ways, all of which depend on the computation of correlations among different measures. Each estimate of reliability differs in terms of what is treated as error variance, that is, measurement of something other than what was intended. Thus, any given test contains both a *true score* component and an *error* component, such that

$$x = t + e \qquad (12)$$

where x refers to variance in the manifest variable, t refers to the true score variance (i.e., underlying latent construct), and e refers to error variance (i.e., any variance not associated with the latent construct, such as test fatigue, exhaustion, and forgetting). According to classical test theory (CTT), reliability may be conceptualized as follows:

$$\rho = t/(t + e) \qquad (13)$$

This general form of reliability only suggests reliability is equal to the amount of true score variance divided by the total variance in a test or measure. Clearly, reliability can only range from 0 (no reliability) to 1 (perfect reliability); values of 0.70 or 0.80 are usually interpreted as minimally acceptable. However, operationally there are several different ways one can use the correlation to estimate reliability, and these various estimates differ in what they consider to be error variance (e.g., Schmidt & Hunter, 1996).

Test–Retest Reliability

First, the most commonsense notion of reliability is that of the stability of measurement. To calculate the stability of a measure, we correlate scores of persons on the same measure given at two different points in time. This is most often called *test–retest* reliability. Implicit in this treatment of reliability is the notion that any changes in the scores of persons over time are errors. Perfect reliability of measurement in this sense would mean individuals would be ordered in identical fashion both times the test was given. There are several reasons why a person's score might change over time, including forgetting, practice, or learning.

Parallel Forms Reliability

A second estimate of reliability can be obtained by correlating scores derived from *parallel forms* of a test. Parallel forms of a test are supposed to measure the same construct with different items; hence, the content of the items is treated as error. Arithmetically, parallel forms are also assumed to have equal means and standard deviations and item intercorrelations (some models of parallel forms reliability relax these assumptions).

The previous estimates of reliability are based on multiple time periods (test–retest) or multiple versions (parallel forms). Whenever we calculate the reliability of a test based only on an analysis of that test at one point in time, we are calculating what is called internal consistency reliability. There are two types of internal consistency reliability: split-half and coefficient alpha.

Split-Half Reliability

Third, a test is sometimes given for which no alternate form is available. In this case, the test may be split in half (randomly assigning items to halves) and *split-half* reliability may be calculated. In this procedure, each individual taking the test is given a score on both halves of the test, and the two sets of half scores are correlated across all test takers. However, reliability is related to test length (such that longer tests are more reliable), and the correlation between half scores represents the reliability of a test half as long as the one on which we will make decisions. Assuming items in the two halves are equivalent in the sense defined previously, we can use the Spearman Brown correction formula to estimate the reliability of the full-length test. This formula reads as follows:

$$r_f = (2r)/(1+r) \tag{14}$$

where r_f represents the reliability of the full-length test, and r is the correlation between the two sets of half scores. For example, if the half scores on a test are correlated to be 0.6, the reliability of the full-length test would be

$$r_f = (2 \times 0.6)/(1 + 0.6) \text{ or } 0.75 \tag{15}$$

Coefficient Alpha Reliability

Fourth, the notion of a split-half reliability led to development of what is perhaps the most popular single way to estimate internal consistency reliability: *coefficient alpha*. Coefficient alpha rests on the assumption that a test can be split in many different ways. If we split a test in all these ways, compute the correlations between all pairs of splits, and average the correlations, we have coefficient alpha. Fortunately, this involved process is not necessary. A relatively simple formula exists:

$$\alpha = n^2 (\text{average } COV_{ij})/SD^2 \tag{16}$$

where n equals the number of items in the test, COV_{ij} is the covariance between the items, and SD^2 is the squared standard deviation of the test, or the test variance. The covariance is an unstandardized correlation or the correlation between items multiplied by both items' standard deviation.

These latter three estimates of reliability—parallel forms, split-half, and coefficient alpha—are reliability estimates that treat content differences in items on the measure as error. Such differences might reflect minor changes in the wording of the items or differences in the way people interpret the items. Because all three involve administration of a measure or two forms of a measure at a single point in time, stability, or lack thereof, is not treated as error.

Incidentally, the general form of the Spearman Brown formula is extremely useful. It reads as follows:

$$r_f = (nr)/(1 + (n-1)(r)) \tag{17}$$

where r_f is a full-length test as before, r is the reliability of the original set of items, and n is the number of times by which a test's length is multiplied. Assume, for example, we have a three-item measure that has an internal consistency reliability of 0.50 and that we can develop six new equally good items (their intercorrelations or covariances are the same as those of the original three) to measure the concept. This nine-item measure will be three times as long as the original, and we want to estimate its reliability. Applying the Spearman Brown formula gives the following numbers:

$$r_f = (3 \times 0.5)/(1 + (3-1)(0.5)) = 0.75 \tag{18}$$

The 9-item test should have a reliability of 0.75. Or, alternatively, we have a 30-item test with a reliability of 0.90, and we want to shorten this measure to 12 items. What would its reliability be? In this case, $n = 12/30$ or 0.4:

$$R_f = (0.4 \times 0.90)/(1 + (0.4 - 1)(0.9)) = 0.78 \tag{19}$$

The short test has a reliability of 0.78, and we have saved time and space associated with 18 items.

These examples reveal an interesting characteristic of these kinds of estimates of the internal reliability of a test or measure: The incremental reliability gained from adding items to a test or measure falls off rather quickly.

To see this, think for a moment of how many questions you would want to have on your final exam so you would feel comfortable that the exam was reliable. One question, 5 questions, 25 questions, 125 questions? Five questions is certainly better than 1, and 15 better than 10; 20 is better than 15, and 25 better than 20. We know intuitively that the incremental reliability in going from 1 to 5 is greater than going from 5 to 10 and so on. Thus, although length always increases internal consistency reliability (when the added length is at least as reliable as what exists), more and more length must be added after a certain point to obtain equivalent benefits. That is, adding more items results in diminishing returns such that there is a point where adding more items shows little increase in reliability.

Parallel Forms With Time Interval Reliability

A fifth approach to reliability is to assess it with *parallel forms with a time interval* between administrations of the forms. The correlation of the two resulting sets of scores will be the lowest estimate of reliability because both stability and content differences can produce differences across scores for the individuals taking the test. This approach is particularly appropriate when we want to assess performance rating criteria used to validate a selection instrument. Because we are interested in predicting the criterion, we must be certain the criterion itself can be measured with some degree of stability.

Interrater Reliability

Often, we ask individuals (e.g., supervisors) to provide ratings of another person's behavior (e.g., employee performance). In such cases, we hope the performance rating is consistent across raters (or stated differently, not affected by any particular rater). This is the same issue as saying an individual's test score does not depend on a particular test or set of items. Therefore, *interrater reliability* can be viewed as the internal consistency of a test. Interrater reliability is the correlation of the ratings from two raters; if we use the composite of those two ratings as our criterion, then we can correct the intercorrelation of the ratings with the Spearman Brown correction to estimate the reliability of the composite. Likewise, coefficient alpha can be computed for three or more raters.

An appropriate applied question would be "What form of reliability should we use?" The answer is that it depends on our intended use of an instrument, or as Cronbach, Gleser, Nanda, and Rajaratnam (1972) indicate, we must decide to what situations we intend to generalize. An appropriate reliability estimate for a performance rating criterion is interrater reliability with an "appropriate time interval" between ratings. The appropriate time interval depends on the period of time over which predictions about the criterion are to be made. For managers, this could be 3 to 5 years; for short order cooks, 2 weeks. When our primary interest is if our test measures a given set of content, then an internal consistency measure is more appropriate. We revisit this issue in chapter 4.

Consequences of Reliability

As we said at the beginning of the discussion of reliability, we are interested in reliability because it places limits on validity—the correlation of our measure with other measures or phenomena of interest. More generally, unreliability attenuates (reduces) effect sizes and makes them appear smaller than they really are. We can use our reliability estimates to estimate what the correlation among various measures would be if they were measured with perfect reliability. This estimate for two variables, x and y, is given as follows:

$$\text{Corrected } r_{xy} = \frac{r_{xy}}{\sqrt{r_{xx}}\sqrt{r_{yy}}} \qquad (20)$$

where r_{xy} is the observed correlation, and r_{xx} and r_{yy} are the reliabilities of the predictor and criterion, respectively. Note that theoretically the maximum true intercorrelation between two variables ought to be 1.00, but it will not be because of imperfectly reliable measures. For example, the highest correlation between two measures, both of which are measured with a reliability of 0.80, would be 0.80. Hence, we say observed correlations are attenuated and the previous formula is referred to as the *correction for attenuation*. Observed validity coefficients (or relationships between manifest variables using terminology introduced earlier in the chapter) are typically corrected for unreliability of measurement in both variables to estimate true validity (or the relationship between latent constructs). Corrections for lack of reliability in the predictor are usually not made in practice, because when using the predictors for selection, we must live with their unreliability. When assessing the relationship between two latent constructs in conducting basic research in test theory or when examining construct validity (chapter 6), corrections for lack of reliability in the measurement of both manifest variables are made. For example, suppose two measures of extraversion have reliabilities of 0.50 and 0.80. If the observed correlation is 0.50, we might conclude there is not much overlap between the two measures. However, after we correct for unreliability, we find this correlation is actually 0.79 (i.e., the square root of 0.50 × 0.80 is 0.63, and 0.50 divided by 0.63 is 0.79). Thus, the "low" correlation is an artifact of the poor reliability of the measures, especially the first variable.

Reliability can be used in still another way—to examine the precision or confidence we can place in the fact that our test scores represent an individual's true level on some measurement. Theoretically, if it were possible to repeatedly give a test to some person and make each observation independently, not all measured values would be the same, but they would distribute themselves around the true value of this person's tested ability. The standard deviation of this distribution is called the standard error of measurement and is given by the following formula:

$$SEM = SD_{test}\left(\sqrt{1 - r_{xx}}\right) \qquad (21)$$

Assume we have a person whose score on some test is 40 and the standard error of measurement is 5.00. The obtained score of 40 could be part of a distribution of scores for true values as different as 30 and 50 (given four standard errors would encompass about 96% of the cases). So, we could say that an obtained score of 40 could be a result of a true score (a person's standing on the latent construct) anywhere between 30 and 50. Examination of similar confidence limits for tests with very high reliabilities will usually indicate that any given score can be evidence of relatively different levels of true score on the part of a given person. The lesson of the standard error of measurement is that we exercise caution in statements about the scores we get on a particular individual and that we are cautious when we make comparisons of persons or scores. These cautionary statements are facilitated by many of the better test manuals now published because they give confidence limits for various score levels.

VALIDITY

Validity is the degree to which the inferences we draw about our test scores are correct. Correlation and regression analysis are used extensively in examining the relationship between the test scores on our instruments and measures of job performance. This approach to validity is called criterion-related validity, and we have already seen how correlation and regression can be used to make job performance predictions. In chapter 6, we extend our discussion of the concept of validity and identify various strategies by which personnel researchers attempt to establish the validity of their procedures. Validity is an important topic in staffing and to fully appreciate it, it is first necessary to understand measurement (this chapter), job analysis (chapter 3), performance (chapter 4), and recruitment (chapter 5). This is why we do not discuss validity in more detail here.

2. MEASUREMENT AND DATA ANALYTIC TOOLS

Correlation and regression are the basic analytic tools used to summarize relationships of interest to staffing researchers. Various more sophisticated analytic tools that build on the basic concepts underlying correlation and regression are also frequently used, particularly in scientific papers. In the next sections of this chapter, we introduce the basic notions underlying the most frequently used of these topics, along with references that may help the interested person gain expertise in the use of these methods.

MODERATED REGRESSION

A frequently proposed hypothesis among staffing researchers is that the relationship between two variables is stronger or weaker for one group of people than another. The variable that separates people in this manner is called a *moderator variable*. In staffing research, the most often discussed and researched moderator variables have been race and gender. The notion that test–performance relationships are different for males and females or Blacks and Whites has generated a great deal of social and political debate and countless legal cases since the mid-1960s. There has also been interest in moderator variables that are not discrete like race and gender. For example, one hypothesis is that the relationship between ability and performance on a job is partly a function of achievement motivation. That is, the hypothesis may be that for people with more achievement motivation, we should find strong ability–performance correlations, whereas for people whose need for achievement is not that high, we should observe lower correlations between ability and performance. In this latter example, achievement motivation can take on many different levels (i.e., we often say that achievement motivation is continuous), whereas in the gender and race examples, the moderator is discrete (i.e., a nominal variable).

To analyze the degree to which some variable moderates a relationship, we use *moderated regression*. In these cases, we have a predictor (ability), an outcome or criterion (rating of overall performance), and a moderator variable (achievement motivation). Evidence for differential prediction, or a moderator variable, is analyzed using the following regression equation:

$$\text{Predicted performance} = b_1(\text{Ability}) + b_2(\text{Ach.mot.}) + b_3(\text{Ability} \times \text{Ach.mot.}) + A_{xy}$$

where b_1, b_2, and b_3 are the regression weights for ability, achievement motivation, and the arithmetic product of ability and achievement motivation (Ach. mot.), respectively. Ability and motivation are entered first in this regression, followed by the product of these two variables (Aiken & West, 1991; Bartlett, Bobko, Mosier, & Hannon, 1978).

If the regression weight for ability times achievement motivation is statistically significant, then there is evidence that the relationship between ability and performance varies as a function of the level of achievement motivation. To provide an interpretation of how the ability–performance relationship varies as a function of achievement motivation, we usually plot the relationship for arbitrarily selected high and low (usually 1 SD above and below the mean) levels of achievement motivation. A hypothetical moderated regression equation and the results of plotting this interaction are shown in Fig. 2.12. The equation is presented at the top of the figure with the means and standard deviations of the two variables. Predicted performance values of high-achieving individuals (those with scores 1 SD above the mean) of low and high ability (1 SD below and above the ability mean) are then calculated. The same is done for low-achieving individuals (i.e., 1 SD below the achievement mean). These four predicted performance values are then plotted in the bottom part of Fig. 2.12. In this case, we see there is evidence to support the hypothesis that ability is related to performance for both high- and low-achieving individuals, but that the relationship is stronger for the high-achievement group.

Moderated regression has been used frequently in staffing research to investigate differential prediction across subgroups usually defined by race or gender. Moderated regression represents the commonly accepted statistical methodology used to examine test bias issues (American Educational Research Association, American Psychological Association, & National Council on Measurement in Education, 1999). Results of studies of differential prediction and the implications for selection of members of different subgroups are presented in chapters 7 to 10.

However, it is not only for different groups of individuals that we might propose potential moderation. For example, it might be hypothesized that the relationship between ability and performance will vary as a function of the kind of leadership under which employees work. That is, we could develop the hypothesis that under democratic leadership the relationship between ability and performance is stronger than under autocratic leadership. This hypothesis would follow from the idea

Moderated Regression Equation
Predicted Performance = 6.0 + 0.6(Ability) + 0.1(AchMot) + 0.5(Ability × AchMot)

$\bar{x}_{Ability} = 10, SD_{Ability} = 4$
$\bar{x}_{AchMot} = 12, SD_{AchMot} = 3$

High-Achievement Individuals (+1 SD = 15)

$-1\ SD_{Ability} = 6.0 + 0.6(6) + 0.1(15) + 0.5(90)$
$= 56.1$
$+1\ SD_{Ability} = 6.0 + 0.6(14) + 0.1(15) + 0.5(210)$
$= 120.9$

Low-Achievement Individuals (−1 SD = 9)

$-1\ SD_{Ability} = 6.0 + 0.6(6) + 0.1(9) + 0.5(54)$
$= 37.5$
$+1\ SD_{Ability} = 6.0 + 0.6(14) + 0.1(9) + 0.5(126)$
$= 78.3$

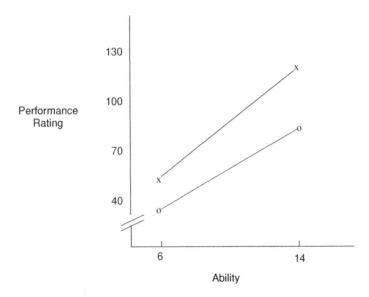

FIG. 2.12. Regression equation and plot of the relationship between an ability test, achievement motivation, and job performance. x = high achievement, o = low achievement.

that individual differences among people are more easily expressed under conditions of democratic leadership. The statistical model that would be used for exploring such a possibility would be identical to that just described using achievement motivation of individuals as the potential moderator. This time, however, the potential moderator is a situational variable (leadership). It turns out to be true that such potential moderator variables, although theoretically interesting regarding both personal characteristics and situational attributes, are difficult to find consistently in the real world (Schneider, 1983).

STRUCTURAL EQUATION MODELING

In many cases, researchers have provided theoretical arguments for much more complex models of work behavior than is represented by a simple ability–performance relationship or even a regression equation that includes several variables that are believed to predict work behavior. One such model is depicted in Fig. 2.13. Several hypotheses are implicit in Fig. 2.13, which represents one explanation of supervisory

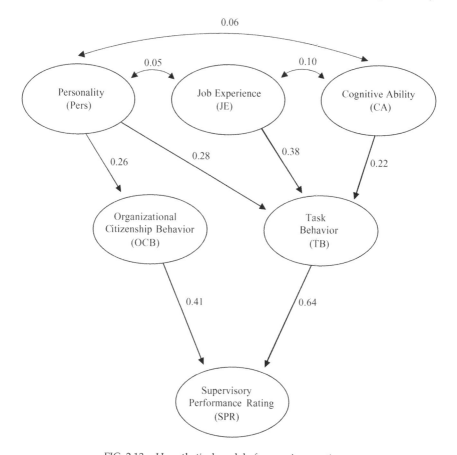

FIG. 2.13. Hypothetical model of supervisory ratings.

Observed Correlation Matrix

	1	2	3	4	5	6
Pers(1)	1.00					
JE(2)	0.05	1.00				
CA(3)	0.06	0.10	1.00			
OCB(4)	0.26	0.03	0.12	1.00		
TB(5)	0.31	0.42	0.28	0.03	1.00	
SPR(6)	0.18	0.25	0.21	0.43	0.65	1.00

Reproduced Correlation Matrix

	1	2	3	4	5	6
Pers(1)	1.00					
JE(2)	0.05	1.00				
CA(3)	0.06	0.10	1.00			
OCB(4)	0.26	0.01	0.02	1.00		
TB(5)	0.31	0.42	0.28	0.08	1.00	
SPR(6)	0.30	0.27	0.18	0.46	0.67	1.00

Examples of Reproduced Correlation and Indirect Effects

$r_{SPR.PERS}$ = $(0.26 \times 0.41) + (0.28 \times 0.64) + (0.05 \times 0.38 \times 0.64) + (0.06 \times 0.22 \times 0.64)$

= 0.1066 + 0.1792 + 0.0121 + 0.0084 = 0.3063

$r_{SPR.JE}$ = $(0.38 \times 0.64) + (0.10 \times 0.22 \times 0.64) + (0.05 \times 0.26 \times 0.41) + (0.05 \times 0.28 \times 0.64)$

= 0.2432 + 0.0141 + 0.0052 + 0.0090 = 0.2715

$r_{SPR.CA}$ = $(0.22 \times 0.64) + (0.06 \times 0.26 \times 0.41) + (0.06 \times 0.28 \times 0.64) + (0.10 \times 0.38 \times 0.64)$

= 0.1408 + 0.0064 + 0.0108 + 0.0243 = 0.1823

FIG. 2.13. (Continued)

performance ratings. First, these ratings should be explained by workers' actual task behavior (i.e., performing the work as defined by one's job description) and behavior that represents good organizational citizenship (cleaning the work area, organizing social events, or helping a worker with problems). The impact of personality, work experience, and cognitive ability on supervisory ratings are all mediated by the workers' behavior—either task related or of a citizenship nature.

The theory represented in fig. 2.13 posits that personality, work experience, and cognitive ability affect supervisory ratings of performance, but only indirectly through their impact on worker task and citizenship behavior. These mediation models are popular in psychology and often specify the process by which variables affect each other. Mediation is distinguished from moderation discussed in the previous section. Moderation specifies that the relationship between two variables is direct,

STRUCTURAL EQUATION MODELING

but that the relationship differs as a function of values of the moderator variable.

In contrast, the model in Fig. 2.13 does not posit that personality, experience, or ability directly affect the supervisors' ratings, but rather that they indirectly affect the ratings through their mediating effects on actual work behavior. Each single-headed arrow in Fig. 2.13 represents a relationship that can be estimated using regression (double-headed arrows indicate covariances). The simultaneous estimation of these regression equations, including the regression coefficients associated with each predictor, is termed *structural equation modeling*. The equations represented in Fig. 2.13 are the following:

Supervisory rating = Task behavior + Citizenship behavior + Error
Task behavior = Ability + Experience + Personality + Error
Citizenship behavior = Personality + Error

Therefore, each single-headed arrow represents a regression coefficient or path from one variable to another. Each variable to which an arrow points represents an outcome variable in a separate regression equation. These paths (as well as the absence of possible paths) represent one possible explanation (the investigator's hypothesized model) of the observed relationship among all six variables in the diagram. That is, the investigator's theory defines how the variables should be related to each other. Given this theoretical model and the observed data, software programs (e.g., AMOS, EQS, LISREL) provide a set of estimates of the regression coefficients. These regression coefficients are then used to reproduce the covariances (or correlations) between the variables in the model. These reproduced covariances are compared with the actual observed covariances, and various measures of the degree to which the reproduced covariances are similar to the actual covariances are computed as estimates of the fit of the hypothesized model to the data. If the theoretical model is correct, the reproduced covariances should be similar to the actual covariances. Structural equation models are always evaluated using covariances, but for simplicity, the model and example represented in Fig. 2.13 consists of standardized covariances, or correlations.

At the bottom of Fig. 2.13, we present the actual observed correlations between the six variables in our model. These data, along with a specification of the regression equations that are believed to explain the relationship among the variables, were input to the LISREL software package, and the estimates of the path coefficients contained in Fig. 2.13 were obtained. The statistical significance of each coefficient is estimated as tests of individual relationships between variables.

These estimated regression coefficients are also used by the program to reproduce the correlation matrix. As a general rule, the reproduced correlations are a sum of the products of the standardized path coefficients that connect two variables. Three examples of the computations and the fully reproduced matrix are presented at the bottom of Fig. 2.13. Reproduced correlations are then used as the basis of testing the overall fit of one's model. As mentioned previously, the difference between the observed and reproduced covariance matrix is used to compute various indices of fit (see Bollen & Long, 1993; Tanaka, 1993).

The examples of reproduced correlations at the bottom of Fig. 2.13 involve the correlations of the three exogenous variables (their determinants are not specified in the model) and the major outcome variable. As such, the different explanations of the correlations between these variables are all indirect or mediated. Each product in the reproduction of these correlations represents a different explanation of the correlation of these predictors of supervisory ratings. For example, the reproduced correlation between supervisory ratings and personality is 0.3063. The product of 0.26 and 0.41 (0.1063) represents the portion of this correlation that is due to the impact of personality on the organizational citizenship behavior mediator. The product of 0.28 and 0.64 is the portion of the correlation between personality and supervisory ratings that is mediated by task behavior. The other two products in this reproduced correlation are relatively tiny and result from the fact that all three original predictors are intercorrelated (see the curved lines and corresponding values at the top of Fig. 2.13).

What we have just illustrated is the analysis of a set of relationships between six observed variables in which our hypothesized model specifies a set of causal relationships between the variables, or a structural model.[3] Structural equation modeling can also be used to test the degree to which latent, or unobserved constructs, provide explanations for a set of relationships among manifest or observed variables. When we are only interested in the degree to which a set of manifest variables represents a latent variable or variables, we are conducting a *confirmatory*

---

[3]Note that the use of the term "causal relationships" does not mean we can prove which variable causes another. Despite some prior claims, SEM models (and HLM discussed next) do not test causality as much as they provide inferences of causality. Longitudinal models with multiple measurement sources or experimental research designs are necessary to provide the strongest inferences of causality.

STRUCTURAL EQUATION MODELING 85

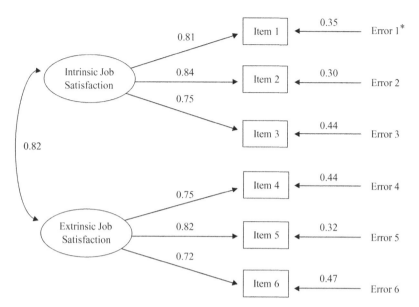

*Numbers for error are actually error variances.

Observed Correlations							Reproduced Correlations					
	<u>1</u>	<u>2</u>	<u>3</u>	<u>4</u>	<u>5</u>	<u>6</u>		<u>1</u>	<u>2</u>	<u>3</u>	<u>4</u>	<u>5</u>	<u>6</u>
Item 1	1.00						Item 1	1.00					
Item 2	0.63	1.00					Item 2	0.67	1.00				
Item 3	0.63	0.64	1.00				Item 3	0.59	0.62	1.00			
Item 4	0.46	0.50	0.41	1.00			Item 4	0.49	0.52	0.46	1.00		
Item 5	0.61	0.62	0.39	0.62	1.00		Item 5	0.54	0.57	0.51	0.61	1.00	
Item 6	0.46	0.53	0.45	0.57	0.57	1.00	Item 6	0.48	0.50	0.45	0.54	0.60	1.00

FIG. 2.14. A confirmatory factor model of six items measuring extrinsic and intrinsic job satisfaction.

factor analysis (CFA) or evaluating a *measurement model*. One such measurement model is depicted in Fig. 2.14. In Fig. 2.14, boxes represent manifest measures, and circles represent latent constructs. According to this model, we are hypothesizing three manifest items of a job satisfaction measure that indicate individuals' status on a latent intrinsic satisfaction construct, whereas three different manifest items index a latent extrinsic satisfaction construct. As in the structural model described previously, Fig. 2.14 specifies a set of regression equations in this case:

Item 1 = Intrinsic satisfaction + Error
Item 2 = Intrinsic satisfaction + Error
Item 3 = Intrinsic satisfaction + Error

Item 4 = Extrinsic satisfaction + Error
Item 5 = Extrinsic satisfaction + Error
Item 6 = Extrinsic satisfaction + Error

These equations indicate that the shared variance in items 1 to 3 is due to intrinsic satisfaction and that the shared variance in items 4 to 6 is due to extrinsic satisfaction. If you recall the discussion of reliability presented earlier, you can see this method maps directly onto the classical test theory approach for estimating reliability, such that each item is composed of a true score component (either intrinsic or extrinsic satisfaction) and an error component, or $x = t + e$.

In addition, the covariance, or correlation, between the two latent constructs is estimated. For each equation, a regression coefficient for the hypothesized predictor (intrinsic or extrinsic satisfaction) is estimated. Model specification (hypothesizing a model that explains the covariances among observed variables) and model estimation (getting estimates of the regression coefficients) is followed as discussed previously by evaluation of model fit. This involves comparing the observed covariance matrix with that reproduced using the estimates of the parameters in our model and computing the various model fit statistics referred to previously. This is done for the factor model at the bottom of Fig. 2.14. As is obvious again, the reproduced correlations are not exactly the same as the observed correlations. This is evidence of a lack of perfect fit of the hypothesized model to the observed data and will be reflected in the various measures of model fit that are used to evaluate models.

A full structural equation model involves the simultaneous estimation of the measurement and structural model. One such model is specified in Fig. 2.15. As was true in the two previous examples, this model specification is followed by estimation of the regression coefficients implied in this model (all variables with arrows pointing to them represent outcomes in a separate equation) and assessment of overall model fit. In these models, the structural model involves estimates that reflect only relationships among the latent variables. Errors in the measurement model that would influence our conclusions about the structural relationships have been corrected for in providing the structural model coefficients. This correction is similar to the correction for attenuation due to unreliability described previously in the reliability section. The end result is estimation of relationships among the latent or true variables in our model. It is hoped that we have provided sufficient detail to

HIERARCHICAL LINEAR MODELING

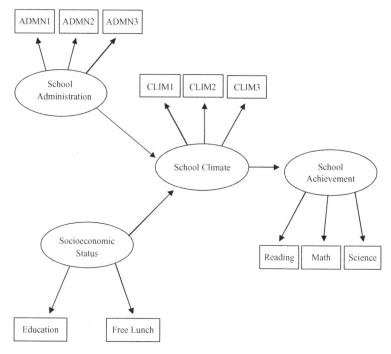

FIG. 2.15. An example of a model that includes both structural and measurement components. Each indicator variable in boxes also has an error variance associated with it, but that error is not depicted.

allow you to read papers that use SEM in evaluating complex models of human performance. Full descriptions of the use of this technique are provided in numerous textbooks (e.g., Bollen, 1989; Kline, 2004).

HIERARCHICAL LINEAR MODELING

In chapter 1, we made a strong case for the idea that we cannot fully understand work behavior unless we consider the context (e.g., group, organization, social unit) in which that behavior occurs (Fig. 1.1). When we use multiple individuals from a single group and their group status has some influence on their work behavior, our observations of these people are not independent of each other. For example, suppose you have to complete a class paper as a group. If one or more of your group members does not do their share of the work, your grade may be affected, and thus your behavior must change to make up for the uncooperative group member or members. The independence of observations (specifically the independence of the errors we make in estimating the outcome variable in a regression equation) is an assumption of the regression-based tools discussed in this chapter. Ignoring violations of this assumption can distort our estimates of their statistical significance, and sometimes, even the estimates themselves. Hierarchical linear modeling (Bliese, 2002; Raudenbush & Bryk, 2002) not only avoids this problem, but also explicitly models the influence of context, thereby providing a much richer explanation of behavior than do single-level models.

Hierarchical linear models (HLMs) are conceptually similar to the regression analyses described previously, but they provide additional parameters to account for the variance in behavior due to levels above the individual level, which comprises the usual unit of analysis in staffing. Such levels most often include the group, department, or organization levels.

Recall that earlier in this chapter we defined a simple regression equation as follows:

$$\text{Predicted } Y = b_1 X + A_{yx} \tag{22}$$

In this equation, b_1 is the slope of the regression line and A_{yx} is referred to as the intercept. We also noted our regression equations also include an error term; that is, we do not predict Y perfectly. When the context or group in which an individual is located influences Y, part of the error in the equation occurs because we have not taken into account the fact that the regression equation is different for different groups of individuals in our analyses. If we are interested in the performance of individuals who work in teams and the behavior of individuals in each team is described by a different regression equation, rather than a single overall equation, then we can have variability in the slopes and/or intercepts of the regression equation. This means every person could have a different slope and/or intercept, and these differences may in part be explained by the group to which they belong. In these cases, a superior model of behavior is represented by the following set of equations:

$$Y_{ij} = b_{1j} X_{ij} + A_{yij} + e_{ij} \tag{23}$$

$$A_{yij} = \gamma_{00} + u_{0j} \tag{24}$$

$$b_{1j} = \gamma_{10} + u_{1j} \tag{25}$$

Notice Eq. (23) looks just like the usual regression Eq. (22), except there are now subscripts to denote individual-level observations (i) and groups (j). However, because intercepts (A) and slopes (b) may differ across groups, Eq. (24) and (25) are necessary to model the extent to which such variation may be explained by between-group differences. Note that the u's in Eq. (24) and (25) refer to group-level errors, or the variance in intercepts and slopes that is not explained by between-group differences. In this manner, we can estimate the amount of variability in intercepts across groups (u_{0j}), and the amount of variability in slopes across groups (u_{1j}). When there are no group differences in slopes and intercepts, these equations reduce to Eq. (23).

So, for example, if one is interested in examining the relationship between ability and performance in different work teams, we begin by using Eqs. (23) to (25), which involve estimating the group average intercept (γ_{00}) and slope (γ_{10}) and the variance in intercepts (u_{0j}) and slopes(u_{1j}). If the intercept variability is significant and large, we know there are differences between teams in their mean performance (i.e., the groups have different average levels of performance). If the slope variability is significant and large, we know that for some teams the relationship between individual ability and individual performance is strong, whereas for other teams, the relationship is weak, zero, or negative. We also know the typical regression equation is too simplistic as a model of work performance in this instance because there are between-group differences in intercepts and slopes. Perhaps an equally interesting question is what predicts the variance in intercepts and slopes. For example, we might hypothesize that the ability–performance slope differences across groups is a function of the leadership style in the teams and that the between-group differences in intercepts is a function of the economy in the areas from which applicants for these work teams are chosen. As you probably guessed, we can also add variables to our regression model to evaluate these hypotheses.

The mechanics of HLM along with an example are well explained in Bliese (2002), and a complete exposition of various HLM models is available in Raudenbush and Bryk (2002). Staffing researchers have only begun to consider the potential hierarchical nature of many of the relationships we study and, as we argue in chapter 1, this should change the way practitioners and scientists think about individual differences in work behavior. At the least, we must be aware that our simple regression equations with one or several predictors may be incomplete. However, if we seek a real understanding of the nature and causes of

2. MEASUREMENT AND DATA ANALYTIC TOOLS

hierarchical effects, we will almost certainly provide more complex and useful explanations. Ployhart and Schneider (2005) described how to conduct and analyze a multilevel staffing study using HLM and related approaches.

ITEM RESPONSE THEORY

The discussion of reliability, the standard error of measurement, and the relationship between test length and reliability in previous sections of this chapter are part of what is often referred to as CTT. In the last few decades, *item response theory* (IRT), which addresses similar problems in constructing and evaluating tests and test scores, has become a widely used alternative to CTT. The basic unit in IRT is the item response curve (IRC), which expresses the relationship between test respondents' underlying ability and the probability that they will answer an item correctly. One such curve is illustrated in Fig. 2.16, and we use it to discuss the parameters in this model of examinee responses. The theory underlying IRT assumes the relationship between ability and the probability of a correct answer is nonlinear and reflected in the normal ogive (s-shaped) curve represented in this figure, and IRCs are scaled accordingly. The commonly used IRT model also assumes

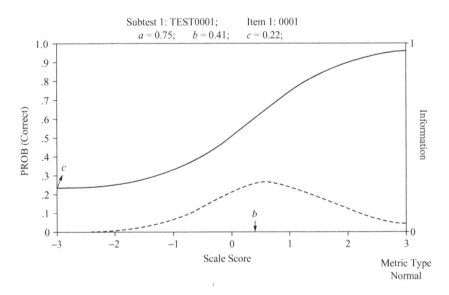

FIG. 2.16. Item response function and item information curve.

ITEM RESPONSE THEORY 91

all items measure a single underlying or latent trait, although multidimensional models are now used in some operational settings (e.g., Reckase, 1997). Therefore, the vertical axis represents the probability of getting a test question correct, whereas the horizontal axis represents the underlying person's true score (note that the true score is placed on a z distribution). In IRT, the true score is denoted by the Greek symbol theta (θ).

The most complex of the IRT models includes three parameters. In Fig. 2.16, the steepness of the curve is referred to as the item discrimination parameter (a). Item discrimination values typically range from 0.5 to 1.5, with higher values indicating a steeper slope and better discrimination within a narrow range of ability. A flat IRC indicating an absence of discrimination would have a parameter of zero. In this case, the probability of a correct answer is the same regardless of the person's ability; clearly, such items do not contain information about the person's level on the underlying trait. The steepness of the curve is measured at the point indicated by b in Fig. 2.16.

The second parameter that provides information about the IRC is referred to as the difficulty parameter, b. The difficulty parameter usually varies from -2.0 to 2.0, reflecting the fact that the difficulty of most items will be within 2 SDs of the mean of zero. The ability, or trait level, of respondents and the difficulty of the items are arbitrarily scaled so they have means of 0 and standard deviations of 1. Difficult items have large positive b values, and the IRC in these cases is shifted to the right. Easy items have large negative b, with IRCs shifted to the left. In the latter case, even persons with low ability have a high probability of getting the item right. The b value is the estimate of 0 at half the distance between the lower asymptote of the IRC and the upper asymptote of the IRC.

The last parameter that describes the IRC is c, which represents the lower asymptote of the IRC as depicted in Fig. 2.16. The value of c is usually 0 to 0.25 and reflects the probability of a person with very low trait values getting an item right or endorsing the item in the scored direction. This parameter is most frequently a nonzero value in multiple-choice exams. For example, with a four-choice item, we would expect c to be approximately 0.25 because there is a 25% chance of a person getting an item correct simply by guessing.

When all three parameters are used to describe an IRC, it is said that we have a three-parameter model. Frequently, the value of c can be assumed or fixed to be equal across all items (e.g., if all items are four-option, multiple-choice items). In this case, all IRCs would have the same asymptote. Two-parameter models in these instances will be adequate to describe item responses.

A single-parameter model results when the item discrimination parameter is also fixed to be equal across all items. In this case, IRCs would be parallel to each other; the only difference being their location on the theta or true score axis. Assuming item discrimination parameters to be equal is equivalent to saying that all items represent a trait equally well. This model, sometimes called the *Rasch* model, occasionally provides an adequate representation of item responses, but its popularity among researchers has waned with the development of more efficient computational methods to estimate the more complex models and with the development of faster computers.

The formula for the probability of a correct answer also contains a value that represents the trait level of the person being measured.[4] Because of the inclusion of the person's trait level in the calculation of the item statistics and vice versa, these parameters are said to be invariant (constant) across samples from a given population. This is not true for CTT; for example, the difficulty value of an item is a function of the ability of the group who took the item. The fact that these parameters are invariant allows us to estimate persons' trait values with different items. The application of this notion is responsible for computer adaptive tests (CATs), which we describe in more detail later. For a more detailed explanation of invariance and its implications for the solution of various measurement problems, see Embretson and Reise (2000) and Hambleton, Swaminathan, and Rogers (1991).

Another important idea in IRT is that items afford differing amounts of information about the person being measured, given that person's standing on the trait and the difficulty of the item being considered. The dashed line at the bottom of Fig. 2.16 represents the information provided by this item for individuals of different standing on the trait. There is, of course, a formula that provides a quantitative index of information (Embretson & Reise, 2000). In this diagram, we can see the item provides the most information close to its difficulty value, b, and at the point where the IRC is the steepest. At low and high levels of the trait, the IRC is relatively flat, indicating that the item

---

[4] $P_{(correct/\theta)} = c + (1-c)/(1 + e^{-1.7a(\theta-b)})$, where θ represents the trait level of the person measured; e is the natural logarithm, approximately 2.72; and $P_{(correct/\theta)}$ is the probability of a person with the trait level equal to θ getting the item correct.

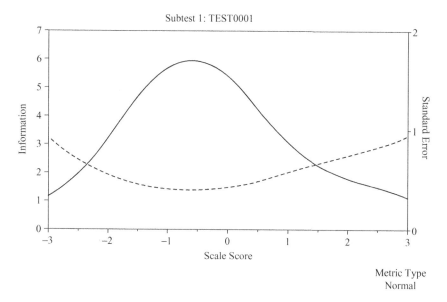

FIG. 2.17. Test information and measurement error.

does not discriminate between persons (i.e., provides no information) in these ranges of scores on the trait. One desirable property of these information curves is that they are additive; by adding the curves, we get a test information curve (see the solid line in Fig. 2.17). This means, that with an appropriate choice of items, we can produce tests that vary in their ability to measure people with different levels of the trait (θ) in question. In Fig. 2.17, we see this test provides little information about those at the extremes of the true score distribution, relative to those in the middle of the true score distribution.

The notion that items and tests provide differing levels of information at different ability levels is another way of saying that they have different standard errors of measurement or accuracy at different trait levels. The dashed line in Fig. 2.17 represents the standard error of the test involved at different levels. Recall that in the previous discussion of CTT, we computed a single standard error for the entire test. An IRT-based test has a different standard error, or differing levels of measurement accuracy, at each level of the trait. Thus, those at the extremes of the true score distribution are measured with less accuracy.

Of course, like other representations of the relationships among variables (e.g., regression or SEM), the advantages of IRT models depend on the degree to which these models actually fit our observations or

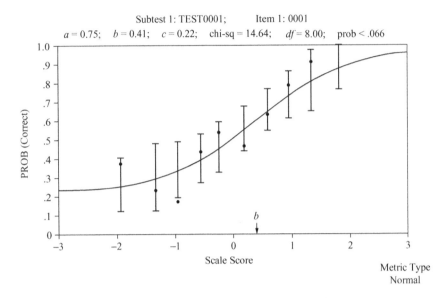

FIG. 2.18. Item response function and observed percent correct.

the responses of a set of examinees. In IRT, the focus has been on the correspondence between IRCs and the actual responses of examinees at different levels of a trait. A graphical depiction of fit is constructed by dividing examinees into groups based on their overall test performance or θ. In each group, one computes the proportion of people who provided the scored response.

These proportions are then plotted next to the IRC. This is done for one item in Fig. 2.18. As can be seen, not all the dots representing the proportion at 10 different levels of θ fall on the IRC; hence, there is some lack of correspondence between actual and observed proportions. This is conceptually similar to the notion of error in multiple regression discussed earlier. There is also a test of the significance of this lack of fit (Mislevy & Bock, 1990). This chi-square value is presented at the top of Fig. 2.18, and the corresponding probability of finding a chi-square this large is printed at the top on the far right. In this case, that probability is 0.066. Usually, the IRC would be considered a poor fit to a set of examinees responses if this probability were less than 0.05 or 0.01.

A fit index for persons, often called *appropriateness fit*, can also be calculated. In the case of person fit indices, we are looking at individuals' response patterns and identifying those whose pattern of endorsements or right–wrong answers do not fit their estimated trait level. So we

TABLE 2.2
Example of a Computer Adaptive Test Item Presentation Sequence

Item Presentation Sequence	Item No.	Item Response	Ability Estimate	Standard Error
1	34	Right	.40	0.85
2	17	Right	.90	0.75
3	31	Right	1.25	0.65
4	8	Right	1.40	0.58
5	4	Right	1.55	0.53
6	19	Right	1.78	0.50
7	45	Right	1.85	0.48
8	7	Right	1.98	0.45
9	13	Wrong	1.80	0.39
10	17	Right	1.85	0.38
11	27	Right	1.95	0.37
12	3	Wrong	1.80	0.34
13	18	Right	1.85	0.34
14	40	Right	1.90	0.34

would be identifying high-trait persons who fail to correctly answer easy items and low-trait persons who get difficult items correct. This information is helpful in identifying people who for whatever reason do not behave like others in the sample. Such people might be those who were careless when completing the test or who tried to misrepresent themselves or fake a test.

There are several important applications of IRT. The use of IRT to develop computer adaptive tests has probably been the most practically useful. When we have calibrated a set of items measuring some trait (i.e., we know their a, b, and c values), we can then use that knowledge to pick items that provide us the most information about an individual with a specified level of the trait in question. An example of the data generated by a set of responses to a CAT is presented in Table 2.2. Before we collect any data on a person, our best estimates of their mean and standard deviation on θ is 0 and 1, respectively, when the trait is standardized. So the best first items to give a new examinee about whom we have no information are items that provide maximum information at the mean of θ, or 0.

In this case, the examinee, who answers items on a computer, answered the first eight items in the keyed direction, so the computer estimate of the person's trait level increased with each item administration. Each new item was chosen to provide maximum information at the new estimate of the person's θ. You should also note that with each new response, the standard error associated with the computer estimate of the person's ability went down, indicating greater accuracy or confidence in the estimate of θ. Each new item is chosen from the set of calibrated items to provide the maximum amount of information about the examinee (see the dotted line in Fig. 2.16). You will notice that as new items are presented, the standard error decreases, but that for the last three items administered, there is no change at all within the two decimal points accuracy provided. At some point, the decrease in standard error becomes so small that the computer (actually the person who programmed the computer) concludes it is not useful to present new items and provides the final estimate of the person's θ (in this case, 1.90 standard deviation units above the mean).

CAT provides several important advantages. For the test taker and administrator, it takes much less time, and fewer items, to get an accurate estimate of the trait level than is usually true when an examinee must take a long test. This is because the examinee is given items that provide maximum information about θ in each step of the examination process. We do not waste time by providing items that are way too easy or difficult given the person's ability. Each examinee can be given different items specific to their trait level, thus minimizing the concern about test security. It should be noted that this is true only when a very large number of items are available at different trait levels. Because of these advantages, several large-scale tests are now available in their CAT form. The one with which students are likely to be most familiar is the Graduate Record Examination or GRE.

In staffing research, IRT is also of great utility in examining the degree to which items are biased for various subgroups of examinees. If IRC's for two or more groups of people are different, this means that people of equal trait levels will have different probabilities of answering an item in the scored direction. This possibility is depicted in Fig. 2.19. In this case, individuals in Group B who are equal in status on θ will have lower probability of responding in the scored direction for all levels of the trait, except at the asymptotes of the IRC's. Obviously, such an item is biased against members of Group B. Identification of biased items is an involved process (Holland & Wainer, 1993), however, and rarely as simple as depicted in Fig. 2.19.

Another practical problem facing test users is the ability to equate people on some trait when their scores on the trait are obtained from different tests. Even when items or tests have been calibrated on samples

SUMMARY

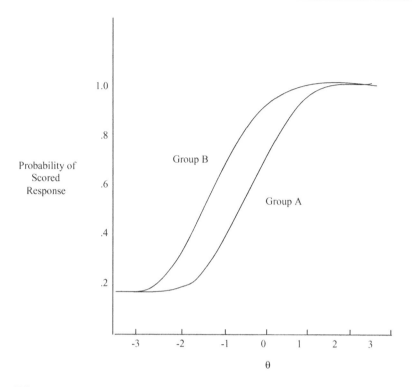

FIG. 2.19. Item characteristic curves on two groups on an item that is differentially difficult for equally able persons from the two groups.

of very different trait levels, it is relatively easy to equate tests using IRT. If the IRT assumptions are met, then the item parameters (i.e., a, b, c) can be transformed linearly by multiplying, adding, or subtracting constants to reflect the calibration group differences. A full discussion of equating item parameters is available in Hambleton et al. (1991).

SUMMARY

We begin this chapter with a discussion of the levels of measurement with which we can assign numbers to persons or objects. Ways of summarizing measurements on many objects and methods of describing the degree of relationship between two or more measured variables (correlation) are described. Regression analysis with a single predictor and multiple predictors are outlined briefly.

Next, we discuss reliability and its relationship to validity of measurement. Various types of reliability information, as well as uses of reliability information, are detailed. Finally, we describe how we could use a relatively simple examination of correlations among scales and items to better describe and understand the measures we take.

Throughout, we become aware of the absolutely central role of the correlation coefficient, r, in understanding the relationships between measures and the characteristics (reliability) of measures. It is critical that the reader grasp the origins of r in data, the degree to which r follows from means and standard deviations in terms of underlying mathematics, and its use as a foundation for much of the discussion in the rest of this book.

We also discuss several much more sophisticated data analytic techniques (i.e., moderated regression, SEM, HLM, IRT). Important and informative analyses of many data-sets do not require the use of these techniques. However, as our models of human behavior become more complex and include considerations of context, these models are necessary. It is hoped that our treatment of these techniques will allow the reader to understand research articles that employ these techniques and that the references we provide will allow access to those who want more information or expertise in the use of these techniques.

REFERENCES

Aiken, L. S., & West, S. G. (1991). *Multiple regression: Testing and interpreting interactions.* Newbury Park, CA: Sage.

American Educational Research Association (AERA), American Psychological Association, & National Council on Measurement in Education. (1999). *Standards for educational and psychological testing.* Washington, DC: AERA.

Baker, B. O., Hardyck, C. D., & Petrinovich, L. F. (1966). Weak measurement vs. strong statistics: An empirical critique of S. S. Stevens' proscriptions on statistics. *Educational and Psychological Measurement, 26,* 291–309.

Bartlett, C. J., Bobko, P., Mosier, S. B., & Hannon, R. (1978). Testing for fairness with a moderated multiple regression strategy: An alternative for differential analysis. *Personnel Psychology, 31,* 233–241.

Bliese, P. D. (2002). Multilevel random coefficient modeling in organizational research. In F. Drasgow & N. Schmitt (Eds.), *Measuring and analyzing behavior in organizations* (pp. 401–445). San Francisco: Jossey-Bass.

Bollen, K. A. (1989). *Structural equations with latent variables.* New York: Wiley.

Bollen, K. A., & Long, J. S. (Eds.). (1993). *Testing structural equation models.* Newbury Park, CA: Sage.

Burket, G. R. (1964). A study of reduced rank models for multiple prediction. *British Journal of Mathematical and Statistical Psychology, 28,* 79–87.

Cronbach, L. J., Gleser, G. C., Nanda, H., & Rajaratnam, N. (1972). *The dependability of behavioral measurements: Theory of generalizability for scores and profiles.* New York: Wiley.

REFERENCES

Embretson, S. E., & Reise, S. P. (2000). *Item response theory for psychologists*. Mahwah, NJ: Erlbaum.
Hambleton, R. K., Swaminathan, H., & Rogers, H. J. (1991). *Fundamentals of item response theory*. Newbury Park, CA: Sage.
Hastie, R., & Dawes, R. M. (2001). *Rational choice in an uncertain world*. Thousand Oaks, CA: Sage.
Holland, P. W., & Wainer, H. (1993). *Differential item functioning*. Hillsdale, NJ: Erlbaum.
Kline, R. B. (2004). *Principles and practice of structural equation modeling*. New York: Guilford Press.
Magnusson, D. (1966). *Test theory*. New York: Addison-Wesley.
Meehl, P. E. (1954). *Clinical versus statistical prediction: A theoretical analysis and a review of the evidence*. Minneapolis: University of Minnesota Press.
Mislevy, R. J., & Bock, R. D. (1990). *BILOG-3: Item analysis and test scoring with binary logistic models*. Mooresville, IN: Scientific Software.
Mosier, C. I. (1951). Problems and design of cross-validation. *Educational and Psychological Measurement, 11*, 5–11.
Ployhart, R. E., & Schneider, B. (2005). Multilevel selection and prediction: Theories, methods, and models. In A. Evers, O. Smit-Voskuyl, & N. Anderson (Eds.), *Handbook of personnel selection* (pp. 495–516). Chichester/London: Wiley.
Raudenbush, S. W., & Bryk, A. S. (2002). *Hierarchical linear models* (2nd ed.). Thousand Oaks, CA: Sage.
Reckase, M. D. (1997). The past and future of multidimensional item response theory. *Applied Psychological Measurement, 21*, 25–36.
Schmidt, F. L., & Hunter, J. E. (1996). Measurement error in psychological research: Lessons from 26 research scenarios. *Psychological Methods, 1*, 199–223.
Schmitt, N., Coyle, B. W., & Rauschenberger, J. M. (1977). A Monte Carlo evaluation of three formula estimates of cross-validated multiple correlation. *Psychological Bulletin, 84*, 751–758.
Schneider, B. (1983). International psychology and organizational behavior. In L. L. Cummings & B. M. Staw (Eds.), Research in Organizational Behavior, Vol. 5 (pp. 1–32). Greenwhich, CT: JAI.
Stevens, S. S. (1946). On the theory of scales of measurement. *Science, 103*, 677–680.
Stine, W. W. (1989). Meaningful inference: The role of measurement in science. *Psychological Bulletin, 105*, 147–155.
Tanaka, J. S. (1993). Multifaceted conceptions of fit in structural equation models. In K. A. Bollen & J. S. Long (Eds.), *Testing structural equation models* (pp.10–39). Newbury Park, CA: Sage.

3

Job Analysis

AIMS OF THE CHAPTER

The orientation of this book, toward staffing organizations, emphasizes first the organization and then the staff. That is, the burden of responsibility for what the staff does *in* the organization is placed *on* the organization.

This chapter provides a basic concept for defining what jobs and organizations are, of what they are comprised, and how they can be described in human terms (i.e., in terms that allow them to be staffed with humans). We attempt to describe the organizations in which people work and the jobs at which they work. We also discuss the many difficulties with describing jobs and worker specifications, and the numerous judgment calls and inferences that must be made.

We recognize up front that there is an important distinction between job analysis and job specifications. As noted by Harvey and Wilson (2000), *job analysis* involves "the process of describing what is done on a job, and the context in which the work activities are performed" (p. 836). *Job specification* is "the process of inferring the human-trait requirements presumed to be necessary for successful job performance" (p. 836). These two activities, comprehensively identifying the nature of work tasks performed on the job and the nature of the individual knowledge, skill, ability, and other (KSAO) requirements to perform those tasks, set the foundation for nearly all major human resources (HR) activities. However, for the sake of simplicity of language, we use job analysis to refer to both the identification of work activities and the development of job specifications, unless otherwise noted.

AIMS OF THE CHAPTER

From a staffing viewpoint, the purpose of job and organizational analysis is really quite simple—to discover the kinds of people the organization needs to be effective. The information collected to achieve the staffing purpose, however, can serve many functions—as a basis for designing training programs and wage and salary (compensation) programs, for job redesign, for organizational renewal and change, and so on. In other words, job and organizational analyses, because they describe the work setting in human terms, can be useful as a basis for any human-related program in organizations (although if done specifically for staffing purposes, additional work will be necessary to have a job analysis be completely relevant for these other purposes).

Our view of the interrelationship among organizations, tasks, and worker qualifications is aptly expressed in Fig. 3.1, influenced by Sidney Fine. When the organization and the work fit the worker qualifications and the motivation of the workers, the result will be optimal in terms of individual and organizational productivity and worker growth and satisfaction. It is worth noting that this broad, multilevel organizational perspective to job analysis is currently the Department of Labor's model of work and workers, as operationalized in the Occupational Information Network or O*NET (discussed later in this chapter).

For staffing purposes, job and organizational analyses serve three major functions:

1. Job analysis is a means for identifying the important job tasks and human behaviors necessary for adequate performance on those tasks. Based on the identification of such behaviors, theories about the kinds of people the job requires (usually in terms of knowledge, skills, abilities, and other characteristics) can be formulated, and procedures (tests, exercises, interviews) for identifying such people can be developed. The procedures can then be submitted to a test of their effectiveness. The process of testing the procedures is, in fact, a test of hypotheses derived from the job and from organizational analyses. This process of testing hypotheses or theories is at the heart of a systematic staffing program and recalls Binet and Simon's comments about "testing the test," that is, about ensuring the hypothesis or theory is valid. The job and organizational analyses on which such programs are based ultimately determine the effectiveness of a staffing program.

Another outcome of identifying necessary human behaviors is the development of standards against which the behavior of workers, and, therefore, the effectiveness of the staffing program, can be evaluated. If

3. JOB ANALYSIS

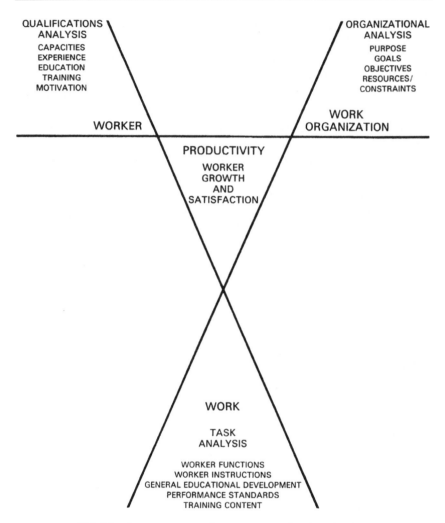

FIG. 3.1. A systems approach to manpower planning: structure.

we know what behaviors are required for adequate performance, we can evaluate performance against those standards (procedures for identifying the worker behaviors necessary for adequate job performance are discussed later in this chapter).

2. Job analysis identifies the rewards the job itself offers to the humans who do the job. Just as there are individual differences in the degree to which people can perform adequately on a job, there are individual differences in the degree to which the rewards that jobs offer will be satisfying to people.

AIMS OF THE CHAPTER

3. Organizational analysis is necessary because of the clear findings regarding the impact of organizational features, aside from the job, on human behavior in the work setting. Such HR practices as leadership and supervision, organizational reward systems, and organizational climate have direct effects on employee behavior, and thus they play a major role in understanding and predicting aggregate levels of performance and productivity.

What this means is that certain conditions of jobs and organizations seem to be capable of generally suppressing or generally enhancing the work-relevant behaviors of people. By generally suppressing or enhancing, we mean that it is in the organization as a whole that the condition has its effect and that different individuals tend to be equally affected (i.e., a "main effect" for the organization).

For example, it is clear now that when organizations establish clearly specified, difficult but attainable goals for employees, and provide employees feedback on their performance, that aggregate performance levels are higher than when such goals do not exist (Locke & Latham, 1990). These specific goals are apparently interpreted similarly by all workers; the goals have a general and pervasive effect on all employees for whom the goals are set.

It is still true, of course, that workers who have more of the qualifications a job requires will outperform people who fail to meet the qualifications. However, in organizations where the goal-setting procedures exist, it is possible for the performance of people with less than average qualifications to be as good as people who have average qualifications in another organization that fails to set specific and difficult goals. Indeed, research on organizational-level HR practices finds that the mere presence of "high-performance" work systems, such as formal reward systems for performance and challenging work, will influence organizational effectiveness without ever considering the nature of the individual differences of employees within these organizations (e.g., Huselid, 1995). So, organizational performance is a function of both the qualifications of workers and the conditions under which they work; it is the combination that is important because both seem to have main effects.

In the following sections, we describe the many issues related to job analysis and job specification. First, we discuss how job analysis sets the foundation for all staffing practice. We then take a step back to examine how the larger environmental and organizational context impacts job analysis. This larger environment must be understood to fully appreciate the job analysis process—its limits and its potential. These first two sections provide the conceptual overview of the job analysis process. Next, we move to the specific procedures for performing job analyses, examining first the historical approaches and then the modern approaches. We conclude with current issues and future challenges.

JOB ANALYSIS IS THE FOUNDATION OF STAFFING

Job analysis sets the foundation for all later staffing procedures and practices. Information missing from the job analysis, or irrelevant material that is included, will impact the later identification of performance and predictor constructs, as well as the nature of the selection battery and decision rules to be used in selection. Because job analysis is the foundation of these later functions, a faulty job analysis foundation will lead to the faulty building of staffing systems.

Figure 3.2 shows the relationship between job analysis, criterion development, and predictor development. The arrows in the figure represent inferences that must be supported to justify the use of a particular performance measure and staffing procedure. In many ways, you can think of these inferences as hypotheses, with the overall figure representing a theory of work behavior. There are many choices for how one can propose and test these hypotheses (Sackett & Laczo, 2003). A key task for the job analyst is to provide evidence (data) to support these inferences and justify the choices. The job analysis methods noted later in this chapter are all concerned with providing this information.

The first step, discussed more in the next section, begins with an understanding of the organization, its strategy and goals, and the external environment in which it exists. These form the bases of understanding whether job analysis (and staffing) is even the appropriate thing to do. If the answer to this question is "yes," then the task of defining the job and determining the staffing procedure commences.

Next, the job analysis method and nature of subject-matter experts (SMEs) will help determine (2) the critical job tasks and (3) the important KSAOs. This corresponds to the job analysis and job specification distinction noted previously. Notice the linkage between critical tasks and critical KSAOs (inference 4) is unidirectional, with important KSAOs being determined by critical tasks. This is intentional because "what" is to be predicted is used to determine "how" we will select.

Once the important tasks and KSAOs are identified, the next step is to develop measures of the KSAO constructs. Thus, criterion development

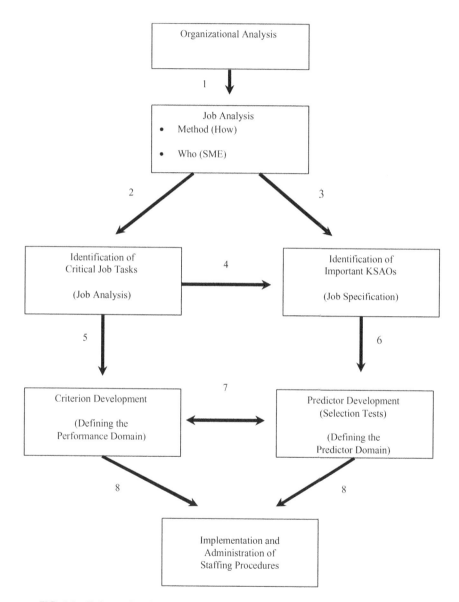

FIG. 3.2. Relationship between organizational analysis, job analysis, and staffing system design.

(inference 5) and predictor development (inference 6) are based directly on the tasks and KSAOs found to be important. This point is a key one because it recognizes that effective staffing practices are not arbitrary; rather, they are based on sound reasoning and data. In contrast to inference 4, inference 7 is a two-headed arrow indicating that the direction of influence moves to and from criteria and predictors. This is because (a) predictors are important to the extent they are related to criteria, and (b) criteria are theoretically "explained" by predictor measures assessing underlying or latent KSAO constructs.

Finally, by defining the criterion and predictor domains, one has the basis for implementing and administering the staffing system (inference 8). The job analysis information is used to identify the types of tests that will be used to assess the KSAOs. For example, will applicants be assessed with interviews, written tests, or simulations? How much weight should each attribute be given in the selection decision? Will multiple tests be used, and if so, when will different tests be administered?

Thus, job analysis provides the basis for defining the performance domain and predictor domain. It also provides a foundation for other HR functions, including training and compensation systems. The role of job analysis in this entire process can be summarized by GIGO—garbage in, garbage out. A poorly done or incomplete job analysis is not better than conducting no job analysis. Although conducting an appropriate job analysis is time consuming and can be expensive and involved, there is currently no better way to ensure the appropriate development of predictors and criteria. However, we do not believe job analysis and the process described in Fig. 3.2 exist in a vacuum; rather, they are highly affected by the larger environment.

THE LARGER ENVIRONMENT

The larger environment in which organizations operate plays an important role in the way organizations design jobs and supervise people, and indeed, in the way people in organizations make decisions. We can see everyday—and frequently dramatic—examples of this all around us. For example, a downturn in the economy, stimulated in part by the terrorist attacks on September 11, 2001, results in workers losing their jobs and organizations restructuring to meet the changing economy. Such restructuring may result in the elimination of jobs, the combination of jobs, or even the creation of new jobs. When such dramatic

THE LARGER ENVIRONMENT

changes occur, HR practices and the definition of the "job" may be altered forever.

It is therefore critical that before exploring the individual-level details of job analysis, we first consider how jobs and job analysis fit into (and are a product in part of) the larger environment. It is clear that the organizational environment can influence the nature of the job. Of course, such influence is not direct but operates through multiple levels: The external environment influences organizational variables, organizational variables influence HR practices, and HR practices influence people and jobs. Jackson and Schuler (1995) articulated a model describing how this occurs. In this model, the external environment operates at local, national, and global levels. External influences include laws and regulations, culture, politics, unions, labor markets, and industry characteristics. These external influences affect the internal organizational context, composed of technology, structure, size, life cycle, and strategy. These internal elements in turn influence organizational decision making, HR practices, and individual and organizational outcomes. We summarize these relationships in Fig. 3.3.

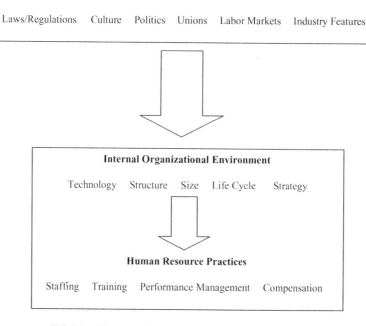

FIG. 3.3. The external environment influences HR practices.
Note: Adapted from Jackson & Schuler, 1995.

TABLE 3.1
Mechanistic and Organic Organizational Forms

Mechanistic	Organic
1. Tasks are broken into very specialized abstract units	1. Tasks are broken down into subunits, but relation to total task of organization is much more clear
2. Tasks remain rigidly defined	2. Adjustment and continued redefinition of tasks through interaction of organizational members
3. Specific definition of responsibility that is attached to individual's functional role only	3. Broader acceptance of responsibility and commitment to organization that goes beyond individual's functional role
4. Strict hierarchy of control and authority	4. Less hierarchy of control and authority sanctions derive more from presumed community of interest
5. Formal leader assumed to be omniscient in all matters	5. Formal leader not assumed to be omniscient in all matters
6. Communication is mainly vertical between superiors and subordinates	6. Communication is lateral between people of different ranks and resembles consultation rather than command
7. Content of communication consists of instructions and decisions issued by superiors	7. Content of communication is information and advice
8. Loyalty and obedience to organization and superiors is highly valued	8. Commitment to tasks and progress and expansion of the firm is highly valued
9. Importance and prestige attached to identification with organization itself	9. Importance and prestige attached to affiliations and expertise in larger enviroment

Source: From G. Zaltman, R. Duncan, and J. Holbek, *Innovations and Organization* (New York: Wiley, 1973), p. 171. Reprinted by permission.

A classic book by Burns and Stalker (1961) illustrates how the structure of an organization may be affected by the external environment. They conceptualized the "style" of organizations to be a function of the larger environment in which the organization operates. They suggested this larger environment determines how an organization accomplishes its goals. They noted that, when the larger environment is stable, a mechanistic type of organization may be most effective. In a changing and unstable environment, an organic type of organization may be best. Burns and Stalker hypothesized that some differences in job and organizational designs may result from differences in the turbulence of the external world. Table 3.1 summarizes some of these differences in mechanistic and organic organizations.

THE LARGER ENVIRONMENT 109

The stability of the external environment seems to be a function of the volatility or perceived predictability of some or all the following systems outside the organization with which organization members must deal:

- The nature of the customer
- The nature of the supplier
- The nature of the competition
- The nature of the sociopolitical environment
- The nature of technology

How an organization perceives its external environment, particularly the organization's strengths, weaknesses, opportunities, and threats, will influence the strategy the organization pursues to compete in this environment. This in turn influences the role of staffing within the organizational context. Whereas staffing has historically been considered a by-product of the strategy process, more recent thinking argues that staffing itself may be a source of strategy. For example, Snow and Snell (1993) noted that when organizations operate in fluid, fast-changing markets, they need to adapt quickly, and thus proactive staffing can ensure sufficient worker competencies for demands that may currently not be needed. As such, staffing becomes the basis of strategy and competitive advantage.

Because strategy and HR practices are related, so too will HR practices influence the nature of jobs and job specifications. HR practices refer to the way an organization recruits, selects, trains, and compensates employees, among other things. One goal of HR practices is to maximize the benefits associated with wise use of an organization's human capital. *Human capital* refers to an organization's composition of KSAOs or the KSAOs that characterize an entire organization. An important contribution to this thinking is provided by Lepak and Snell (1999, 2002). They argued that within a single organization there are differences in human capital varying on dimensions of strategic value and uniqueness. Depending on the strategic value and uniqueness of particular configurations of human capital, different HR practices will be necessary to maximize their effectiveness.

For example, KSAOs reflective of knowledge-based employment (e.g., top management) have high uniqueness and high strategic value, whereas KSAOs of job-based employment (e.g., engineers) have high strategic value but are not unique. HR practices must follow a commitment-based configuration with knowledge workers because their human capital is central to the organization and a source of competitive advantage (because it is unique). Commitment-based HR practices might include more decision-making responsibility and autonomy, development, and high job security. Staffing is more likely to be focused on identifying employees with aptitudes rather than task-specific skills (Snell & Dean, 1992). In contrast, a productivity-based configuration will be most useful for job-based workers because their capital is relatively more common in the labor market. A productivity-based configuration might involve a more generic approach to the position, more emphasis on short-term job performance, and staffing based on job-specific KSAOs (rather than potential; Lepak & Snell, 2002).

Figure 3.4 shows how job and organizational analysis fits within the organizational context. We can see that job and organizational analysis influences the staffing process, but that is not all. Job analyses are also translated into several additional processes:

1. *The Industrial Relations Process*, including labor/management negotiations, wage and salary administration and so on.
2. *The Training Process*, including skills training, management training and so on.
3. *The Organization Development Process*, including organizational renewal, job enrichment and enlargement, structural redesign and so on.

Application of job analysis information helps these other HR functions ensure employees remain competent, well satisfied, and well compensated. Therefore, the sustained effectiveness of staffing procedures is a function of the outcomes of this interaction with other HR practices. Poorly designed jobs that result in poor performance, high turnover (as a result of dissatisfaction), or alienation cannot be compensated for by better staffing procedures. Organizational diagnoses may show that revised or new training procedures are required, job enlargement is necessary, or the emphasis should be on staffing procedures. Organizations may effectively move toward achieving their goals by many routes, only one of which is through the implementation of various staffing policies and procedures. The contribution of the staffing procedures to the prediction of performance must be compared with what may be expected from other sources (i.e., job redesign, changes in management philosophy, training) and various HR efforts must be consistent with each other.

THE LARGER ENVIRONMENT 111

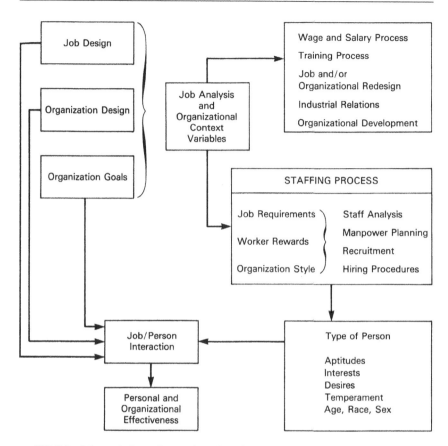

FIG. 3.4. Schematic for understanding the role of job and organizational analysis in the staffing process.

Guion (1965) proposed that about half the behavior we observe in organizations is due to individual (person) characteristics, and the other half to situational variables (job and organizational characteristics). It is hoped that the reader of this book will become a believer in regarding both people and situational factors as explanations for behavior. This point of view sets up the situation so organizations and the people in them will profit maximally. Job and organizational analyses can help an organization understand the nature of its situation and thus the nature of the staffing problem.

Thus, the external environment, HR practices, and strategy are intricately related. This is a relationship (rather than a one-way street) in the sense that the external environment affects strategy, organization management philosophy, and job design, whereas the success of HR practices in effectively meeting the requirements of the external world will influence the external environment. Thus, customer demands, when met, result in less customer volatility (Schneider & White, 2004). Less customer volatility, in turn, may result in a more mechanistic form of organization and job design, which again may spur a subsequent stimulation of customer volatility and so on.

We address the larger environment issue to emphasize that jobs and organizations do not exist independently; they are caused by understandable phenomena. We can conclude that organizations operating in different kinds of larger environments will develop different kinds of jobs because, for example, they will have different kinds of customers, work with different kinds of technology, and be subject to different kinds of market forces. It follows that these jobs will require (and attract) different kinds of people. It is not surprising that, as noted earlier, organizations differ from each other as a result of the kinds of people who work there. Hospitals attract, recruit, and select people who are different from those found in advertising companies. In fact, Schneider, Smith, Taylor, and Fleenor (1998) found that different organizations had different personality profiles, suggesting average personality *within* organizations is one way of describing differences *between* organizations. This idea of the affinities of people for being attracted to, selected by, and staying with particular organizations will be referred to throughout the book (Schneider, 1987). Thus, understanding jobs can only occur through an understanding of the organizational context.

Our concern now turns to understanding what jobs are, how they can be described, and how such descriptions are used to infer worker characteristics. There are a variety of approaches that represent different theories of work behavior and organizational structure. We briefly review key past approaches and then consider more recent innovations, but all must be considered because there is no one single type of job analysis appropriate for all situations. To facilitate a comparison of these approaches, Sackett and Laczo (2003) suggested contrasting different job analysis methods in terms of the following:

- Activity (what a person does at work) versus attribute (worker characteristics)
- General versus specific information
- Qualitative versus quantitative information
- Applying a taxonomy of tasks/KSAOs versus using a "blank slate"

HISTORICAL PERSPECTIVES ON JOB ANALYSIS

- Collecting information from informants (job incumbents) versus trained observers
- Focusing on knowledge, skills, and abilities (KSAs) versus KSAOs
- Examining a single job versus making comparisons across multiple jobs
- The purpose is descriptive versus prescriptive

As you read through the following methodologies, it is helpful to keep these distinctions in mind.

HISTORICAL PERSPECTIVES ON JOB ANALYSIS

Over the history of the staffing enterprise, a number of procedures have been devised for describing the demands that jobs make on people—for describing what the job requires in terms of individual attributes. Once carried out, these procedures allow for the relatively unambiguous specification of acceptable performance and, as a result, help an organization identify the kinds of individuals it should recruit, select, develop, promote, and retain.

All such procedures ultimately attempt to quantify a set of inferences that must be made about the job, the characteristics of the people working in the job, and how they are related. These are the inferences discussed earlier and shown in Fig. 3.2 (in particular, inferences 2, 3, and 4). In this section, we consider historical approaches for validating these inferences.

Functional Job Analysis

A major breakthrough in job analysis occurred when the 1977 edition of the *Dictionary of Occupational Titles* (DOT; U.S. Department of Labor, 1977) was issued by the U.S. Employment Service. The publication of the DOT was important for two reasons. First, the DOT included information on approximately 12,000 jobs useful for determining job requirements. Second, and more important, the DOT presented a system for describing worker attributes and jobs. Although the DOT has been phased out and replaced by the O*NET system, it is still instructive to briefly describe the functional job analysis (FJA) methodology created by Sidney Fine and his colleagues because it contributed to many job analysis practices still in use.

Perhaps the major contribution of Fine's work was the standardized manner in which task statements were written. This standardized form has been widely adopted by job analysts; its purpose was to clearly articulate the nature of different jobs to supervisors, trainers, recruiters, and other users of job information.

Specifically, task statements should include five types of information:

1. The subject of the task statement is assumed to be the worker or job incumbent.
2. The action performed requires an explicit, concrete verb. Verbs that point to a process such as "develops," "prepares," or "assesses" are undesirable.
3. The immediate objective of the work should be defined.
4. A task statement should identify the tools, equipment, or work aids a worker uses.
5. The task statement should include the type of instructions given to the worker.

An example of a complete edited task statement is shown here:

> Asks customer questions, listens to responses, and transcribes relevant details on computer, exercising discretion as to sequence of questions, in order to record basic information.

In this task statement, "asks customer questions, listens to responses, and transcribes relevant details" are the actions required. The immediate objective of the action is the recording of "basic information," and the tools, equipment, and work aids are a computer. The instructions include using the computer but exercising discretion with respect to the sequence of questions asked.

When a task statement includes these pieces of information, staffing personnel have the bases on which they can infer the worker qualifications needed to perform the task. The information required to write task statements is gathered by reviewing existing materials concerning the job and by interviewing SMEs such as job incumbents and their supervisors who are familiar with the job and the KSAOs required to perform it effectively. Trained job analysts, however, write the task statements and provide the ratings discussed shortly. An example of a completed task analysis form for the job of an industrial/organizational psychologist is presented in Fig. 3.5. Note that the completed task analysis form includes information on the training and performance standards relevant to the task and a series of ratings of this task along the top.

FJA is also a conceptual system used to define the level and orientation of worker activity. Fine assumes the tasks people do can be organized along three dimensions that reveal the extent of their involvement with data, people, and things. The actions on each of these three dimensions are classified by the three worker function scales depicted in Fig. 3.6 (see Fine & Wiley, 1974). As can be seen in Fig. 3.6, the actions or functions in each scale are actually examples of behavior ranging from the simple to the more complex. There are two ways of comparing and measuring the behavioral requirements of any task in a job. Jobs can be described in terms of their level of involvement with data, people, and things—ranging from the relatively complex at the top of the scales in Fig. 3.6 to the relatively simple at the bottom of the same scales. The orientation of a worker with respect to data, people, and things can be described by the percent of time spent on each dimension. When a given task has been assigned functional level and orientation scores for data, people, and things, the worker's total involvement with the specific facets of that task—mentally (data), physically (things), and interpersonally (people)—is obtained. Similarly, a job comprised of many tasks can be given functional level and orientation ratings.

Position Analysis Questionnaire

Beginning in the 1950s, Dr. Ernest J. McCormick and his colleagues at Purdue University began a program of research that led to the development of a job analysis instrument called the Position Analysis Questionnaire (PAQ). Instead of requiring a job analyst to produce a narrative description of tasks for the PAQ, the job analyst or expert job incumbent makes ratings of a job on a number of already prepared descriptors. These descriptors are elements of work activity, and the analyst judges the degree to which the element is present in the job being observed.

The quantitative nature of the responses to the PAQ permits the application of a statistical procedure called factor analysis (see chapter 2) to be used in isolating the major underlying dimensions of job behavior across a large number of jobs. Factor analyses allow us to answer

TASK CODE:

WORKER FUNCTION LEVEL AND ORIENTATION						WORKER INSTRUCTIONS	GENERAL EDUCATIONAL DEVELOPMENT		
DATA	%	PEOPLE	%	THINGS	%		REASONING	MATH	LANGUAGE
3B	85	1A	5	1A	10	3	4	4	5

GOAL:

OBJECTIVE:

TASK: Writes report of validation study, outlining the procedures followed, the results obtained, and the conclusions drawn using standard reporting procedures and some discretion in the way the material is presented in order to comply with organization's reporting requirements.

(To Perform This Task)

PERFORMANCE STANDARDS | TRAINING CONTENT

DESCRIPTIVE:
1. Writes clearly and legibly.
2. Writes on a level appropriate to the audience.
3. Report is complete and accurate.
4. Format of report follows the standard reporting procedures.

NUMERICAL:
1. Report includes all of the following: procedures, validity, reliability, norms, differential validity, test bias.
2. Report is completed within X amount of time from completion of validation study.
3. X% feel level of report is understandable.

(To These Standards)

FUNCTIONAL:
1. How to write report of validation study.
2. How to express oneself in non-technical terms.
3. How to interpret results of a validation study.

SPECIFIC:
1. Knowledge of validation procedures used and results obtained.
2. Knowledge of standard reporting procedure.

(Worker Needs This Training)

FIG. 3.5. Functional job analysis task statement and ratings.

DATA	PEOPLE	THINGS
0 Synthesizing	0 Mentoring	0 Setting Up
1 Coordinating	1 Negotiating	1 Precision Working
2 Analyzing	2 Instructing	2 Operating-Controlling
3 Compiling	3 Supervising	3 Driving-Operating
4 Computing	4 Diverting	4 Manipulating
5 Copying	5 Persuading	5 Tending
6 Comparing	6 Speaking-Signaling	6 Feeding-Offbearing
	7 Serving	7 Handling
	8 Taking Instructions-Helping	

FIG. 3.6. Summary chart of worker functions.
Source: Adapted from Figure 1 of Fine and Wiley (1974), p. 10.

the question, "What kinds of behavior characterize work behavior?" The FJA and DOT answer this question with data, people, and things, but these are very general behavioral categories. Factor analyses of the ratings of 536 different jobs in relation to the 194 elements in the PAQ indicated that jobs tend to differ from each other on five important dimensions:

1. Having decision-making/communication/social responsibilities
2. Performing skilled activities (use of tools and eye–hand coordination)
3. Being physically active/related environmental conditions
4. Operating vehicles/equipment
5. Processing information

The PAQ is important to know about because it had such an early impact on modern approaches to job analysis methods (e.g., O*NET, to be described later). The idea behind the PAQ was to derive a set of behaviors that could be used to describe all jobs so one job could be compared with another, so job families could be prepared in which seemingly different job titles actually revealed the need for very similar task behaviors and similar KSAOs, and to permit the clustering of jobs into sets that would likely require more similar than different staffing procedures and approaches.

We should note that, in addition to the PAQ, there is a newer job analysis form called the Professional and Managerial Position Questionnaire (PMPQ), which focuses on these different categories of jobs. The PMPQ covers planning and scheduling activities, processing information and ideas, exercising judgment, communication, interpersonal activities and relationships, and technical activities, as well as the education, training, and experience required to perform these activities.

Databases covering thousands of jobs for the PAQ and the PMPQ are maintained by PAQ Services, Inc. These techniques have been found useful by many organizations as they prepare to design staffing procedures, especially when whole organizations want to standardize the job analysis procedures being used. Their Web page (www.PAQ.com) provides more information about these job analysis procedures.

Critical Incident Technique

A method that is still frequently used for gathering information on the tasks and KSAOs required to do those tasks is one developed by John Flanagan (1954) called the critical incident technique (CIT).

In the typical use of the CIT, SMEs are asked to focus on the job of interest and describe the major functions of the job. For example, the major functions of a supermarket bakery department manager may be to train subordinates, interact with customers, schedule employees, place orders for merchandise, and display merchandise. Then, the SMEs are asked to think of an incident in which unusually good (or poor) performance was observed in carrying out each function. For each incident, they are asked a series of questions:

1. What were the circumstances leading to the incident?
2. What did the manager do that made you think she was a good, average, or poor performer?
3. What were the consequences of the manager's behavior in the critical incident?
4. How effective do you feel this example of performance was (circle one)?

   1   2   3   4   5   6   7
   Extremely   About   Extremely
   ineffective   average   effective

If the SMEs are successful in completing these steps, they should have a list of incidents that are behavioral and observable, a list of the situational factors involved, and judgments concerning the criticalness and importance of the incidents.

Since the mid-1960s, the critical incident methodology has been adapted to develop performance measures as well (e.g., Latham & Wexley, 1981; Smith & Kendall, 1963). This adaptation represents the most widely used form of critical incidents methodology and is described in chapter 4. It has also been used as a way to develop work simulations and situational judgment tests, as discussed in chapter 10.

The strength of the critical incidents approach lies in the emphasis placed on the observation of behavior that is critically important to successful job performance, that is, behaviors that discriminate effective from ineffective performers. In contrast, several deficiencies have been noted in the use of the critical incidents approach in establishing the KSAO requirements of a job. Most basic is the fact that nothing in the methodology leads to an explicit statement of ability requirements; one must still make a judgment concerning what KSAs indicate individual differences in the capacity of a job applicant to perform an effective critical incident and to avoid ineffective critical incidents. In addition, job tasks that everyone masters and that they spend most of their time doing may never appear in a list of critical incidents. This criticism is not really relevant to personnel selection because such tasks by definition can have nothing to do with individual differences in ability to perform the job. Finally, emphasis on incidents may lead to a fragmented view of the job in which one is unable to ascertain how various elements or tasks fit together.

MODERN PERSPECTIVES ON JOBS AND JOB SPECIFICATIONS

In this section, we now turn to an examination of modern perspectives on job analysis and job specifications.

The Occupational Information Network

In an effort to replace the outdated DOT, the Department of Labor sponsored a large research project to develop a better and more inclusive system. This effort resulted in what has become known as O*NET (www.onetcenter.org). O*NET was designed to achieve four principles:

1. Create multiple task descriptors and categories.
2. Use descriptors based on a common language applicable to all jobs and occupations.

3. Ensure the occupational descriptors are organized according to taxonomies and hierarchies within taxonomies.

4. Design the system to be comprehensive, yet reduce the number of occupational categories from about 12,000 in the old DOT to about 1,100. (Peterson et al., 2001)

What makes the O*NET system such a major departure from all previous job analysis methods is its comprehensiveness. It considers six major content domains as necessary and important to describe work and workers. Rather than relying on job analysts as the DOT required, the O*NET system relies primarily on job incumbents to provide the information necessary to build the information database. This simple change alone is likely to result in a vast savings in terms of human and monetary resources, relative to the DOT. The system cuts across many of the perspectives noted in historical job analysis methods, taking both a work- and a worker-oriented focus. These content domains include the following:

1. *Worker Characteristics* include major individual difference domains that tend to be rather stable and enduring. They include abilities (cognitive, psychomotor, physical, sensory), work styles (achievement orientation, social influence, interpersonal orientation, adjustment, conscientiousness, independence, practical intelligence), and occupational values and interests (realistic, investigative, artistic, social, enterprising, conventional). The content within each domain is structured hierarchically, such that more specific abilities, styles, and interests are clustered within the highest-order factors. For example, cognitive ability is composed of verbal ability, idea generation and reasoning, quantitative ability, memory, perceptual ability, spatial ability, and attentiveness. In turn, each of these more specific abilities is composed of even more specific elements (e.g., verbal ability is composed of oral comprehension, written comprehension, oral expression, and written expression).

2. *Worker Requirements* reflect more general attributes such as knowledge and skills and formal educational accomplishments. As a result of their development through education and experience, worker requirements tend to be more changeable than worker characteristics. For example, the skills taxonomy describes two major skill sets (content, such as science and math, and process, such as learning and critical thinking) and five cross-functional skill sets (problem solving, technical, resource management, social, systems).

MODERN PERSPECTIVES ON JOBS 121

3. *Experience Requirements* involve the kinds of experiences a worker must have to be successful on the job. These may include both on-the-job and off-the-job training, as well as more formal licensure and certification requirements.

4. *Occupational Requirements* describe the content and context of the work to be performed in the job. It includes several categories of requirements. Generalized work activities are tasks and activities that are expected to cut across most, if not all, occupations. These may include information input, mental processes, work output, and interactions with others. The work context is also described in terms of the interpersonal, physical, and structural characteristics in which the job occurs. Finally, the organizational context is considered, making reference to the type of industry, structure, culture, roles, and leadership of the organization. This latter point is particularly interesting because it recognizes the important role organizational factors may have on the structure of jobs and roles.

5. *Occupation-Specific Requirements* reflect the knowledge, skills, abilities, characteristics, and tasks that are specific to a given job and occupation. Thus, despite the existence of generalized work activities, it is obvious there may be job-specific requirements (an example would be minimum vision requirements for airline pilots). By definition, the content of the occupation-specific requirements are expected not to generalize beyond the job in question.

6. *Occupation Characteristics* refer to more macro and economic factors that may influence jobs and organizations. These include such factors as labor market conditions, labor supply projections, and related labor market information contained in national databases.

It is clear that the O*NET system is an ambitious attempt to unify many diverse perspectives that independently or additively influence jobs and work. It simultaneously considers features of both the worker (KSAOs) and work (tasks and activities). An impressive feature of the O*NET system is its explicit consideration of multiple levels of analysis. For example, it clearly recognizes the level of the individual worker (KSAOs) is nested within a particular job and work context, which in turn is nested within an organizational context, which in turn is nested within a particular economic context and labor market.

Although an impressive system, O*NET is just starting to be used, and data collection to complete the system has just begun. There are currently several limitations with the system. As noted by Peterson et al. (2001), one of the most serious is the small number of incumbents who have responded to the job analysis surveys. Although an advantage of using incumbents is that they are most familiar with the job in question, the disadvantage is that researchers must find ways to ensure adequate sampling of incumbents; otherwise, the data may not accurately reflect the realities of the job.

Despite potential limitations, O*NET is already slated to be the system and model for describing jobs and work for all major government agencies, and it is likely to be adopted by most educational and vocational agencies as well (Peterson et al., 2001). A complete description of the development and system can be found in Peterson, Mumford, Borman, Jeanneret, and Fleishman (1999). The system can be accessed on the Internet free of charge, making it particularly attractive to small employers who do not have the resources available to conduct a comprehensive job analysis. Box 3.1 gives an example of how a small business owner can use O*NET.

Box 3.1 Example of a Small Business Owner Using O*NET to Obtain Job Analysis and Job Specification Information

The beauty of the O*NET system is that it is freely available and accessible on the Internet. This makes it an ideal resource for small business owners who cannot devote the resources for conducting a full-scale job analysis. Let us take the perspective of an owner of a small restaurant (employing less than 50 people). Suppose this individual wants to develop an interview to hire waitstaff and knows enough about staffing to realize the interview questions should be based on a job analysis. Rather than conduct such an analysis, the owner goes to O*NET's Web site (www.onetcenter.org). The owner then clicks on the link to find occupations, typing in the keywords "waiter and waitress." The O*NET system responds by producing eight job categories in the O*NET system that reflect variations of the waiter/waitress occupation.

The owner chooses the occupation labeled "waiters and waitresses." In doing so, the owner can examine detailed job analysis information that has already been collected for this occupation. This includes information about critical tasks, the KSAOs needed to perform the job, work activities and work context, and related

information. Shown below is a sample of the information obtained from the Web site:
- Critical Tasks
  - Check patrons' identification in order to ensure that they meet minimum age requirements for consumption of alcoholic beverages.
  - Check with customers to ensure that they are enjoying their meals and take action to correct any problems.
  - Escort customers to their tables.
  - Explain how various menu items are prepared, describing ingredients and cooking methods.
  - Inform customers of daily specials.
- Knowledge
  - Customer and personal service; English language
- Skills
  - Service orientation; active listening; writing
- Abilities
  - Oral expression; memorization; manual dexterity
- Work context
  - Indoors, environmentally controlled; spend time standing; dealing with unpleasant or angry people

Now, based on this information, our small business owner knows what KSAOs to target with interview questions. This helps ensure the appropriate selection procedure and the hiring of those capable of performing the job.

Homegrown Job Analysis Procedures

Often, companies want to develop their own procedures for studying jobs. As noted by Sackett and Laczo (2003), there are no "one size fits all" job analysis methods, and so frequently it is best to use methods most appropriate to a given situation. Some reasons for this include the belief that a specific job is so different in the tasks to be performed and the KSAOS required to perform them that existing methods such as the PAQ or O*NET will not do it justice. Prior to modern job analysis techniques, this is what the military did, and the procedures developed in the military were used as guides in the development of O*NET, just as the earlier DOT processes were built on for FJA.

One begins this type of job analysis with a review of existing job descriptions and training manuals, as well as with interviews and observations of the job incumbents. The job analyst then meets with a group of five to eight SMEs to generate a list of task statements. First, any special environmental conditions (shift work, toxic chemicals, heat, noise) are noted. Then, the job analyst provides a definition of a task and examples of the tasks generated from a preliminary review of the job, and the group is asked to generate additional tasks. These task statements can be done in written form, but for job incumbents who are not used to writing, it can be done orally, with the job analyst or an assistant recording the tasks. The analyst then edits, deletes duplicate tasks, and groups the tasks into major dimensions. This process can be repeated with two or three groups of SMEs or until new groups fail to add novel tasks. A final group meets to edit the combined lists of tasks (tasks are best written to conform with the functional job analysis statements described earlier) and review the grouping of tasks into behavioral dimensions.

The next step involves SMEs rating each task on three aspects of the task: time spent, task difficulty, and criticality. Because this is a written task (literally, a task inventory), a larger group of job incumbents can be included, although accurate ratings can sometimes be obtained with a relatively small number of task raters (5–10). Examples of the three scales are shown in Table 3.2.

Raters should be instructed to make their ratings based on the job in general rather than on their own particular position. For each task, an overall task importance value is computed by multiplying difficulty times criticality ratings and adding time spent ratings. Although this method of combination seems to make logical sense (i.e., a task must have some degree of criticality or difficulty), other variations may be appropriate in different situations. The job analyst can then produce a list of tasks organized by major functional categories in order of importance.

The next phase focuses on generating the human attributes required for performing each task or major functional dimension. Human attributes are generally defined as KSAOs, although the distinction is occasionally ambiguous:

- Knowledge is usually defined as the degree to which a job incumbent is required to know certain technical material.

TABLE 3.2

Samples of Time Spent, Importance, and Criticality Task Ratings

Time Spent: How much time is spent performing this task?
0 = Never perform
1 = Very low amount
2 = Low amount
3 = Moderate amount
4 = High amount
5 = Very high amount

Difficulty: How difficult is it to perform this task relative to other tasks on this job?
1 = One of the easiest tasks
2 = Easier than most tasks
3 = About the same as most other tasks
4 = Harder than most tasks
5 = One of the most difficult tasks

Criticality: How serious are the consequences for failing to perform this task?
1 = Very minor consequences
2 = Minor consequences
3 = Moderate consequences
4 = Major consequences
5 = Very major consequences

- A skill indicates adequate performance on tasks requiring the use of tools, equipment, and machinery.
- Abilities are physical and mental capacities to perform tasks not requiring the use of tools, equipment, or machinery.
- Other characteristics include personality, interest, or motivational attributes that indicate a job incumbent will perform certain tasks, rather than whether they can perform those tasks.

An example of KSAOs generated for the staff selection, evaluation, and development dimension of school administrator performance is shown in Table 3.3. KSAOs are usually generated in small group meetings of four to eight SMEs. When the group is satisfied, a relatively complete list of KSAOs has been generated, the job analyst edits, removes duplicates, and has the list typed. This list is then rated by the SMEs, again on a series of scales such as those in Table 3.4.

If one uses these scales to determine personnel selection procedures, a clear majority of the SMEs should indicate that an item is necessary for newly hired workers and that it would be practical to expect that newly hired workers possess a given KSAO. These are called the

TABLE 3.3

Example of Tasks and Associated KSAOs for the Staff Selection, Evaluation, and Development Dimension

A. Two Tasks in Staff Selection, Evaluation, and Development Dimension of Job Performance
  1. Observes teachers' classroom performance to evaluate their performance and provide feedback to teachers.
  2. Confers with other principals and/or district personnel to coordinate educational programs across schools.
B. Four KSAOs Written for the Staff Selection, Evaluation, and Development Dimension of Job Performance
  1. Knowledge of curricula in various subject matter areas in own school as well as the district.
  2. Knowledge of appropriate instructional behavior in various subject matter areas.
  3. Sensitivity in dealing with observational instructions in classrooms.
  4. Knowledge of school and district organizational chart and programs/facilities/personnel.

TABLE 3.4

Example of a Form for Rating KSAOs of a Personnel Specialist

KSAOs	NECESSARY FOR NEW WORKERS YES or NO (Circle One)	PRACTICAL TO EXPECT YES or NO (Circle One)	EXTENT OF TROUBLE LIKELY 1 = Very Little or None 2 = To Some Extent 3 = To a Great Extent 4 = To a Very Great Extent 5 = To an Extremely Great Extent (Enter Your Rating)	DISTINGUISH SUPERIOR WORKER FROM AVERAGE 1 = Very Little or None 2 = To Some Extent 3 = To a Great Extent 4 = To a Very Great Extent 5 = To an Extremely Great Extent (Enter Your Rating)
Knowledge of item construction	(YES) or NO	(YES) or NO	3	3
Ability to communicate with angry/confused examinees	(YES) or NO	(YES) or NO	4	4
Knowledge of item analysis statistics, difficulty indices, and item-total correlations	(YES) or NO	(YES) or NO	2	3

Source: Adapted from Levine, (1983).

selection KSAOs. There must also be some consensus that lack of a particular KSAO would produce trouble for the workers, their coworkers, or the organization. That is, the mean of the "trouble likely scale" should be at least 1.5. Given the KSAO is necessary, practical, and, if ignored, could produce trouble, then the most useful KSAOs from a selection perspective are those that receive high superior ratings. These are the KSAOs on which the SMEs see the greatest individual differences; hence, informed selection can produce the greatest change in employee effectiveness. Items and tests would be selected and/or constructed to measure those KSAOs that pass the criteria mentioned here for the first three items and are highest on the scale, indicating that the KSAO distinguishes superior workers from average workers.

For example, if we have the pattern of ratings for the three KSAO items for the job of personnel specialist shown in Table 3.4, we would develop three items or groups of items for any test (or other kind of predictor) because all three are necessary for newly hired workers. All three items are possessed by many applicants, so it is practical to select on these items, all would produce some trouble if ignored, and there are differences among current employees' performance on these items. We would expect that the ability to communicate with angry/confused examinees would be the most valid predictor of effectiveness because of the perception that the range of individual differences is greatest for this item.

In constructing items for an exam, we should return to the task analysis part of the job analysis to determine the content of the exam. Frequently, a KSAO will be relevant for several tasks; in this case, test items are written using the tasks that are judged most important. Consideration of the tasks is critical when we use a content validation strategy (see chapter 6). We must use items that are representative of and linked to important job tasks.

If we want to use the job analysis results for training, we need to know if new workers needed to have a KSAO. If not and this item also distinguishes superior workers from average workers, then training on the KSAO would be appropriate. If it is practical to require a KSAO in the labor market, it may be cost effective to select for the characteristic. However, selecting on these attributes may eliminate otherwise excellent applicants who lack particular educational or job experience. Again, tasks associated with KSAOs that are considered important for training would be used as the basis for determining training content.

Other uses of a job analysis conducted in this manner are discussed in Brannick and Levine (2002).

Campion

Campion (1988, 1989) proposed an interdisciplinary model of work and job design. In this model, four basic approaches to work design are considered, with each stemming from a different theoretical tradition. First, a *mechanistic* approach focuses on improving the efficiency of job tasks. Such job design features include task simplification and specialization. Second, a *motivational* approach focuses on enhancing the motivational properties of jobs, such as increasing autonomy. Third, a *perceptual* approach focuses on improving the information processing requirements of jobs. These interventions are drawn from human factors research and include such manipulations as enhancing workplace layout and improving lighting. Finally, a *biological* approach focuses on reducing stressors and strains in the workplace, both mental and physical. Such methods include the presence of work breaks and reducing noise.

One implication of this multidimensional framework is that it identifies the trade-off between making work more efficient versus making work more satisfying and motivating. Morgeson and Campion (2003) showed that when work is designed with only one of these perspectives in mind, the trade-off is observed. However, when both perspectives are considered simultaneously, the trade-off can be avoided.

Competency Models

Since the early 1990s, a major trend in HR has been the development and implementation of competency models. In general terms, competency models describe the core set of KSAOs that are judged to be important for an organization. They tend not to be specific to a particular job or level, but rather they focus on what is common across jobs and levels. Competency models have a certain appeal to managers because they are presented in a language and style with which organizational decision makers are familiar. For example, a competency model might identify the leadership skills necessary for successful performance in jobs throughout an entire organization, thereby communicating the organization's values and culture. This is in stark contrast to the very job- and level-specific nature of a traditional job analysis.

However, questions remain as to whether competency modeling is a useful practice and how competency models differ from a traditional job analysis.

To answer this question, the Society for Industrial and Organizational Psychology created a Job Analysis and Competency Modeling Task Force. This task force surveyed leading experts in different areas of consulting, business, government, and academics, asking them to describe the science and practice of competency modeling and how they compare to job analysis.

As summarized in Schippmann et al. (2000), there are some important differences between traditional job analysis and competency modeling. Job analysis methods tend to be more rigorous, psychometrically sound, and well documented than competency models. In contrast, competency models are more strongly linked to the organization's strategy and goals than traditional job analysis. Not surprisingly, competency models are conceptualized and operationalized at more broad or general levels of specificity than would be true with a traditional job analysis approach. They tend to be general to an entire occupational category, level within an organization, or even the entire organization. They also tend to focus more on personality and values than does a traditional job analysis.

Such breadth makes competency models appealing to management, but their lack of specificity with regard to KSAOs and tasks makes them unlikely to provide a sound basis for developing a staffing plan for a specific job. Because KSAOs are broadly defined, it becomes difficult to link these KSAOs to important tasks for a job and to develop predictor batteries assessing the KSAOs. The lack of specificity in competency models makes them unlikely to provide a sufficient form of content validity evidence (discussed later in this chapter and in chapter 6), and as such, they may be more useful for developmental purposes than staffing purposes. Of course, this assumes competency models and job analysis must be used in an either–or fashion. There is little reason why an organization could not use both forms of information because both may serve different purposes and thus be differentially effective. For example, an organization may decide that everyone hired who will interact with customers as part of their job must be service oriented, so all hiring for all jobs might include procedures for assessing service orientation. Therefore, people hired for a hotel chain in such different jobs as call center representative, front desk clerk, waitperson, or

> **Box 3.2 Sample Competency Analysis for a Series of Positions in the Snack Food Delivery Industry**
>
> | Delivery Salesperson | Supervisor | Region Manager |
> |---|---|---|
> | * Customer orientation
* Planfulnes
* Arithmetic competencies
* Social skills
* Conscientiousness
* Computer knowledge | * Customer orientation
* Planfulness
* Arithmetic competencies
* Social skills
* Conscientiousness
* Computer knowledge
* Coordination skills
* Influencing skills
* Adaptability
* Sales territory knowledge | * Customer orientation
* Planfulness
* Arithmetic competencies
* Social skills
* Conscientiousness
* Computer knowledge
* Coordination skills
* Influencing skills
* Adaptability
* Region knowledge
* Leadership skills |
>
> For each position, there are many of the same competencies, but these competencies will obviously manifest themselves in different ways. For example, computer knowledge competencies are required by all positions, but the level and kind of knowledge will vary. This is where the traditional job analysis focus on KSAOs becomes appropriate.

hotel manager would all be assessed for generic service orientation. In addition, of course, assessments would be made of the specific KSAOs required for effective performance in the different jobs, thus combining the organizational competency model approach with the traditional job analysis approach with regard to a specific job.

One of the authors recently participated in a project with a company that manufactures and delivers snacks to retail outlets (e.g., supermarkets, drugstores). The issue the company wanted to have explicated was the nest of competencies that different positions associated with delivery of the snacks required. Thus, not one job, but the nest of jobs, was the target: delivery person, delivery person supervisor, region manager, and so on. Box 3.2 presents the results of this project.

Summary

Obviously, a variety of techniques has been developed for understanding the task behaviors and the KSAOs required to perform them. We must again emphasize that there is no single best approach for all situations, and the particular job analysis method used should best match the purpose for conducting the study. However, do not believe that the application of job analysis is merely a matter of picking the most appropriate method. Job analysis in the 21st century is facing a variety of new challenges.

CURRENT CHALLENGES FOR JOB ANALYSIS

Is the job dead? For a time in the mid-1990s, several articles and books in the popular press touted "the end of the job." The idea was that technological advancements such as e-mail and the Internet, the shift to service-oriented work, the increasing use of team- or contract-based work, and the growth of small businesses had destroyed the job as we previously knew it. The job was simply a concept that had outlived its usefulness. Its replacement was work based entirely on the type of tasks or projects the organization needed accomplished, without formal job descriptions. This would supposedly result in more "freelance" or contract-based work, where workers were more like autonomous external contractors to larger organizations, not unlike the craftsmen of a century before (Bridges, 1994).

Did the job actually die? For the vast majority of organizations and occupations, jobs are still alive and well and are likely to be for many years to come. However, the realities of modern work have certainly changed the nature of jobs and work, and as a result, pose new challenges to the science and practice of job analysis. In this section, we discuss several of these challenges.

The Social Setting for Job Tasks

It is clear that modern work is increasingly interpersonal, service oriented, and team based. Job analysis methodologies therefore need to consider the social context of work. When teams become the basic organizing structure of work, the job must necessarily change because an individual's behavior is influenced by teammates. Furthermore, the work being performed may not be successfully performed by any one person; that is, the job may require interdependencies for effective performance—think basketball as a metaphor. It is not that individuals and their KSAOs are unimportant, but it is that a focus exclusively on such individual attributes ignores the team issues and interdependencies that make for effective performance of the task.

132 **3. JOB ANALYSIS**

For example, suppose a 12-person project team works to develop a new product. Members on the team have distributed expertise; some are marketing experts, some are engineers, some are production specialists, and others are experts in distribution. No one person can accomplish the tasks of another team member, and no one person can successfully complete the team task. How then will the "job" be defined? If we define it at the individual level, we miss the coordination and synergy that results from having a coordinated well-functioning team. Only some of the tasks are truly individual-level tasks, and by ignoring the collective nature of performance, important tasks and thus KSAOs might not be identified.

Research is only starting to tackle the difficult issues associated with job analysis in team-based environments. What will be required is a truly multilevel conceptualization of job analysis. Early work by Fleishman and Zaccaro (1992) developed a taxonomy of team performance functions. Many of the individual behaviors that emerge in looking at team performance functions, such as activity pacing and monitoring and adjusting to other team members, would not likely be identified in a traditionally focused individual job analysis. Klimoski and Jones (1995) identified different variations of job analysis that may be necessary when work is conducted in team environments. They note the need to consider broader definitions of job requirements that incorporate interdependencies among team members. Further, they propose more SMEs will need to be sampled because each team may have slightly different norms and working styles. Finally, the importance of considering team composition (the team's mix of individual KSAOs) and making linkages from individual to team KSAOs will be critical for accurately identifying selection KSAOs in team settings. Stevens and Campion (1994) took this issue one step further and identified the KSAOs necessary for successful individual performance within team contexts (i.e., teamwork KSAOs). These KSAOs included interpersonal (e.g., conflict resolution, collaborative problem solving, communication) and self-management (e.g., goal setting, planning, task coordination) dimensions.

One example of how job analysis might be conducted in multilevel settings is provided by Schmitt (2002). He gave examples of how rating scales for task and KSAO statements could be altered to incorporate multilevel issues. For example, a task rating scale might be "Rate the degree to which this knowledge, skill, or ability distinguishes a highly effective work group from a group that is just barely acceptable." The key difference here is that the reference is at the group level rather than the usual individual level, a change in referent that can produce large changes in responses (e.g., Klein, Conn, Smith, & Sorra, 2001).

Beyond basic changes in the structure of tasks in team-based settings, another important consideration is that of roles. Roles concern the expectations of others for behavior that includes but extends beyond formal task behavior. Because these expectations can emerge from different sources (coworkers, customers, professional standards), roles are said to be emergent in that individual employees implicitly negotiate what their role behaviors will be vis-à-vis different others. Ilgen and Hollenbeck (1991) proposed a provocative theory called *job role differentiation* that explored the boundaries of jobs and roles. The central thesis was that both job tasks and roles exist for employees, but tasks tend to be formally defined, whereas roles are more emergent properties of jobs and negotiated between an employee, coworkers, customers, and a supervisor. Thus, roles encompass job tasks and responsibilities, as well as the (mostly unwritten) expectations of others.

One important consequence of this conceptualization is that job analysis methodologies may be missing an important part of this broader work context by focusing only on the job-specific tasks. Another consequence is that roles, being more emergent properties, may be more idiosyncratic across employees, and thus not easily examined using the traditional estimates of agreement and reliability that are used in job analysis. Indeed, when discussing the nature of roles, Ilgen and Hollenbeck (1991) noted "emergent task elements are by definition subjective, personal, dynamic, and specified by a variety of social sources other than the prime beneficiaries" (p. 174). If only the formal (established) elements of work are analyzed, much of what a person does on the job may not be built into a staffing battery. Given that roles may be determined more by social and interpersonal factors than task-specific KSAOs, it is possible that many selection batteries do not tap the interpersonal constructs that are actually important for successful performance (at least to the extent that role-consistent behavior is an important determinant of performance).

We believe the link between roles and tasks is an important issue in need of further consideration. For example, in customer service settings, a service provider often must deal with considerable role conflict: between a supervisor and customer, between coworkers and customers, and between a supervisor and coworkers. As a function of different relationships with these constituencies, different employees may very well have different roles, and thus, more or less conflict. A traditional job analysis that misses this "nontask" information may be missing a major portion of what it means to be an effective service provider.

Job Analysis for Dispositional Constructs

We noted previously how competency models tend to focus more on personality and values than traditional job analysis methods. Worker-oriented job analysis methods, such as the PAQ, tend to be closer to competency models in their consideration of noncognitive KSAOs (e.g., personality). However, most traditional job analysis methods are largely deficient in their explicit consideration of such KSAOs. As a result, key personality constructs might not be identified in a job analysis, leading to their absence in predictor batteries.

In an attempt to better identify important personality constructs, Raymark, Schmit, and Guion (1997) developed the Personality-Related Position Requirements Form (PPRF). This instrument was designed to aid in the identification of personality constructs based on the Five Factor Model (FFM) of personality, currently the dominant personality framework (the FFM is discussed in more detail in chapter 9). Behavioral statements reflecting the constructs of neuroticism, extraversion, openness to new experiences, agreeableness, and conscientiousness are administered to job experts, who rate each item on an "Effective performance in this position requires the person to..." "$0 =$ not required, $1 =$ helpful, $2 =$ essential" scale. Sample items for neuroticism and conscientiousness are, respectively, "Keep cool when confronted with conflicts" and "Work until task is done rather than stopping at quitting time."

Whether the PPRF leads to the identification of personality constructs that are missed by traditional job analysis methods remains to be seen. It is possible that traditional methods can identify personality constructs in sufficient detail, but a large-scale comparison between traditional methods and the PPRF (or similar personality-based methods) has not been conducted.

Demographic Attributes of SMEs and Job Analysis Ratings

There has been a fair amount of research examining the relationships between a variety of demographic attributes (e.g., race, gender, tenure) and job analysis ratings (primarily task importance, difficulty, and/or time spent). The issue here is whether people with different backgrounds or experiences perceive the job differently. If so, it becomes difficult to unambiguously define "the job" because the job means different things to different people. Another consequence is that if there are demographic differences in perceptions of jobs, and the job analyst does not incorporate all such perspectives into the SME sampling strategy, then the final job analysis information may look good from a quantitative perspective (e.g., high agreement) but not fully represent the opinions and perceptions of all job experts. The implication would be the design of incomplete—or inappropriate—staffing procedures.

Consider an example situation where the job of firefighter is being analyzed. Assume only 10% of the force is comprised of women. If participants are sampled randomly and all participate, then only about 10% of the sample will be women (if 200 people were in the sample, only 20 would be female). What might be the consequence of this? Figure 3.7 shows one possible consequence, where men and women differ with respect to how they would carry 80 lb of equipment. Men, having more physical strength, indicate they would simply lift the equipment and carry it; women, having less physical strength, indicate they would drag the equipment. If carrying the equipment was a correct behavior and dragging the equipment was a incorrect behavior, then it means women could not do the job. However, if both behaviors are correct, what then? A random sampling strategy shown in Fig. 3.7 illustrates how the results of the job analysis would lead to the conclusion that an important task is lifting 80 lb of equipment. Developing a physical strength test that requires applicants to lift 80 lb is likely to result in men being hired at a disproportionate rate.

In contrast, if this job analyst was to oversample women, he or she might realize that one may either lift or drag the equipment, and the task statement would be edited appropriately. This could result in a change to the physical ability test, such that lifting 80 lb would not be required, but rather one must simply be able to drag 80 lb. This would lead to a physical test requiring applicants to drag 80 lb, and hence, not produce such severe differences between men and women. We have constructed this example to make the point; rarely in practice is the issue so clear-cut. However, this illustrates how and why it is important to consider all perspectives in a job analysis; it is only by identifying the varieties of ways effective performance can be accomplished that the variety of KSAOs that make for effective performance can be identified.

Despite this suggestion, empirical research has been mixed about the extent to which demographic factors are related to job analysis ratings,

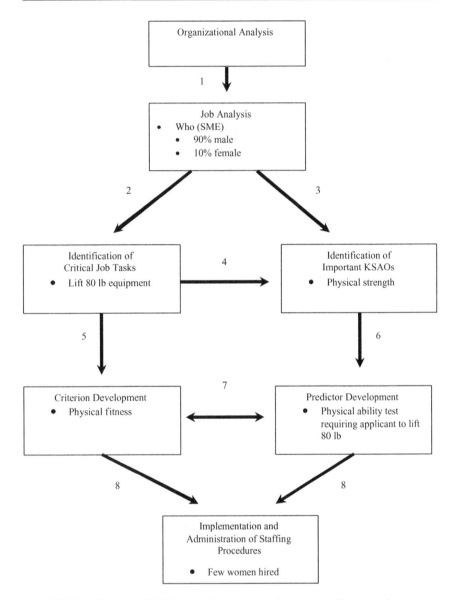

FIG. 3.7. Example of SME composition in determination of staffing procedures.

and the effect sizes for these relationships have tended to be small. A few studies have found small effects for gender and even smaller effects for race (e.g., Landy & Vasey, 1991; Schmitt & Cohen, 1989). Job experience appears to be the more consistent—and stronger—correlate of job analysis ratings (Borman, Dorsey, & Ackerman, 1992; Landy & Vasey,

1991). Interestingly, some research has even linked job analysis ratings to indices of effectiveness. Borman et al. (1992) found that salesperson's time-spent ratings were related to individual sales, such that better salespeople reported spending different amounts of time on different tasks. Lindell, Clause, Brandt, and Landis (1998) similarly found time-spent and difficulty ratings were related to supervisors' evaluations of unit effectiveness. However, not all research has been supportive of a job analysis rating effectiveness link (Conley & Sackett, 1987; Wexley & Silverman, 1978).

These inconsistencies and small effects across studies suggest that demographic factors may not contribute substantially to job analysis ratings. However, we believe such factors should still be important considerations in the planning and administration of job analysis procedures, if for no other reason than to enhance perceptions of fairness and inclusiveness from job incumbents and management. It is also important to recognize that legal considerations make such issues extremely important, so an organization may best protect itself by making a good faith effort to ensure all constituencies are represented. Finally, relatively little research has examined this issue and with a narrow range of jobs. Borman et al. (1992) noted the key role of job autonomy as a reason why they found relations between time-spent ratings and sales performance (see also Lindell et al., 1998). To the extent that jobs allow flexibility and autonomy in work behavior, stronger relations between job analysis ratings, demographic attributes, and effectiveness might be found.

If one is to formally incorporate various constituencies into a job analysis, a question becomes how to best compose SME panels. In our experience, a random sampling strategy where SMEs are asked to volunteer to participate is only useful if evidence can show the SMEs are demographically similar to the job incumbent population. In instances where an organization may be under a court-ordered consent decree to increase minority representation (to address past instances of discrimination), the job analysis should possibly oversample the discriminated minority group to ensure complete coverage (as alluded to in Fig. 3.7). In situations where the organization may have jobs in multiple branch locations distributed throughout various geographic locations, the sampling strategy should incorporate SMEs from these different locations because there may be local differences in how people perform the job. This is particularly likely when the organization operates in multiple countries because there are large cultural differences in how work is performed. The answer to all such sampling questions comes from the reason the job analysis was implemented in the first place.

The Legal Necessity of Job Analysis

We discuss legal issues more fully in chapter 11, but for now it is important to recognize that job analysis helps comprise the legal defensibility of a staffing system. It is also true that many legal battles are reduced to questions about the adequacy of the job analysis (Harvey, 1991). In our experience, the following features are most often challenged:

- Number of SMEs and adequacy of sampling strategy
- Comprehensiveness of task and KSAO statements
- Linkage between tasks and KSAOs
- Content overlap between job analysis and selection tests
- Adequacy and comprehensiveness of criterion dimensions
- SME accuracy and accountability

As we noted earlier in Fig. 3.2, job analysis sets the foundation of staffing, but this foundation is based on many inferences. The rigor needed to document and provide evidence for such inferences is something that many organizations find frustrating and try to avoid. For example, an organization may not want to require SMEs to participate in a survey, but if this organization is legally challenged, this sampling strategy will be one of the first things that is attacked. The legal battlefield is often won or lost on the job analysis front.

A slightly different concern is when an organization adopts a content validation strategy. Discussed in more detail in chapter 6, content validity involves a systematic examination of the content of an assessment instrument (e.g., interview) with the KSAO content of the job (as defined through a job analysis). Thus, a content-oriented validation strategy depends on an adequate job analysis; however, the link is not only a physical one, but also a psychological one. That is, physical fidelity (does the test look like what is required on the job?) is important but secondary to psychological fidelity (does the test assess the important KSAOs required on the job?).

Goldstein, Zedeck, and Schneider (1993) described the procedures and inferences necessary to establish content validity. Because the foundation of content validity rests on a sound job analysis, a comprehensive job analysis is conducted following procedures noted throughout this chapter. After the identification of critical tasks and KSAOs, and the establishment of predictor measures, the content validity of the measures is assured by having a group of experts review the test items and determine the extent to which each item taps an important KSAO. Weighting KSAOs by different amounts to reflect their contribution to task performance is also possible under content validation strategies. Many public sector employers, such as police and fire departments, use content validity studies to support their use of knowledge exams. In such cases, every item will be linked to a particular KSAO, and the number of items for each KSAO will reflect the percentage importance of the KSAO for successful task performance.

Accuracy and Job Analysis

A more recent controversy among job analysis researchers involves the extent to which job analysis ratings are accurate and reflect a "true score," that is, the true description of the job. In fact, the issue of whether a true score exists in job analysis ratings has been raised (Sanchez & Levine, 2000). As we noted earlier, different demographic groups and those with different amounts of experience may see the same job differently. The typical solution to this situation is to average task or KSAO ratings across raters and compute interrater reliability and/or agreement indices. Provided there is minimally sufficient interrater reliability or agreement, the thought is that averaging across raters removes idiosyncratic perceptions and biases.

The issue for job analysis is whether these differences should be averaged out, and whether the differences actually reflect biased perceptions (Sanchez & Levine, 2000). If these differences reflect real differences in jobs, then how can a comprehensive staffing system (or any HR system, for that matter) be developed? Harvey and Wilson (2000) noted that if job analysis procedures are based on relatively observable behaviors and rated using unambiguous scales, objective job analysis information can be obtained (see also Ilgen & Hollenbeck, 1991). However, given the many issues raised earlier, including the difference between tasks versus roles, the social context of work, and differences in job tasks across organizational units, one must wonder to what extent high agreement on job tasks and specifications should be found.

To further complicate matters, Morgeson and Campion (1997) identified a host of cognitive and perceptual factors that make the "objective" collection of job analysis information difficult. Such perceptual bases of inaccuracy include social sources (e.g., conformity pressures), self-presentation processes (e.g., impression management), limitations in information processing (e.g., information overload), and biases in information processing (e.g., carelessness). They maintained a distinction must be made between inaccuracy and actual differences in ratings and that perceptual inaccuracies can make differences in job analysis ratings appear larger (or smaller) than they really are. Furthermore, they noted the statistical artifacts that such inaccuracies can cause, including less dimensionality, inflated ratings, and higher (or lower) interrater agreement. Empirical support is now gathering to support such claims (Morgeson, Delaney-Klinger, Mayfield, Ferrara, & Campion, 2004).

These concerns lead to potential differences due to the *source* of the job analysis ratings. Sources refer to whether job analysis information is obtained from job incumbents (either high or moderate performing), supervisors, expert job analysts, or similar people. The position taken by the developers of O*NET is that job incumbents are the best source of job information (Peterson et al., 2001). However, there may be important differences between employees and supervisors, such that employees tend to report what is done, whereas supervisors tend to report how the job should be done. Goldstein et al. (1993) found that supervisors are better for providing information about KSAOs, whereas employees are better for providing information about tasks. Another issue is the composition of the SME panel, or even whether a panel will be used (Sanchez & Levine, 2002).

The expectation that SMEs have many difficulties making accurate job analysis ratings is not unique; it is an issue with all judgmental ratings (as discussed in chapter 4). However, the important issue is that one may always be able to question the accuracy of any given piece of job analysis information. Sanchez and Levine (2000) suggested the consequences of the job analysis information should be used to determine accuracy; if the job analysis information leads to the development of effective staffing procedures, then the information was reasonably accurate. In contrast, Harvey and Wilson (2000) argued for the use of formal "expert accuracy reviews," which consist of panels of job experts who are highly motivated and accountable for ensuring the accuracy of the ratings. Perhaps the safest conclusion that can be made at this point is that multiple sources should be used whenever possible, and the SMEs should be held accountable for their ratings.

Cognitive Task Analysis

In contrast to the traditional job analysis, which examines a comprehensive set of tasks, a cognitive task analysis is a methodology that seeks to identify and isolate the cognitive processes associated with a particular task. For example, a cognitive task analysis might examine the cognitive operations involved in observing, interpreting, and responding to customer e-mail requests. Whereas in a job analysis this task would be one of several hundred others, in a cognitive task analysis this one task would be the focus of study, and the goal would be to understand how individuals perceive, encode, process, and respond to the information provided.

Cognitive task analysis can be described as a general set of methods that decompose tasks into the basic cognitive processes that are required to perform those tasks (DuBois & Shalin, 1995). Traditional job analysis focuses on what gets done, but cognitive task analysis focuses on the mental operations underlying how this work gets done. Cognitive task analysis uses a variety of data collection methods perfected in the more basic cognitive science literature, including verbal protocol analysis, process tracing, communication analysis, and reaction time measures.

Cognitive task analysis has been used more frequently in training design than in staffing. In fact, we are not familiar with any application of cognitive task analysis in the development of a staffing measure. However, we believe this represents an important avenue for future research, and in particular, to explore why different groups of people might perform differently on the same task or selection test. We discuss racial, gender, and age subgroup differences in job performance and various selection constructs and measures in later chapters, but we raise the possibility here that cognitive task analysis has the potential to unlock why such differences exist. As such, it represents a rich potential source of information that staffing researchers have not used.

Job Analysis for Jobs That Do Not Exist

One issue that has not been addressed so far concerns the specification of KSAOs for jobs that do not yet exist or for jobs that exist in a particular form today but will change tomorrow, perhaps then requiring a person with somewhat different KSAOs. One of your authors has been involved in a project of the latter type. The job of telephone order taker was going to change to telephone order taker and salesperson. The job analysis was conducted with SMEs who were familiar with the present job and involved in planning the future job and its objectives. This same group of people provided estimates of the KSAOs required to carry out both the old and the new set of tasks. Based on those KSAOs, a set of predictors was developed to predict performance on the new job prior to its actual existence. The U.S. Army is currently going through a similar exercise, called the Select21 Project. The expectation is that soldiers will have to use more judgment in the future because the nature of their missions will be more decentralized (e.g., interacting with friendly and hostile civilians). Given the changing context, the Army is asking SMEs to make predictions about which KSAOs will be important for these future assignments.

Schneider and Konz (1989) discussed many issues with the specification of KSAOs for job tasks that have not yet been created. They outlined a general strategy through which job analysis may still be conducted. First, a comprehensive job analysis should be conducted on the current job. Second, a group of SMEs with insight into what the future job might involve are asked to think of likely tasks and KSAOs. Third, a new set of task and KSAO statements are prepared and administered to experts. Finally, the data are analyzed to identify the critical—but as yet unmeasured—future tasks and KSAOs.

A related issue to thinking about jobs that do not yet exist is the concept known as strategic job modeling. Schippmann (1999) provided a comprehensive treatment of this issue, making heavy reliance on broad general competencies and linking these to business strategy. The development and use of broad competencies provides one solution to this problem. For example, it is probably safe to say the job of manager will require interpersonal skills, but what specific kinds of skills may be harder to identify. Methods designed to predict requirements (e.g., selection, training) are still not well developed. Given rapidly changing technology, these methods are certainly necessary. Brannick and Levine (2002) provided a more detailed treatment of these issues.

Understanding Jobs From an Organizational Perspective

Another major issue left unexplored in traditional job analysis procedures concerns the following question: Given two organizations that are in the same industry (so they have the same kinds of jobs) hiring

CURRENT CHALLENGES FOR JOB ANALYSIS 143

people with equivalent KSAOs, why does one organization consistently outperform the other? In other words, are the KSAOs of employees the only determinant of organizational effectiveness? Is there an "organizational main effect" on organizational effectiveness, independent of the human capital within the organization? Answering this question requires an approach to job analysis that includes specifying the organizational conditions in which the job will be performed.

In chapter 4, we see how organizational conditions come into play when evaluating the performance of individual workers; the concern there is for how one person in a job may work under different conditions than another person in the same job, and thereby perform differently (better or poorer). Here we are concerned with a different level of the problem: how some organizations create generally superior or inferior conditions for workers as a group, keeping the general level of effectiveness higher or lower than would be expected given the qualifications (KSAOs) of those who are hired.

This is an important issue because staffing procedures are designed, ultimately, to produce organizations that can compete and survive. It is, then, one thing to hire tomorrow more competent people than are being hired today, but it may be another thing to be competitive and survive. We know how to hire more competent people—that is what this book is about—but we also know that hiring people with more competence to do particular jobs does not assure organizational effectiveness (Harris, 1994).

Perhaps Fig. 3.8 will make the issue clear. The top of Fig. 3.8 shows the distribution of the ability of workers in organizations A and B. The bottom of Fig. 3.8 shows, however, that the distributions of productivity are not equal, that productivity in organization B is superior to productivity in organization A. Our contention is that certain organizational attributes, such as HR practices, have an impact on people that is essentially independent of the impact of their competence level as reflected in their KSAOs.

The tendency of staffing researchers and practitioners has not been to examine HR practices of organizations as much as to focus on jobs. However, jobs do not exist in a vacuum. Because job–person interaction is the ultimate focus of staffing decisions, it is important to understand external conditions that may affect their interaction. Thus, the basic staffing model is to assess job and person characteristics and predict the outcome of their interaction. However, if HR practices (aside from the immediate attributes of the job) cause workers to behave in particular

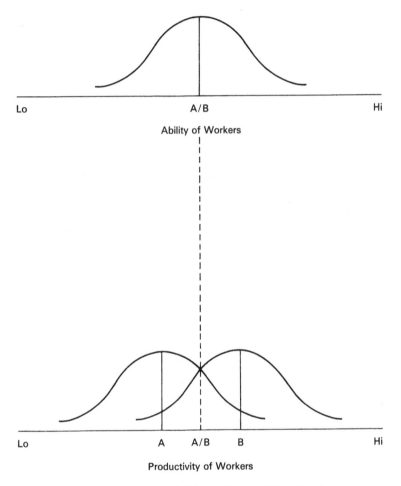

FIG. 3.8. Distributions of ability and productivity of workers.

ways, then the staffing process may work more or less effectively than would be predicted on the basis of worker attributes alone.

In more recent years, effective use of HR practices has become known as "high-performance work systems." We touched on these issues earlier when considering the external environment, noting how HR practices are driven by the external environment and the organization's strategy, goals, and culture. The key issue here is that staffing is just one HR practice, and other HR practices are important and must be considered in combination (or HR "bundles"; MacDuffie, 1995).

CURRENT CHALLENGES FOR JOB ANALYSIS 145

What are the HR practices that contribute to performance? Clearly *staffing* is one such practice, and researchers have shown how organizations that use job analysis methods, formal recruiting procedures, validation studies, structured interviews, cognitive tests, and biographical data tests, show higher levels of organizational profitability and sales growth (e.g., Huselid, 1995; Koch & McGrath, 1996; Terpstra & Rozell, 1993). Likewise, organizations that employ formal *training* programs and invest in *employee development* tend to show higher levels of effectiveness (Huselid, 1995).

Compensation systems also have an influence on the effectiveness of workers. The use of bonuses, attractive benefits, pay-for-performance plans, incentive programs, stock options, and related systems are present in more high-performing organizations (Huselid, 1995). Also relevant is the presence, nature, and administration of formal performance appraisal systems, and whether these systems are linked to various rewards. Two theories dominate present thinking along these lines—the organizational justice literature (e.g., Greenberg, 1990) and whether rewards are tied to performance (Lawler, Mohrman, & Ledford, 1998). In chapter 5, we discuss the implications of these reward systems for employee recruitment.

Organizational culture may also influence the way individuals behave in a setting and the degree to which their individual attributes are reflected in their individual performance. For example, consider the idea that Organization A has established a culture that values innovation and creativity. The organization establishes this culture through the kinds of behavior it rewards, the kinds of people it hires, the emphases management places on innovative and creative behaviors at performance appraisal time, the kinds of myths and stories people share about who gets promoted and why, and so forth (Martin 2002; Schein, 1992).

Now, take Organization B where the emphasis is on monitoring costs, establishing efficient and routine ways to get the work done, outsourcing the labor pool, and so forth. We might make the prediction as to when individuals of equal competency skill sets will be differentially productive merely as a function of in which organization, A or B, they work.

When the HR practices are aligned with each other and the organization's strategy, external environment, and goals, positive effects on performance at all levels are likely to occur. The truly amazing finding is that a sizeable portion of an organization's profitability may stem directly from HR practices (Gerhart, 1999). More often than not, however, such alignment is not present, and the HR practices that are implemented to improve performance may actually work against each other. For example, consider an organization that makes hiring decisions based on cognitive ability. If those with more ability are hired, but the reward system is such that the use of cognitive ability is not valued, lower job performance and increased turnover are likely to result. In such an instance, the HR practices are fighting against each other, and success in one practice is failure for another.

A similar point of view has been expressed by Peters and O'Connor (1980), namely, that organizations can do much to facilitate or constrain the performance of employees. They suggested that systematic differences in work situations (organizations or units within an organization) exist and determine whether employees can be maximally effective. Among these situational resource variables are materials and supplies, budgetary support, time availability, and the physical work environment.

If these kinds of changes yield improved productivity from people of equivalent levels of ability or competence or if they serve as important constraints on productivity, then we can infer that organizations that are more like the changed organizations are going to be more effective. If this is true, then merely finding competent people may not result in the kinds of effectiveness that were hoped for.

We believe the following:

1. More competent people on a job will outperform less competent people on the job.
2. Performance of the same jobs in two different organizations with people of the same levels of competence may produce or result in different levels of effectiveness.
3. It is important for organizations to ask themselves if their lack of effectiveness is attributable to the competence of workers or to the way the organization is managed.

Because of the potentially great impact organizations may have on the outcomes of staffing decisions, it is important for an organization to determine its own stance on various "style" factors. Unfortunately, we are not able to be as specific about these factors as we were with respect to job analysis. Leadership practices, reward systems, general interpersonal relationships, and organizational goals are all important targets of analysis. All these factors affect potential success or failure in explaining and/or predicting how well people actually will do their jobs and how satisfied they will be with what they do.

SUMMARY

In this chapter, we try to show that jobs do not exist in a vacuum; the kinds of jobs in an organization are a function of its goals, strategies, HR practices, and the larger environment. We discuss this larger external environment to place a context for our review of different job analysis procedures, concluding with more recent advances in theories of work, such as O*NET and competency modeling, and more recent challenges to job analysis. We stress the idea that understanding jobs and job context is necessary for developing the staffing process and for evaluating the potential contribution to the accomplishment of organizational goals. The importance of organizational goals and measurement of the degree to which goals are accomplished are the focus of the next chapter.

REFERENCES

Bliese, P. D. (2002). Multilevel random coefficient modeling in organizational research: Examples using SAS and S-PLUS. In F. Drasgow & N. Schmitt (Eds.), *Modeling in organizational research: Measuring and analyzing behavior in organizations* (pp. 401–445). San Francisco: Jossey-Bass.
Borman, W. C., Dorsey, D., & Ackerman, L. (1992). Time-spent responses as time allocation strategies: Relations with sales performance in a stockbroker sample. *Personnel Psychology, 45*, 763–777.
Brannick, M. T., & Levine, E. L. (2002). *Job analysis*. Thousand Oaks, CA: Sage.
Bridges, W. (1994). The end of the job. *Fortune, 130*, 62–74.
Burns, T., & Stalker, G. M. (1961). *The management of innovation*. London: Tavistock.
Campion, M. A. (1988). Interdisciplinary approaches to job design: A constructive replication with extensions. *Journal of Applied Psychology, 73*, 467–481.
Campion, M. A. (1989). Ability requirement implications of job design: An interdisciplinary perspective. *Personnel Psychology, 42*, 1–24.
Colihan, J., & Burger, G. K. (1995). Constructing job families: An analysis of quantitative techniques used for grouping jobs. *Personnel Psychology, 48*, 563–586.
Conley, P. R., & Sackett, P. R. (1987). Effects of using high- versus low-performing job incumbents as sources of job-analysis information. *Journal of Applied Psychology, 72*, 434–437.
Cranny, C. J., & Doherty, M. E. (1988). Importance ratings in job analysis: Note on the misinterpretation of factor analyses. *Journal of Applied Psychology, 73*, 320–322.
DuBois, D., & Shalin, U. L. (1995). Adapting cognitive methods to real-world objectives: an application to job knowledge testing. In P. D. Nichols, S. F. Chipman, &

R. L. Brennan (Eds.), *Cognitively diagnostic assessment* (pp. 189–220). Hillsdale, NJ: Erlbaum.
Dunnette, M. D. (1966). *Personnel selection and placement*. Belmont, CA: Wadsworth.
Fine, S. A., & Wiley, W. W. (1974). An introduction to functional job analysis. In F. A. Fleishman & A. R. Bass (Eds.), *Studies in personnel and industrial psychology* (3rd ed., pp. 6–13). Homewood, IL: Irwin.
Flanagan, J. C. (1954). The critical incident technique. *Psychological Bulletin, 51*, 327–355.
Fleishman, E. A., & Zaccaro, S. J. (1992). Toward a taxonomy of team performance functions. In R. W. Swezey & E. Salas (Eds.), *Teams: Their training and performance* (pp. 31–56). Norwood, NJ: Ablex.
Gerhart, B. (1999). Human resource management and firm performance: Measurement issues and their effect on causal and policy inferences. In P. M. Wright, L. D. Dyer, J. W. Boudreau, & G. T. Gerhart (Eds.), *Research in personnel and human resource management* (Suppl. 4, pp. 31–51). Greenwich, CT: JAI Press.
Goldstein, I. L., Zedeck, S., & Schneider, B. (1993). An exploration of the job analysis-content validity process. In N. Schmitt & W. C. Borman (Eds.), *Personnel selection in organizations* (pp. 3–34). San Francisco: Jossey-Bass.
Greenberg, J. (1990). Organizational justice: Yesterday, today, and tomorrow. *Journal of Management, 16*, 399–432.
Guion, R. M. (1965). *Personnel testing*. New York: McGraw-Hill.
Harris, D. H. (1994). *Organizational linkages: Understanding the productivity paradox*. Washington, DC: National Academy Press.
Harvey, R. J. (1991). Job analysis. In M. D. Dunnette & L. M. Hough (Eds.), *Handbook of industrial and organizational psychology* (pp. 71–163). Palo Alto, CA: Consulting Psychologists Press.
Harvey, R. J., & Wilson, M. A. (2000). Yes Virginia, there *is* an objective reality in job analysis. *Journal of Organizational Behavior, 21*, 829–854.
Huselid, M. A. (1995). The impact of human resource management practices on turnover, productivity, and corporate financial performance. *Academy of Management Journal, 38*, 635–672.
Ilgen, D. R., & Hollenbeck, J. R. (1991). The structure of work: Job design and roles. In M. D. Dunnette & L. M. Hough (Eds.), *Handbook of industrial and organizational psychology* (Vol. 2, pp. 165–207). Palo Alto, CA: Consulting Psychologists Press.
Jackson, S. E., & Schuler, R. S. (1995). Understanding human resource management in the context of organizations and their environments. *Annual Review of Psychology, 46*, 237–264.
Klein, K. J., Conn, A. B., Smith, D. B., & Sorra, J. S. (2001). Is everyone in agreement? An exploration of within-group agreement in survey responses. *Journal of Applied Psychology, 86*, 3–16.
Klimoski, R., & Jones, R. G. (1995). Staffing for effective group decision making: Key issues in matching people to teams. In R. Guzzo, E. Salas, & Associates (Eds.), *Team effectiveness and decision making in organizations* (pp. 291–332). San Francisco: Jossey-Bass.
Koch, M. J., & McGrath, R. G. (1996). Improving labor productivity: Human resource management policies do matter. *Strategic Management Journal, 17*, 335–354.
Landy, F. J., & Vasey, J. (1991). Job analysis: The composition of SME samples. *Personnel Psychology, 44*, 27–50.
Latham, G. P., & Wexley, K. N. (1981). *Increasing productivity through performance appraisal*. Reading, MA: Addison-Wesley.
Lawler, E. E., III, Mohrman, S. A., & Ledford, G. E., Jr. (1998). *Strategies for high performance organizations: Employee involvement, TQM, and reengineering programs in Fortune 1000 companies*. San Francisco: Jossey-Bass.

REFERENCES

Lepak, D. P., & Snell, S. A. (1999). The human resource architecture: Toward a theory of human capital allocation and development. *The Academy of Management Review, 24,* 34–48.

Lepak, D. P., & Snell, S. A. (2002). Examining the human resource architecture: The relationships among human capital, employment, and human resource configurations. *Journal of Management, 28,* 517–543.

Lindell, M. K., Clause, C. S., Brandt, C. J., & Landis, R. S. (1998). Relationship between organizational context and job analysis ratings. *Journal of Applied Psychology, 83,* 769–776.

Locke, E. A., & Latham, G. P. (1990). *A theory of goal setting and task performance.* Englewood Cliffs, NJ: Prentice Hall.

MacDuffie, J. P. (1995). Human resource bundles and manufacturing performance: Organizational logic and flexible production systems in the world auto industry. *Industrial and Labor Relations Review, 48,* 197–221.

Martin, J. (2002). *Organizational culture: Mapping the terrain.* Thousand Oaks, CA: Sage.

Morgeson, F. P., & Campion, M. A. (1997). Social and cognitive sources of potential inaccuracy in job analysis. *Journal of Applied Psychology, 82,* 627–655.

Morgeson, F. P., & Campion, M. A. (2003). Work design. In W. C. Borman, D. R. Ilgen, & R. Klimoski (Eds.), *Handbook of psychology. Volume 12: Industrial and organizational psychology* (pp. 423–452). Hoboken, NJ: Wiley.

Morgeson, F. P., Delaney-Klinger, K., Mayfield, M. S., Ferrara, P., & Campion, M. A. (2004). Self-presentation processes in job analysis: A field experiment investigating inflation in abilities, tasks, and competencies. *Journal of Applied Psychology, 89,* 674–686.

Peters, L. H., & O'Connor, E. J. (1980). Situational constraints and work outcomes: The influences of a frequently overlooked construct. *Academy of Management Review, 5,* 391–397.

Peterson, N. G., Mumford, M. D., Borman, W. C., Jeanneret, P. R., & Fleishman, E. A. (1999). *An occupational information system for the 21st century: The development of O*NET.* Washington, DC: American Psychological Association.

Peterson, N. G., Mumford, M. D., Borman, W. C., Jeanneret, P. R., Levin, K. Y., Campion, M. A., Mayfield, M. S., Morgeson, F. P., Pearlman, K., Gowing, M. K., Lancaster, A. R., Silver, M. B., & Dye, D. M. (2001). Understanding work using the Occupational Information Network (O*NET): Implications for practice and research. *Personnel Psychology, 54,* 451–492.

Raymark, P. H., Schmit, M. J., & Guion, R. M. (1997). Identifying potentially useful personality constructs for employee selection. *Personnel Psychology, 50,* 723–736.

Sackett, P. R., & Laczo, R. M. (2003). Job and work analysis. In W. C. Borman, D. R. Ilgen, & R. Klimoski (Eds.), *Handbook of psychology. Volume 12: Industrial and organizational psychology* (pp. 21–37). Hoboken, NJ: Wiley.

Sanchez, J. I., & Levine, E. L. (2000). Accuracy or consequential validity: Which is the better standard for job analysis data? *Journal of Organizational Behavior, 21,* 809–818.

Sanchez, J. I., & Levine, E. L. (2002). The analysis of work in the 20th and 21st centuries. In N. R. Anderson, D. S. Ones, H. K. Sinangil, & C. Viswesvaran (Eds.), *Handbook of industrial, work & organizational psychology. Volume 1: Personnel psychology* (pp. 71–89). Thousand Oaks, CA: Sage.

Schein, E. A. (1992). *Organizational culture and leadership* (2nd ed.). San Francisco: Jossey-Bass.

Schippmann, J. S. (1999). *Strategic job modeling.* Mahwah, NJ: Erlbaum.

Schippmann, J. S., Ash, R. A., Battista, M., Carr, L., Eyde, L. D., Hesketh, B., Kehoe, J., Pearlman, K., Prien, E. P., & Sanchez, J. (2000). The practice of competency modeling. *Personnel Psychology, 53,* 703–740.

Schmitt, N. (2002). A multi-level perspective on personnel selection: Are we ready? In F. J. Dansereau & F. Yamarino (Eds.), *Research in multi-level issues. Volume 1: The many faces of multi-level issues* (pp. 155–164). Oxford, UK: Elsevier Science.
Schmitt, N., & Cohen, S. A. (1989). Internal analyses of task ratings by job incumbents. *Journal of Applied Psychology, 74*, 96–104.
Schneider, B. (1987). The people make the place. *Personnel Psychology, 40*, 437–454.
Schneider, B., & Konz, A. M. (1989). Strategic job analysis. *Human Resource Management, 28*, 51–63.
Schneider, B., Smith, D. B., Taylor, S., & Fleenor, J. (1998). Personality and organizations: A test of the homogeneity of personality hypothesis. *Journal of Applied Psychology, 83*, 462–470.
Schneider, B., & White, S. S. (2004). *Service quality: Research perspectives.* Thousand Oaks, CA: Sage.
Smith, P., & Kendall, L. M. (1963). Retranslation of expectations: An approach to the construction of unambiguous anchors for rating scales. *Journal of Applied Psychology, 47*, 149–155.
Snell, S. A., & Dean, J. W., Jr. (1992). Integrated manufacturing and human resources management: A human capital perspective. *Academy of Management Journal, 35*, 467–504.
Snow, C. C., & Snell, S. A. (1993). Staffing as strategy. In N. Schmitt & W. C. Borman (Eds.), *Personnel selection in organizations* (pp. 448–478). San Francisco: Jossey-Bass.
Stevens, M. J., & Campion, M. A. (1994). The knowledge, skill, and ability requirements for teamwork: Implications for human resource management. *Journal of Management, 20*, 503–530.
Terpstra, D. E., & Rozell, E. J. (1993). The relationship of staffing practices to organizational level measures of performance. *Personnel Psychology, 46*, 27–48.
U.S. Department of Labor. *Selected characteristics of occupations defined in the dictionary of occupational titles.* (1981). Washington, DC: Author.
U.S. Department of Labor. *Dictionary of occupational titles.* (1977). Washington, DC: Author.
Wexley, K. N., & Silverman, S. B. (1978). An examination of differences between managerial effectiveness and response patterns on a structured job analysis questionnaire. *Journal of Applied Psychology, 63*, 646–649.

4

Performance and Criterion Development

AIMS OF THE CHAPTER

Staff appraisal is concerned with the conceptualization, measurement, and analysis of how well people do their jobs and how satisfied they are in their work setting. More generally, staff appraisal is the process of gaining information about what individual workers do and feel. As noted by Randell, Packard, Shaw, and Slater (1974), this information is collected so organizations can make improved decisions regarding

1. Evaluation: to enable the organization to share the money, promotions, and perquisites "fairly"
2. Auditing: to discover the work potential, both present and future, of individuals and departments
3. Constructing succession plans: for manpower, departmental, and corporate planning
4. Discovering training needs: by exposing inadequacies and deficiencies that could be remedied by training
5. Motivating staff: to reach organizational standards and objectives
6. Developing individuals: by advice, information, and attempts at shaping their behavior by praise or punishment
7. Checking: the effectiveness of personnel procedures and practices

We can see that just as job and organizational analyses are useful for multiple purposes, so is staff appraisal. Staff appraisal issues must be considered before hiring the staff; organizations must decide on what bases and how they will assess staff before meaningful programs for selection, training, or motivation can be developed. Without knowing the standards against which people will be evaluated, procedures for getting people who will meet those standards are only guesswork. It is for this reason that staff appraisal should be based on the results of the job analysis. That is, what the organization does and values is formally articulated through the staff appraisal process. To the extent staff are evaluated on factors unrelated to what the organization values, the appraisal process will not accomplish organizational objectives.

A critically important point for staffing, one that is often forgotten, is that predictors derive their importance from criteria (Wallace, 1965). That is, we use a particular test for selection not because we think it is useful or appropriate, but because it has a demonstrated relationship with job performance. For example, if we administer a battery of personality and cognitive measures, identify which specific personality and cognitive measures predict job performance, and then only use those measures related to the job to hire applicants, it is obvious the criterion is the determinant of which measures are actually used and who is hired. It is essential that the conceptual and measurement issues surrounding job performance—that which determines the predictors used—are fully understood. Not to belabor the point, but this logic is why we discuss staff appraisal and performance criteria before we discuss the predictors of those criteria.

This chapter is also about the importance of specifying standards and goals for understanding why people behave the way they do in organizations. A number of issues connected with how organizations go about conducting staff appraisals and with what issues the organization focuses on serve as clues to employees about what their organization stands for, what is important, and where their efforts are required (Schein, 1992). Indeed, one may think of staff appraisal as the operational definition of what the organization values. If the organization involves employees in the design of the appraisal process, it tells employees their opinions are valued. This is especially true if the organization also evaluates how satisfied employees are and holds managers responsible for taking action on this kind of information. If the organization evaluates employees regarding their rejects and faulty products, then employees have one kind of feeling about the organization. Contrast that organization with another that focuses exclusively on employees' total quantity of productivity. Because we think the nature of organizational goal emphasis and the social context are so important for understanding behavior in organizations, we turn to this topic before discussing the details of staff appraisal.

THE IMPORTANCE OF ORGANIZATIONAL CONTEXT AND GOALS

In earlier chapters, we emphasize the nested, hierarchical nature of organizations, such that staffing is embedded within these organizational levels. This multilevel perspective is perhaps most obvious when considering the nature of organizational goals and how they relate to performance. Whether formalized through a mission statement or informally conveyed through daily rewards and punishments, organizations exist for a particular purpose and tend to structure themselves in ways they believe lead to goal attainment. Although an organization's goals may often be poorly understood and ambiguous, they become operationalized through the types of behaviors, results, and indices of effectiveness the organization supports or discourages.

Indeed, perhaps the best way to think about staffing appraisal is to conceptualize it as the formal definition of what the organization values. For example, an organization that values innovation is likely to evaluate employees on the extent to which they develop new products or contribute to the development of new ideas. The organization may reward department managers to the extent they can allocate scarce resources for product development or by the number of new products they propose each year.

This leads to a recognition that the development of a staff appraisal system must begin with an understanding of superordinate organizational goals and of the subordinate goal hierarchies that may exist to achieve these superordinate goals. Staff appraisal operationalizes these goals at multiple levels. A full understanding of staff appraisal therefore requires a firm understanding of the social, organizational, economic, and political context within which the organization and staff operate.

A rich appreciation of the contextual nature of staff appraisal is offered by Murphy and Cleveland (1995). Whereas much traditional performance appraisal research focused on the cognitive processes and biases of individual raters, Murphy and Cleveland argued that such performance judgment processes cannot be cleanly separated from context. Figure 4.1 summarizes their arguments by identifying the contextual factors of most relevance. Note that Murphy and Cleveland lay these factors on a continuum from proximal to distal. Figure 4.1 shows that distal factors include the economic, societal, technical, legal, and physical environments, whereas proximal factors include organizational goals, the purposes of the appraisal system, workgroups, task characteristics, and related factors (see also Katz & Kahn, 1978). An important point to recognize is that the effects of distal environmental factors do not directly influence individual-level staffing processes; rather, they operate indirectly through influencing staff appraisal factors (these are mediated relationships; chapter 2). Altogether, these distal, proximal, and intermediate factors jointly influence individual-level staff appraisal behavior.

Of particular importance in the organizational factors is the role of climate and culture. People decide what the general or guiding practices and procedures of an organization are by observing many little events, the hundreds of work experiences they have, and what practices and procedures the organization seems to emphasize. In a real sense, they form concepts of their organization's priorities (climate; Schneider, 1975) and the deeply rooted beliefs and values (culture) that senior management holds (Schein, 1992).

People form concepts about the climates that exist in their organization with reference to many issues. Thus, Tom Smith may have concepts about the organization he works for with respect to what its goals are, how it rewards people, the kind of supervisors it promotes, its policies on accidents and safety, and how it relates to the world external to the organization. A single climate does not exist in an organization; many climates are created by the various practices and procedures in which an organization engages (Schneider, 1980).

The important thing about organizational climates is that employees use them as benchmarks or guidelines for behavior. We have all had the experience of trying to decide how and whether to do something at work. How does the decision get made? It is made on the basis of what we sense is most appropriate for the situation.

Think of people weighing how to best accomplish a particular work task. On what basis do they decide how to accomplish their objective? They determine where they are, try to locate themselves in the larger environment, and, after locating themselves, chart the course necessary to achieve the goal. Employees use their concepts of the organization's priorities and values as frames of reference for locating themselves and deciding how to behave (Martin, 2002).

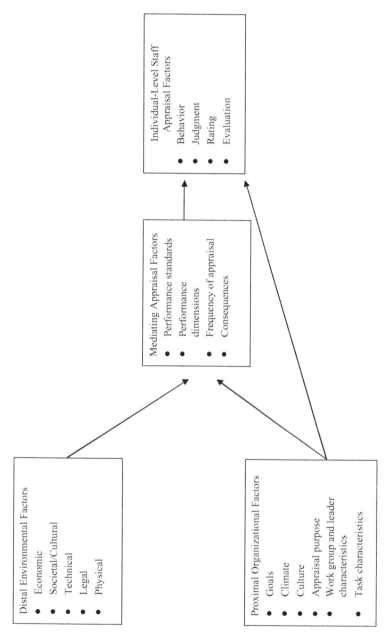

FIG. 4.1. Contextual influences on staff appraisal. *Source:* Adapted from Murphy & Cleveland, 1995.

If the concepts people have of their organization help guide their behavior, then the stated goals and the subsequent standards of excellence an organization establishes give employees clear cues about what is important and where their efforts are required and valued. When translated into specific standards for evaluating employees, the goals or objectives of an organization lead to particular employee behavior.

Going back to the issue of organizational goals; they are important for reasons beyond gaining an understanding of the bases of behavior among employees. They are also important because they dictate the nature of the entire staffing process. That is, if company policy determines the definition of what a superior employee is, then the organization's staffing procedures will be targeted toward those same goals and objectives. Particular kinds of employees will be attracted to the organization, recruited and selected for the organization, stay in the organization, be evaluated highly by the organization, and be promoted within the organization (Schneider, 1987).

A final point concerning the contextual nature of staff appraisal needs to be made. When developing a staffing system, the impact of national culture is almost taken for granted. In the United States, which is a very individualistic culture, the concept of evaluating each individual employee makes common sense. However, in collectivistic cultures that value group harmony (e.g., many Asian cultures), the concept of individual staff appraisal is barely comprehensible. As organizations increasingly move into multinational operations and become global, they face a conflict where the climate, goals, culture, and values of an organization in the host country clash with those from other nations and cultures. Thus, the nature of what is appropriate staffing appraisal practice may be culture specific. Indeed, more recent research into the European community has not clearly identified a common set of staffing practices even within Western Europe (see Brewster, 1995; Brewster, Larsen, & Mayrhofer, 2000; Larsen & Brewster, 2000a, 2000b; see also Ryan, McFarland, Baron, & Page, 1999). We return to more specific issues of context and performance measurement in later sections of this chapter. We now turn to a consideration of the historical difficulty with defining and measuring performance.

CRITERIA DEFINED

When we speak of job performance, what exactly do we mean? Consider trying to define the job performance of a waiter: Should performance be what the waiter does (e.g., friendliness and promptness), how large a tip the waiter receives, or how much total sales the waiter makes in a day? One might be tempted to argue that we should focus on sales—the real bottom line. However, what if the waiter works an early afternoon shift when there are few customers or if the restaurant is in a poor location. One could focus on tips, but many people tip a flat 15%, regardless of service. The quality of the food (which is beyond the waiter's control) can also influence the size of the tip. Perhaps then we should focus solely on the behavior of the employee; however, what if the supervisor has little interaction with and little chance to observe the employee?

This simple example demonstrates the difficulty of defining criteria and job performance. The criteria that determines performance often depend on who you ask, what they value, and what information is available to them. Staffing researchers have dealt with this issue for a long time. It is commonly known as the "criterion problem"; a problem that stems from an inability to comprehensively conceptualize and measure the job performance domain. Indeed, Austin and Villanova (1992) noted the historical record of the criterion problem goes back to at least 1917. Perhaps a quote by Campbell, McCloy, Oppler, and Sager (1993) summed it up best:

> If the dependent variable is the variable of real interest and if performance is perhaps our most important dependent variable, then more often than not we simply may not know what our real interest is. If we want to accumulate knowledge about how to measure, predict, explain, and change performance but have no common understanding of what it is, then building a cumulative research record is difficult to impossible and industrial and organizational (I/O) psychology is in for continued unfavorable comparisons with other sciences. (p. 35)

When discussing performance and criteria, staffing researchers often speak only of a *criterion* or *criteria*. A simple definition of a criterion is something in which you are interested. A definition closer to the use of the word in the staffing literature is a standard of excellence. Campbell et al. (1993) restricted their definition of performance solely to employee behavior: "Performance is what the organization hires one to do, and do well. Performance is not the consequence or result of action, it is the action itself" (p. 40). This builds on a distinction raised by Smith (1976), between behavior (what a person does), results (outcomes of these behaviors), and organizational effectiveness (effects of behavior and results on indices of organizational effectiveness). The definition we find most suitable for our purposes is the following:

> Criteria are those behaviors and outcomes at work that competent observers can agree constitute necessary standards of excellence to be achieved in order for the individual and the organization to both accomplish their goals.

Criteria are those work behaviors and outcomes that we try to predict with our staffing procedures. Criteria are the behaviors and outcomes against which employees are judged, and on which organizations base various administrative decisions (e.g., promotion, wages and salaries, training). Criteria are those factors that constitute doing the job well.

Cutting across all types of criteria, including behavior, results, and effectiveness, are issues relating to criterion measurement. At this point, it is informative to consider a classic distinction between the "ultimate criterion" and the "actual criterion." Definition and articulation of organizational and individual goals provide the *ultimate criteria*. For example, an organization's ultimate goals might include (a) investor confidence that allows them to generate money in the amounts and at the time it is needed, (b) having a reputation for honesty and integrity, and (c) loyal and motivated customers. At the personal level, the ultimate goal may be to believe we are living satisfying and productive lives both in work and nonwork spheres. Clearly, these are desirable objectives, but they are often, if not always, difficult to measure. Our actual measures of the degree to which we meet these ultimate criteria represent proxies to some extent and have been labeled *actual criteria*. They include measures of production, absenteeism, accidents, performance ratings, customer satisfaction, and employee attitudes.

The goal of criterion development efforts is to increase the degree to which actual and ultimate criteria overlap. One way to think about this is in terms of latent constructs and manifest measures (see chapter 2). Latent constructs represent the ultimate criterion (what we theoretically want to measure), and the measures we use (e.g., performance ratings, accidents) represent manifest indicators of latent performance. Although this distinction, and the discussion in general, may sound abstract, it is extremely useful in describing the type of problems staffing researchers are interested in and discuss when they speak of the criterion problem.

Consider Fig. 4.2, which depicts the relationship between an actual (manifest measure depicted by a box) and ultimate (latent construct

CRITERIA DEFINED 159

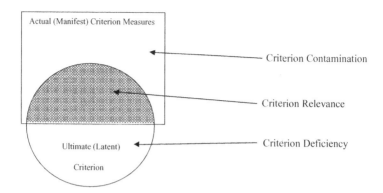

FIG. 4.2. Relationship between the ultimate criterion and an actual criterion. The box represents actual (manifest) criterion measures; the circle represents ultimate (latent) criteria.

represented by a circle) criterion. The degree to which the actual and ultimate criterion overlap is the degree to which we say that the actual criterion is *relevant*. The portion of the ultimate criterion that does not overlap with the actual criterion is referred to as *criterion deficiency*, or that part of the ultimate criterion that we are not measuring but should be. It is through systematic job and organizational analyses such as those described in chapter 3 that we want to minimize criterion deficiency and maximize criterion relevance. The third area in the diagram, called *criterion contamination*, represents variance in the actual criterion that has nothing to do with the ultimate criterion. Traditionally, discussions of criterion contamination include a distinction among three major types:

1. *Opportunity bias* occurs when some employees have advantages that are not available to other workers. A salesperson in a wealthy region has more opportunity to sell than a salesperson in a poorer region.

2. *Group characteristic bias* occurs when measures of an individual's job performance are partly a function of the work or social group to which they belong. In such instances, the behavior of the person is constrained or affected by the group interactions. An otherwise effective pilot who must work with a poorly trained flight crew may not be able to fly the plane as well as his ability allows (e.g., Tziner & Eden, 1985).

3. *Predictor bias* occurs when a supervisor knows the skill levels of employees, either through standardized tests or the recommendations

(formal or informal) of colleagues. Indeed, Eden (1990) showed that prior information supervisors have about new subordinates has dramatic effects on how those subordinates are treated and large effects on the performance of subordinates. So, if the supervisor also supplies the criteria in the form of performance ratings, such knowledge may directly affect the ratings of certain individuals, compounding the bias or criterion contamination. We consider other specific forms of contamination at the appropriate sections throughout the remainder of this chapter.

Both deficiency and contamination represent significant problems to the staffing researcher or practitioner who wants to use measurements of job performance when making various personnel decisions. A critically important point is that both contamination and deficiency are present to some degree in every criterion measure; a key job for staffing researchers is to maximize the amount of criterion relevance. However, "proving" how much variance in a criterion measure is relevant is not something that is easily accomplished. Because we rarely, if ever, have data on any ultimate criteria, we must rely on the judgments of "experts" concerning the appropriateness of the actual criteria we measure. Clearly, an important first step for assessing performance is knowing what our ultimate criterion should look like. This requires a consideration of the latent structure of job performance.

MODELS OF THE JOB PERFORMANCE DOMAIN

A major advancement for addressing the criterion problem was made by John Campbell and colleagues as part of their work on Project A. Project A was a large-scale selection and classification project funded by the U.S. Army; it resulted in a complete and comprehensive assessment of criteria and predictors for entry-level Army positions. A key outcome of Project A was the development of a model of job performance that described its latent structure and determinants. This work represents perhaps the most important advancement in criterion development in the last 50 years. For our purposes, we simply refer to this as the "Campbell performance model," but the actual summaries of the model can be found in Campbell (1990; Campbell, Gasser, & Oswald, 1996; Campbell et al., 1993).

As noted earlier, Campbell et al. (1993, 1996) argued that our focus should be on performance behavior. Taking this perspective, the model

MODELS OF THE JOB PERFORMANCE DOMAIN

hypothesizes there are eight, and only eight, latent dimensions of performance that comprehensively define the performance domain for all jobs. These dimensions are as follows (Campbell et al., 1996):

1. Job-specific task proficiency: Performance on technical, core tasks specific to the job
2. Non–job-specific task proficiency: Performance on technical, core tasks generic to all similar jobs
3. Written and oral communication task proficiency: Performance on writing and oral communication tasks
4. Demonstration of effort: The direction, persistence, and intensity of performance behavior
5. Maintenance of personal discipline: Avoidance of such negative actions as tardiness, absence, drug abuse, and rule violations
6. Facilitation of peer and team performance: Helping colleagues, coworkers, and peers in achieving their performance objectives
7. Supervision/leadership: Influencing, directing, and monitoring employees or coworkers
8. Management/administration: Behaviors involved with organization, implementation, and obtaining resources

To claim that these eight factors comprehensively define the performance domain does not preclude the relevance of more specific forms of criteria. The model proposed is hierarchical, such that these eight dimensions comprise the highest (most broad) level in the hierarchy, with more specific dimensions being subsumed within them. The model also does not propose all eight factors will be present in every job (Campbell et al., 1996). Rather, the most important implication of the model is that it attempts to define the latent taxonomy of ultimate criteria relevant in some form for all jobs.

Beyond creating a taxonomy of performance factors, the model further proposes the KSAO determinants of each factor. There are three determinants: *declarative knowledge* (knowing what to do), *procedural knowledge* (knowing how to do it), and *motivation* (choice, level of effort, and persistence). Thus, performance on any given performance dimension is a function of these three determinants, although their relative importance may differ across the eight dimensions.

As noted by Motowidlo (2003), the Campbell performance model defines the latent structure in terms of performance content. In contrast, a different performance model proposed by Borman and Motowidlo (1993, 1997) defines the latent structure in terms of the organizational value or consequences for the performance behaviors. Specifically, Borman and Motowidlo (1993, 1997) distinguished between task and contextual performance. *Task performance* generally reflects the behaviors directly responsible for producing the products the organization values, such as putting together components on an assembly line, writing programming code for a software application, or completing accounting forms. Many of the dimensions in Campbell's model, such as the job-specific and generic forms of task proficiency, relate to technical performance.

In contrast, *contextual performance* reflects those behaviors that help support these core task responsibilities. Motowidlo (2003) noted that contextual performance is defined "in terms of behavior that contributes to organizational effectiveness through its effects on the psychological, social, and organizational context of work" (p. 44). Contextual performance helps support the production of organizational goods and services by providing a supportive contextual environment. Borman and Motowidlo (1993, 1997) described five major contextual performance behaviors:

1. Volunteering to complete tasks not formally part of one's job
2. Persisting with extra effort and enthusiasm
3. Helping and cooperating with others
4. Following company rules, procedures, and policies
5. Supporting and defending the organization

Motowidlo and colleagues (Motowidlo, 2003; Motowidlo & Schmit, 1999) noted that both task and contextual performance contribute to the achievement of organizational goals. However, they differ in their means: Task performance contributes to organizational goals by accomplishing core, technical tasks, and contextual performance contributes to organizational goals by developing a supportive social, psychological, and organizational context surrounding this technical core. Contextual performance is very much linked to unit-level constructs; contextual performance collectively aggregates to higher levels and provides a climate, culture, and work environment conducive to getting task responsibilities accomplished (or not accomplished, as the case may be). Indeed, relations between contextual performance and aggregate performance are to be expected (Organ & Ryan, 1995).

MODELS OF THE JOB PERFORMANCE DOMAIN 163

Another distinction between task and contextual performance is that they differ in their antecedents. Because task performance reflects the technical core, it is predicted primarily by individual differences in cognitive ability. Because contextual performance is more discretionary and/or interpersonal in nature, it is predicted primarily by individual differences in personality (e.g., Borman, Penner, Allen, & Motowidlo, 2001). As is seen in later chapters, this has enormous practical implications because different predictor batteries will be used to the extent criteria reflect task versus contextual dimensions.

It is also worth mentioning a concept closely related to contextual performance—*organizational citizenship behavior* (OCB). Organ (1988) defined OCB as discretionary behavior that is not formally rewarded but in the aggregate contributes to organizational effectiveness. Two key features of this definition are that OCBs represent discretionary behaviors, and the effects of OCBs are expected to occur at an aggregate level of analysis. The focus on discretionary (not job-prescribed) behaviors causes some concerns for considering OCBs in staffing, because if such behaviors are not formally part of the job, they cannot be used as criteria for staffing. Similarly, the expectation of OCBs having effects at an aggregate level of analysis suggests the traditional individual-level validation model may have difficulty incorporating them as a criterion. With that said, more recent research on OCBs has resulted in some assimilation to the conceptualizations underlying contextual performance (e.g., Organ, 1997). Although some important theoretical differences between the two types of behavior remain unresolved (Motowidlo, 2003), we believe behaviors such as OCB, contextual performance, and other role behaviors (Ilgen & Hollenbeck, 1991) not traditionally considered in staffing research need to be considered in the future. Contextual performance and OCB are revisited numerous times in this book because of their central role in organizational life.

A final issue concerning contextual performance is whether it is distinguishable from a class of performance behaviors known as *counterproductive work behaviors* (CWBs). In contrast to the positive features of contextual performance, CWBs represent negative and destructive behaviors to the organization and other people within the organization (Sackett & DeVore, 2001). Sackett (2002), citing an unpublished dissertation by Gruys (1999), notes 11 types of CWBs:

1. Theft and related behavior (theft of cash or property; giving away of goods and services; misuse of employee discount).

2. Destruction of property (deface, damage, or destroy property; sabotage production).
3. Misuse of information (reveal confidential information; falsify records).
4. Misuse of time and resources (waste time, alter time card, conduct personal business during work time).
5. Unsafe behavior (failure to follow safety procedures; failure to learn safety procedures).
6. Poor attendance (unexcused absence or tardiness; misuse sick leave).
7. Poor quality of work (intentionally slow or sloppy work).
8. Alcohol use (alcohol use on the job; coming to work under the influence of alcohol).
9. Drug use (possess, use, or sell drugs at work).
10. Inappropriate verbal actions (argue with customers, verbally harass co-workers).
11. Inappropriate physical actions (physically attack co-workers, physical sexual advances toward co-worker). (Sackett, 2002, p. 6)

Although these behaviors and outcomes represent a diverse number of domains, they may have a common underlying set of determinants. That is, some of the antecedents of CWBs may cut across all 11 types of behaviors, with other antecedents being more specific to a particular CWB. Sackett (2002) noted that these behaviors, when examined in the aggregate, correlate -0.60 with contextual performance. Thus, a question for future research is whether CWBs and contextual performance represent the same set of constructs at different sides of the same continuum (e.g., Sackett, 2002) or different classes of behaviors (e.g., Kelloway, Loughlin, Barling, & Nault, 2002). Now that some common measures of CWB have become available (e.g., Bennett & Robinson, 2000), such research is likely to accumulate rapidly.

A final question with contextual performance is whether the task performance/contextual performance distinction is contrary to the Campbell performance model. More recent writing on this topic suggests that the various approaches are quite complementary (Campbell, 1999). Indeed, many of the specific dimensions of contextual performance may be considered more specific factors of the eight proposed by Campbell. For example, the contextual performance dimension "persisting with extra effort and enthusiasm" is essentially the same as Campbell's "demonstrating effort" dimension. In contrast, the distinction between task and contextual performance may occur at a higher level than Campbell's eight dimensions, such that each of the eight dimensions may be subsumed within general task and contextual performance higher-order dimensions.

Future research will need to resolve some important theoretical and conceptual distinctions between these and related performance taxonomies, such as those with CWBs. Such research is not likely to result in a single performance model applicable to all contexts and for all purposes because the structure of any taxonomy is a function of its purpose. With that said, these taxonomies represent an enormous theoretical advancement toward an understanding of the latent structure of job performance and some solutions for the criterion problem. Indeed, these models represent impressive attempts to define the ultimate criterion—a definition many have believed impossible. Despite such theoretical advancements, one must always remember that a key issue facing the staffing specialist is also to develop appropriate measures of the ultimate criterion to maximize relevance and minimize deficiency and contamination. We must now consider characteristics of criteria that make them useful for research and practice.

DESIRABLE ASPECTS OF CRITERIA

When listing aspects of criteria that are particularly desirable, most authors usually state relevance, discriminability, reliability, and practicality.

Relevance

Of the four aspects of criteria, relevance is, of course, indispensable. Researchers used two major approaches in developing criteria that are maximally relevant. Muchinsky (1983) referred to these two approaches as inductive and deductive; other writers have referred to them as objective and subjective. *Inductive* (objective) procedures are not based on any prior theory or conceptual framework; rather, they tend to be a data-driven approach to demonstrating criterion relevance. That is, criteria are relevant to the extent that they relate to other important organizational outcomes. In *deductive* (subjective) approaches to criterion development, the researcher begins with a careful and rational identification of the possible criterion variables of interest. This is usually accomplished through job analyses, literature reviews, review of training program goals, or even a review of organizational goals. The relative importance of these various criteria is best deduced by rational judgments of a panel of judges. As we saw in chapter 3, several SMEs can be interviewed in groups to generate a list of issues on which staff performance can be judged. These issues are then grouped into major dimensions, and their relative importance is judged by the same or a different group of experts. Once the important dimensions are identified, then actual measures of each are developed.

Whether one employs an inductive or deductive approach to the identification of criteria, remember that, as a cluster or set, they define effectiveness for us. Or, as in selection, we use criteria as the standard(s) against which we evaluate the usefulness of hiring procedures. What this discussion of objective and subjective criterion development shows is that relevant issues or facets of criteria that need to be assessed are identifiable through empirical analysis and/or informed judgments. When feasible, it would be useful to employ both strategies. It is important to use at least one strategy for the development of criteria, or the criteria against which staffing procedures are validated may not be relevant. The prediction and/or understanding of nonrelevant criteria is a useless exercise, to say the least.

Discriminability

Another desirable aspect of criteria is that they be useful in discriminating among staff members. In chapter 2, we make the point that the purpose of measurement is discrimination; this also applies to job performance measurement. It may be important that workers come to work. Yet, if all employees in a company are absent the full amount of time allowed under union contract, or if none of them miss any work, then absenteeism does not differentiate among employees and is of no use in making hiring decisions based on this criterion. It may be desirable to introduce programs to cut down the general level of absenteeism when all are missing a large constant amount, but no basis for differential treatment of staff is justified. The point is that, for measures to be useful as criteria in personnel selection, there must be differences in individual performance levels. We see shortly that various rater errors and biases are problematic precisely because they reduce the discriminability of criteria.

Discriminability is important as a "criterion for criteria" in personnel selection, but it may not be important when raising the issue of overall unit or overall organizational effectiveness. That is, if we can assume no one individual's poor performance will bury a unit or the

DESIRABLE ASPECTS OF CRITERIA

organization, then for most practical purposes it is how well people, in the aggregate, are performing that is the question of interest. In other words, in general, are employees performing effectively? Answering this question requires an examination of average unit performance; hence, agreement will be critical for demonstrating the mean value accurately represents the unit's performance.

However, any time a decision needs to be made about a particular employee—send to training, promote, dismiss, give a raise—then a procedure for measuring performance that has good discriminability is critical.

Reliability

We discuss various concepts relating to reliability in chapter 2. Based on this discussion, it should be clear that if criteria are to be useful, they must be measurable in a consistent manner. Several types of reliability can be assessed, and determining which is appropriate depends on the actual measure itself and the use to which we want to adapt the criterion.

The first kind of reliability is test–retest reliability or stability. That is, our performance measures must show some degree of consistency over a reasonable period of time. For example, if an employee's behavior does not change and nothing in the environment changes, then we would expect the evaluation of this performance to be the same at both time periods. Sometimes it is not, such as when output fluctuates according to equipment failures, sales increase or decrease due to changes in the economy, or the employee disengages from the job and exerts less effort. The degree to which criteria are unstable over time will directly affect whether they can be predicted. Criterion-related validity cannot exceed the square root of stability.

A second manner in which reliability is assessed is referred to as *internal consistency reliability*. For example, we may have several kinds of measures of a single dimension of the criterion, such as when task performance is assessed with three items relating to speed, quantity, and productivity. In this case, those separate measures should yield similar information, and thus, internal consistency estimates of reliability are appropriate. We should also note high internal consistency is desirable only when we want to make the case that various indices are measures of the same aspect of performance. When the actual measures are perceived of as measures of different performance dimensions, then they might yield different values for a person's performance and thus rank order people differently (hence, low internal consistency reliability).

Internal consistency estimates of reliability are interesting because they focus on the many different ways of analyzing particular facets of performance. For example, suppose we wanted to assess the quality of bank tellers' performances at work. We could do one or more of the following:

1. Ask their coworkers about the quality of their performance.
2. Track their balances at the end of the day (i.e., are they consistently over or under?).
3. Ask their customers about the quality of their performance.
4. Ask their managers to rate their quality.

It is hoped that the different kinds of measures of quality across a group of tellers would yield similar data, allowing us to rank order the tellers in terms of the quality of their performance. When the different kinds of measures yield similar rank orderings, we can infer we have internal consistency reliability—the different kinds of measures of quality are consistent with each other.

There is a different kind of reliability that is also important for criteria, that is, interrater reliability. *Interrater reliability* reflects whether two or more raters give similar rank orders to ratees. For example, high interrater reliability occurs when the numbers assigned to represent employee performance by two or more supervisors end up rank ordering employees in consistent ways across raters. Interrater reliability should not be confused with interrater agreement, however. *Interrater agreement* focuses on the absolute ratings assigned by two or more raters. For example, high agreement would occur when two or more raters provide the same employee similar numbers. Thus, one may think of interrater reliability as reflecting consistency in rank orders and interrater agreement as reflecting consistency in absolute scores. Of the two, interrater reliability is usually the more important estimate when making decisions about individuals. Unfortunately, meta-analytic research suggests that ratings show extremely poor levels of interrater reliability, with 0.52 to 0.60 being common (Rothstein, 1990; Viswesvaran, Ones, & Schmidt, 1996). We return to the implications of this estimate in a

DESIRABLE ASPECTS OF CRITERIA

later section when we discuss statistical corrections for unreliability. For present purposes, it is sufficient to emphasize the following:

1. Reliability is an important criterion attribute.
2. Depending on the situation, one or more kinds of reliability are important: stability, interrater, and/or internal consistency.
3. Reliability is not agreement.

Finally, although reliability of criteria is an important issue, it should always be secondary to relevance. This has not always been true, and it has led to the problem of emphasizing what we can measure rather than what we should measure.

Practicality

Obviously, an actual measure of any criterion must be available (or collection must be possible) and generalizable across units of an organization. It must be a criterion that organizational members accept as an appropriate measuring index. We can develop accurate and sophisticated performance assessments, but if they are so expensive and time consuming to administer that no one will use them, they are not going to be practically feasible. Most organizations are interested in the dollar contribution of employees to the organization. This remains a difficult criterion to measure, particularly for complex managerial and technical jobs. For people in service industries, we are probably most interested in how satisfied clients or customers are and how often they return for service (or how often they do not return) because of the way they have been helped. The practical problems in collecting these types of data can be significant, although in the case of customer satisfaction measurement great progress has been made, including estimates of the dollar value of customer satisfaction to the organization (Schneider & White, 2004). If a criterion is judged relevant, however, effort must be made to overcome practicality problems.

Summary

Four aspects of a good criterion, in rank order of importance, are relevance, discriminability, reliability, and practicality. Shun practicality as a major determinant whenever it interferes with any of the other three.

OBJECTIVE CRITERION MEASURES

We now move to a discussion of the various objective measures of criteria, and then turn to a discussion of judgmental criteria in the following section. Objective criteria (sometimes also called "hard criteria") earn their name because their measurement is not based on human judgment. For example, sales dollars are what they are and require no interpretation (unless one puts a value on a particular amount of sales). With that said, there is often judgment present with objective criteria, it just plays a less pivotal role in the measure. On the surface, one may think objective criteria would be preferable to judgmental criteria; after all, they reflect the "bottom line." However, as we will see, objective criteria are often more deficient and contaminated. There are a variety of such objective criteria, including production of output (quantity), quality, trainability, absenteeism, lateness, withdrawal and turnover, promotion rate, salary increase rate, and safety, accidents, and errors.

Production of Output

Measures of the amount produced (e.g., quantity, sales) appear to have obvious relevance. The amount of life insurance sold, the amount of machinings per day, the number of letters typed, the number of legal briefs written—all these measures seem at first glance to have undeniable utility. Indeed, they sometimes do. However, all too often, problems of contamination essentially invalidate production indices as criterion measures. Some salespeople by chance have "richer" sales territories than others. One machinist appropriates the newest lathe, whereas others must do their work on an outmoded heap that operates slowly, if at all. One secretary is assigned difficult letters and papers because of a reputation for quality work, whereas another receives simpler material. Thus, the production figures may not really reflect individual differences in value to the organization. Sometimes it is possible to make statistical corrections for contaminants of this sort. That is, one can assess the contaminating factors and remove them from the criterion. Yet, such adjustments can be fairly difficult and are sometimes made impossible by the unavailability of the information necessary to make the corrections.

Quality

For any job, the old adage that quality is more important than quantity is almost certainly true. One important patented invention by a research engineer is apt to be more valuable to the organization than a number of minor ones. A smaller number of automobiles properly repaired by a mechanic may contribute more to shop success than a larger number improperly "repaired." Quality of output may often be difficult to assess, but it can be assessed in many situations. Illustrative examples include percentage of work products that pass (or fail) inspection, error rates, scrap rates, the percent of insurance policies sold that are later renewed, or the number of legal clients who later return for further services. Like measures of quantity, quality measures must be closely scrutinized for contamination. The quality of machinings, for example, will not accurately reflect individual differences in employee contributions if there are wide variations in the condition of machines used. Likewise, the secretary who is assigned the more difficult papers may have more errors per paper despite her superior value to the organization. Furthermore, if the best secretaries are assigned the most difficult material, contamination of this latter kind may result in the following paradox: If the test is valid, the best secretaries will tend to have the highest test scores and the lowest quality of work scores. This produces a negative relationship between test scores and performance that is entirely spurious.

Training Success and Work Samples

Trainability, as defined by Wexley (1984), is a person's ability to acquire the skills, knowledge, or behavior necessary to perform a job at a given level and to achieve these outcomes in a given time. Such criteria are important in many public safety occupations such as police, firefighters, and the military, where training is continuous and must be successfully completed before one is allowed to work on the job. *Work sample tests* (sometimes known as performance tests, see chapter 10) are often used as training success criteria. A work sample is a test developed where the individual performs actual tasks that will be (or are) performed on the job (Green & Wigdor, 1991). For example, the secretary will type legal documents, and the machinist will actually produce the requested part. Work sample tests are often objective measures in that scores comprise amount, quantity, quality, or related indices (although they may also be scored using the rating procedures described in the next section). Use of training criteria in personnel selection is attractive because criteria such as training time and cost can usually be calculated fairly objectively, and savings from a selection program that reduces training time and cost can be fairly easily documented. This, in turn, makes it easy to promote a good selection program to management. However, measures of training success, particularly work samples, may be expensive to develop and administer, and yet still be deficient (Borman, 1991). Another objection against using training success as a staffing criterion is that there may be little relationship between training success and on-the-job performance; perhaps because everyone must pass training to work on the job (and thus no variance). This means that if a test is related to training success, it may not be related to performance on the job. Thus, training success criteria should only be used when they are meaningfully related to job performance.

Withdrawal: Lateness, Tardiness, Absenteeism, and Turnover

We have included these various behaviors, lateness, tardiness, absenteeism, and turnover, into a general withdrawal category because influential research by Hulin and colleagues (summarized in Hulin, 1991; Hulin, Roznowski, & Hachiya, 1985) has argued that all such behaviors are adaptations to unacceptable or dissatisfying work arrangements (see also the previous discussion of CWBs; Sackett, 2002). That is, they are all reasonable reactions to a job that is no longer satisfying to the employee.

Hulin (1991) proposed a model of organizational/adaptation/ withdrawal that was interesting because it integrated many different types of withdrawal behaviors and linked these to a set of common determinants. A simple overview of this model is shown in Fig. 4.3, illustrating how dissatisfaction with the job contributes to four classes of withdrawal behavior. *Increasing outputs* represent attempts to make a dissatisfying job more equitable, such as when someone feeling underappreciated steals office supplies to "even the deal." *Psychological withdrawal* refers to attempts to mentally make the job more tolerable (e.g., excessive daydreaming). *Behavioral withdrawal* results in active attempts to remove oneself from the situation (e.g., tardiness). *Changing work roles* are ways to change the nature of the job, such as asking for

OBJECTIVE CRITERION MEASURES 173

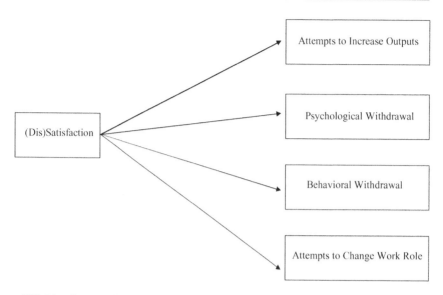

FIG. 4.3. Organization adaptation/withdrawal model. *Note.* Adapted from Hulin, 1991.

a transfer. The model also proposed various stages, such that at early stages of dissatisfaction the employee may simply start showing up late. This in turn might lead to tardiness and absenteeism, followed by actual turnover—but only if alternative jobs are available (the theory assumes behavior is volitional). Of course, the progression need not go in a specific sequence, but it is clear that each behavior does carry a greater consequence for the person and organization. The Hulin (1991) model is similar to the potential interrelationships among the CWBs discussed above, and one may perhaps integrate the 11 types of CWBs into the Hulin (1991) model to identify an even wider range of adaptations to dissatisfaction at work. For example, an employee may withdraw effort, take extra breaks, or steal from the organization. Keeping in mind the various withdrawal behaviors may share common antecedents, we briefly discuss each type of withdrawal behavior.

First, *lateness and tardiness* refer to behaviors associated with showing up to work tasks late, if at all. They are problematic in that they require other employees to do more work, may decrease morale, and cost the organization time and money. Blau (1994) developed a taxonomy of lateness behavior, demonstrating how each type has different antecedents and consequences. *Increasing chronic lateness* is intentional lateness behavior that becomes increasingly longer in duration and is caused by negative job attitudes. An example is when employees purposely show up to work late because they do not like the job. *Stable periodic lateness* is intentional but stable amounts of lateness behavior and may be caused by the worker making work–nonwork trade-offs. For example, an employee who must balance child care with work may be consistently late on certain days of the week. *Unavoidable lateness* is unintentional and random, due to factors beyond the person's control. Examples include illness, being caught in traffic, or inclement weather. Lateness behaviors may be measured in terms of number of instances one is late, in the amount of time one missed work, or some combination of the two. An implication of this taxonomy is that we cannot assume all lateness behaviors are the same; only increasing chronic and perhaps stable periodic lateness behaviors are the targets for potential staffing solutions.

Second, *absenteeism* refers to being gone from work for extended periods of time. One way to distinguish absenteeism from lateness is that absenteeism behaviors involve the person not showing up for work at all, in contrast to showing up late. Absenteeism is an important criterion because it is costly to the organization; it is relevant to staffing because it can be predicted and is associated with job satisfaction. The association of low job satisfaction with above average absence rates has repeatedly been confirmed (Mowday, Porter, & Steers, 1982; Steers & Rhodes, 1978). However, different indices of absenteeism apparently differ in reliability. Huse and Taylor (1962) examined the reliability of four different measures over a 1-year period for 393 truck drivers. They found "attitudinal absences" (frequency of one day absences) were correlated 0.52 during a 1-year time period. "Absence frequency" (the total number of times absent) had slightly higher reliability, 0.61. The other two absence measures were much lower in reliability. "Absence severity" (total number of days absent) and "medical absences" (total number of 3 days or longer absences) had reliabilities of 0.23 and 0.19, respectively. For selection purposes, attitudinal absence or absence frequency would then be the most appropriate. As noted by Hulin (1991), higher reliabilities may require measurement over even longer time periods or aggregation across different forms of absenteeism.

Finally, *turnover* occurs when people terminate employment with their current organization and choose employment with another. Sometimes turnover is functional, such as when an employee poorly suited for the job quits. However, in staffing we are most often concerned with dysfunctional turnover, such as when a valued and competent employee leaves the organization. Turnover may also be voluntary and involuntary. *Voluntary* turnover occurs when the employee chooses to leave the job; *involuntary* turnover occurs when the employee did not make this choice (e.g., terminated by the organization, the organization goes broke, spousal relocation). Clearly, voluntary turnover is of more importance to staffing and is important for organizations to predict because it is so costly. Turnover has been hard to predict, probably because of its dichotomous nature (people either quit or stay) and the fact that longitudinal research is difficult because of the necessity to wait long periods of time until sufficient numbers of people quit. Further, as the Hulin (1991) model indicates, turnover may be contingent on alternate job opportunities. Not so incidentally, the stability of turnover is not knowable because once employees leave we do not have an opportunity to reassess them on turnover. Turnover at the unit level is often quite unreliable in that high turnover in a department this month leads to hiring next month and likely low turnover. So, although a practical and important issue, predicting turnover based on staffing practices can be difficult.

A general solution for analyzing turnover data (or absenteeism or any data where the criterion is dichotomous) is to consider the issue as a longitudinal process, using appropriate data analytic methods such as growth modeling, survival analysis, or event history analysis (e.g., Harrison, 2002). Although each model is slightly different, it can be used to predict the likelihood of turnover as a function of various predictors and how much time has passed. For example, if we know that people with lots of alternative job offers are likely to leave an organization, we can predict that as more time passes, the likelihood of the person leaving increases as well.

Such a longitudinal perspective is certainly consistent with the theory proposed by Hulin (1991). It is also consistent with more recent refinements. For example, Lee and Mitchell (1994) proposed an "unfolding" model of turnover, such that individuals make a variety of effortless and effortful choices to determine whether to stay or leave. What is important to predict in such a model is not just the final stay–quit decision, but the variety of choices and processes that occur prior to the final decision. In this way, obviously, we can obtain an estimate of the stability of the causes of turnover, even though the stability of turnover itself is not knowable.

Safety and Accidents

Worker safety and accidents can also have important consequences for individuals and organizations. For example, Hofmann, Morgeson, and Gerras (2003) noted the number of workplace accidents is in the millions, with the costs totaling into the tens of billions of dollars. To the extent safety and accident rates are affected by individual differences, they become possible targets for staffing. The difficulty becomes recording these behaviors because employees (and organizations) are often penalized when accidents occur; thus, there is a strong urge to hide such incidents. Likewise, accident counts may be contaminated because the accident may not have been under the person's control. In contrast, safety behaviors may be more volitional, frequent, and hence more able to be measured. For example, employees will often show more individual differences in wearing safety equipment, such as hardhats and safety glasses, than they will in actual injuries and accidents. One of the authors worked in manufacturing settings where safety equipment was only kept around for rare visits from supervisors; interestingly, no supervisors found it unusual that the workers' clothes were dirty but the hardhats perfectly clean. Climate perceptions are also important, such that safety climate can contribute to safety behavior above and beyond the characteristics of individuals (e.g., Hofmann & Stetzer, 1996).

Other Counterproductive Work Behaviors

Beyond CWBs relating to withdrawal and safety, research on other types of CWBs, such as theft, shrinkage (percentage of missing inventory due to theft), reports of alcohol and drug use, and number of verbal or physical complaints, may also comprise objective measures. As noted earlier, these measures may show moderate and consistent intercorrelations, arguing for an underlying latent factor accounting for all their shared covariance (Hulin, 1991; Sackett, 2002). For example, Viswesvaran (2002) demonstrated absenteeism true score correlations of -0.21 and -0.48 with records of productivity and quality, respectively. This confirms a reasonable expectation that when one is gone from work, work outcomes suffer. Of course, issues that plague other negative outcome measures, such as withdrawal and accidents, are also likely to cause reluctance to report the occurrence, low base rates, and skewed distributions. Research is just starting to tackle these various kinds of criteria, mapping out their underlying relationships. Robinson and Bennett (1995) indicated that two underlying dimensions may be found: *organizational deviance*, which refers to destructive behaviors directed toward the organization, and *interpersonal deviance*, which refers to destructive behaviors directed toward other coworkers and individuals within the organization. Sackett (2002) questioned whether these two dimensions may themselves be composed of an underlying latent common factor. Thus, CWBs follow a hierarchical model of specific CWBs (theft, physical abuse) nested within organizational and interpersonal latent factors, which in turn are nested within an overall CWB factor.

Rate of Advancement

Rate of advancement or promotion can be taken as a reflection of the value the organization places on the services of the individual. If those promoted faster have more organizational value, it is logical to attempt to increase the number of people capable of rapid advancement. The relevance of this criterion, as of all others, is based ultimately on judgment. In addition to number of promotions per time period, measures of rate of salary increase and time to first promotion have also been used. Contamination in the form of opportunity bias, where not all employees have access to the same opportunities, is obviously a potential problem and should be watched for carefully. As with other objective indices, longitudinal studies of the rate of advancement necessitate a long time perspective because of the relative infrequency of advancements in most organizations and jobs. Rate of advancement is most frequently used as a criterion in studies of managerial or management trainee selection. Rate of advancement has often been used as a "real world" criterion to demonstrate the societal value of cognitive ability and personality tests (Boudreau, Boswell, & Judge, 2001; Gottfredson, 1997).

Summary of Objective Criteria

Objective criteria have an obvious appeal because they are represented in the kinds of outcomes about which organizations care (and track for administrative purposes). However, objective measures are often highly deficient and contaminated. For example, with objective criteria, the nature of the task, job, equipment/resources, or external environment may all influence the criterion measure. Alternatively, with judgmental ratings, the supervisor can consider this information when providing the evaluation. Yet, although judgmental ratings are perhaps better at reducing contamination and deficiency, they are by no means perfect measures of performance.

JUDGMENTAL CRITERION MEASURES

Given that objective criterion measures tend to be both contaminated and deficient, and that as psychologists our interventions are primarily behavioral in nature, it should come as no surprise that supervisory judgments of employee performance are the most common type of criterion measurement method (e.g., Landy & Farr, 1980). There are many ways one may judge another's performance, including methods requiring absolute judgments, methods requiring employee comparisons, and methods requiring reports of behavior. In this section, we review these various types of judgmental criteria and outline their relative advantages and disadvantages. There are many variations on these basic themes that can be found in Borman (1991) and Bernardin and Beatty (1984).

Methods Requiring Absolute Judgments

Graphic Rating Scales

Formats differ widely among graphic rating scales. Figure 4.4 is an illustration of some of the many variations. At first glance, it would appear that the formats in Fig. 4.4 vary along the degree of structure provided to the rater. The formats at the top of the page provide little information to the rater on the dimension to be rated and no definition of the meaning of different points on the scale. At the other extreme, the formats at the bottom of the page attempt to give some idea of what different points on the scale along the dimension mean. It seems reasonable that the formats providing more information should produce "better" ratings, that is, ratings with higher reliabilities and fewer rating errors. The surprising finding, however, is that this is often not the case. Indeed, rating formats providing even more information than any in Fig. 4.4 do not, in general, produce better ratings (Wexley & Klimoski, 1984).

The formats in Fig. 4.4 also differ in number of rating categories. That is, they differ in the fineness of discriminations required. Although earlier research concluded number of rating categories does not make

JUDGMENTAL CRITERION MEASURES

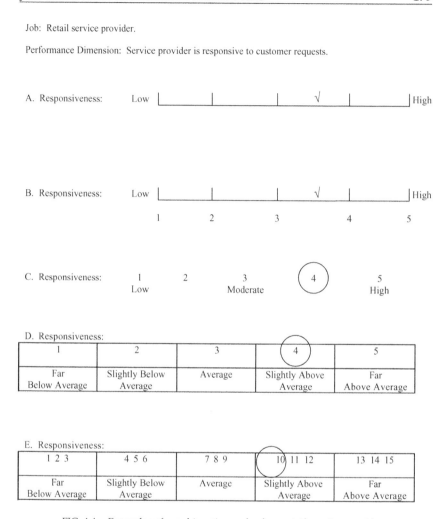

FIG. 4.4. Examples of graphic rating scales for a retail service provider.

much of a psychometric difference (provided at least five categories were used; Landy & Farr, 1980; Matell & Jacoby, 1972), more recent studies have suggested the number of categories may influence rater motivation and hence accuracy (Bretz, Milkovich, & Read, 1992). For example, Bartol, Durham, and Poon (2001) found ratings based on five (rather than three) categories produced higher rater self-efficacy and goals for accuracy. Thus, it would appear that five categories has a number of positive psychometric and motivational features (e.g., greater goals for accuracy and self-efficacy).

Behaviorally Anchored Rating Scales

The idea behind behaviorally anchored rating scales (BARS) was that there were two primary reasons graphic rating scales typically showed low reliability and poor discriminability. The first was that the performance dimensions included in the scales were vague and lacked meaning to raters. It was believed that dimensions such as "overall performance," "quality of output," and "human relations ability," which were chosen by psychologists or other personnel researchers, were not the dimensions used by supervisors when they viewed job performance. As a result, different supervisors read different meanings and intentions into the same dimension. The second reason was that scale points on the dimensions were vague and undefined, again hypothetically leading to different interpretations by different raters. For example, does "satisfactory" mean "about average," or does it mean something closer to "in the top two-thirds"?

In an attempt to overcome these two perceived deficiencies, Smith and Kendall (1963) proposed that the dimensions to be used be derived by raters who would actually use the scale, and that different points on each dimension be anchored by statements describing actual job behavior that would illustrate specific levels of performance. There are variations in how these two concepts are applied in developing BARS, but typically at least five steps are included:

1. *Generation of Descriptive Statements.* Persons with a thorough knowledge of the job (typically supervisors or job incumbents) are assembled in a group and asked to describe specific behaviors that indicate effective or ineffective job performance. What is really done is close to a critical incidents job analysis as described in chapter 3. The following are two examples taken from a study of retail managers:

> Holds daily meetings with associates to ensure they are familiar with new products/promotions.
> Speaks with customers to resolve questions or problems.

2. *Determination of Performance Dimensions.* The same expert group is asked to define the major dimensions of job performance. Actually, this step can be carried out first but has been moved to second in a number of studies to focus the experts on specific job behaviors before moving to the broader dimensions.

3. *Assignment of Behaviors to Dimensions.* Members of the group then independently assign each behavior from step 1 to the various dimensions from step 2.

4. *"Retranslation" of Behavior Statements.* A new group of job knowledgeable judges then "retranslates" or reallocates each behavioral statement to the performance dimension to which they think it belongs. If a certain specified percentage (usually 50%–80%) of the judges in this group reallocate a statement to its dimension as established in step 3, the statement is retained.

5. *Scaling of Behavior Statements.* This group or another similar group then rates each statement as to the level of performance it represents on its performance dimension. The statements finally chosen to anchor each performance dimension are those showing low standard deviations across judges on these ratings (indicating good rater agreement) and covering the range of each performance dimension as evenly as possible.

6. *Final Rating Scale.* The final instrument consists of 5 to 10 performance dimensions, each anchored by about 5 to 7 behavioral statements. An example of one such anchored performance dimension—resolves customers' disputes by a retail manager—is shown in Fig. 4.5.

Although BARS require much effort and time to construct, they surprisingly do not result in scales with better measurement properties (e.g., Borman & Dunnette, 1975; Keaveny & McGann, 1975; Schwab, Heneman, & Decotiis, 1975). However, there are other advantages that may make BARS attractive. There is some evidence that the greater participation that workers and supervisors have in developing these scales leads to all employees taking the rating process much more seriously, and this in turn generates positive motivational outcomes. This result may occur with greater employee participation regardless of what type of scale is used (Friedman & Cornelius, 1976). A second possible advantage may be that the BARS themselves communicate performance expectations more explicitly than other formats. Finally, the BARS format is much more acceptable from a legal standpoint because it is behaviorally based (Bernardin & Beatty, 1984).

Behavioral Observation Scales

An approach to rating format that represents something of a blend between graphic rating and a checklist as defined here is the behavioral

Job: Retail service manager.

Performance Dimension: Resolve customers' disputes.

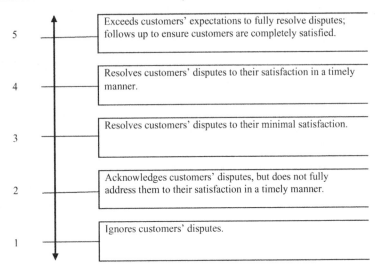

FIG. 4.5. Example of BARS scale for a retail manager for the "Resolves customers' disputes" performance dimension.

observation scales (BOS; Latham & Wexley, 1994). An example of these scales is given in Fig. 4.6. BOS retain the behavioral specificity of BARS but require the rater to indicate how often each behavior occurred. No evaluation of the behavior or the person is required, although in developing these scales some behaviors are recognized as effective and others as ineffective regarding their contribution to total job performance. The major advantage of BOS relative to BARS may be that BOS provides more data for performance feedback and counseling. Memory requirements of BOS, however, may be such that the frequency ratings become evaluations (Murphy, Martin, & Garcia, 1982) rather than reports of the frequency of particular behaviors.

Mixed Standard Scales

Another attempt to construct rating scales with a behavioral format is the mixed standard scales (MSS) proposed by Blanz and Ghiselli (1972). MSS are similar to checklists in that the rater checks which one of three statements in a set is most like the person rated. One of several of these triads that might be used in a rating instrument is presented in

Job: Retail service manager.

Performance Dimension: Resolve customers' disputes.

1. Questions customers to understand the nature of their problems.

    Almost Never   1   2   3   4   5   Almost Always

2. Upon resolution of dispute, will again ask customers if they are satisfied.

    Almost Never   1   2   3   4   5   Almost Always

3. Follows up with customers after the dispute to ensure they are still satisfied.

    Almost Never   1   2   3   4   5   Almost Always

4. Interacts with customers in an empathetic manner.

    Almost Never   1   2   3   4   5   Almost Always

5. Interacts with customers in a respectful manner.

    Almost Never   1   2   3   4   5   Almost Always

6. Resolves customers' disputes in a timely manner.

    Almost Never   1   2   3   4   5   Almost Always

FIG. 4.6. Example of BOS scale for a retail manager for the "Resolves customers' disputes" performance dimension.

Fig. 4.7. Triad statements are developed from critical incidents describing especially effective and ineffective behavior. Within each triad, the three statements describe good, average, or poor performance along a single dimension. These statements are then randomly ordered on the form, and for each statement the rater indicates whether the ratee's performance is better than, worse than, or equal to, that described by the statements. Scoring procedures for each of these triads are described at the bottom of Fig. 4.7.

"Errors" Associated With Absolute Ratings

A variety of errors may occur when ratings are made using absolute formats. Rating errors occur when raters are either unwilling or unable

1. He/she is a self-starter. Always takes the initiative. Superior never has to stimulate.
2. While generally he/she shows initiative, occasionally superior has to prod to get work done.
3. He/she has a tendency to sit around and wait for directions.

Scoring Mixed Standard Scale Triads

1. Assign numerical values to the statements in a triad: 1, 2, 3 according to their favorability.
2. Assign similar values to the response options; that is, Better than = 3, Equal to = 2, Worse than = 1.
3. The response to each statement is then scored as the product of the statement and response values as follows:

	Response Value		
	Worse than	Equal to	Better than
Statement Value	(1)	(2)	(3)
Low (1)	1	2	3
Medium (2)	2	4	6
High (3)	3	6	9

4. The resulting scale value for the triads range from 6–18 because the total sum for the triad is the sum of the Response Value–Statement Value product for each of the three items.

FIG. 4.7. Examples of a mixed standard scale triad used to measure employee initiative.

to make distinctions between employees. There are three major types of this error:

1. *Leniency error* refers to the fact that some raters give higher ratings than do others. Most students are familiar with faculty members who have a reputation for giving all As and Bs, whereas others rarely give grades above a C. Leniency error becomes a problem when performance ratings are compared across raters or when students who took similar courses from different professors are to be compared.

2. *Central tendency error* occurs when raters are reluctant to use the extreme points on a rating scale. Such a rater would be giving ratings that are biased against the truly superior performers and unfairly advantageous to those employees whose performance is really inadequate.

3. *Halo error* refers to the fact that a rater does not discriminate among the different facets of performance. A rater who is impressed by an employee's punctuality gives that person high ratings on other dimensions of performance such as the quality of work, the amount of work completed, and the degree to which the person's behavior contributes to

the social atmosphere of work, in spite of evidence that performance on these latter dimensions is not as superior as their punctuality.

These errors are problematic because they reduce the magnitude of individual differences in performance between employees, and hence, reduce the ability of the performance measures to distinguish between high- and low-performing individuals. As a result of this reduced variability, relationships with predictors will necessarily be lower. However, it is important to keep in mind that these errors may not be errors in terms of psychological processes. Indeed, such "errors" may serve a useful purpose for the rater. For example, Murphy and Cleveland (1995) noted how it may be functional for managers to ignore distinctions among employees (leniency or halo) because it could promote workgroup harmony and reduce their personal anxiety with making such distinctions. Or, consider the manager who may live in the same neighborhood as the employee; this manager knows that a poor performance evaluation could have personal consequences for their family. So even though we refer to these as errors, we need to keep a broader perspective on this issue and consider the social context within which the ratings are made, as well as the purposes they seek to achieve for the organization and rater. We return to this issue later, when we discuss the appropriateness of corrections for interrater unreliability.

Attempts to understand the psychological processes contributing to rating error led to an extensive program of research examining social-cognitive theories of the rating process. Several influential studies in the early 1980s, such as DeNisi, Cafferty, and Meglino (1984), Feldman (1981), Landy and Farr (1983), and Wherry and Bartlett (1982), basically directed rating research to adopt social-cognitive theories of psychological processes involved in providing ratings. This research generated a number of interesting concepts now central to performance rating theory. For example, it led to the recognition that raters may hold "folk theories" about performance, which are supervisors' implicit theories about what the job requires and what it means to perform well in the job. This research also documented the importance of having an opportunity to observe ratees, such that having more opportunities to observe employee behavior positively influences ratings (Judge & Ferris, 1993; Rothstein, 1990).

Many of these more cognitively oriented models were adapted to field settings and tested using SEM techniques (see chapter 2). An early study by Hunter (1983) showed how cognitive ability determined job knowledge and work sample performance, which in turn influenced supervisory ratings. This basic model has been extended in several ways by Borman, White, and Dorsey (1995); Borman, White, Pulakos, and Oppler (1991); Pulakos, Schmitt, and Chan (1996); and Schmidt, Hunter, and Outerbridge (1986), to name just a few. In each model, different cognitive, affective, or behavioral antecedents of performance ratings are examined (to help understand what such models look like, please see chapter 2, Fig. 2.13). For example, Borman et al. (1995) examined a model for peer ratings and a model for supervisory ratings. The models extended Hunter (1983) by also including technical proficiency, dependability, friendliness, and obnoxiousness of the ratee constructs (note the latter two constructs would represent contamination). In both models, technical proficiency and dependability were strong predictors of ratings. However, although there was no significant effect for friendliness and obnoxiousness with supervisory ratings, these effects were present for peer ratings. Thus, Borman et al. (1995) showed how supervisory ratings tend to be less contaminated by these factors than peer ratings (see also Oppler, Campbell, Pulakos, & Borman, 1992).

Research Comparing Absolute Rating Formats

Research to explore differences among rating scale formats in their ability to eliminate halo, central tendency, and leniency indicates that the more involved attempts (BARS, BOS, MSS) yield little improvement in measurement (Wexley & Klimoski, 1984). In fact, Landy and Farr (1980) noted less than about 4% of the variance in ratings is due to rating format. As long as the rating format is well developed and raters use it appropriately (see later section on rater training), various formats tend to produce similar results. However, in terms of performance feedback and counseling, communicating the importance of high standards of performance and performance appraisal, communicating to employees what aspects of performance are most relevant, and enhancing the motivation and goals of accuracy for raters, behaviorally based ratings represent a significant improvement (e.g., Bretz et al., 1992; Guion & Gibson, 1988).

Methods Requiring Employee Comparisons

Judgments about performance that require the rater to directly compare various employees are called *employee comparisons* and consist of

JUDGMENTAL CRITERION MEASURES 187

three types: rank ordering, paired comparison, and forced distribution. Employee comparisons are highly reliable; apparently, comparative judgments are easier to make than absolute judgments (Guilford, 1954). Of the rating errors, central tendency and leniency cannot occur because in an ordering system someone must be highest and someone else lowest. Further, halo is not usually a problem because, for practical reasons, employee comparison systems are limited to assessing only overall performance (comparing large numbers of employees only once is tedious; comparisons along several dimensions would be extremely time consuming). One of your authors once required the supervisors of a group of prison guards to pair-compare their subordinates along three performance dimensions. The resulting correlations among performance dimensions exceeded 0.80, indicating employees were rank ordered almost identically on all three dimensions. Finally, rank ordering people in separate supervisory groups and comparing those rank orders across groups assumes equal performance across groups. So, a person working in a group in which all employees are above average would be at a disadvantage when compared with another person working in a group in which all employees were below average.

Methods of Rank Order

One of the simplest approaches to job performance measurement is to have supervisors rank order their employees. The ranking can be on overall job performance only or on separate dimensions of job performance. Ranking becomes more difficult as the number to be ranked increases; thus, a technique called alternate ranking can be helpful in this connection. The supervisor is given a stack of cards, each one containing the name of an employee. The supervisor sorts through the stack and picks out the two individuals who are highest and lowest on the dimension being ranked. Then, the supervisor goes through the remaining cards and selects the highest and the lowest. This process continues until the ranking is complete. With the available computer technology, the process is even more simplified and easier to administer.

Method of Paired Comparisons

In using this method, one first constructs all possible pairings of employee names. For each pair of names, raters then choose the one they believe is highest on the dimension being evaluated. For each ratee, we then compute the percentage of comparisons in which that ratee was preferred over the other person in the pair. These percentages are then converted into their corresponding standard scores. Tables are available for converting to standard scores (see chapter 2) directly from frequency data. The information needed is the number of times the ratee was preferred over others and the total number of ratees (Lawshe, Kephart, & McCormick, 1949). These standard scores are most useful if one is interested in making cross-group comparisons of individuals.

The paired comparison technique is capable of producing highly reliable ratings with a minimal number of judges. The major drawback is the fact that the number of pairs of judgments one must make increases rapidly as n, the number rated, increases. The number of pairs of employees to be compared can be computed by the following formula:

$$\text{Number of comparisons} = (n(n-1))/2$$

So, to pair compare 20 employees, one must make 190 comparisons; with 30 employees, this number becomes 435.

More recently, Borman, Buck, et al. (2001) provided an interesting variation on the paired comparison technique. Using principles from adaptive testing and IRT (chapter 2), they developed a method called the *computerized adaptive rating scale* (CARS). This format uses the computer to administer pairs of behavioral statements that have been shown to differ in their effectiveness, with the first comparison having one statement being only somewhat more effective than the other. The rater is asked to consider an employee and decide which of the two statements best describes the employee's performance. If the rater chooses the more effective statement, another statement that is even more effective is given and the rater is asked to choose between the two. In contrast, if the rater chose the less effective statement, an even worse behavior is given and the rater is asked to choose between the two. By iterating through continually more or less effective behaviors until all relevant comparisons have been made, the computer ultimately settles on an estimate of the ratee's performance that is "just right." This process is repeated for each performance dimension. Borman, Buck, et al. (2001) showed how the CARS format was superior to BARS and graphic rating scales in terms of reliability, validity, and accuracy. Note this procedure does not pair-compare individuals, but rather behavioral statements about individuals, so the end result does not have the same psychometric limitations of other employee comparison methods. Future research will need to further establish the effectiveness and practicality of the CARS format, but it holds a great deal of promise.

JUDGMENTAL CRITERION MEASURES

Forced Distribution Method

This employee comparison system is actually an approximate ranking method that allows numerous ties. Typically, there are five categories (and thus five "ranks") into which raters must sort ratees in specified proportions. These proportions are usually chosen to approximate frequencies in the normal distribution. As an aid to the rater, the number and the percentage of individuals to be assigned to each category is given. The scale might appear as follows for ranking 50 workers:

Lowest	Next	Middle	Next	Highest
10%	20%	40%	20%	10%
5 names	10 names	20 names	10 names	5 names

Typically, a list of names of ratees is presented along with this information. The forced distribution method is usually used to measure along a single dimension of overall performance, but this need not be the case. The forced distribution method is often useful when the number to be rated is large, making ranking and pair comparison difficult, and when there are at least several raters, which ensures the reliability of the final composite. When the added condition holds that the raters would find it difficult to make fine differentiations between the ratees, we have the ideal set of circumstances for use of the forced distribution. In combining ratings across judges, the five categories can be assigned numbers (e.g., ranging from 1 through 5) and values averaged across judges. Incidentally, if raters are equal in their ability to rate, it is always better to have more raters because the reliability of the composite rating increases as a function of the number of raters, just like in a test where reliability increases with the number of items.

Forced distribution ratings are superior to graphic rating scales in that there can be no leniency or central tendency errors. Halo, however, is not necessarily reduced. Reliability of forced distribution ratings, although usually superior to that of graphic rating scales, will typically be lower than that of pair comparisons or ranking, given the number of raters is constant. Differences in reliability will be greater the smaller the number of raters; conversely, differences should be negligible with large numbers of raters. An attractive feature of forced distributions is that they yield a normal distribution of ratees, always an attractive measurement feature.

All three employee comparison methods discussed are highly reliable and have worked well when a researcher is interested only in a unidimensional judgment of employees' overall organizational contribution. This might be the case when we want to determine pay raises or promotions, or when the judgments are to be used as standards against which to evaluate some personnel staffing function, such as selection. However, there are two distinct limitations. First, these systems are of no use for performance feedback or counseling and may be most difficult to explain or justify to questioning employees. Second, the use of employee comparisons for pay raises or promotions will most certainly increase competition among the members of a work group, which may produce significant negative long-term outcomes (see Schein, 1980).

Methods Requiring Reports of Behavior

Checklists

The idea behind checklists is that the quality of information obtained from supervisors about employee work performance can be improved by eliminating the evaluative function and turning the supervisor into a mere describer or reporter of observed behavior. This distinction often appears somewhat strained. For example, a rater reporting the frequency with which the ratee displays a desirable work behavior (on a scale where, say, 5 = "always" and 1 = "never") is likely making an evaluative judgment. For this reason, we included our discussion of BOS and MSS with our discussion of graphic ratings. However, one checklist technique, forced choice, may be more truly descriptive than others.

The Forced Choice Technique

This method was developed by Wherry in the late 1940s for use in rating officers in the U.S. Army. The major purpose in substituting forced choice scales for the older graphic rating scales was reduction in leniency: Military personnel consistently found that the vast majority of ratees were landing in the top two rating categories. Although they believed their officer corps was good, they knew it was not that good. The forced choice procedure was designed to reduce leniency by making it difficult, if not impossible, for a rater to deliberately assign a high rating to an undeserving ratee. In this method, the rater who, it will be remembered, is cast as an observer-reporter rather than an evaluator, is typically presented with blocks of two or more (usually four) statements, each statement within a block having been carefully

JUDGMENTAL CRITERION MEASURES

equated with the others on some "attractiveness" index. All statements within a block thus look equally flattering, but, based on prior analyses of the items, it is known that they differ in ability to differentiate high from low employees. Only those that differentiate the high from the low are counted when the form is scored. The rater-reporter's task is to select the statements in each block that are most (and sometimes least) descriptive of the ratee. The theory is that even a rater who wants to fake will not be able to do so under these circumstances.

A number of research studies has been done on the appropriate format of forced choice scales. The conclusion of these studies is that the attractiveness of items should be judged on the basis of "importance to the job," that forced choice items should be arranged in groups of four equally attractive items, and that the rater-reporter be asked to pick the two most descriptive of a ratee. An example of forced choice blocks developed for the rating of state troopers is presented as Fig. 4.8.

There are three potential problems with the use of forced choice techniques. First, in developing these scales, we frequently find the

> Select two phrases which best describe the trooper.
> Remember to treat each set of statements independently.
>
> 1
> 1. He makes good contacts with both the general public and public officials.
> 2. He leaves a very good impression of the department with the younger generation.
> 3. He does not accept or solicit gifts or services from the public.
> 4. He knows the criminal element in the post area.
>
> 2
> 1. His attitude toward the job is one of sincerity and belief that the job is important.
> 2. He has pride in the department and himself.
> 3. He keeps the firearms that he carries clean and in proper working order.
> 4. He never compromises a principle or writes a ticket just to be number one on the activity sheet.
>
> 3
> 1. He seems to know when the letter of the law should be discarded in favor of the spirit of the law.
> 2. He practices good first aid.
> 3. He takes advantage of resource materials department-wide.
> 4. His reports convey meaning without the use of superficial and excessive language.
>
> 4
> 1. He does not present a false front to command officers and fellow workers.
> 2. He is diplomatic with the public.
> 3. He knows his limitations.
> 4. He knows the trouble areas in traffic and he works them.

FIG. 4.8. Examples of forced choice tetrads.

discrimination index (the ability of the item to separate high- from low-performing individuals) and the job importance index (or attractiveness index) computed for each item are highly correlated. This means that it is impossible to arrange items in groups of four equally attractive items, two of which discriminate between good and poor workers and two others that fail to discriminate. Second, the technique is not popular with supervisors who want to know what rating they are giving subordinates. Third, the scales cannot be used for feedback because once one tells supervisors and employees what the scored items are, the scale is no longer useful in eliminating leniency—the very reason for its development. The latter two reasons were instrumental in deciding to discontinue using a forced choice scale for evaluating state troopers. If the latter two concerns are organizationally relevant, we would not recommend using forced choice rating scales.

Rater Training

Regardless of the judgmental rating format used, it is critical to train raters before implementing the staff appraisal system. Rater training refers to a variety of methods used to teach raters how to provide ratings of better quality. Pulakos (1997; see also Smith, 1986) noted effective rater training programs have four common features: They involve explanations for why rating accuracy is important, they allow raters to practice, they allow group interaction to discuss rating issues, and the raters are provided with feedback to improve the rating process. Although there are a variety of rater training programs (e.g., Hauenstein, 1998), two are dominant. *Rater error training* attempts to teach raters to identify the common errors and mistakes (e.g., leniency) and how to reduce them (e.g., Latham & Wexley, 1994). *Rater accuracy training* (sometimes also called frame of reference training) attempts to teach raters how to correctly observe and report performance behavior using the rating format. In both types of training, raters may observe videos of employee performance that differ from ineffective to effective. Raters will make their ratings, and then the staffing expert will discuss the ratings and compare them to the "true" ratings obtained by a group of knowledgeable job experts. Raters will see how and why their ratings deviated form the true ratings, and this process will continue until all raters provide accurate ratings.

Both types of rater training should routinely be used, with the appropriate type of training matched to the rating format and appraisal process (Hauenstein, 1998; Pulakos, 1984). However, although there is consistent evidence that rater training effectively reduces rating errors, there are two problems. First, the reduction of these errors is not large even when statistically significant. This may in part be caused by individual differences in rater ability and motivation (Hauenstein, 1998). Second, in the absence of information about actual performance levels (Schmitt & Lappin, 1980), we do not know whether reduction in these errors (as defined in the experimental studies) results in more accurate ratings (Borman, 1978). In other words, raters can agree with each other and be wrong, or all employees in a group can be excellent performers, in which case all the ratings in that group should be good. Accurate ratings for such a group may, in fact, look like leniency errors have occurred. Some research (Pulakos, 1984) suggests that training to reduce errors may have no effect on accuracy even though ratings seem to have improved.

The main conclusion to be drawn from this research on rating errors is that rater training should familiarize raters with possible errors, the meaning of the rating dimensions, and the defined levels of effective and ineffective performance. It should also give them practice at rating and use that practice to demonstrate errors and ways to correct those errors. Finally, raters must believe that the organization places importance on securing accurate and carefully documented performance appraisals.

Appraisal Purpose

The purpose of the appraisal system, or at least the collection of performance judgments, can have a fundamental impact on the quality of the performance ratings. That is, why one provides the ratings will influence the way they are made. There are three major purposes, and each purpose will affect the quality of the ratings (Murphy & Cleveland, 1995). Ratings collected for administrative purposes, such as promotion, termination, and salary decisions, tend to suffer from the types of errors and social-cognitive influences noted earlier. The reason is because such ratings carry real and important consequences for people, so it should come as no surprise that raters are hesitant to make drastic discriminations between employees. Ratings collected for research purposes, such as in the conduct of a validation study, tend to show much better discrimination among employees because the ratings have no direct consequence for the employee or rater. Between these two extremes are ratings collected for developmental purposes. As we discuss later in the section on multisource feedback, these ratings may or may not be affected by various rater errors, depending on whether raters are anonymous. One implication of these differences is that if one wants to conduct a validation study, one should be sure to collect ratings for research-only purposes. This is easily accomplished by explaining to supervisors that the study has no relation to their current or future employment (or that of their ratees), using confidential data collection methods done through a neutral third party (e.g., a consulting agency), and having employees complete the performance appraisal form during paid company time. We give examples of many such studies in chapters 7 to 10.

Summary of Judgmental Criteria

Judgmental criteria have a nice advantage over objective criteria in that they can more directly focus on performance behavior; what many argue is the appropriate target for staffing (e.g., Campbell et al., 1993; Motowidlo, 2003). Despite this, there remain several continuing difficulties with collecting high-quality ratings (e.g., leniency). Although there are many different ways of assessing performance judgments, the differences across these approaches have traditionally been small. Rather, as long as the performance appraisal forms are well developed; raters are trained, willing, and able to provide accurate ratings; and raters have the opportunity to observe employee performance, it does not make a great deal of difference what format is used. Of course, the purpose of the appraisal system does matter, such that administrative ratings tend to suffer from more rating errors than ratings collected for research or developmental purposes. Further, if the ratings are to be used not only as criteria in selection validity projects, but also as developmental feedback, the choice of the rating procedure used should err on the side of comprehensiveness of the performance domains covered.

A COMPARISON OF OBJECTIVE AND JUDGMENTAL CRITERIA

Beyond the obvious differences between objective and judgmental criteria, there is another important reason to consider them as distinct. Specifically, meta-analyses have shown the two types of criteria are not interchangeable. Heneman (1986); Bommer, Johnson, Rich, Podsakoff, and MacKenzie (1995); and Rich, Bommer, MacKenzie, Podsakoff, and Johnson (1999) reported the corrected correlation between objective and judgmental criteria ranges only from about 0.28 to 0.45. There are many explanations to account for the low correlations, including the various kinds of contamination, deficiency, and related considerations noted earlier. However, when considering criteria that represent the same construct, convergence between objective and judgmental ratings should be better (Bommer et al., 1995).

So which criterion should staffing specialists use? Given their low correlations, it would generally not be appropriate for researchers to treat this question arbitrarily. Hoffman, Nathan, and Holden (1991) demonstrated this fact by showing the same predictor composite had substantially different relationships with objective and judgmental measures. The advice of researchers such as Campbell and Motowidlo is to focus only on performance behaviors because we study human behavior, and thus the usefulness of our behavioral interventions should be assessed via employee behavior. In contrast, objective criteria are more closely aligned to the organization's goals and objectives and are more easy to convey to management. Although there are no data on this issue, we propose that if judgmental criteria were aggregated to unit levels, such as average department performance or average work group performance, one would find judgmental criteria could relate to important organizational "hard" criteria such as profitability and return on investment because these relationships would exist at the same level of analysis.

CRITERION MEASUREMENT CONSIDERATIONS

In this section, we outline three major issues staffing personnel confront when developing or measuring criteria. First, how do we combine information on various aspects of performance? Second, does performance change over time? Third, what characterizes typical from maximum performance? In a sense, all three involve the problem of including irrelevant information in our criterion measure; hence, all are aspects of the contamination problem depicted in Fig. 4.2. However, they deserve such careful consideration because they may have a dramatic effect on the measurement of both objective and subjective criteria, and because they may influence the relationships with predictor constructs.

Specific Versus General Dimensions of Performance

Among applied psychologists, a controversy has waged for years between those who maintain measures of different dimensions of job performance should be combined into a single overall composite measure (e.g., Nagle, 1953) and those who believe that measures of performance should be kept separate and used independently of each other (Dunnette, 1963; Guion, 1961). In more recent years, there appears to be some resolution to this issue; we have already seen how models of job performance make important distinctions between different dimensions. A question becomes, however, what one should do with these dimensions in practice—do we attempt to predict each dimension separately, or should we combine them into an overall dimension? Does it even make sense to speak of "overall job performance"?

The importance of this issue can perhaps be best illustrated with an example. If task performance is primarily driven by cognitive ability and contextual performance is primarily driven by personality, the potential ways we may combine these dimensions into a single index of overall performance could have serious consequences. Hattrup, Rock, and Scalia (1997) demonstrated this effect with minority hiring rates and adverse impact (i.e., the ratio of minority to majority group hires). They showed that if the composite criterion was weighted more by contextual performance, personality was a stronger predictor and thus contributed to smaller racial subgroup differences (because personality measures exhibit smaller differences than cognitive ability). Alternatively, if task performance was weighted more heavily, cognitive ability was the stronger predictor and subgroup differences were greater. The point is that when predictors have different relations to different criteria, how the criterion dimensions are formed into a composite will have implications for the nature of predictive relationships.

Murphy and Shiarella (1997) showed another consequence of criterion weighting, demonstrating predictive validities could differ dramatically as a function of criterion and predictor weights. In fact, they illustrated how validities (correlations) could range from 0.20 to 0.78 simply as a function of these weightings. This means the magnitude of validities one might find for predictors is affected by how the criteria are weighted and formed into a composite.

However, the fact that criterion weighting is an important issue does not address the question of whether one should use multiple dimensions or combine the dimensions into a single overall composite.

Guion (1961) attempted to solve this question by first demonstrating the validity of each predictor separately for each criterion dimension (a "research" question), and then laying out the set of predictive relationships for management and letting them determine how they want to combine everything into a staffing decision (an "administrative" question). Schmidt and Kaplan (1971) extended this thinking by proposing the selection of multiple or composite criteria should depend on the intended use. If the goal is to make practical decisions about staff members (as in making salary recommendations), then computation of some weighted composite is essential. Weights are determined by considering the contribution of each performance facet to the total contribution of the job to the organization. These weights are multiplied by the measured performance of people on corresponding job dimensions to come to a value that represents the overall worth of a worker. However, if the goal is to understand the dimensions of job performance and how they contribute to job success or how they are determined by various job experiences, training, or employee attitudes, then multiple criteria should be used. A meta-analysis by Viswesvaran, Schmidt, and Ones (2005) challenged this thinking. Analyzing 90 years worth of performance data, they argued that a general factor accounts for approximately half of the total variance in job performance ratings. Perhaps more important, they argued that this general factor represents a psychologically meaningful construct because the performance dimensions themselves are positively correlated.

If management decides they want to weight criterion dimensions by their value to the organization, a methodology known as *policy capturing* is extremely useful. In policy capturing, one would give supervisors all combinations of the different dimensions (e.g., high on contextual and task performance, high on contextual performance but low on task performance) and ask them to provide an overall evaluative performance judgment. One could then statistically determine how much weight supervisors give to each dimension. However, in this policy capturing approach, supervisors are implicitly, not explicitly, providing the weights. An approach similar in concept was used by Rotundo and Sackett (2002). They used policy capturing to show some supervisors tended to weight task performance more than contextual and counterproductive performance, another group weighted counterproductive performance more heavily, and a third group weighted task and counterproductive performance equally. This finding suggests the importance of discussion and a coming to consensus over weights for specific jobs so some uniformity in how overall performance indices for employees in jobs may be obtained.

Overall, the basis for resolution of the controversy lies in the use we intend to make of the criteria. If the interest is in promoting the best person from a group, then a composite criterion is essential. This will be true for most predictor validation studies. If we want to train or counsel staff people concerning their job performance, then multiple criteria are most useful. The composite criterion notion, however, is not useful when high job performance on one dimension cannot compensate for high job performance on another job facet. Consider the difference between quality and quantity in an assembly job; both are necessary for effective overall performance. The idea that lack of good performance in one dimension cannot be compensated for by high performance in other dimensions works for most, but not all, jobs. When combining criterion dimensions to create an overall composite, average the dimensions if there is no compelling rationale for weighting some more than others; otherwise, weight the dimensions by value to the organization.

Maintenance Versus Transition Stages of Performance: The Dynamic Criterion Issue

To predict performance, one must be sure performance will be stable over some reasonable amount of time. If it is not, then the performance prediction problem becomes considerably more difficult because one must not only identify the types of performance to predict, but also when to predict them. To take an extreme example, what if the performance of managers during the first year on the job was nearly uncorrelated with performance in subsequent years—using a criterion assessed within that first year would not be useful for predicting and enhancing long-term performance. Thus, as noted earlier, the stability of the validity (correlation) coefficient depends on the stability of the criterion (once again demonstrating that predictors derive their importance from criteria).

The question of whether criteria change over time has been a heated one in applied psychology. That criteria were likely to be dynamic and changing over time was a long-standing assumption noted by Guion (1961) and Ghiselli (1956; Ghiselli & Haire, 1960). However, a critical review by Barrett, Caldwell, and Alexander (1985) challenged this assumption. Their review of the literature found little support for criterion change over time, as assessed via mean changes in performance,

CRITERION MEASUREMENT CONSIDERATIONS

changes in validity, and changes in correlations between criteria. A number of researchers then demonstrated empirical evidence showing both criterion change over time (Henry & Hulin, 1987, 1989; Hulin, Henry, & Noon, 1990) and criterion stability (Ackerman, 1989; Barrett & Alexander, 1989; Hanges, Schneider, & Niles, 1990).

A key factor contributing to the debate is the difficulty in answering questions about change. In fact, an understanding of criterion change requires a consideration of several interrelated substantive, methodological, and statistical issues. Many of these issues have been addressed only within the last decade.

Substantive advances have been made in terms of delineating how and why criteria might change over time. Building from research in skill acquisition (primarily Ackerman, 1987), Murphy (1989) proposed two general stages to performance change. The first is called a transition stage, which occurs when people must learn new tasks and skills; performance during this stage is primarily determined by cognitive ability. The second is called a maintenance stage, which occurs after people have learned the new skills; here the focus is on continuing to perform already learned behaviors, and hence, performance is primarily determined by personality and motivation. Because the transition stage requires learning new tasks, it is likely to occur for shorter periods of time than maintenance stages. In general, these predictions appear to be supported empirically. Using a growth curve methodology whereby individual performance is plotted over time, several researchers found performance tends to change rapidly during the early stages of learning a job, and then approach an asymptote where performance becomes more stable (e.g., Hofmann, Jacobs, & Baratta, 1993; Hofmann, Jacobs, & Gerras, 1992; Ployhart & Hakel, 1998). A typical representation of this performance curve is shown in Fig. 4.9. In Fig. 4.9, performance over time increases rapidly during the early phases of performance, but starts to level off and approach an asymptote around time period 2. Individual differences in performance also increase as time goes on, meaning that the variability in performance is greater later (in the maintenance stage) than earlier (in the transition stage).

Methodological advances have contributed to our understanding of how performance change must be studied. A better understanding of the types of designs one must adopt to study dynamic criteria has been proposed (e.g., Willett & Sayer, 1994). For example, one must study at least three time periods to determine if performance changes in nonlinear ways (such as shown in Fig. 4.9) because two time periods can only

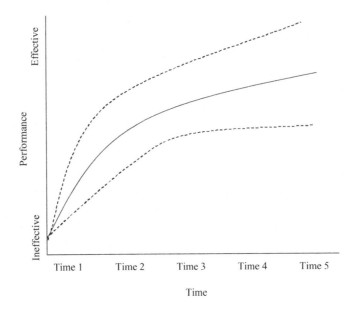

FIG. 4.9. Function describing performance change over time. Solid line indicates average performance; dashed lines indicate variability (individual differences) in average performance.

examine a straight line (Rogosa, 1995). Furthermore, one needs to make distinctions between individual performance change and average performance change. As shown in Fig. 4.9, the solid line represents the average rate of change for all employees, but this average curve does not adequately capture the amount of individual differences in average performance reflected by the dashed lines. Simply examining correlations between criteria over time, as was so common in past research, is unlikely to enhance our understanding of how criteria change. This is because the rank order of people over time can remain fairly constant even when there have been significant changes in the levels of the data. When the rank order of people is constant, it looks like there has been no change even though change in actual levels of the criteria may be significant. Likewise, ignoring missing data (something common in longitudinal designs) can sometimes make it difficult to interpret the results (Sturman & Trevor, 2001).

Statistical advances have been made by adopting models developed in other areas of science for the explicit consideration of change over time. These models are loosely termed "growth models" because they were adapted from developmental psychology (e.g., McArdle &

Epstein, 1987). These models examine the nature of both within- and between-person performance change over time. A powerful component in these models is the ability to not only identify the functional form of performance change over time (e.g., the solid line in Fig. 4.9), but also to estimate and explain individual differences in such change (denoted by the dashed lines in Fig. 4.9). Several studies have now used these models and shown that, although performance may be dynamic, it is still explainable and predictable. For example, Hofmann et al. (1992, 1993) and Deadrick, Bennett, and Russell (1997) used random coefficient growth curve models to show that performance follows a nonlinear trend such as that shown in Fig. 4.9: Performance level improves quickly and then approaches an asymptote where it starts to stabilize.

An extension to this work has been studies showing how, even though criteria may change over time, they are sufficiently stable to allow one to predict the performance variability—that is, predicting who changes faster, maintains more stability, or declines in performance over time. Ployhart and Hakel (1998) used latent growth curve modeling to demonstrate the nonlinear trend found in previous research, but also showed that the dynamic performance variance could be predicted by personality (self-assessed empathy and persuasive ability) and a biodata form. Their results demonstrated that the biodata form predicted average performance midway through the time period studied, whereas empathy and persuasion predicted rate of performance improvement. Those salespeople with more empathy and persuasive ability improved sales performance more quickly. Stewart (1999) also showed how different personality traits predict performance during maintenance and transition stages: Order predicted performance in the transition stage, achievement predicted performance in the maintenance stage, and conscientiousness was predictive of both.

In sum, research tends to suggest performance is dynamic over time. To a certain extent, this means we need to think carefully about when we want to gather the criterion data used in validity studies. We know performance is likely to be more variable when one first starts a job, so often we will want performance to move to maintenance stages before using it as a criterion for validation studies. This may not always be true, however, because transition times may be important criteria in some organizations (e.g., in police departments when one moves from a cadet position to a patrol position). Yet, even when performance is dynamic, there is still sufficient systematic variance for it to be predicted by individual difference constructs. That is, different types of performance change—increasing, decreasing, stable—may occur for different people, and may be explained by individual difference constructs such as cognitive ability and personality. An implication here is that individual differences predict performance in maintenance and transition stages in unique ways, again arguing for a need to think first about our criteria before our predictors (Wallace, 1965).

Typical Versus Maximum Measures of Performance

A third important consideration in the measurement of criteria is whether the measure is assessed under typical or maximum performance conditions. Sackett, Zedeck, and Fogli (1988) noted three features—time frame, instructions, and evaluation—that distinguish between the two types of measures. *Maximum performance* is defined by conditions where performance occurs over relatively short periods of time so one can exert total effort, there are accepted instructions to exert total effort, and the context is evaluative in nature. In contrast, *typical performance* is defined by conditions where performance occurs over long periods of time, there are no explicit instructions to exert total effort, and the context may not be evaluative (at least to the extent of being observed continuously). An example from a college course might show maximum performance occurs when taking a test, whereas typical performance might involve how people prepare for the test throughout the semester. Interestingly, Sackett et al. (1988) showed that both types of performance were practically uncorrelated with each other. Research conducted by McCloy, Campbell, and Cudeck (1994) and Vance, MacCallum, Coovert, and Hedge (1988) showed results similar to the typical and maximum performance distinction.

The importance of the distinction between typical and maximum performance is most apparent when examining predictive relationships. This reflects an old distinction first noted by Cronbach (1949). He argued maximum performance represents "can do" measures and is driven primarily by individual differences in cognitive ability. In contrast, typical performance represents "will do" measures and is driven primarily by individual differences in motivation and personality. Research has generally supported this expectation. DuBois, Sackett, Zedeck, and Fogli (1993) found maximum performance measures were predicted more strongly by cognitive ability than personality.

In addition, Mount and Barrick (1995) meta-analyzed personality measures and segmented the criteria into "can do" and "will do" dimensions. They generally found conscientiousness predicted "will do" performance more strongly than "can do" performance. Ployhart, Lim, and Chan (2001) demonstrated the typical/maximum distinction is present at both the manifest and latent levels, and that the distinction moderates the validity of the Five Factor Model (FFM) of personality. Typical performance was predicted by emotional stability, maximum performance was predicted by openness to new experiences, and both types of performance were predicted by extraversion. Finally, Lim and Ployhart (2004) found the typical/maximum performance distinction was present at the team level of analysis, with personality and transformational leadership predicting typical team performance, but only transformational leadership predicting maximum team performance. However, the determinants between typical and maximum performance may not be so clear because research suggests motivation can influence performance in maximum performance contexts when theoretically it should be constant (Kirk & Brown, 2003).

Altogether, these results indicate the distinction between typical and maximum performance is important because even when both types of performance assess the same content, they are unlikely to be highly related. This means when thinking about purposes for criteria, one must consider whether the interest is in typical performance, maximum performance, or both. For example, police officers may spend the majority of their time attending to mundane tasks such as writing accident reports (typical performance), but clearly when the situation demands maximum performance (apprehending a violent criminal), effective performance is critical. In such instances, the determinants of performance may differ between the two contexts, so a validation study using only one type of performance measure will miss an important aspect of the criterion space—and failure to understand the criterion dimensionality issue will leave some of the predictor space also unattended.

Summary and Integration of Criterion Measurement Issues: Mapping the Criterion Space

To this point, we have discussed the three criterion measurement issues as though they were independent. In fact, issues of criterion dimensionality, stability, and context are interrelated. We can see some similarity,

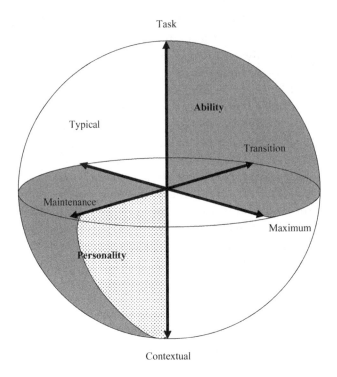

FIG. 4.10. Integrative content, stability, intensity model of job performance measures.

for example, between typical performance and maintenance performance, but these two aspects of criteria are not identical. If one treats each criterion measurement issue as a different axis, one may map the three-dimensional nature of the criterion space such as that shown in Fig. 4.10.

In Fig. 4.10, notice there are three underlying dimensions cutting across all types of criteria (e.g., objective, judgmental). The first is the *content dimension*, roughly corresponding to the single–multiple dimension issue presented earlier. That is, at the extremes, do criteria correspond to task (technical) performance, contextual performance, or some composite combination in the middle? The second is the *stability dimension*, corresponding to the difference between maintenance and transition stages. The third is the *intensity dimension*, representing the distinction between typical and maximum performance contexts. If one plots these dimensions in accordance with Fig. 4.10, we see that the integration prompts a deeper understanding of criteria by allowing any criterion measure to be put into a three-dimensional criterion space.

For example, Campbell's job-specific task performance would clearly fall into the upper part of the sphere, but where does it fall along the typical/maximum and maintenance/transition continuums?

Perhaps more important, the dimensionality portrayed in Fig. 4.10 allows a better understanding of the determinants of performance, and thus a means to compare and contrast performance measures. In the shaded area in the upper right, cognitive ability is the dominant determinant of performance because performance is technically oriented, in transition stages, and in maximum performance contexts. In the polar opposite extreme in the lower left, performance is driven primarily by personality because it is contextual, in maintenance stages, and in typical performance contexts. However, the real value of this dimensionality can be seen as one moves away from these extremes, where the determinants of performance become less straightforward. For example, why should task performance be driven primarily by ability when it is typical performance in a maintenance stage? Might personality actually matter more in this instance? Likewise, would maximum contextual performance be affected by ability and personality? The point is that we have only collected data for the shaded areas in Fig. 4.10, comprising only about 25% of the entire criterion space. The remainder of the criterion space remains relatively uncharted territory, one for which future research is sorely needed and to which practitioners must pay attention to gain for their employing organizations some otherwise unattended potential for competitive advantage.

CURRENT CHALLENGES FOR PERFORMANCE AND CRITERION DEVELOPMENT

To this point, we have discussed the definition, structure, and mechanics of criterion measurement. In this final section, we raise a number of current and special issues regarding employee performance appraisal.

The Legal Context of Staff Appraisal

Fair employment complaints reflect a dramatic increase in those cases in which the relevance of job performance appraisals is the central issue. Examples include promotional disputes, merit-based layoff decisions, terminations of older employees, and the employment of disabled persons, as well as selection decisions. Bernardin and Beatty (1984), Malos

(1998), and Pulakos (1997) noted legally defensible appraisal methods will tend to share the following characteristics:

- They are based on sound job analyses.
- Criteria are communicated to the employee.
- They assess specific dimensions of performance.
- They focus on behaviors rather than traits.
- Behaviors are under the employee's control.
- They use trained raters.
- Procedures are well documented.
- Procedures are standardized.
- A formal appeals procedure exists.

Malos (1998) provided a summary of various laws influencing the practice of staff appraisal. We discuss many of these laws in chapter 11; the important point is that laws and legal concepts form the boundaries of acceptable staff appraisal. Developing appraisal practices that meet these standards can be difficult, but fortunately, criterion measures and procedures that meet scientific rigor will also tend to meet legal rigor. Regardless of whether we want to admit it, the legal context probably improves staffing appraisal practice more often than it hurts it. In general, demonstrating the minimization of deficiency and contamination is the key to a successful defense of appraisal methods.

Potential Bias in Staff Appraisals

An issue intertwined within the legal context of staff appraisal is the issue of bias. Bias most commonly refers to instances where the criterion measure is contaminated by factors unrelated to performance. For example, if male employees are rated higher than women at least in part due to their gender, then gender is contaminating the ratings. However, it is important to recognize that bias only occurs when the difference observed in the ratings is not due to real differences in performance. If in fact men perform better than women, the difference in ratings is not bias but reflects true differences in performance. In practice, it is extremely difficult to determine whether subgroup differences in criteria reflect true differences, bias, or some other factor (Rotundo

& Sackett, 1999). Therefore, studies have sought to examine the magnitude of subgroup differences in ratings, where any differences must usually only be interpreted as potentially indicative of bias.

In general, there is little indication that mean differences—and hence the potential for bias—in criteria are a pervasive problem. Effect sizes for race (White–Black, White–Hispanic) and gender tend to be relatively small, although same race raters tend to give members from the same racial group slightly higher ratings (Ford, Kraiger, & Schechtman, 1986; Kraiger & Ford, 1985; Oppler et al., 1992; Pulakos, White, Oppler, & Borman, 1989; Waldman & Avolio, 1991). Differences between majority and minority incumbents are about one third of a standard deviation, favoring majority group members, both for judgmental and objective measures. However, a more recent and comprehensive meta-analysis by Roth, Huffcutt, and Bobko (2003) found subgroup differences for objective performance measures were similar to or greater than judgmental ratings. Consistent with Campbell's model of performance, the criterion measures with the largest racial differences corresponded to the measures most strongly determined by cognitive ability, such as knowledge tests and work samples. In this context ability measure differences approach one full standard deviation so the differences in performance do not seem to be the same as the differences in measures of cognitive ability. This makes sense given that performance at work is based on more than ability and these other measures (e.g., measures of personality) do not reveal these large differences. This logic is consistent with Waldman and Avolio (1991), who demonstrated how controlling for cognitive ability, experience, and education essentially reduced the Black–White performance difference to zero. Further, the less behavioral and thus more inferential the nature of the ratings, the more potential for bias. Although rater training helps reduce such bias, it does not eliminate it under most conditions (Hoyt, 2000; Hoyt & Kerns, 1999).

Thus, current data find some but not much potential for bias in measures of performance; most differences appear to occur as a function of subgroup differences in the determinants of performance (of course, if predictors are biased in the same manner as the criteria, our ability to detect criterion bias becomes nearly impossible). An important condition on this finding is that these studies have primarily been conducted using supervisory ratings; peer ratings appear to be more prone to greater contamination (Borman et al., 1995; Oppler et al., 1992).

The Meaning of Interrater Unreliability

When obtaining judgmental performance information, supervisory ratings remain the most common source. However, we noted earlier how different supervisors may provide different ratings contributing to lower interrater reliability and lower interrater agreement. Likewise, different sources—ratee, supervisory, peers, subordinates, customers—can all provide unique insight into an individual's performance, and as a result, show low interrater reliability and agreement (Conway, Lombardo, & Sanders, 2001). Indeed, considerable systematic variance is often due to who the rater is (Mount, Judge, Scullen, Sytsma, & Hezlett, 1998) and their level (e.g., supervisory, peer), such that ratings tend to be highly idiosyncratic (e.g., Conway & Huffcutt, 1997; Greguras & Robie, 1998; Hoyt & Kerns, 1999; Scullen, Mount, & Goff, 2000).

At least in terms of supervisory ratings, the most common type of rating in staffing research and practice, meta-analytic research suggests interrater reliability is only approximately 0.52 (Viswesvaran et al., 1996). This has two important implications. First, it means the variance shared by two supervisors tends to be no greater than about 28%, a somewhat low level of interrater agreement given that the two raters are supposedly trained and observing the same employee. Second, it means the unreliability of performance measures in criterion-related validity studies serves to produce serious underestimates of predictor–criterion relationships. Estimates of what the validity would be with a perfectly measured criterion where differences between raters did not exist will be much larger than the observed validity. These two implications are intertwined and have far-reaching consequences for staffing research and practice.

If one follows some more recent suggestions (Schmidt & Hunter, 1996), assumes the differences between ratings are due to error, and thus uses the 0.52 estimate of unreliability in the correction for attenuation formula, one will see a dramatic increase in the validity coefficient. However, the question is whether differences in supervisory ratings are due to error. In a published debate on this issue, Murphy and DeShon (2000a, 2000b) argued that this is unlikely. Differences in what the raters observe, their opportunities to observe, and real differences in their implicit theories of performance are not sources of error in the sense that they are unsystematic. Rather, they may often occur for purposeful and explicit political purposes.

For example, Kozlowski, Chao, and Morrison (1998) argued performance appraisal is frequently a political process and discussed several factors that may contribute to rater distortion of ratings. These include appraisal system factors (e.g., lack of monitoring, competing goals) and organizational system factors (e.g., a climate supporting politics, informal socialization and training).

Murphy and Cleveland (1995) similarly note how rater goals and motivation influence rating distortion and accuracy, such that formal rewards can influence motivation to provide accurate ratings with potentially negative consequences (e.g., detrimental effects on group cohesion). Such systematic sources of variance that contribute to differences in ratings are perhaps better construed as substantive and important in their own right. Indeed, Murphy, Cleveland, Skattebo, and Kinney (2004) found that measures of rater goals correlated with their ratings. This may be particularly true when comparing ratings from different sources because supervisors are likely to see different behaviors than peers or customers (e.g., Cheung, 1999). All of this suggests differences in ratings should not be treated as error and thus corrected in attenuation formulas (although see Schmidt, Viswesvaran, & Ones, 2000, for a contrasting view).

This issue currently remains unresolved, but the consequences are potentially great because using the wrong attenuation correction estimate can dramatically influence the magnitude of validity estimates (not to mention result in over- or underestimates of subgroup differences, when the differences are also corrected for unreliability). To address the issue, research needs to incorporate theories for why sources should have different perspectives (e.g., implicit theories), and then assess these hypothesized constructs using statistical models that include measures of relevant sources of systematic variance (DeShon, 1998; Hoyt, 2000; Murphy & DeShon, 2000a).

Staff Appraisal in Team Contexts

As we saw in chapter 3, assessing individual behavior becomes difficult when the job occurs within a team context because the individual's behavior is no longer independent of the team members. In such contexts, staff appraisal systems must likewise consider the embedded nature of the individual's behavior. Ratings of performance are expected to focus primarily on the behavior that is under the individual's control, but such observations may be difficult to separate from the team's interactions.

This is likely to be most true when supervisors do not have sufficient opportunity to observe the individual's behavior (indeed, peer ratings may be critical here). However, perhaps a more important question is whether in such team environments staff appraisal should focus only on individual behavior. Perhaps such individuals should be evaluated in terms of their contribution to the team—do they facilitate team functioning or detract from team interactions and so on? Although the behavior being evaluated may still be a person's, the context will influence the interpretation of that behavior (Murphy & Cleveland, 1995). For example, suppose two individuals from different teams challenge their group's purpose and goals. Is such behavior equally negative? Not necessarily because one team may be a marketing team trying to develop a new project (and hence independence of thought is an asset), the other may be an army tank team in combat (and hence independence of thought could get the team killed). There is essentially no research on this topic from the perspective of staff appraisal, but it is reasonable to expect appraisal of staff behavior in team environments cannot be separated from the team context.

It is also important to realize that assessment of team performance may occur at the team level of analysis. For example, supervisors may be asked to observe and evaluate team-level processes, such as coordination, interpersonal interactions, communication, and related activities (e.g., Fleishman & Zaccaro, 1992; McIntyre & Salas, 1995; Reilly & McGourty, 1998). In such instances, it is not the individual's behavior that is assessed, but rather the behavior of team members interacting with each other. Thus, if a team is evaluated in terms of "coordination," "communication," and "performance regulation," each member of a given team would receive the same score on each dimension (as opposed to each team member potentially receiving a different score). Such direct measures of team processes and performance are common in research settings, but we have seen few applications to real world practical staffing contexts (although see Reilly & McGourty, 1998, for some examples).

Staff Appraisal in Service Contexts

In a different but no less social context, assessing staff behavior in face-to-face customer service settings can be difficult. The provision of customer service is frequently characterized by three features not present in more "traditional" forms of work: Service quality is more

CURRENT CHALLENGES FOR PERFORMANCE 211

dependent on the delivery experience and thus intangible, the customer is frequently involved in the coproduction of the service, and service production and consumption are more simultaneous (Bowen & Schneider, 1988; Schneider & Bowen, 1985). Unlike more traditional types of work, supervisors are often in a poor position to evaluate employee service provision (Bowen & Waldman, 1999). Furthermore; customer satisfaction with service delivery is inherently within the eye of the customer and supervisory ratings of customer service behavior may not be reliable indicators of what the customers themselves are actually experiencing. Given that supervisors cannot observe and evaluate the majority of an employees' service behaviors when the employee is face to face with a customer, HR practices are critical such that employees who are "service oriented" are attracted, selected, and retained (e.g., Schneider & Bowen, 1985, 1992). It is also important that a climate in which service excellence is rewarded, supported, and expected is created because the supervisor is not present when the actual service behavior is delivered to customers (Schneider & White, 2004).

However, as a consequence, the measurement of staff behavior in service contexts can be difficult. I/O research squarely addressing this issue is nearly nonexistent, in stark contrast to the marketing literature, which has devoted an enormous amount of attention to the issue (see Ryan & Ployhart, 2003). Although many conceptualizations exist, perhaps the dominant marketing perspective on service performance measurement is based on the service quality (SERVQUAL) performance scale developed by Parasuraman, Zeithaml, and Berry (1985; see also Parasuraman, Berry, & Zeithaml, 1991; Parasuraman, Zeithaml, & Berry, 1988). The dimensions and brief definitions of the revised SERVQUAL instrument are shown here. The instrument is used such that customers make a comparison between the type of service they expected and the type of service they received:

1. *Tangibles*: perceptions based on the appearance of store personnel, and the physical layout of the store and service setting
2. *Responsiveness*: behaviors that include attentiveness, readiness, and ability to provide fast service for the customer
3. *Assurance*: behaviors demonstrating competence in the service provision; behaviors that enhance customers' trust and confidence in the service
4. *Empathy*: behaviors demonstrating the service provider understands and is attentive to the customer's expectations, needs,

and concerns; provides individualized treatment and consideration for the customer

5. *Reliability*: extent to which the service provider exhibits consistent service provision

It is important to recognize that the SERVQUAL instrument (and marketing research in general) takes the position that service performance should be assessed by the customers themselves, not management. Therefore, the SERVQUAL instrument would be completed by customers. Yet, as noted by Schneider and Bowen (1995) and Ryan and Ployhart (2003), the approach primarily taken by I/O psychologists has been to ignore customer perceptions and rely on supervisor ratings of service performance. Thus, marketing research argues for a customer perspective, whereas I/O research argues for a supervisor perspective. Further, although the marketing research makes many distinctions between different types of service behaviors, the I/O literature tends to use a single dimension of "service behavior" or simply evaluate overall performance via multiple dimensions that reflect a mix and match of service and nonservice behaviors. This is interesting because job analyses conducted in service occupations tend to find dimensions similar to the SERVQUAL dimensions (e.g., Hogan, Hogan, & Busch, 1984).

If it is true that customers have the best perspective on employee service behaviors, a question becomes whether customer ratings of service should be used as criteria for staffing. There is essentially no research on this issue, but in practice both customer reports of their experiences and third-party monitors of service people, especially in call centers, has become the norm. Weekley and Jones (1997) used customers to help develop and assess their criteria for test validation, but this has been the exception as reported in the research literature. Although we would support the use of customers as a source of performance data, we suspect that there are several problems with relying solely on customer perceptions of service as criteria for staffing. First, customer service expectations and perceptions may be idiosyncratic. This would be true because different customers present different service delivery problems to employees and thus may receive different kinds and levels of service in return. This would make interrater reliability and interrater agreement low. Second, customers will not have received either rater accuracy or rater error training (why would they?), and thus may make a variety of self-serving biases and rating errors. Third, customers' ratings may be contaminated. For example, an older customer may perceive poor service performance from a teenager wearing a nose ring, even though the teenager may have performed well. Thus, we see little legal or scientific possibility for relying solely on customer ratings as criteria. However, we do see the potential for customer ratings to supplement supervisory ratings in a more research-based manner. As each source provides unique information, using multiple criteria from different sources could help better evaluate the validities of predictor constructs. The difficulty arises in defending the legal basis of customer ratings for staffing decisions (because we are unaware of cases where these customer ratings have had to be defended), so these ratings may perhaps be most useful for research or developmental purposes. With service occupations comprising the majority of the U.S. workforce, research on this issue is sorely needed.

Performance Adaptability

There have obviously been a large number of political, societal, technological, and economic changes taking place in business since the mid-1980s (Cascio, 2003). Ilgen and Pulakos (1999) argued seven changes have occurred that require a shift in performance assessment and management:

- Interweaving of technology and jobs
- Changes in the design of jobs
- Integration of contingent workers into the workforce
- Increased emphasis on continuous learning
- External control of performance standards by customers
- Limitations on leadership and supervision
- Changes in the structure of work from individuals to teams

A common theme among many of these developments is a need for employees to continually learn, change, and adapt to new tasks and environments. This has led to a vigorous interest in employee adaptability, defined loosely as the extent to which an employee can change to succeed with changing work demands (Chan, 2000). Research directly focused on adaptability is just starting to appear. In terms of adaptive performance requirements, Pulakos and colleagues (Pulakos, Arad, Donovan, & Plamondon, 2000; Pulakos et al., 2002) conducted a comprehensive review of the literature to understand the structure of adaptability. They then collected several thousand critical incidents from 21 different jobs to identify and test an eight-factor structure of adaptive performance. Support was found for this structure, with the final dimensions shown as follows (Pulakos et al., 2000, p. 617):

- Handling emergency or crisis situations
- Handling work stress
- Solving problems creatively
- Handling uncertain and unpredictable situations
- Learning new tasks and procedures
- Interpersonal adaptability
- Cultural adaptability
- Physical adaptability

Although this research shows considerable promise for understanding the meaning of adaptable performance, how it is different from other forms of performance needs to be better understood. It is not clear, for example, whether adaptability comprises different dimensions of performance (perhaps adding to Campbell's or contextual performance models), or whether it is a type of criterion measurement similar to transition and/or maximum performance conditions noted earlier. Research to date has tended to treat adaptability as a separate dimension of performance, but one must question whether any performance dimension (e.g., technical proficiency, effort) could be adaptable. For example, research has measured adaptability in terms of performance to changing tasks (Kozlowski et al., 2001; LePine, Colquitt, & Erez, 2000). Likewise, how adaptability is similar to or different from more well-established constructs such as coping skills (e.g., Judge, Thoresen, Pucik, & Welbourne, 1999) and change self-efficacy (e.g., Wanberg & Banas, 2000) will need to be established because they share a number of conceptual similarities. Finally, considering adaptability in light of the transition stage of performance, or maximum performance contexts, needs to be considered. For example, is adaptability similar to or different from transition performance? However adaptability fits within the nomological network of performance and individual difference constructs, it is clear that research must begin to examine these important questions because adaptability will become a criterion measure for most jobs. In Box 4.1, one of the authors describes his work on performance adaptability.

Box 4.1 Development of a Cross-Occupation Adaptability Model

The research on adaptability by Pulakos et al. (2000) in your text was a thorough examination of the types of situations to which employees must adapt in various occupations. These researchers began by conducting workshops with supervisors and incumbents in 21 different jobs in 11 different organizations. The jobs included service, technical, support, law enforcement, military, and supervisory jobs.

The workshops were oriented to the generation of critical incidents (an especially effective or ineffective example of work performance) that included a brief description of the situation, what the employee did or failed to do in this situation, and what resulted. They were also taught to generate these incidents across all aspects of their job. A total of 9,462 incidents were generated across the 21 jobs.

Five I/O psychologists identified those incidents that involved adaptive performance defined as a situation in which individuals modified their behavior to meet the demands of a new situation, event, or changed environment. Adaptive behavior was believed to characterize 1,311 of these incidents. After consideration of the content of these incidents and several attempts to categorize them into major dimensions of adaptive behavior, the psychologists then decided that they represented the eight dimensions of adaptive behavior listed in the text. Another group of five psychologists then sorted each incident into one of these eight dimensions. Incidents that were assigned to the same dimension by three of the five judges were retained as examples of that dimension. Seven hundred sixty-seven unique incidents were retained as examples of adaptive behavior in one of these eight areas.

Using a subset of these dimensions, the authors developed a job adaptability inventory that has been used to assess adaptive behavior in 24 different jobs. They have also developed an experience inventory, which assesses the degree to which job applicants have experience in any of the adaptive behavior dimensions, and a performance evaluation instrument tied to these eight dimensions. Finally, they have developed training programs to give employees some experience in coping with the demands of the situations that require the various forms of adaptive behavior.

Cross-Cultural Issues With Staff Appraisal

As organizations continue to globalize and operate in different countries and cultures, they must adapt their existing staffing approaches to work effectively in other cultures. For example, the notion of individual performance assessment so common in the United States is often difficult to transport into cultures more collectivistic and group centered, such as in Asian countries (e.g., Fulkerson & Schuler, 1992). There can be considerable resistance to adopting the staff appraisal process if it conflicts with the culture of the raters. Indeed, there has been quite a debate in Europe over whether a common or generic set of HR practices can be applied consistently within the European Community (e.g., Brewster & Larsen, 2000), and research suggests considerable differences in practices across cultures beyond those in Europe (Ryan et al., 1999).

Beyond such macro-level considerations that reflect the acceptance of staff appraisal practices in different cultures, another issue is whether appraisal practices that are applied in different cultures have the same meaning. Stated differently, when measuring performance in different cultures, a key issue is whether the meaning, interpretation, and measurement of performance is equivalent across cultures. To date, only a handful of studies have addressed this issue with judgmental ratings (there is no such data to our knowledge with objective measures). Farh, Earley, and Lin (1997) examined organizational citizenship behavior and found it had both universal and culture-specific aspects. Lam, Hui, and Law (1999) similarly found that ratings of organizational citizenship behavior demonstrated similar internal consistency reliability and relations to other variables across Australia, Hong Kong, Japan, and the United States. A direct test of the cross-cultural equivalence of judgmental ratings was performed by Ployhart, Wiechmann, Schmitt, Sacco, and Rogg (2003). They examined whether supervisory ratings of customer service, technical proficiency, and teamwork were equivalent at the manifest and latent levels across Canada, South Korea, and Spain. Although their results showed supervisors could use the rating scales and thus evaluate the performance of employees in an equivalent fashion, the latent performance constructs were affected by culture to a degree.

Considerably more research will be needed to examine both the ability and motivation of raters to provide accurate ratings across cultures. Given the strong effects of context found primarily within the United

States (Murphy & Cleveland, 1995), it is likely that strong contextual effects will influence criterion measurement across cultures. For example, the research on rating process models described earlier identified such factors as implicit theories of performance as influencing ratings; one would expect such implicit theories to also be affected by cultural values.

Staff Appraisal and Aggregate Effectiveness

We have talked about the assessment of performance in terms of the behavior, results, and effectiveness of individuals. Even when there is a team context, staff appraisal research has been primarily individual in orientation. However, a question remains how individual performance contributes to unit, and ultimately, organizational performance and effectiveness. For example, is department performance the simple sum of individual performance? This seems highly unlikely (DeNisi, 2000), and indeed, is typically not supported. Harris (1994) discussed an issue known as the "performance paradox"; how does performance from lower levels contribute to performance at higher levels? This is a new variation of the old criterion problem.

DeNisi (2000) noted a variety of ways through which performance may be related at different levels of analysis. Moreover, he argued that organizations typically care more about unit performance than the performance of an individual; therefore, research must start to address the processes through which individual-level performance contributes to organizational-level effectiveness. Toward that end, DeNisi (2000) provided a number of directions for establishing such relationships. Schneider, Smith, and Sipe (2000) similarly provided arguments for understanding how performance aggregates into unit-level effectiveness. Ployhart and Schneider (2002) built on this earlier work to lay out a multilevel framework illustrating why individual performance is unlikely to aggregate directly to organizational effectiveness. However, they also provide propositions for indirect effects through which individual behavior contributes to organizational effectiveness.

A different way to think about unit-level criteria is to consider relationships among criteria assessed entirely at the unit level of analysis (e.g., department, organization). Often referred to as *linkage research*, this approach seeks to identify relationships between aggregate perceptions, such as employee attitudes, service climate perceptions, and aggregate customer satisfaction perceptions. This research has found such aggregate employee perceptions may predict aggregate customer satisfaction (e.g., Ryan, Schmit, & Johnson, 1996; Schmit & Allscheid, 1995; Schneider & Bowen, 1985; Schneider, Parkington, & Buxton, 1980) and organizational profitability indices (e.g., Schneider et al., 1980; Schneider, White & Paul, 1998). However, this research has not examined aggregate ratings of staff appraisal, except insofar as they have used aggregate ratings by customers of the service quality they receive (e.g., Schneider et al., 1998). An interesting question would be whether aggregate ratings of performance by supervisors would likewise contribute to higher levels of customer satisfaction and organizational profitability, or whether aggregate KSAOs relate to aggregate performance.

Employee Development

Staff appraisal serves many purposes, and one of the most important is the development of existing employees. We note in several places how work is continually changing, and employees must therefore continually improve and learn new skills to meet these changing demands. Using the staff appraisal process as one form of employee development can help accomplish these goals (London & Mone, 1999).

Whether the appraisal is done primarily for monitoring and evaluation purposes or developmental purposes, employees will want to know how they are doing and what information is being collected and retained about their performance. It seems performance appraisal often means the obligation to feed information back to the person appraised. This latter purpose requires a different kind of philosophy. For staff development, analyses must be conducted with the heart and mind of a counselor rather than treated as an accounting problem. Staff analysis for developmental purposes must involve the person whose performance is being analyzed. Although such a requirement seems obvious, research suggests companies often believe that they have involved their staff in an appraisal, whereas the staff may not even be aware of the fact that they have been appraised. Hall and Lawler (1969) referred to this phenomenon as the "vanishing performance appraisal."

The performance appraisal that most involves the appraised person focuses on the development of the individual and on goal setting for specific needs. A joint appraiser–appraisee conference may be held aside from a more traditional evaluation session. Both conferences may

CURRENT CHALLENGES FOR PERFORMANCE 219

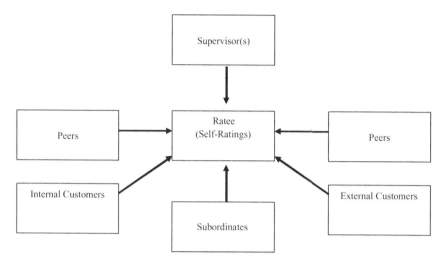

FIG. 4.11. Sources of ratings in multisource feedback systems.

use the same data as input to discussion, but in one session the appraiser plays the role of evaluator, in the other the role of developer.

Since the mid-1990s, practitioners have increasingly adopted multisource rating and feedback mechanisms as a way to promote employee development (Dunnette, 1993; Tornow, 1993). Often termed *360-degree* or *multirater feedback performance appraisals,* these systems obtain performance feedback from several sources—supervisors, peers, the employee, subordinates, customers—who each have a unique perspective on the person's job behavior. Figure 4.11 shows an example of such sources in the multisource appraisal context. These sources each evaluate the person, and then a coach or development personnel work with the person to clarify strengths and weaknesses, set goals and timelines for meeting these goals, and develop a way of assessing progress for the goals. For this process to work, it is critical that all involved realize the system is for developmental purposes only; no administrative decisions (e.g., promotion, termination) can be based on such information or the entire process is likely to fail (Dalessio, 1998; Murphy & Cleveland, 1995). With that said, it is also important for participants to be accountable for their ratings, evaluations, and behavior, even though they may be anonymous (Dalessio, 1998; London & Mone, 1999). Another important consideration is to use enough raters to obtain sufficient interrater reliability; current practice is inadequate (Greguras & Robie, 1998).

Obviously, a key component of any staff appraisal or development effort is the provision of performance feedback. To ensure the feedback, particularly negative feedback, is useful and accepted, it must be focused on specific behaviors. Being critical of the person, rather than the behavior, evokes a different and much more negative reaction, with less likelihood of acceptance (London, 1995). A meta-analysis by Kluger and DeNisi (1996) found that although feedback interventions generally enhanced performance by about 0.41 standard deviation units, these effects were quite variable and one third of the time performance actually got worse after feedback. They reviewed a considerable amount of research on this topic and proposed a general model of feedback intervention effects. It remains the definitive review on the topic.

Satisfaction, Well-Being, and Performance

From the literature reviewed in this chapter, it is probably obvious that job performance tends to be the key target of staffing research and practice. However, if organizations are structured and operate in such a way as to make employees dissatisfied with their work and lives, or contribute to stress and burnout, what good is effective performance? With reference to organizational philosophy, the emphasis on individual performance alone is "an unworkable model for social organization":

> Extreme emphasis on performance as a criterion [of status] may foster an atmosphere of raw striving that results in brutal treatment of the less able, or less vigorous, or less aggressive; it may wantonly injure those whose temperament or whose values make them unwilling to engage in performance rivalries; it may penalize those whose undeniable excellences do not add up to the kinds of performance that society at any given moment chooses to reward. (Gardner, 1961, p. 18)

Interestingly, worker satisfaction has served as an important criterion for evaluating the effectiveness of vocational counseling, organizational change, human relations training, and task redesign, but it is difficult to find satisfaction used as a criterion of staffing procedures. Indeed, worker attitudes of any kind have rarely been used as criteria for evaluating the effectiveness of staffing procedures (interestingly, as we see in chapters 5 and 7, a more recent change has occurred in that the attitudes of applicants, but not of incumbent employees, are being considered). The selection of workers who will have more positive job attitudes (be more satisfied, be more committed, experience less alienation) has not been a goal of staffing researchers (see Schneider, Hall, & Nygren, 1972).

Yet, research on job satisfaction, stress, and well-being is entirely consistent with thinking that staffing and staff appraisal will have implications for such employee attitudes. For example, early work by Pulakos and Schmitt (1983) and Schmitt and Pulakos (1985) demonstrated job satisfaction was predictable over a period of up to 2 years. Judge, Heller, and Mount (2002) performed a meta-analysis linking the FFM of personality to job satisfaction and found all five personality traits were related to satisfaction, with the primary correlates being emotional stability and extraversion. Such findings are consistent with a variety of research studies finding job satisfaction has a strong dispositional, perhaps even genetic, component (Arvey, Bouchard, Segal, & Abraham, 1989; Brief, 1998; Staw & Ross, 1985).

Thus, it seems satisfaction may be a relatively stable and general aspect of certain individuals that is a function of particular personality characteristics and/or an inclination toward interpreting various situations in a favorable manner (e.g., Judge & Larsen, 2001). This has a number of implications for staffing. If staffing practices focus solely on predicting job performance and personality is related to such performance, then organizations may also unintentionally be hiring individuals of relatively stable (dis)satisfaction dispositions. Fortunately, the pattern of correlations among satisfaction, performance, and personality tend to be positive, meaning organizations hiring better-performing employees will also be hiring employees who are more dispositionally satisfied. However, if for some reason the relationship between a given trait and performance was positive, but the trait was negatively related to satisfaction, the organization may inadvertently be hiring a dissatisfied employee.

Of course, one might directly ask the question, "Are task performance and satisfaction related?" There has been a debate between researchers both supporting and refuting the relationship, but in general the historical record has not been very supportive. However, research has added a few new twists to this old question. First, linkage research has found meaningful relationships between aggregate employee satisfaction perceptions and aggregate criteria such as customer satisfaction and performance (e.g., Ostroff, 1992; Ryan et al., 1996; Schmit & Allscheid, 1995). For example, Schmitt, Colligan, and Fitzgerald (1980) found little evidence of a job attitude–job stress relationship at the individual level, but they found meaningful differences among organizations. Attitude–stress relationships were much stronger when organizations were treated as subjects. More recently, Harter, Schmidt, and Hayes (2002) found aggregate job satisfaction and engagement related to, among other things, unit-level profit, turnover, and accident rates. Schneider, Hanges, Smith, and Salvaggio (2003) found such aggregated employee attitudes to be related to organizational-level market and financial performance. However, what is missing is a correlation of aggregate individual job performance and aggregate individual job satisfaction. Second, job satisfaction may also be related to a variety of other individual-level criteria such as absenteeism, commitment, turnover, and withdrawal (Hulin, 1991; Lee & Mitchell, 1994; Mobley, Horner, & Hollingsworth, 1978; Porter & Steers, 1973); physical health (Burke, 1970); mental health (Kornhauser, 1965); and life satisfaction (Iris & Barrett, 1972). Third, a more recent meta-analysis on the job satisfaction–performance relationship found a stronger effect ($r = 0.18; r_{corrected} = 0.30$) using newer statistical corrections (Judge, Thoresen, Bono, & Patton, 2001). They also delineated seven different ways job satisfaction and performance may theoretically be related. Finally, broadening our conceptualization of performance to include contextual performance does produce stronger relationships with satisfaction, particularly in the aggregate (e.g., Organ & Ryan, 1995). A comprehensive review of the job satisfaction literature can be found in Brief (1998).

Although the relationship between satisfaction and performance is being clarified, the relationship between performance and stress remains unclear. As summarized by Jex (1998) and Sonnentag and Frese (2003), laboratory studies tend to show fairly consistent and strong debilitating effects of stress on performance, but results from field studies are more inconclusive. For example, Beehr, Jex, Stacy, and Murray (2000) found both positive and negative effects between stress and various measures of job performance. Likewise, although stress may relate to absenteeism (Farrell & Stamm, 1988) and turnover (Griffeth, Hom, & Gaertner, 2000), the results may be small and are often inconclusive (Sonnentag & Frese, 2003). There are several reasons for these inconclusive findings, but the presence of multiple moderators appears to be the most likely explanation.

There are then a number of good reasons for organizations to evaluate employees on how satisfied they are in addition to evaluating their behavior and the outcomes from their behavior. Evaluating work

CURRENT CHALLENGES FOR PERFORMANCE 223

stress may also be important, but the data thus far do not suggest a consistent pattern of relationships with performance criteria. We thus do not argue for their inclusion as criteria until more data are collected. Note we are not arguing job satisfaction must be considered as a criterion in staffing research and practice to the exclusion of job performance. Rather, we are suggesting that job satisfaction is an inherently worthwhile criterion to consider as a supplement to task and contextual performance.

The Criterion–Predictor Relationship

Earlier in this chapter we used Fig. 4.2 to illustrate the relationship between the actual and ultimate criterion and to define criterion deficiency, contamination, and relevance. Now that we have seen some of the strengths and potential limitations of various criteria, let us return to Fig. 4.2, but with an added complexity so as to point out the importance of criterion measurement as it relates to employee selection. In Fig. 4.12, we illustrate the actual–ultimate criterion relationship and add a predictor of job performance (as we said at the beginning of the chapter, a major purpose of criterion measurement is to assess the

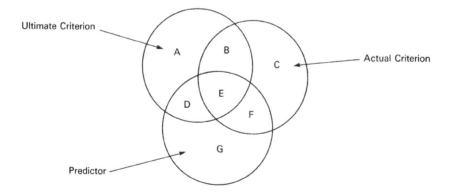

A: Criterion *deficiency* which is uncorrelated with the predictor.
B: Criterion *relevance* which is uncorrelated with the predictor.
C: Criterion *contamination* which is uncorrelated with the predictor.
D: Criterion *deficiency* which is correlated with the predictor.
E: Criterion *relevance* which is correlated with the predictor.
F: Criterion *contamination* which is correlated with the predictor.
G: Portion of predictor which is uncorrelated with either actual or ultimate criterion.

FIG. 4.12. Relationship among actual and ultimate criterion and a predictor variable. *Source:* Adapted from Blum and Naylor (1998).

appropriateness of staffing decisions). Theoretically, at least, this produces seven different areas as identified at the bottom of Fig. 4.12. We have no knowledge of the ultimate criterion, so the relative sizes of A, B, D, and E are unknown. In an actual study performed to evaluate the relationship between a test or some other selection instrument, areas E and F represent the validity of the test as estimated in the study (see chapter 6 for a definition and discussion of validity issues), whereas areas D and E represent the "true" validity of the test as measured against the ultimate criterion. Because we do not know the relative sizes of areas D and F, we do not know whether our actual study validity over- or underestimates the "true" validity of our predictor.

Perhaps the area of most concern in the diagram is area F, which is that part of the actual criterion we can predict that is unrelated to the ultimate criterion. If this area is large, then our selection instrument is actually predicting contamination or bias. Our validity study will yield high-validity estimates, but certainly not for the right reason. An example of this kind of bias may occur in the following fashion. Suppose we use supervisory ratings as a criterion, and those ratings are largely contaminated by how much the rater likes the employee (areas C and F). If our predictor is a measure of agreeableness, then agreeableness may predict ratings (area F) even though it is unrelated to actual performance.

If area D, in contrast, is very large, then a potentially useful predictor will be discarded or its utility seriously underestimated. The size of area E, then, is what one wants to maximize. Unfortunately, in practice, one has no way of knowing whether the actual estimate of a test's validity is an over- or underestimate because the relative size of areas D and F are unknown. However, chapter 2 and this chapter are devoted to minimizing the problems of criterion deficiency and contamination and maximizing relevance and, hence, area E in Fig. 4.12.

SUMMARY

We begin this chapter discussing the role of staff appraisal in organizations and how organizational goals are translated into measures of individual excellence called criteria. We note the social, political, and global context within which staff appraisal takes place. We say that a desirable criterion should be relevant, reliable, show discriminability, and be practical. Attempts to define the latent criterion domain are examined. We discuss alternative ways of developing criteria and provide some examples. Problems in the contamination of criteria are noted. We then describe underlying issues present with the measurement and use of criteria and propose a latent dimensionality that would allow one to compare and contrast various criteria. We end this chapter with discussions of many current issues in staff appraisal.

The importance of criterion issues to staffing is obviously one that is fundamental to the design of effective staffing programs. We conclude by reiterating the insight provided by Wallace (1965): Predictors derive their importance from criteria.

REFERENCES

Ackerman, P. L. (1987). Individual differences in skill learning: An integration of psychometric and information processing perspectives. *Psychological Bulletin, 102,* 3–27.
Ackerman, P. L. (1989). Within-task intercorrelations of skilled performance: Implications for predicting individual differences? (A comment on Henry & Hulin, 1987). *Journal of Applied Psychology, 74,* 360–364.
Arvey, R. D., Bouchard, T. J., Segal, N. L., & Abraham, L. M. (1989). Job satisfaction: Environmental and genetic components. *Journal of Applied Psychology, 74,* 187–192.
Austin, J. T., & Villanova, P. (1992). The criterion problem: 1917–1992. *Journal of Applied Psychology, 77,* 836–874.
Barrett, G. V., & Alexander, R. A. (1989). Rejoinder to Austin, Humphreys, and Hulin: A critical reanalysis of Barrett, Caldwell, and Alexander. *Personnel Psychology, 42,* 597–612.
Barrett, G. V., Caldwell, M. S., & Alexander, R. A. (1985). The concept of dynamic criteria: A critical reanalysis. *Personnel Psychology, 38,* 41–56.
Bartol, K. M., Durham, C. C., & Poon, J. M. L. (2001). Influence of performance evaluation rating segmentation on motivation and fairness perceptions. *Journal of Applied Psychology, 86,* 1106–1119.
Beehr, T. A., Jex, S. M., Stacy, B. A., & Murray, M. A. (2000). Work stressors and coworker support as predictors of individual strain and job performance. *Journal of Organizational Behavior, 21,* 391–405.
Bennett, R. J., & Robinson, S. L. (2000). Development of a measure of workplace deviance. *Journal of Applied Psychology, 85,* 349–360.
Bernardin, H. J., Alvarez, K. M., & Cranny, C. J. (1976). A recomparison of behavioral expectation scales to summated scales. *Journal of Applied Psychology, 61,* 554–560.
Bernardin, H. J., & Beatty, R. W. (1984). *Performance appraisal: Assessing human behavior at work.* Belmont, CA: Wadsworth.
Blanz, F., & Ghiselli, E. E. (1972). The mixed standard scale: A new rating system. *Personnel Psychology, 25,* 185–200.
Blau, G. (1994). Developing and testing a taxonomy of lateness behavior. *Journal of Applied Psychology, 79,* 959–970.
Bommer, W. H., Johnson, J. L., Rich, G. A., Podsakoff, P. M., & MacKenzie, S. B. (1995). On the interchangeability of objective and subjective measures of employee performance: A meta-analysis. *Personnel Psychology, 48,* 587–605.

Borman, W. C. (1978). Exploring the upper limits of reliability and validity in job performance ratings. *Journal of Applied Psychology, 63*, 135–144.
Borman, W. C. (1991). Job behavior, performance, and effectiveness. In M. D. Dunnette & L. M. Hough (Eds.), *Handbook of industrial and organizational psychology* (2nd ed., Vol. 2, pp. 271–326). Palo Alto, CA: Consulting Psychologists Press.
Borman, W. C., Buck, D. E., Hanson, M. A., Motowidlo, S. J., Stark, S., & Drasgow, F. (2001). An examination of the comparative reliability, validity, and accuracy of performance ratings made using computerized adaptive rating scales. *Journal of Applied Psychology, 86*, 965–973.
Borman, W. C., & Dunnette, M. D. (1975). Behavior-based versus trial-oriented performance ratings: An empirical study. *Journal of Applied Psychology, 60*, 561–565.
Borman, W. C., Eaton, N. K., Bryan, J., & Rosse, R. L. (1983). Validity of Army recruiter behavioral assessment: Does the assessor make a difference? *Journal of Applied Psychology, 68*, 415–419.
Borman, W. C., & Motowidlo, S. J. (1993). Expanding the criterion domain to include elements of contextual performance. In N. Schmitt & W. C. Borman (Eds.), *Personnel selection in organizations* (pp. 71–98). San Francisco: Jossey-Bass.
Borman, W. C., & Motowidlo, S. J. (1997). Task performance and contextual performance: The meaning for personnel selection research. *Human Performance, 10*, 99–109.
Borman, W. C., Penner, L. A., Allen, T. D., & Motowidlo, S. J. (2001). Personality predictors of citizenship performance. *International Journal of Selection and Assessment, 9*, 52–69.
Borman, W. C., White, L. A., & Dorsey, D. W. (1995). Effects of ratee task performance and interpersonal factors on supervisory and peer performance ratings. *Journal of Applied Psychology, 80*, 168–177.
Borman, W. C., White, L. A., Pulakos, E. D., & Oppler, S. H. (1991). Models of supervisory job performance ratings. *Journal of Applied Psychology, 76*, 863–872.
Boudreau, J. W., Boswell, W. R., & Judge, T. A. (2001). Effects of personality on executive career success in the United States and Europe. *Journal of Vocational Behavior, 58*, 53–81.
Bowen, D. E., & Schneider, B. (1988). Services marketing and management: Implications for organizational behavior. *Research in Organizational Behavior, 10*, 43–80.
Bowen, D. E., & Waldman, D. A. (1999). Customer-driven employee performance. In D. R. Ilgen & E. D. Pulakos (Eds.), *The changing nature of performance* (pp. 154–191). San Francisco: Jossey-Bass.
Bretz, R. D., Jr., Milkovich, G. T., & Read, W. (1992). The current state of performance appraisal research and practice: Concerns, directions, and implications. *Journal of Management, 18*, 321–352.
Brewster, C. (1995). Towards a "European" model of human resource management. *Journal of International Business Studies, 26*, 1–21.
Brewster, C., & Larsen, H. H. (2000). *Human resource management in northern Europe: Trends, dilemmas, and strategy.* Oxford: Blackwell.
Brewster, C., Larsen, H. H., & Mayrhofer, W. (2000). Human resource management: A strategic approach? In C. Brewster & H. H. Larsen (Eds.), *Human resource management in northern Europe: Trends, dilemmas, and strategy* (pp. 39–65). Oxford: Blackwell.
Brief, A. P. (1998). *Attitudes in and around organizations.* Thousand Oaks, CA: Sage.
Burke, R. J. (1970). Occupational and life strains, satisfaction, and mental health. *Journal of Business Administration, 1*, 35–41.
Campbell, J. P. (1990). Modeling the performance prediction problem in industrial and organizational psychology. In M. D. Dunnette & L. M. Hough (Eds.), *Handbook of*

REFERENCES

industrial and organizational psychology (2nd ed., Vol. 1, pp. 687–732). Palo Alto, CA: Consulting Psychologists Press.
Campbell, J. P. (1999). The definition and measurement of performance in the new age. In D. R. Ilgen & E. D. Pulakos (Eds.), *The changing nature of performance* (pp. 399–429). San Francisco: Jossey-Bass.
Campbell, J. P., Gasser, M. B., & Oswald, F. L. (1996). The substantive nature of job performance variability. In K. R. Murphy (Ed.), *Individual differences and behavior in organizations* (pp. 258–299). San Francisco: Jossey-Bass.
Campbell, J. P., McCloy, R. A., Oppler, S. H., & Sager, C. E. (1993). A theory of performance. In N. Schmitt & W. C. Borman (Eds.), *Personnel selection in organizations* (pp. 35–70). San Francisco: Jossey-Bass.
Cascio, W. F. (2003). Changes in workers, work, and organizations. In W. C. Borman, D. R. Ilgen, & R. J. Klimoski (Eds.), *Handbook of psychology (Vol. 12*, pp. 401–422). Hoboken, NJ: Wiley.
Chan, D. (2000). Conceptual and empirical gaps in research on individual adaptation at work. In C. L. Cooper & I. T. Robertson (Eds.), *International review of industrial and organizational psychology* (Vol. 15, pp. 143–164). New York: Wiley.
Cheung, G. W. (1999). Multifaceted conceptions of self-other ratings disagreement. *Personnel Psychology, 52*, 1–36.
Conway, J. M., & Huffcutt, A. I. (1997). Psychometric properties of multisource performance ratings: A meta-analysis of subordinate, supervisor, peer, and self-ratings. *Human Performance, 10*, 331–360.
Conway, J. M., Lombardo, K., & Sanders, K. C. (2001). A meta-analysis of incremental validity and nomological networks for subordinate and peer rating. *Human Performance, 14*, 267–303.
Cronbach, L. J. (1949). *Essentials of psychological testing.* New York: Harper.
Dalessio, A. T. (1998). Using multi-source feedback for employee development and personnel decisions. In J. W. Smither (Ed.), *Performance appraisal: State of the art in practice* (pp. 278–330). San Francisco: Jossey-Bass.
Deadrick, D. L., Bennett, N., & Russell, C. J. (1997). Using hierarchical linear modeling to examine dynamic performance criteria over time. *Journal of Management, 23*, 745–757.
DeNisi, A. S. (2000). Performance appraisal and performance management: A multilevel analysis. In K. J. Klein & S. W. J. Kozlowski (Eds.), *Multilevel theory, research, and methods in organizations: Foundations, extensions, and new directions* (pp. 121–156). San Francisco: Jossey-Bass.
DeNisi, A. S., Cafferty, T. P., & Meglino, B. M. (1984). A cognitive view of the performance appraisal process: A model and research propositions. *Organizational Behavior and Human Decision Processes, 33*, 360–396.
DeShon, R. P. (1998). A cautionary note on measurement error corrections in structural equation modeling. *Psychological Methods, 3*, 412–423.
DuBois, C. L. Z., Sackett, P. R., Zedeck, S., & Fogli, L. (1993). Further exploration of typical and maximum performance criteria: Definitional issues, prediction, and Black–White differences. *Journal of Applied Psychology, 78*, 205–211.
Dunnette, M. D. (1963). A note on the criterion. *Journal of Applied Psychology, 47*, 251–254.
Dunnette, M. D. (1993). My hammer or your hammer. *Human Resource Management, 32*, 373–384.
Eden, D. (1990). *Pygmalion in management: Productivity as a self-fulfilling prophecy.* Lexington, MA: Lexington Books.
Farh, J. L., Earley, P. C., & Lin, S. C. (1997). Impetus for action: A cultural analysis of justice and organizational citizenship behavior in Chinese society. *Administrative Science Quarterly, 42*, 421–444.

Farrell, D., & Stamm, C. L. (1988). Meta-analysis of the correlates of employee absence. *Human Relations, 41*, 211–227.

Feldman, J. M. (1981). Beyond attribution theory: Cognitive processes in performance appraisal. *Journal of Applied Psychology, 66*, 127–148.

Fleishman, E. A., & Zaccaro, S. J. (1992). Toward a taxonomy of team performance functions. In R. W. Swezey & E. Salas (Eds.), *Teams: Their training and performance* (pp. 31–56). Norwood, NJ: Ablex.

Ford, J. K., Kraiger, K., & Schechtman, S. L. (1986). Study of race effects in objective indices and subjective evaluations of performance: A meta-analysis of performance criteria. *Psychological Bulletin, 99*, 330–337.

Friedman, B. A., & Cornelius, E. T., III. (1976). Effect of rater participation in scale construction on two psychometric characteristics of two rating scale formats. *Journal of Applied Psychology, 61*, 210–216.

Fulkerson, J. R., & Schuler, R. S. (1992). Managing worldwide diversity at Pepsi-Cola International. In S. S. Jackson (Ed.), *Diversity in the workplace: Human resource initiatives* (pp. 248–276). New York: Guilford Press.

Gardner, J. N. (1961). *Excellence: Can we be equal and excellent too?* New York: Harper.

Ghiselli, E. E. (1956). Dimensional problems of criteria. *Journal of Applied Psychology, 40*, 1–4.

Ghiselli, E. E., & Haire, M. (1960). The validation of selection tests in light of the dynamic character of criteria. *Personnel Psychology, 13*, 225–232.

Gottfredson, L. S. (1997). Why g matters: The complexity of everyday life. *Intelligence, 24*, 79–132.

Green, B. F., & Wigdor, A. K. (1991). Measuring job competency. In A. K. Wigdor & B. F. Green (Eds.), *Performance assessment for the workplace* (Vol. 2, pp. 53–74). Washington, DC: National Research Council.

Greguras, G. J., & Robie, C. (1998). A new look at within-source interrater reliability of 360-degree feedback ratings. *Journal of Applied Psychology, 83*, 960–968.

Griffeth, R. W., Hom, P. W., & Gaertner, S. (2000). A meta-analysis of antecedents and correlates of employee turnover: Update, moderator tests, and research implications for the next millennium. *Journal of Management, 26*, 436–488.

Gruys, M. (1999). *The dimensionality of deviant employee performance in the workplace.* Unpublished doctoral dissertation, University of Minnesota: Minneapolis.

Guilford, J. P. (1954). *Psychometric methods.* New York: McGraw-Hill.

Guion, R. M. (1961). Criterion measurement and personnel judgments. *Personnel Psychology, 14*, 141–149.

Guion, R. M., & Gibson, W. M. (1988). Personnel selection and placement. *Annual Review of Psychology, 39*, 349–374.

Hall, D. T., & Lawler, E. E., III. (1969). Unused potential in research and development organizations. *Research Management, 12*, 339–354.

Hanges, P. J., Schneider, B., & Niles, K. (1990). Stability of performance: An interactionist perspective. *Journal of Applied Psychology, 75*, 658–667.

Harris, D. H. (1994). *Organizational linkages: Understanding the productivity paradox.* Washington, DC: National Academy Press.

Harrison, D. A. (2002). Structure and timing in limited range dependent variables: Regression models for predicting if and when. In F. Drasgow & N. Schmitt (Eds.), *Measuring and analyzing behavior in organizations: Advances in measurement and data analysis* (pp. 446–497). San Francisco: Jossey-Bass.

Harter, J. K., Schmidt, F. L., & Hayes, T. L. (2002). Business-unit-level relationship between employee satisfaction, employee engagement, and business outcomes: A meta-analysis. *Journal of Applied Psychology, 87*, 268–279.

REFERENCES

Hattrup, K., Rock, J., & Scalia, C. (1997). The effects of varying conceptualizations of job performance on adverse impact, minority hiring, and predicted performance. *Journal of Applied Psychology, 82*, 656–664.
Hauenstein, N. M. A. (1998). Training raters to increase the accuracy of appraisals and the usefulness of feedback. In J. W. Smither (Ed.), *Performance appraisal: State of the art in practice* (pp. 404–442). San Francisco: Jossey-Bass.
Heneman, R. C. (1986). The relationship between supervisory ratings and results-oriented measures of performance: A meta-analysis. *Personnel Psychology, 39*, 811–826.
Henry, R. A., & Hulin, C. L. (1987). Stability of skilled performance across time: Some generalizations and limitations on utilities. *Journal of Applied Psychology, 72*, 457–462.
Henry, R. A., & Hulin, C. L. (1989). Changing validities: Ability–performance relations and utilities. *Journal of Applied Psychology, 54*, 365–367.
Hoffman, C. C., Nathan, B. R., & Holden, L. M. (1991). A comparison of validation criteria: Objective versus subjective performance measures and self versus supervisor ratings. *Personnel Psychology, 44*, 601–619.
Hofmann, D. A., Jacobs, R., & Baratta, J. E. (1993). Dynamic criteria and the measurement of change. *Journal of Applied Psychology, 78*, 194–204.
Hofmann, D. A., Jacobs, R., & Gerras, S. J. (1992). Mapping individual performance across time. *Journal of Applied Psychology, 77*, 185–195.
Hofmann, D. A., Morgeson, F. P., & Gerras, S. J. (2003). Climate as a moderator of the relationship between leader–member exchange and content specific citizenship: Safety climate as an exemplar. *Journal of Applied Psychology, 88*, 170–178.
Hofmann, D. A., & Stetzer, A. (1996). A cross-level investigation of factors influencing unsafe behaviors and accidents. *Personnel Psychology, 49*, 307–339.
Hogan, J., Hogan, R., & Busch, C. M. (1984). How to measure service orientation. *Journal of Applied Psychology, 69*, 167–173.
Hoyt, W. T. (2000). Rater bias in psychological research: When is it a problem and what can we do about it? *Psychological Methods, 5*, 64–86.
Hoyt, W. T., & Kerns, M. D. (1999). Magnitude and moderators of bias in observer ratings: A meta-analysis. *Psychological Methods, 4*, 403–424.
Hulin, C. L. (1991). Adaptation, persistence, and commitment in organizations. In M. D. Dunnette & L. M. Hough (Eds.), *Handbook of industrial and organizational psychology* (Vol. 2, pp. 445–506). Palo Alto, CA: Consulting Psychologists Press.
Hulin, C. L., Henry, R. A., & Noon S. L. (1990). Adding a dimension: Time as a factor in the generalizability of predictive relationships. *Psychological Bulletin, 107*, 328–340.
Hulin, C. L., Roznowski, M., & Hachiya, D. (1985). Alternative opportunities and withdrawal decisions: Empirical and theoretical discrepancies and an integration. *Psychological Bulletin, 97*, 233–250.
Hunter, J. E. (1983). A causal analysis of cognitive ability, job knowledge, job performance, and supervisory ratings. In F. J. Landy, S. Zedeck, & J. Cleveland (Eds.), *Performance measurement and theory* (pp. 257–266). Hillsdale, NJ: Erlbaum.
Huse, E. F., & Taylor, E. K. (1962). Reliability of absence measures. *Journal of Applied Psychology, 46*, 159–160.
Ilgen, D. R., & Hollenbeck, J. R. (1991). The structure of work: Job design and roles. In M. D. Dunnette & L. M. Hough (Eds.), *Handbook of industrial and organizational psychology* (Vol. 2, pp. 165–207). Palo Alto, CA: Consulting Psychologists Press.
Ilgen, D. R., & Pulakos, E. D. (1999). Introduction: Employee performance in today's organizations. In D. R. Ilgen & E. D. Pulakos (Eds.), *The changing nature of performance* (pp. 1–18). San Francisco: Jossey-Bass.

Iris, B., & Barrett, G. V. (1972). Some relations between job and life satisfaction and job importance. *Journal of Applied Psychology, 56*, 301–304.
Jex, S. M. (1998). *Stress and job performance: Theory, research, and implications for managerial practice.* Thousand Oaks, CA: Sage.
Judge, T. A., & Ferris, G. R. (1993). Social context of performance evaluation decisions. *Academy of Management Journal, 36*, 80–105.
Judge, T. A., Heller, D., & Mount, M. K. (2002). Five-factor model of personality and job satisfaction: A meta-analysis. *Journal of Applied Psychology, 87*, 530–541.
Judge, T. A., & Larsen, R. J. (2001). Dispositional affect and job satisfaction: A review and theoretical extension. *Organizational Behavior & Human Decision Processes, 86*, 67–98.
Judge, T. A., Thoresen, C. J., Bono, J. E., & Patton, G. K. (2001). The job satisfaction–job performance relationship: A qualitative and quantitative review. *Psychological Bulletin, 127*, 376–407.
Judge, T. A., Thoresen, C. J., Pucik, V., & Welbourne, T. M. (1999). Managerial coping with organizational change: A dispositional perspective. *Journal of Applied Psychology, 84*, 107–122.
Katz, D., & Kahn, R. L. (1978). *The social psychology of organizations* (2nd ed.). New York: Wiley.
Keaveny, T. J., & McGann, A. F. (1975). A comparison of behavioral expectation scales and graphic rating scales. *Journal of Applied Psychology, 60*, 695–703.
Kelloway, E. K., Loughlin, C., Barling, J., & Nault, A. (2002). Self-reported counterproductive behaviors and organizational citizenship behaviors: Separate but related constructs. *International Journal of Selection and Assessment, 10*, 143–151.
Kirk, A. K., & Brown, D. F. (2003). Latent constructs of proximal and distal motivation predicting performance under maximum test conditions. *Journal of Applied Psychology, 88*, 40–49.
Kluger, A. N., & DeNisi, A. (1996). The effects of feedback interventions on performance: A historical review, a meta-analysis, and a preliminary feedback intervention theory. *Psychological Bulletin, 119*, 254–284.
Kornhauser, A. W. (1965). *Mental health of the industrial worker: A Detroit study.* New York: Wiley.
Kozlowski, S. W. J., Chao, G. T., & Morrison, R. F. (1998). Games raters play: Politics, strategies, and impression management in performance appraisal. In J. W. Smither (Ed.), *Performance appraisal: State of the art in practice* (pp. 163–205). San Francisco: Jossey-Bass.
Kozlowski, S. W. J., Gully, S. M., Brown, K. G., Salas, E., Smith, E. M., & Nason, E. R. (2001). Effects of training goals and goal orientation traits on multidimensional training outcomes and performance adaptability. *Organizational Behavior and Human Decision Processes, 85*, 1–31.
Kraiger, K., & Ford, J. K. (1985). A meta-analysis of ratee race effects in performance ratings. *Journal of Applied Psychology, 69*, 56–65.
Lam, S. S. K., Hui, C., & Law, K. S. (1999). Organizational citizenship behavior: Comparing perspectives of supervisors and subordinates across four international samples. *Journal of Applied Psychology, 84*, 594–601.
Landy, F. J., & Farr, J. L. (1980). A process model of performance rating. *Psychological Bulletin, 87*, 72–107.
Landy, F. J., & Farr, J. L. (1983). *The measurement of work performance: Methods, theory, and applications.* New York: Academic Press.
Larsen, H. H., & Brewster, C. (2000a). Human resource management in northern Europe. In C. Brewster & H. H. Larsen (Eds.), *Human resource management in northern Europe: Trends, dilemmas, and strategy* (pp. 1–23). Oxford: Blackwell.

REFERENCES

Larsen, H. H., & Brewster, C. (2000b). On the road again. In C. Brewster & H. H. Larsen (Eds.), *Human resource management in northern Europe: Trends, dilemmas, and strategy* (pp. 219–227). Oxford: Blackwell.

Latham, G. P., & Wexley, K. N. (1994). *Increasing productivity through performance appraisal* (2nd ed.). Reading, MA: Addison-Wesley.

Lawshe, C. H., Kephart, N. C., & McCormick, E. J. (1949). The paired comparison technique for rating performance of industrial employees. *Journal of Applied Psychology, 33,* 69–77.

Lee, T. L., & Mitchell, T. R. (1994). An alternative approach: The unfolding model of voluntary employee turnover. *Academy of Management Review, 19,* 51–89.

LePine, J. A., Colquitt, J. A., & Erez, A. (2000). Adaptability to changing task contexts: Effects of general cognitive ability, conscientiousness, and openness to experience. *Personnel Psychology, 53,* 563–593.

Lim, B. C., & Ployhart, R. E. (2004). Transformational leadership: Relations to the five factor model, typical and maximum performance, and leader and team effectiveness. *Journal of Applied Psychology, 89,* 610–621.

London, M. (1995). Giving feedback: Source-centered antecedents and consequences of constructive and destructive feedback. *Human Resource Management Review, 3,* 159–188.

London, M., & Mone, E. M. (1999). Continuous learning. In D. R. Ilgen & E. D. Pulakos (Eds.), *The changing nature of performance* (pp. 119–153). San Francisco: Jossey-Bass.

Malos, S. B. (1998). Current legal issues in performance appraisal. In J. W. Smither (Ed.), *Performance appraisal: State of the art in practice* (pp. 49–94). San Francisco: Jossey-Bass.

Martin, J. (2002). *Organizational culture: Charting the territory.* Thousand Oaks, CA: Sage.

Matell, M. S., & Jacoby, J. (1972). Is there an optimal number of alternatives for Likert-scale items? Effects of testing time and scale properties. *Journal of Applied Psychology, 56,* 506–509.

McArdle, J. J., & Epstein, D. (1987). Latent growth curves within developmental structural equation models. *Child Development, 58,* 110–133.

McCloy, R. A., Campbell, J. P., & Cudeck, R. (1994). A confirmatory test of a model of performance determinants. *Journal of Applied Psychology, 79,* 493–505.

McIntyre, R. M., & Salas, E. (1995). Measuring and managing for team performance: Lessons from complex environments. In R. A. Guzzo, E. Salas, & Associates (Eds.), *Team effectiveness and decision making in organizations* (pp. 9–45). San Francisco: Jossey-Bass.

Mobley, W. H., Horner, S. O., & Hollingsworth, A. T. (1978). An evaluation of precursors of hospital employee turnover. *Journal of Applied Psychology, 63,* 408–414.

Motowidlo, S. J. (2003). Job performance. In W. C. Borman, D. R. Ilgen, & R. J. Klimoski (Eds.), *Handbook of psychology* (Vol. 12, pp. 39–53). Hoboken, NJ: Wiley.

Motowidlo, S. J., & Schmit, M. J. (1999). Performance assessment in unique jobs. In D. R. Ilgen & E. D. Pulakos (Eds.), *The changing nature of performance* (pp. 56–86). San Francisco: Jossey-Bass.

Mount, M. K., & Barrick, M. R. (1995). The big five personality dimensions: Implications for research and practice in human resources management. In G. Ferris & J. Martocchio (Eds.), *Research in personnel and human resource management* (Vol. 13, pp. 153–200). Greenwich, CT: JAI Press.

Mount, M. K., Judge, T. A., Scullen, S. E., Sytsma, M. R., & Hezlett, S. A. (1998). Trait, rater and level effects in 360-degree performance ratings. *Personnel Psychology, 51,* 557–576.

Mowday, R. T., Porter, L. W., & Steers, R. M. (1982). *Employee–organizational linkages: The psychology of commitment, absenteeism and turnover.* New York: Academic Press.

Muchinsky, P. M. (1983). *Psychology applied to work*. Homewood, IL: Irwin.
Murphy, K. R. (1989). Is the relationship between cognitive ability and job performance stable over time? *Human Performance, 2*, 183–200.
Murphy, K. R., & Cleveland, J. N. (1995). *Understanding performance appraisal: Social, organizational, and goal-based perspectives*. Thousand Oaks, CA: Sage.
Murphy, K. R., Cleveland, J. N., Skattebo, A. L., & Kinney, T. B. (2004). Raters who pursue different goals give different ratings. *Journal of Applied Psychology, 89*, 158–164.
Murphy, K. R., & DeShon, R. P. (2000a). Interrater correlations do not estimate the reliability of job performance ratings. *Personnel Psychology, 53*, 873–900.
Murphy, K. R., & DeShon, R. P. (2000b). Progress in psychometrics: Can industrial and organizational psychology catch up? *Personnel Psychology, 53*, 913–922.
Murphy, K. R., Martin, C., & Garcia, M. (1982). Do behavioral observation scales measure observations? *Journal of Applied Psychology, 67*, 562–567.
Murphy, K. R., & Shiarella, A. H. (1997). Implications of the multidimensional nature of job performance for the validity of selection tests: Multivariate frameworks for studying test validity. *Personnel Psychology, 50*, 823–854.
Nagle, B. F. (1953). Criterion development. *Personnel Psychology, 6*, 271–289.
Oppler, S. H., Campbell, J. P., Pulakos, E. D., & Borman, W. C. (1992). Three approaches to the investigation of subgroup bias in performance measurement: Review, results, and conclusions. *Journal of Applied Psychology, 77*, 201–217.
Organ, D. W. (1988). *Organizational citizenship behavior: The good soldier syndrome*. Lexington, MA: Lexington Books.
Organ, D. W. (1997). Organizational citizenship behavior: It's construct clean-up time. *Human Performance, 10*, 85–97.
Organ, D. W., & Ryan, K. (1995). A meta-analytic review of attitudinal and dispositional predictors of organizational citizenship behavior. *Personnel Psychology, 48*, 775–802.
Ostroff, C. (1992). The relationship between satisfaction, attitudes, and performance: An organizational level analysis. *Journal of Applied Psychology, 78*, 569–582.
Parasuraman, A., Berry, L. L., & Zeithaml, V. A. (1991). Refinement and reassessment of the SERVQUAL scale. *Journal of Retailing, 67*, 420–451.
Parasuraman, A., Zeithaml, V. A., & Berry, L. L. (1985). A conceptual model of service quality and its implications for future research. *Journal of Marketing, 49*, 41–50.
Parasuraman, A., Zeithaml, V. A., & Berry, L. L. (1988). SERVQUAL: A multiple-item scale for measuring consumer perceptions of service quality. *Journal of Retailing, 64*, 12–40.
Ployhart, R. E., & Hakel, M. D. (1998). The substantive nature of performance variability: Predicting interindividual differences in intraindividual performance. *Personnel Psychology, 51*, 859–901.
Ployhart, R. E., Lim, B. C., & Chan, K. Y. (2001). Exploring relations between typical and maximum performance ratings and the five factor model of personality. *Personnel Psychology, 54*, 809–843.
Ployhart, R. E., & Schneider, B. (2002). A multi-level perspective on personnel selection research and practice: Implications for selection system design, assessment, and construct validation. In F. J. Dansereau & F. Yammarino (Eds.), *Research in multi-level issues. Volume 1: The many faces of multi-level issues* (pp. 95–140). Oxford, UK: Elsevier Science.
Ployhart, R. E., Wiechmann, D., Schmitt, N., Sacco, J. M., & Rogg, K. (2003). The cross-cultural equivalence of job performance ratings. *Human Performance, 16*, 49–79.
Porter, L. W., & Steers, R. M. (1973). Organizational, work, and personal factors in employee turnover and absenteeism. *Psychological Bulletin, 80*, 151–176.

REFERENCES

Pulakos, E. D. (1984). A comparison of two rater training programs: Error training versus accuracy training. *Journal of Applied Psychology, 69,* 581–588.

Pulakos, E. D. (1997). Ratings of job performance. In D. L. Whetzel & G. R. Wheaton (Eds.), *Applied measurement methods in industrial psychology* (pp. 291–317). Palo Alto, CA: Davies-Black.

Pulakos, E. D., Arad, S., Donovan, M. A., & Plamondon, K. E. (2000). Adaptability in the workplace: Development of a taxonomy of adaptive performance. *Journal of Applied Psychology, 85,* 612–624.

Pulakos, E. D., & Schmitt, N. (1983). A longitudinal study of a valence model approach for the prediction of job satisfaction of new employees. *Journal of Applied Psychology, 68,* 307–312.

Pulakos, E. D., Schmitt, N., & Chan, D. (1996). Models of job performance ratings: An examination of ratee race, ratee gender, and rater level effects. *Human Performance, 9,* 103–119.

Pulakos, E. D., Schmitt, N., Dorsey, D. W., Arad, S., Hedge, J. W., & Borman, W. C. (2002). Predicting adaptive performance: Further tests of a model of adaptability. *Human Performance, 15,* 299–323.

Pulakos, E. D., White, L. A., Oppler, S. H., & Borman, W. C. (1989). Examination of race and sex effects on performance ratings. *Journal of Applied Psychology, 74,* 770–780.

Randell, G. A., Packard, P. M. A., Shaw, R. L., & Slater, A. M. (1974). *Staff appraisal.* London: Institute of Personnel Management.

Reilly, R. R., & McGourty, J. (1998). Performance appraisal in team settings. In J. W. Smither (Ed.), *Performance appraisal: State of the art in practice* (pp. 244–277). San Francisco: Jossey-Bass.

Rich, G. A., Bommer, W. H., MacKenzie, S. B., Podsakoff, P. M., & Johnson, J. L. (1999). Apples and apples or apples and oranges? A meta-analysis of objective and subjective measures of salesperson performance. *Journal of Personal Selling & Sales Management, 19,* 41–52.

Robinson, S. L., & Bennett, R. J. (1995). A typology of deviant workplace behaviors: A multidimensional scaling study. *Academy of Management Journal, 38,* 555–572.

Rogosa, D. R. (1995). Myths and methods: "Myths about longitudinal research" plus supplemental questions. In J. M. Gottman (Ed.), *The analysis of change* (pp. 3–66). Mahwah, NJ: Erlbaum.

Roth, P. L., Huffcutt, A. I., & Bobko, P. (2003). Ethnic group differences in measures of job performance: A new meta-analysis. *Journal of Applied Psychology, 88,* 694–706.

Rothstein, H. R. (1990). Interrater reliability of job performance ratings: Growth to asymptote level with increasing opportunity to observe. *Journal of Applied Psychology, 75,* 322–327.

Rotundo, M., & Sackett, P. R. (1999). Effect of rater race on conclusions regarding differential prediction in cognitive ability tests. *Journal of Applied Psychology, 84,* 815–822.

Rotundo, M., & Sackett, P. R. (2002). The relative importance of task, citizenship, and counterproductive performance to global ratings of job performance: A policy-capturing approach. *Journal of Applied Psychology, 87,* 66–80.

Ryan, A. M., McFarland, L. A., Baron, H., & Page, R. (1999). An international look at selection practices: Nation and culture as explanations for variability in practice. *Personnel Psychology, 52,* 359–391.

Ryan, A. M., & Ployhart, R. E. (2003). Customer service behavior. In W. C. Borman, D. R. Ilgen, & R. J. Klimoski (Eds.), *Handbook of psychology* (Vol. 12, pp. 377–397). Hoboken, NJ: Wiley.

Ryan, A. M., Schmit, M. J., & Johnson, R. (1996). Attitudes and effectiveness: Examining relations at an organizational level. *Personnel Psychology, 49,* 853–882.

Sackett, P. R. (2002). The structure of counterproductive work behavior: Dimensionality and relationships with facets of job performance. *International Journal of Selection and Assessment, 10,* 5–11.
Sackett, P. R., & DeVore, C. J. (2001). Counterproductive behaviors at work. In N. Anderson, D. S. Ones, H. K. Sinangil, & C. Viswesvaran (Eds.), *Handbook of industrial, work, and organizational psychology: Personnel psychology* (Vol. 1, pp. 145–164). London: Sage.
Sackett, P. R., Zedeck, S., & Fogli, L. (1988). Relations between measures of typical and maximum job performance. *Journal of Applied Psychology, 73,* 482–486.
Schein, E. H. (1980). *Organizational psychology.* Englewood Cliffs, NJ: Prentice Hall.
Schein, E. H. (1992). *Leadership and organizational culture* (2nd ed.). San Francisco: Jossey-Bass.
Schmidt, F. L., & Hunter, J. E. (1996). Measurement error in psychological research: Lessons from 26 research scenarios. *Psychological Methods, 1,* 199–223.
Schmidt, F. L., Hunter, J. E., & Outerbridge, A. N. (1986). Impact of job experience and ability on job knowledge, work sample performance, and supervisory ratings of job performance. *Journal of Applied Psychology, 71,* 432–439.
Schmidt, F. L., & Kaplan, L. B. (1971). Composite vs. multiple criteria: A review and resolution of the controversy. *Personnel Psychology, 24,* 419–434.
Schmidt, F. L., Viswesvaran, C., & Ones, D. S. (2000). Reliability is not validity and validity is not reliability. *Personnel Psychology, 53,* 901–912.
Schmit, M. J., & Allscheid, S. P. (1995). Employee attitudes and customer satisfaction: Making theoretical and empirical connections. *Personnel Psychology, 48,* 521–536.
Schmitt, N., Colligan, M. J., & Fitzgerald, M. (1980). Unexplained physical symptoms in eight organizations: Individual and organizational analyses. *Journal of Occupational Psychology, 53,* 305–317.
Schmitt, N., & Lappin, M. (1980). Race and sex as determinants of mean and variance of performance ratings. *Journal of Applied Psychology, 65,* 428–435.
Schmitt, N., & Pulakos, E. D. (1985). Predicting job satisfaction from life satisfaction: Is there a general satisfaction factor? *International Journal of Psychology, 20,* 155–168.
Schneider, B. (1975). Organizational climate: Individual preferences and organizational realities revisited. *Journal of Applied Psychology, 61,* 459–465.
Schneider, B. (1980). The service organization: Climate is crucial. *Organizational Dynamics, Autumn,* 52–65.
Schneider, B. (1987). The people make the place. *Personnel Psychology, 40,* 437–453.
Schneider, B., & Bowen, D. E. (1985). Employee and customer perceptions of service in banks: Replication and extension. *Journal of Applied Psychology, 70,* 423–433.
Schneider, B., & Bowen, D. E. (1992). Personnel/human resources management in the service sector. In G. R. Ferris & K. M. Rowland (Eds.), *Research in personnel and human resources management* (Vol. 10, pp. 1–30). Greenwich, CT: JAI Press.
Schneider, B., & Bowen, D. E. (1995). *Winning the service game.* Boston: Harvard Business School.
Schneider, B., Hall, D. T., & Nygren, H. T. (1972). Self-image and job characteristics as correlates of changing organizational identification. *Human Relations, 24,* 397–416.
Schneider, B., Hanges, P. J., Smith, D. B., & Salvaggio, A. N. (2003). Which comes first: Employee attitudes or organizational financial and market performance? *Journal of Applied Psychology, 88,* 836–851.
Schneider, B., Parkington, J. J., & Buxton, V. (1980). Employee and customer perceptions of service in banks. *Administrative Science Quarterly, 25,* 252–267.
Schneider, B., Smith, D. B., & Sipe, W. P. (2000). Personnel selection psychology: Multilevel considerations. In K. J. Klein & S. W. J. Kozlowski (Eds.), *Multilevel*

REFERENCES

theory, research, and methods in organizations: Foundations, extensions, and new directions (pp. 91–120). San Francisco: Jossey-Bass.

Schneider, B., & White, S. S. (2004). *Service quality: Research perspectives.* Thousand Oaks, CA: Sage.

Schneider, B., White, S. S., & Paul, M. C. (1998). Linking service climate and customer perceptions of service quality: Test of a causal model. *Journal of Applied Psychology, 83,* 150–163.

Schwab, D. P., Heneman, H. G., III, & Decotiis, T. A. (1975). Behaviorally anchored rating scales: A review of the literature. *Personnel Psychology, 28,* 549–562.

Scullen, S. E., Mount, M. K., & Goff, M. (2000). Understanding the latent structure of job performance ratings. *Journal of Applied Psychology, 85,* 956–970.

Smith, D. E. (1986). Training programs for performance appraisal: A review. *Academy of Management Review, 11,* 22–40.

Smith, P. C. (1976). Behaviors, results, and organizational effectiveness: The problem of criteria. In M. D. Dunnette (Ed.), *Handbook of industrial and organizational psychology* (pp. 745–776). Chicago: Rand McNally.

Smith, P. C., & Kendall, L. M. (1963). Retranslation of expectations: An approach to the construction of unambiguous anchors for rating scales. *Journal of Applied Psychology, 47,* 149–155.

Sonnentag, S., & Frese, M. (2003). Stress in organizations. In W. C. Borman, D. R. Ilgen, & R. J. Klimoski (Eds.), *Handbook of psychology* (Vol. 12, pp. 453–491). Hoboken, NJ: Wiley.

Staw, B. M., & Ross, J. (1985). Stability in the midst of change: A dispositional approach to job attitudes. *Journal of Applied Psychology, 70,* 469–480.

Steers, R. M., & Rhodes, S. R. (1978). Major influences on employee attendance: A process model. *Journal of Applied Psychology, 63,* 391–407.

Stewart, G. L. (1999). Trait bandwidth and stages of job performance: Assessing differential effects for conscientiousness and its subtraits. *Journal of Applied Psychology, 84,* 959–968.

Sturman, M. C., & Trevor, C. O. (2001). The implications of linking the dynamic performance and turnover literatures. *Journal of Applied Psychology, 86,* 684–696.

Tornow, W. W. (1993). Perceptions of reality: Is multi-perspective measurement a means or an end? *Human Resource Management, 32,* 221–229.

Tziner, A., & Eden, D. (1985). Effects of crew composition on crew performance: Does the whole equal the sum of its parts? *Journal of Applied Psychology, 70,* 85–93.

Vance, R. J., MacCallum, R. C., Coovert, M. D., & Hedge, J. W. (1988). Construct validity of multiple job performance measures using confirmatory factor analysis. *Journal of Applied Psychology, 73,* 74–80.

Viswesvaran, C. (2002). Absenteeism and measures of job performance: A meta-analysis. *International Journal of Selection and Assessment, 10,* 12–17.

Viswesvaran, C., Ones, D., & Schmidt, F. L. (1996). Comparative analysis of the reliability of job performance ratings. *Journal of Applied Psychology, 81,* 557–574.

Viswesvaran, C., Schmidt, F. L., & Ones, D. (2005). Is there a general factor in ratings of job performance? A meta-analytic framework for disentangling substantive and error influences. *Journal of Applied Psychology, 90,* 108–131.

Waldman, D. A., & Avolio, B. J. (1991). Race effects in performance evaluations: Controlling for ability, education, and experience. *Journal of Applied Psychology, 76,* 897–901.

Wallace, S. R. (1965). Criteria for what? *American Psychologist, 20,* 411–417.

Wanberg, C. R., & Banas, J. T. (2000). Predictors and outcomes of openness to changes in a reorganizing workplace. *Journal of Applied Psychology, 85,* 132–142.

Weekley, J. A., & Jones, C. (1997). Video-based situational testing. *Personnel Psychology, 50,* 25–49.
Wexley, K. N. (1984). Personnel training. *Annual Review of Psychology, 35,* 519–551.
Wexley, K. N., & Klimoski, R. (1984). Performance appraisal: An update. In K. M. Rowland & G. R. Ferris (Eds.), *Research in personnel and human resources* (Vol. 2, pp. 35–79). Greenwich, CT: JAI Press.
Wherry, R. J., & Bartlett, C. J. (1982). The control of bias in ratings: A theory of rating. *Personnel Psychology, 35,* 521–555.
Willett, J. B., & Sayer, A. G. (1994). Using covariance structure analysis to detect correlates and predictors of individual change over time. *Psychological Bulletin, 116,* 363–381.

5

Recruitment: Retention and Attraction

AIMS OF THE CHAPTER

In chapters 3 and 4, we provide a description of the job that confronts a worker and the knowledge, skills, and abilities required of workers (chapter 3) and then the standards by which we judge how effectively individuals and organizations perform (chapter 4). In this chapter, we turn our attention to considering how organizations keep current employees and attract new employees to work at these jobs.

The chapter begins with a discussion of the role of internal recruitment (filling positions with current employees). In this context, we discuss the role of job satisfaction and organizational commitment in retaining a pool of competent internal recruits. That is, to have sufficient qualified people internal to the organization to fill openings to which they might be promoted, there must be good reasons for people to stay—job satisfaction and organizational commitment are two good reasons. To this end, we also present a brief review of the determinants of satisfaction, commitment, and turnover. Then we describe procedures for internal recruitment, such as job posting and career pathing and planning, to include succession planning.

In discussing external recruitment, a central thesis is that both the organization and the individual are making choices. From the individual's point of view, we discuss career, occupational, and organizational choice. From the organization's point of view, we discuss what individuals want in an organization, how best to present the organization and job to prospective employees, and the availability and effectiveness of various recruitment techniques. We also discuss the interplay between applicants and organizations—in particular, how organizational practices influence applicant perceptions and reactions.

Before we begin, it is important to elaborate three important premises for the chapter:

1. Most openings in companies are filled with internal persons; entry-level jobs are the ones most likely to be filled from the outside. The two major sections in this chapter, then, deal with internal recruitment and external recruitment. For internal recruitment, the central issues are employee retention and the kinds of experiences organizations provide for their employees to prepare them for jobs other than those for which they were initially hired. So, the focus in this section is on understanding the job and organizational conditions that individuals find attractive enough in their current organization to want to be a candidate for available jobs, and creating experiences (training, job assignments) by which incumbents attain the KSAOs required by future job possibilities and alternatives. For external recruitment, the concern is for the ways people and organizations become attractive to and seek out each other and for the reactions applicants for jobs in organizations have to the recruitment experience.

2. Recruitment really is a two-way street, with both the individual and the organization having a series of decisions to make. Many organizations tend to focus on their decisions and forget about applicant decisions. In this chapter, we pay considerable attention to applicants' decisions: what kind of job in what kinds of organization they want to join and the experiences they have as they venture through the recruitment process.

3. Finally, there is the issue of who actually employs the people who work in organizations. Increasingly, the people who work in an organization are employed by a contracting firm who pays them, provides their benefits, and so forth. Contract employees will not be the focus of the chapter although it seems clear to us that, because contracting firms also employ people, the same issues regarding retention and attraction also apply to them. In addition, of course, if the contracting organization and the organization in which employees work are different, there are issues surrounding organizational loyalty and commitment that require consideration (Korman, 1999).

INTERNAL RECRUITMENT

If the organization is not an attractive place to work, internal recruits will not be available for job openings as they occur. More important, if the organization is not attractive to the best current employees, the only incumbents available for new openings are the least desirable, and the organization must go outside the company for candidates. Clearly, the employees of any organization most attractive to other organizations are the best ones, precisely those an organization should attempt to retain. A brief review of the correlates of employee job satisfaction, commitment, and turnover is presented as relevant background information regarding what makes an organization attractive.

Employee Turnover, Job Satisfaction, and Commitment

There are many models and conceptualizations of employee turnover, and they have been reviewed in a number of places (see Griffith, Hom, & Gaertner, 2000, for a comprehensive review). We touched on these issues in chapter 4, but the focus was on using withdrawal and turnover as criteria. Here, we expand on the meaning of turnover and its implications for internal recruitment. In all the models of turnover, five major categories of issues emerge as important elements in turnover decisions (Baysinger & Mobley, 1983):

1. *Attraction of the present job*. The issue of job satisfaction captures most of the research on the attraction of the present job. Because that is the major issue for this chapter, a summary of the job satisfaction literature is presented shortly.

2. *Future attraction of the present job*. Although Baysinger and Mobely (1983) include this as a separate category, our review of the job satisfaction literature suggests this is an issue incumbents include in their thinking when evaluating their present job satisfaction. In fact, one of the most frequently used measures of job satisfaction, the Job Descriptive Index (JDI; Smith, Kendall, & Hulin, 1969), includes an assesssment of promotion opportunities. Schneider, Gunnarson, and Wheeler (1992) even argued that opportunity is the great satisfier; it is not what you have but what you know you have the opportunity to have that leads to present satisfaction. They use the example of people who live in an attractive tourist location (e.g., New York City, Washington, DC, or

San Francisco) who love living there precisely because of the attractions tourists come to see, but they never visit the tourist sites themselves. They have the opportunity to visit them so they do not have to.

3. *Perceived external alternatives.* Many models of turnover include the issue of alternative possibilities as an important component in understanding why, or under what circumstance, incumbents are likely to leave (Maertz & Campion, 1998). In tight labor markets, where jobs are scarce, turnover is lower than when the economy is booming. At the individual level, what this means is that, other things being equal (which they rarely are), the more employees believe there are alternatives that are more attractive than the present job, the more likely they are to leave (Hulin, Roznowski, & Hachiya, 1985). March and Simon (1958), who provided an early conceptual discussion of turnover in organizations, gave this perception the commonsense label of "perceived ease of movement." It remains an assumption in more recent models such as Hulin (1991).

4. *Economic and psychological investments.* Monetary investments include issues such as being "vested" in a retirement plan. Employees who are vested are those who have stayed long enough with a company so, when they retire, they will be able to collect on the company's retirement package. Most companies require that employees work some period of time before they are vested—anywhere from 6 months to 10 years. Vesting refers to all kinds of benefits, not just retirement. Sample benefits include participation in bonus plans, life insurance policies, stock ownership, and so forth. Sometimes, these packages can be quite lucrative and leaving them, therefore, can be costly. When employees choose to remain with a company because of monetary benefits to which they are entitled only if they stay, these benefits are called "golden handcuffs." On the psychological side, investments can take the form of commitment to the organization (Meyer & Allen, 1997). For example, someone who has been with an organization since its beginning can become psychologically committed to the organization. Leaving it would be emotionally painful even when it looks like the "rational" thing to do. The issue of commitment is reviewed in more detail later.

5. *Nonjob factors.* This category includes issues such as family responsibilities and the compatibility of job and perceived nonjob responsibilities. For example, Kraut (1992) presented several case studies of corporate research that concerned work–family issues and the

propensity for employees to quit a job rather than relocate for a promotion as the company wanted them to do. As women, who are still the primary homemakers, assume management and executive positions in organizations, such issues can be expected to increase in frequency for dual-career couples, especially those with children (Brett, Stroh, & Reilly, 1992).

In the aggregate, these nonjob factors plus two additional issues (fit and sacrifice) have been called causes of *job embeddedness* (Mitchell, Holtom, Lee, Sablynski, & Erez, 2001). Job embeddedness is the result of people feeling linked to facets of their nonwork worlds that are important to them—of feeling that their jobs and their communities are a good fit to other aspects of their life—and the sacrifices they would have to make to leave a job. Mitchell et al. (2001) showed job embeddedness contributes to an understanding of turnover that is over and above the contribution of job satisfaction and organizational commitment. Presumably, job embeddedness is tapping into a kind of psychological attachment to a job that lies beyond the attributes of the job or organization and thus takes into account more of the total life space of people. In short, when people think about leaving a job or organization, they consider factors other than the job or organization—they consider their total life space. Although this notion makes good sense and has intuitive appeal, the fact is that most theory and research on understanding turnover have focused on the job and the organization as if they were the only considerations. What is critical to understand here is that organizations only have so much impact on whether people will stay or leave, so this makes it doubly important for organizations to focus greatly on things over which they do have control.

One last important point to make about turnover before we focus on some antecedents of turnover is that turnover from a job and organization is obviously not something that typically happens overnight. Rather, it is an outcome of a decision process, with people beginning to think about such a possibility, perhaps by reading an article in a newspaper about employment rates, and then moving through a behavioral and psychological process involving searching for possibilities that might eventuate in turnover. Steele (2002) did a nice job of describing this process of how people move through the different stages of job search that might eventuate in turnover. His depiction of turnover as a job search process reinforces our own idea that turnover must be intimately connected to recruitment in organizations because both turnover from an organization and attraction to an organization involve job search processes.

The two issues that have received the most attention in the study of turnover are job satisfaction and organizational commitment, listed previously as the attraction of one's present job and investments (monetary and psychological), respectively. These are issues over which organizations have some control, so actions taken to enhance job satisfaction and organizational commitment can have important payoffs in retention.

What Is Job Satisfaction?

Job satisfaction concerns individuals' attitudes about the work and work setting that reflect their feelings about what happens to them and around them. It is most frequently thought of as some comparison people make between what they desired/expected/valued/wanted/hoped for/was important to them (alone or in various combinations) and what they perceive actually is happening to and around them. Measures typically only assess the perceptions people have under the assumption that perceptions include the implicit framing against desires, expectations, and so forth.

There have been literally thousands of studies of job satisfaction, and these studies have been conducted from a variety of theoretical vantage points. These studies have yielded a complex and interesting set of findings:

1. *Job satisfaction is a multifaceted construct.* Because many elements comprise job satisfaction, it is not very useful to speak about "job satisfaction" in a general sense (Scarpello & Campbell, 1983). It is more useful to speak about satisfaction with something. The "somethings" that seem to recur most frequently in the studies of job satisfaction are satisfaction with pay, the opportunities for promotion, supervision (leadership), coworkers (interpersonal relationships), and the work itself (Smith et al., 1969). Other facets of the work setting that have received some attention are fringe benefits (or job security), working conditions, and recognition (Spector, 1997). We need not go into the specifics of the measurement of job satisfaction here (see Spector, 1997, for a review). It is important to know, however, that global measures of job satisfaction, although useful for predicting turnover, do not identify the facets of the job that can be changed to yield improvements in job satisfaction.

For example, if one knows that in a particular organization there is a strong relationship between satisfaction with opportunities for promotion and overall job satisfaction, but a weak relationship between satisfaction with coworkers and overall job satisfaction, then a focus on improving opportunities for promotion is necessary, not a focus on coworkers. Just knowing the overall job satisfaction level would not reveal this fact.

2. *Job satisfaction and productivity are only weakly to moderately related.* Until the mid-1950s, it was assumed by both researchers and managers that a satisfied worker was a productive worker. In 1955, a series of reviews of this assumption began to appear (e.g., Brayfield & Crockett, 1955; Iaffaldano & Muchinsky, 1985; Locke, 1976; Vroom, 1964), and each indicated that the assumption of the satisfied productive worker was overly simplistic.

We think the assumption of a relationship between satisfaction and productivity was derived from observations of researchers and managers that, in organizations that were effective, morale of the workers (the sum of job satisfaction scores from all workers) seemed to be higher than in ineffective organizations. The conclusion that this relationship held for measures of individual job satisfaction and individual job productivity appears to have been somewhat in error. It was somewhat in error because raw productivity is but one facet of overall work performance of individuals, and it is the cumulation of individuals' overall work performance that contributes to overall organizational performance.

We must make a distinction here then, as we did in chapter 4, between raw productivity and overall job performance because the latter does seem to be related to job satisfaction (Judge, Thoresen, Bono, & Patton, 2001). The former is an issue of units produced per hour, and the latter concerns a combination of units produced per hour plus the other behaviors employees must engage in if the organization is to be effective. These other behaviors have come to be called OCBs (Organ, 1988) or contextual performance (Motowidlo, Borman, & Schmit, 1997). These refer to voluntary acts on the part of employees that have to do with being conscientious (coming to work on time, not taking longer breaks than they are supposed to), and being helpful (assisting coworkers and customers, although this may not be written into their job descriptions). OCBs are seen as the kind of social glue that keeps the organization running efficiently because not every action that needs doing can be written down (Podsakoff & MacKenzie, 1997, list many ways that OCBs are useful for organizational performance). In addition to OCBs, job satisfaction also seems to be negatively related to absenteeism, CWBs (stealing, vandalism), and physical health issues (headaches, upset stomach, and stress) (Spector, 1997).

The point is that job satisfaction appears to be reflected in a number of different facets of job performance and behavior at work—even when it may not be strongly related to productivity. It is also worth noting that in the aggregate (e.g., bank branches, business units), job satisfaction relates meaningfully and consistently to several unit level indicators of effectiveness, including turnover, customer satisfaction, profit, and even productivity (Harter, Schmidt, & Hayes, 2002). A caution here is that at the aggregate level of analysis the direction of causality—whether satisfaction causes performance or vice versa—is not always clear (Schneider, Hanges, Smith, & Salvaggio, 2003). Of more relevance to the present case, job satisfaction is strongly related to intentions to leave, increased levels of job search behavior, and finally turnover.

3. *Job satisfaction is a function of personal variables, as well as the result of what organizations do to or for people.* For the most part, job satisfaction has been studied as if its only determinant is what happens to people after they get on the job. This approach makes the assumption that all people tend to be satisfied or dissatisfied by the same things, but this assumption is not necessarily true. Of course, there is a core of organizational conditions that are generally satisfying to a broad cross-section of workers, but at least one of these conditions appears to be the worker's own self-esteem. Locke (1976), in an important earlier summary of the literature on job satisfaction, presented the list of conditions conducive to job satisfaction as follows:

   a. Mentally challenging work with which the individual can cope successfully
   b. Personal interest in the work itself
   c. Work that is not too physically tiring
   d. Rewards for performance that are just, informative, and in line with the person's aspirations
   e. Working conditions that are compatible with the individual's physical needs and that facilitate the accomplishment of work goals
   f. High self-esteem on the part of the employee

INTERNAL RECRUITMENT 245

g. Agents in the workplace who help employees attain job values such as interesting work, pay, and promotions, whose basic values are similar to their own, and who minimize role conflict and ambiguity.

This listing by Locke shows that, for virtually each condition, an interaction of both a personal and a situational condition accounts for the satisfaction the person will derive. So, for mentally challenging work, the person needs to be able to cope, that is, have the skills and abilities required by the job; and the rewards must be in keeping with the person's aspirations. In other words, what an individual brings to a setting in the way of needs, values, interests, abilities, and self-esteem will also determine the degree to which the individual will experience satisfaction there.

On top of interactions between specific facets of the job and personal attributes as important antecedents of job satisfaction, it further appears to be true that some people will experience high and some people will experience low levels of job satisfaction almost regardless of the job at which they work. In other words, there appears to be a disposition to be satisfied or dissatisfied at work (Staw, 2004). This is interesting for many reasons but, from an historical vantage point, it is particularly interesting. So, in his book, *Industrial Psychology*, Viteles (1932) presented job dissatisfaction as a worker maladjustment problem, placing satisfaction or dissatisfaction in the heads of workers. Beginning in the late 1930s and through the mid-1980s, satisfaction and dissatisfaction were taken out of the heads of workers and put into the design of jobs and the management of organizations. In a series of papers published since the early 1990s (see Judge, 1992; Judge, Heller, & Mount, 2002), it now seems safe to argue that (a) people who have negative attitudes toward their life and their life situations early in life will experience higher levels of dissatisfaction later in life, and (b) people's levels of job satisfaction are reliable over long periods of time (decades), even when they change jobs.

There are several lessons to be learned from this foray into the personal correlates of job satisfaction. First, organization and management have some control, but clearly not complete control, over the levels of job satisfaction experienced by employees and thus the likely turnover of employees. Second, an organization that wants to retain employees who will experience higher levels of job satisfaction would do well to hire people who are dispositionally oriented to higher levels of job satisfaction. We discuss this latter point in more detail in chapter 9.

4. *Assessments of job satisfaction yield useful information for organizations, especially when accomplished on a continuing basis.* When broadly conceptualized as quality of work life (QWL), the assessment of job satisfaction has become as important for organizations as other kinds of resource or expense audits. A useful way to think about assessing job satisfaction is to picture it as an audit of the feelings of people about the various facets of organizational life. Many organizations have ongoing programs of assessment, using the data as diagnostic information and as input into corporate and managerial decision making (Kraut, 1996). Indeed, some companies hold supervisors accountable for the data generated by their work units. In this kind of assessment model, supervisors have the same kind of responsibility for the QWL of employees as they do for unit performance of the more traditional kind.

The measurement of job satisfaction is both a science and an art. The science concerns developing reliable and valid procedures for questioning organizational members; the art deals with tapping into the issues to be assessed in ways that make organizational members believe they are contributing to their own benefit, as well as to the organization's interests. One just does not put together a few questions and send them out to employees as a way of assessing the general feelings of employees about the work setting. Just as with the development of tests for assessing skills and abilities for making personnel selection decisions, the development of measures of job satisfaction and QWL is best left to professionals (cf. Kraut, 1996).

As one example of the degree to which such employee surveys are used and found to be valid, consider the work done by the Gallup Organization over the past decade or more. Gallup is probably better known for its public opinion polling, but in reality its greatest source of income is its employee survey program. Over the years, Gallup has isolated what they call a measure of "employee engagement," 12 survey items that in the aggregate represent employee well-being at work (see Harter, Schmidt, & Hayes, 2002, for a description of the items and their validity). What is interesting is not just that these 12 items correlate with other measures of job satisfaction but that employee responses to these items also correlate with important group-level outcomes such as group productivity, profit, accidents, customer satisfaction, and turnover. What is impressive is that Harter et al. documented these relationships based on administration of the measure of employee engagement to a total of almost 200,000 employees in 36 companies representing almost 8,000 distinct business units. The bottom line is that employee surveys are big business, and employee survey data are reflected in important business unit outcomes such as turnover. Companies that want to retain their best and brightest talent need to pay attention to these kinds of findings.

Summary

We try to summarize an unbelievably voluminous amount of literature with a few summary statements. In a real sense, it may be presumptuous of us to detail so much of the relevant findings in a few bullet points; yet, the complexity of the issue seems to come through when it is presented this way. The message is that the work experiences of employees with respect to how satisfied they are in their jobs is multidetermined and multifaceted and has multiple consequences at multiple levels of analysis. Therefore, when attempting to influence employee job satisfaction, remember that no one facet of the situation can be expected to affect employee feelings of job satisfaction. Entire systems and procedures regarding the multifacets of organizational life, such as the 12 facets of employee engagement in the Gallup measure, must be used to obtain significant effects on some global idea such as satisfaction. In addition, it must be understood that employee job satisfaction is not totally under the control of organizations. On the one hand, and as we noted, there are dispositional facets of job satisfaction (Staw, 2004). On the other hand, there are extrawork factors that also influence job satisfaction. So, just as job embeddedness has an impact on turnover in organizations, relationships outside the organization can influence the satisfaction employees experience within the organization and vice versa (Brief, 1998).

In this section, we argue that job satisfaction is important because it is one determinant of continued accessibility to competent internal recruits (i.e., job satisfaction is a consistent and significant predictor of turnover). However, the issue is a bit more complex than simply looking at job satisfaction because there are other reasons why people stay or leave an organization, such as organizational commitment.

What Is Organizational Commitment?

Organizational scientists, it must be observed, speak a convoluted language. By this, we mean that a literal interpretation of the title of this section is that organizations are committed; of the last section, it is that jobs are satisfied. The real purpose of this section, however, is to understand the behavioral commitment of people to their employing organization. Commitment is an important concept because it says something about what people are likely to do when the chips are down, when there is not an obvious reward for doing something, or when no one really asks them to behave a particular way. Katz and Kahn (1978) spoke of this kind of behavior as innovative and spontaneous behavior; performance beyond role requirements for accomplishment of organizational functions:

1. Cooperative activities with fellow members
2. Actions protective of system or subsystem
3. Creative suggestions for organizational improvement
4. Self-training for additional organizational responsibility
5. Creation of favorable climate for organization in the external environment

Katz and Kahn (1978) are quite explicit about the need for these kinds of behaviors in organizations, behaviors not usually addressed by either theories of work motivation or theories of job satisfaction:

> The organizational need for actions of an innovative, relatively spontaneous sort is inevitable and unending. No organizational plan can foresee all contingencies within its own operations, or can control perfectly all human variability. The resources of people for innovation, for spontaneous cooperation, for protective and creative behavior are thus vital to organizational survival and effectiveness. An organization that depends solely on its blueprint of prescribed behavior is a very fragile social system. (pp. 403–404)

The kinds of behaviors described by Katz and Kahn as exemplifying the organizationally committed person have many important components. One component is obviously of a motivational origin because the behaviors are directed at particular kinds of activities, and the desired outcome seems to be more valuable to the organization than to the individual. So, Katz and Kahn were really speaking about behaviors that are indicative of commitment to the organization. Research by Organ (1988, 1997) and by Borman and Motowidlo (1997) has developed the construct of OCB (Organ, 1988) and contextual performance (Borman & Motowidlo, 1997) to capture this inclination on the part of people to engage in behaviors that are supportive of others in the work environment and the organization itself. What is interesting about this research is that it focuses on less tangible features of job performance, than do typical indicators of task performance, and research shows that job satisfaction is a consistent predictor of such helpful behaviors (Podsakoff, MacKenzie, Paine, & Bachrach, 2000).

Another way to think about commitment is from the perspective of individuals and their feelings. This kind of commitment has been called *attitudinal commitment* in contrast to the behavioral commitment just described (Meyer & Allen, 1997). Attitudinal commitment most frequently concerns some variant of the theme that the goals of the individual and the organization are congruent (Morrow, 1983). That is, the individual who views the organization as sharing goals and outcomes will be the one who will behave as Katz and Kahn describe (Bateman & Organ, 1983).

When the behaviors specified by Katz and Kahn are viewed as the outcome of a congruence between individual and organizational goals, commitment becomes a two-way process. The outcome of the two-way process is that individuals feel committed to providing organizations with behaviors that will support the long-term viability of the organization, and organizations feel committed to providing individuals with work environments that satisfy their needs both monetarily and psychologically. Rousseau and her colleagues (e.g., Rousseau, 1995) termed this reciprocity the *psychological contract*. It is a psychological contract because it refers to implicit understandings or assumptions each party has about what the other party will do, with organizations making assumptions about the extra effort employees will provide and employees making assumptions about what the organization will in turn provide them.

It is useful to recognize that there is more than one base for organizational commitment (i.e., it is not only when people and organizations share values that people reveal commitment to the organization). Meyer and Allen (1997) called it *affective commitment* when people have good feelings about their organization, such as when they perceive their own goals are shared with the goals of their employing organization. *Continuance commitment* concerns the costs of leaving, such that the higher the perceived costs associated with leaving an organization, the stronger the continuance commitment. Continuance commitment shares some features with the job embeddedness construct (Mitchell et al., 2001) introduced earlier in that costs associated with leaving are

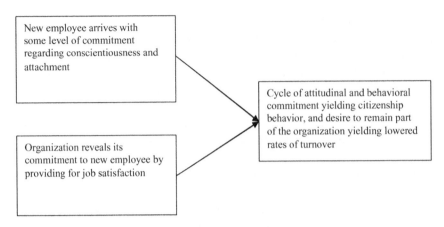

FIG. 5.1. Processes through which commitment emerges.

not restricted to monetary costs and not restricted to job or organizational characteristics. Finally, *normative commitment* refers to peoples' feelings of obligation to remain in the organization. These feelings can concern morality ("it is not right to quit anything"), obligation ("this organization was really good to me when the chips were down"), and so forth. Again, as was true in the earlier discussion on job embeddedness, this presentation of the multiple facets of commitment shows us that organizations only have so much control over things, so when they do they should exercise it.

Figure 5.1 presents an integrated model of the kind of process by which commitment emerges for people in organizations. First, the figure shows that the commitment of people to organizations begins to occur when they make a decision to join an organization. By the act of deciding, they have eliminated other alternatives in favor of the position available in the employing organization. This act of choice indicates a personal investment in the organization, leading to a desire to see the organization survive. However, it also leads to a high value being placed on the organization itself. Our hypothesis then is that organizational employees begin their tenure in an organization with relatively high levels of commitment because of the process we just described. Yet, we believe this high commitment is very tenuous because it has not been tested through actual participation in the organization and because it has not been reinforced by the organization or the individual through many cycles of behavior and attitudes. This suggests that, as far as commitment is concerned, early experiences with the

INTERNAL RECRUITMENT 251

organization are absolutely critical—critical because of the very tenuousness of the commitment new employees bring with them to the organization. In our discussion of external recruitment, this issue of the importance of early experiences will receive more attention.

We hypothesize that these feelings and behaviors on the part of new employees will only be maintained if the organization, in turn, displays its commitment to the employee, literally organizational commitment. Organizations display their commitment to employees in many ways, and these ways are neatly summarized by the facets of job satisfaction already discussed earlier in this chapter. In other words, for the typical employee to be committed in both attitudes and behavior, the employee must have behavioral evidence that the organization is committed, too. Without this behavioral evidence, the attitude–behavior cycle will not begin or be maintained, and the early levels of commitment with which employees arrive will dissipate.

Here again, the concept of the psychological contract (Rousseau, 1995) is a useful one, this time for understanding the reciprocal interaction of the individual and the organization and the role of mutual commitments in facilitating the integration of the individual and the organization. The psychological contract is the set of mutual, implicit agreements between employees and employing organizations that govern the expectations each has of the other. So, for example, the employee expects that for coming to work and working, the organization will provide pay, good supervision, worthwhile work, and opportunities for bettering oneself. In turn, the organization expects the worker to come to work on time, to work hard during the day, to exercise initiative for the organization, and to be loyal to it.

It must be emphasized, prior to leaving this topic, that supporting and maintaining a mutually beneficial psychological contract requires continual attention by both parties. This is true because the expectations of each party will constantly evolve as they work together in achieving organizational goals. It is important to view the psychological contract as being in a kind of constant renegotiation because today's committed employee may be tomorrow's alienated employee due to a change in expectations. Similarly, employees to whom the organization is committed may lose this status because of a failure to identify any new ground rules for behavior that may be associated with changes in the organization, such as those due to new supervisors, new technology, mergers, and so forth. A final consideration here, and one we raised earlier, concerns the new employment contracts that are emerging in the workplace, employment contracts where people work for a contracting organization while being employed within another firm. To whom will these employees likely be committed? Korman (1999) argued forcefully for the notion that workers will become increasingly committed to themselves so they can protect their identity and avoid the potential conflicts and ambiguities associated with splitting their loyalties between the firm in which they work and the firm that employs them under contract. This is an interesting idea, and it fits well with Hall's (2002) notion that in the new world of work people will increasingly have to take control of their own careers, a topic to which we return later in the chapter.

Summary

What is important about this summary of the literature on job satisfaction and organizational commitment is that people who are more satisfied and committed are those who will not only exert extra effort in supporting organizational effectiveness, but who will also be available for openings that emerge in the organization. Of course, our discussion has rested on the assumption that it is better to recruit internally than externally, a topic to which we turn next.

The Benefits of Internal Recruitment

We have been writing about the importance of job satisfaction and commitment as correlates of turnover under the assumption that it is better to retain than recruit (from outside). One of the reasons behind this assumption is that current employees do not have to go through the kinds of socialization experiences new employees require before they can feel adjusted to the (new) job. So, as noted in the section on external recruitment, an important element in recruiting new employees is concern for the entire new employee entry process.

This does not mean that internal recruitment allows organizations to ignore socialization issues. It only means that the problems of adjustment are likely to be less severe for a move involving a current employee than for one involving a new employee. Some of the "problems" we are addressing here are as follows:

1. *Start-up time.* Current employees, especially in smaller organizations, know a lot about the organization and how it functions, so they do not have to go through this learning process. They know about

INTERNAL RECRUITMENT 253

how and when they are paid, what their fringe benefits are, what the goals and objectives of the company are, who holds the power and who wields it, who to avoid and who to try and become friends with, and so forth. These kinds of informal, get along issues will not require so much time; therefore, fewer surprises will exist than would be true for a newcomer (Louis, 1990). Fewer surprises that require attention permit getting up to speed faster.

2. *Probability of success.* An organization that has a satisfactory staff appraisal process bases the choice of an internal candidate on much more information than we have in choosing someone from the outside. There is no substitute for living with a person to know them, and working with someone requires living with them. If the best predictor of future performance is past performance, then observations of persons today provide a lot of the information required for predicting their performance tomorrow. However, this all depends on the appropriateness of the performance appraisal information available at the time internal candidates are recruited, especially with respect to the similarity of the present job to the future job. This means that we can take an excellent salesperson and make her a terrible manager because the past performance, used as the predictor, does not fit the job to which she is moved.

3. *Cost.* It is certainly less costly to the organization to recruit internally than to go outside for people. This would be especially true in management and professional ranks where the costs of recruitment can be very high indeed (see section on executive recruitment). The costs not only include the actual costs of recruitment, but also costs associated with selection, training, start-up time (see Boudreau, 1991), and the psychological advantage organizations can gain from having a promotion from within policy. That is, if one of the issues that is important to employees as they make turnover decisions is the future they may have in the company, then a way for a company to reduce turnover would be to promote from within.

It is useful to note here that internal recruitment is not entirely a bed of roses. If everything is going well on the retention of the best people for the jobs that open, then all is well. However, as noted earlier, people can stay in an organization for many reasons, and some of those are not as healthy for the company as others. For example, consider the case of continuance commitment (Meyer & Allen, 1997) described earlier. Here, people stay because of the costs of leaving, not because of their positive feelings. These costs can be as tangible as financial costs or as intangible as the fact that they derive all their social and emotional support from being part of the organization. We can predict that such people are unlikely to engage in the citizenship behaviors that support organizational health (Meyer & Allen, 1997). Indeed, such people can become a burden on the organization, especially in higher ranks, because their entire being is wrapped up in the organization, and they can become unwilling to take risks that might jeopardize their job. Price (1977), for example, has shown that organizations with lower turnover rates tend to be less innovative. When people are unwilling to take risks that might jeopardize their job, a company that requires people to take risks can be in a great deal of trouble. The point here, and the caution, is that as in all things behavioral and psychological, too much of a seemingly good thing (e.g., organizational commitment) needs to be carefully monitored for potential unintended negative consequences.

Internal recruitment also puts significant demands on the organization's performance appraisal process. Employees must perceive the fairness of internal recruitment; hence, the organization's methods of promotion must be accepted by the employees. People who believe they are unfairly passed over for promotion can create significant problems in the organization and/or leave. This is an interesting point because, as Hulin (1991; Hulin et al., 1985) showed, explicit turnover from an organization is only one way people reveal their dissatisfactions with an organization, especially when the unemployment rate is high and jobs are scarce.

Finally, an organization that promotes from within must be concerned about the explicit policies and practices it has in place to enhance the competencies and loyalties of employees. It cannot generate expectations for promotion that are impossible to meet, and it must plan realistically with employees what opportunities will be available and how the employee will be best prepared to take advantage of those opportunities.

Models and Methods of Internal Recruitment

As strange as it seems, there is not much research literature available on internal recruitment. Some books on recruitment do not mention internal recruitment (Barber, 1998), and some books on personnel selection barely mention recruitment, much less internal recruitment and the issues surrounding turnover, satisfaction, and commitment (e.g., Gatewood & Field, 1994). Others have extensive treatment of the administrative details involved in internal recruitment but not the issues surrounding turnover and so forth (Heneman & Judge, 2004). It is clear from what we have presented that if employees are not satisfied and committed, we are in trouble because incumbent employees provide the potential supply of candidates for job openings, especially higher-level jobs requiring complex sets of skills. It is once we understand that it is important to retain employees that we can become concerned with the mechanics of internal recruitment. That is, administrative procedures for internal recruitment will only be as good as the candidates available for recruitment.

Models of Internal Recruitment

We can segment models of internal recruitment into those that focus on the total career experiences of employees and those that focus on organizational needs for specific competencies at a given point in time.

First, in the *total career experiences* model, the emphasis is on planning the experiences employees require for them to assume increasingly responsible positions in the firm. These experiences begin at entry to the organization (or even earlier in external recruitment) and include identification of socialization activities, assignment to a buddy or mentor, formal training, and exposure to a variety of upward and lateral job assignments that together build competencies and prepare individuals for successively demanding positions in a company. Career planning at one time was interpreted as upward mobility in organizations, but research has shown that lateral moves and occasionally even downward moves can provide employees with the breadth of experiences they may require at later stages in the organization (Hall, 2002). These kinds of carefully developed total career systems are facilitated by developments in information systems, with most companies now having some form of human resource information system (HRIS) that cumulates available information and data on each employee in the firm. The challenge in managing such a total career system is having a series of possible goals for an employee in mind and then articulating the actual experiences required to achieve the goals.

An interesting question, of course, is who should identify the goals and who should identify the experiences. Hall (2002) argued persuasively that from the standpoint of performance outcomes (how effective a person is for the organization), the organization has a vested interest

TABLE 5.1

Types of Relationships of Importance for Career Development: Functions, Characteristics, and Impact

Type	Main Function	Characteristics	Impact
Mentor/protégé	Develops protégé's self-knowledge and awareness	Cross-generational information sharing	Enhanced self-image
Coach/sponsor	Increases protégé's capacity for responsibility; sponsor facilitates learning	Mentor increasingly poses questions and builds capacities	Enhanced capacity for responsibility
Support group	Emotional and instrumental support	Maintaining and enhancing networks	Connection to others; identity
Supervisor/coworker	Task completion and skill development	Intense contact	Growth in capacities
Project team/task force	Cross-boundary collaboration with growth in group skills	Frequent contact	Growth in task capacities
Training program	Increases skills and capacities for thought and action	Knowledge transmission	Learning of ideas and skills
Role model	Provides models for behavior	Observational learning	Self-comparison

Source: Adapted from Hall and Kahn (2001).

in identifying the goals to be achieved. However, from the standpoint of the individual, it is the individual who should articulate the goals desired. The ideal situation is one in which there is mutual sharing and a chance for both parties to explicate their desires—thus changing the psychological contract into an explicit one.

Hall and Kahn (2001) developed a model of the various influences on the ways organizations and people conceptualize their careers in organizations. The framework shows that there are numerous informal and formal relationships that can yield the total systems careers of interest to us here. The model is shown in Table 5.1, which indicates that there is a variety of relationships that serve different functions for people and thus have different kinds of impacts on career development and progress. The more of these kinds of relationships that are promoted in today's more turbulent and unpredictable organizations, the more employees will gather the competencies they require for future growth and development, the more organizations are likely to have in existing employees the competencies they require, and the more likely it is that employees will repay organizations for the investments the organization makes in them.

The second model, *meeting immediate organizational needs*, is the one we usually think of when discussing internal recruitment: An existing position becomes available and/or a new position is created, and we seek someone who has the KSAOs required by the position. In this instance, the HRIS can prove useful, but in addition there are formal processes and procedures that can be put in place to ensure organizations make the most optimal match possible given the KSAOs of existing employees. Several methods for making this match happen exist.

Methods of Internal Recruitment

Heneman and Judge (2004) do an excellent job of presenting the contingencies on the methods used by organizations for identifying candidates for a position. By focusing on contingencies, they and we acknowledge there is no one best way to conduct internal recruitment. However, it depends on whether all openings are posted for all employees or openings are known only to those who have the opening and candidates designated by HR as having the required KSAOs. Obviously, in the open system anyone can apply for a position if they believe they meet the requirements associated with it. In the "closed" system, HR identifies possible candidates, and then the manager who has the opening makes a final decision. Of course, it is possible to combine these approaches into what Heneman and Judge call a targeted system, wherein both the opened and the closed approaches are used to maximize the possibility of obtaining the ideal candidate for the opening. In an era when many employees in most companies have access to e-mail, the available openings and the requirements those openings demand can be made known to the broadest cross-section of employees.

Various formal mechanisms for announcing job openings may be good for employee morale because they are fair, provide opportunity, and may do a good job of matching skills to openings. However, there are also some potential drawbacks to formal posting (Schuler, 1984):

1. *Time*: Posting can increase the time from having an opening to filling it.
2. *Conflict*: A person who believed she would get the job does not because a more qualified person from another unit, for example, posts for the opening.
3. *Stress*: The supervisor who must make the decision can be confronted with two or three equally attractive candidates.
4. *Turmoil*: Subordinates in a unit who constantly bid on any job opening that arises may be viewed by their supervisor in a negative way. This can result in turmoil between superior and subordinate.
5. *Awareness of problems*: Units in which there are frequent job openings that get posted can come to be seen as units that have problems.

Regardless of the method chosen, there are still certain key requirements for ensuring the best possible candidate is eventually chosen:

1. The requirements for the opening must be clearly specified in terms of KSAOs required. In this sense, as in others, internal recruitment is no different from external recruitment.
2. The procedures for making application for openings must be clearly specified so both applicants and their current supervisors know each party's responsibilities. Indeed, it may be required that supervisors be trained to see that development of employees is an important part of their job so when an excellent employee wants to move to another job it will not be something the employee resists.
3. If the system for making application is not user friendly, it is unlikely to be used. Such systems must be designed with input from those people likely to make use of them, or all other good planning will be for naught.
4. Nomination should be encouraged. Some employees may not believe they have the requirements for an opening, but their supervisor does believe they are ready. The supervisor should

INTERNAL RECRUITMENT 259

be encouraged to nominate the employee. One organization in which we have worked requires that a candidate obtain a nomination from the current supervisor as a first step in the process, but this will not always yield a candidate who has the KSAOs but who does not realize it.

5. In larger organizations where significant position openings periodically occur, it is wise to have a pool of ready possible candidates rather than to be scrambling and unprepared. This also applies to firms that make acquisitions on a regular basis—it is wise to have a staff of excellent people ready to move to the acquired firm to introduce those employees to the ways of doing things in the new company.

6. Choose from the pool of available internal candidates wisely, using the same procedures for making selection decisions discussed thus far and discussed in the remainder of this book.

7. Where labor unions are involved, agreements between labor and management must be honored (e.g., seniority agreements).

8. Any time a decision is made about one person over another, a selection decision has been made and is subject to all the legal issues we may usually think of as concerning only selection decisions when external candidates are concerned.

Of course, there are still many organizations that fill internal (and external) positions through informal networks. The probabilities that this approach will yield optimal candidates is low because only a narrow range of candidates is available for consideration. In contrast, one thing that we do know about recruitment from the external world, and it is also likely to be true for internal recruitment, is that suggestions of potential people for openings that come from current valued employees typically yield competent people who are not likely to leave (Jackson & Schuler, 2003). Although the informal network may work effectively in small organizations, larger organizations require more sophisticated HRISs. These systems maintain the kinds of data generated by the job analysis and staff appraisal processes. By maintaining the data according to competencies, requests for persons can specify the desired attributes in the new person, and a computer search can turn up the most appropriate candidates. In today's world of sophisticated information processing and data retrieval systems, there is no good reason why an organization's HR capabilities should not be as well documented as are other features (e.g., investments, production capabilities).

The data collected on employees through these kinds of career planning programs, combined with the information obtained through the staff appraisal process, provides organizations with the potential to understand the nature of their internal human resources, including what their talents and desires are, as well as the kinds of external persons they may have to recruit. The nature of the external recruits needed is obtained by comparing projections of the positions to be filled with the kinds of internal persons who will be available.

Projections of the kinds of positions that will have to be filled is the job of the HR planner. HR planning is a problem that we do not present in detail because the mathematics involved are somewhat complex. In the abstract, the problem is one of tracking past experiences regarding the movement of people into, through, and out of the various positions in the organization combined with projections for growth of the organization and changes in the way the organization functions (e.g., due to technological changes). Early models of HR planning were deficient because they failed to consider strategic issues such as the last two noted previously. More recently, HR planners have engaged in strategic HR planning; that is, they have included long-range corporate strategy in their projections, and corporate strategists have included human resources issues in their own planning (Jackson & Schuler, 2003).

Summary

Our goal in the extensive discussion of internal recruitment is to ensure we adequately expressed the idea that to have internal recruits, the organization needs to provide a satisfying work environment and gain employee commitment to its continued growth and development. The logic here is that most nonentry jobs are filled from within, and organizations must be aware of and take control of the factors that promote the availability of candidates with excellent competencies to fill those internal positions. Organizations can do this by providing a high QWL, by promoting a total career systems approach to employee development, and by having in place an accessible system for making openings known to existing employees. A second major emphasis in the section was the repeated message that organizations must retain adequate information on the available pool of internal recruits to make wise internal choices. Such information is most useful if it is based on a carefully developed staff appraisal and career management system. Therefore, the procedures specified in chapter 4 extend not only beyond criteria for staffing, but also beyond internal recruiting.

Unspoken in this section was the following thought: An organization that creates the kind of work environment that elicits citizenship behaviors revealing attitudinal and behavioral commitment and job satisfaction is also likely to be an organization that will have access to the most talented pool of external candidates. This is important because the kinds of selection procedures and programs that constitute the majority of this book will only be as good as the pool of applicants on whom they are used.

EXTERNAL RECRUITMENT

An organization decides to pursue recruits from the outside when it concludes that no qualified people inside are available. That is, when specification of the openings available and the KSAOs required fails to yield a match among job incumbents, then a search begins for an appropriate recruit.

Actually, at the same time an organization decides to go outside, people outside are looking, exploring, and choosing. In fact, potential candidates for jobs are doing the same things organizations are doing—they are looking for a match.

In this section of the chapter, we first explore some of what we know about career development and choice and then examine what we know about how people become attracted to organizations. Thus, we think that people first implicitly or explicitly conclude something about the kind of job (vocation, occupation) they want and then go looking for that kind of job in specific kinds of organizations. Finally, we present information about what we know with regard to recruitment practices.

Career Development and Choice

This brief section reviews some theories and literature that indicate that people do not just walk into an organization through some random selection and ask to join the staff. This is an important idea because it confirms that the people who comprise the working staff of an organization have prior histories, live outside their organization, and have futures that may not include the organization. To understand the behavior of people at work requires that we place those people in their larger context.

The evidence suggests that particular kinds of people choose certain careers or vocations and even specific occupations within those careers. There is less evidence regarding the circumstances and conditions governing organizational choice, but we explore some hypotheses about the conditions by which people choose one organization over another.

Like jobs, people do not spring full-blown and ready to go. Like jobs, particular kinds of people exist in organizations as a function of the larger environment, the conditions that exist in the world around them, and the nature of the goals they may be trying to achieve. Unlike jobs, people can choose whether they want to be a part of an organization. Also, unlike jobs, people go through different stages as they develop their career plans and aspirations. Choosing a career and an organization is part of a continuum whereby persons first work through the stages of career development, make their career decision, search for the appropriate context in which the career should be carried out, and then choose the organization (Holland, 1997). In terms of career development, Holland persuasively demonstrated that organizational environments may be distinguished in terms of six personality types: realistic, investigative, artistic, social, enterprising, and conventional. People tend to choose careers that are a good fit between their own personality, interests, and values. This means, for example, people who work in accounting departments will tend to be different from people who work in marketing departments.

Figure 5.2 illustrates, for those considering a degree in I/O psychology, how individuals narrow their career and organizational choices through various decisions made over time. Of course, this is not an entirely explicit process, but rather a mix of conscious choices and less conscious desires, wants, and preferences. It is also frequently nonlinear because career interests change over time, and certainly careers themselves can be quite variable. For example, in the late 1980s many mid-level managers found themselves out of work as a result of mergers, acquisitions, and downsizing. Rather than go to work for another large organization, many of these individuals wanted more control over their lives and leveraged their expertise by starting small businesses—leading to the dramatic increase in small business we see today.

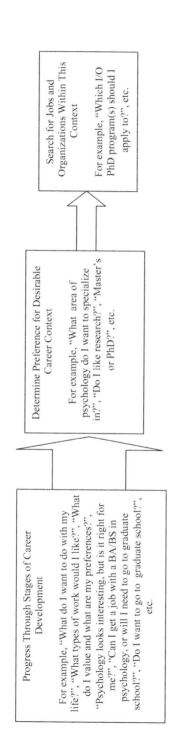
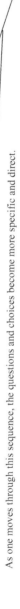

FIG. 5.2. Continuum of career and organizational choice processes. The examples represent career choices for an I/O psychologist.

More than 50 years ago, Super (1953) stated the following nine propositions as a theory of career development:

1. People differ in their abilities, interests, and personalities.
2. They are qualified, by virtue of these characteristics, each for a number of occupations.
3. Each occupation requires a characteristic pattern of abilities, interests, and personality traits, with tolerances wide enough, however, to allow some variety of occupations for each individual and some variety of individuals in each occupation.
4. Vocational preferences and competencies, the situations in which people live and work, and, hence, their self-concepts, change with time and experience (although self-concepts are generally fairly stable from later adolescence until late maturity), making choice and adjustment a continuous process.
5. This process may be summed up in a series of life stages characterized as those of growth, exploration, establishment, maintenance, and decline, and these stages may in turn be subdivided into (a) the fantasy, tentative, and realistic phases of the exploratory stage; and (b) the trial and stable phases of the establishment stage.
6. The nature of the career pattern (i.e., the occupational level attained and the sequence, frequency, and duration of trial and stable jobs) is determined by the individual's parental socioeconomic level, mental ability, and personality characteristics, and by the opportunities to which the individual is exposed.
7. Development through the life stages can be guided, partly by facilitating the process of maturation of abilities and interests and partly by aiding in reality testing and in the development of the self-concept.
8. The process of vocational development is essentially that of developing and implementing a self-concept: It is a compromise process in which the self-concept is a product of the interaction of inherited aptitudes, neural and endocrine makeup, opportunity to play various roles, and evaluations of the extent to which the results of role playing meet with the approval of superiors and fellows.
9. The process of compromise between individual and social factors, between self-concept and reality, is one of role playing,

whether the role is played in fantasy, in the counseling interview, or in real life activities such as school classes, clubs, part-time work, and entry jobs. Work satisfactions and life satisfactions depend on the extent to which individuals find adequate outlets for their abilities, interests, personality traits, and values. These satisfactions depend on establishment in a type of work, a work situation, and a way of life in which individuals can play the kind of role that growth and exploratory experiences have led them to consider congenial and appropriate. (pp. 189–190)

Between the time of this statement and the present, considerable research evidence and more detailed theoretical positions have appeared regarding vocational development and choice (see Feldman, 2002; Hall, 2002; Holland, 1997). However, although Super's (1953) position was quite general, it indicated areas in which more specific information should be collected if we are to understand eventual career decisions. For example, Super noted that the socioeconomic level of parents and an individual's opportunity to obtain occupational information both have an effect on career development. In addition, more personal attributes such as mental ability and interests seem to be related to vocational development and choice.

Figure 5.3 can help summarize the various effects on career development and occupational choice. Figure 5.3 is also a convenient beginning for our discussion of how individuals decide to choose an organization in which to carry out their occupational choice.

Conceptually, socioeconomic status, race, gender, environment, and intelligence/aptitudes serve as limitations to the range of occupations available to individuals. Interests operate to focus preferences. Although not previously discussed, interests also have antecedents, most probably in the psychosocial upbringing of the individual (Roe, 1957).

The successful resolution of conflicts between individuals' perceptions of available alternatives and their preference for alternatives results in a vocational choice. More specific alternatives lead to a clearer specification of choice (i.e., the occupation within the career). This further specification necessarily limits the range of organizations in which the occupation exists. A comparison of specific desires and what potential organizations offer results in an organizational choice.

Note that this conceptualization places the heavy emphasis of a person's eventual career/occupation/organization choice on preexisting

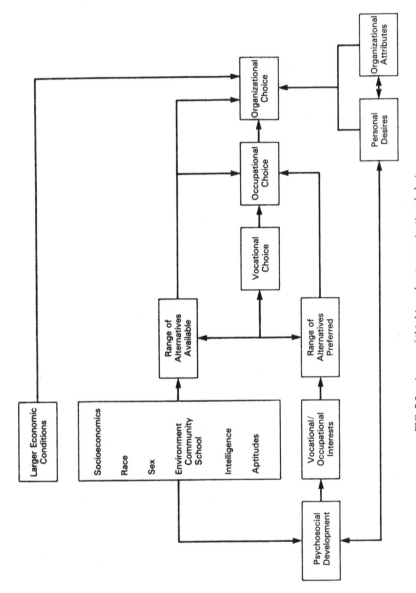

FIG. 5.3. A way of thinking about organizational choice.

conditions in the person's larger environment. Indeed, one may speak realistically of these choices as being, in many cases, the result of a series of externally imposed compromises. That is, the fewer the alternatives available, the more choice is directed or imposed by what is available rather than by what is preferred (cf. Blau, Gustad, Jesson, Fames, & Wilcox, 1956; March & Simon, 1958; Mobley, 1982).

Perhaps the phrase "externally imposed compromise" requires some elaboration. Individuals' vocational maturity may be a composite of their knowledge and ability: (a) the ability to resolve conflicts between the factors in vocational choice; (b) the ability to plan the steps to vocational choice; (c) knowledge about occupations and careers; (d) self-knowledge; and (e) the ability to make realistic vocational choices given appropriate information (Crites, 1978). Vocational maturity then signifies the ability of individuals to synthesize career-relevant knowledge with self-knowledge. The range of alternatives available to some people is restricted by many externally imposed conditions. These people have fewer opportunities to test the world and discover their true capabilities and interests. That is, a career decision will be immature to the extent that people's environment limits their career relevant knowledge and self-knowledge. Holland (1997) showed that early choices about jobs by people in a free and open marketplace are not always the jobs that they best fit. What happens, however, is that people keep changing jobs until they find one that they do fit.

So we are not suggesting a "womb to tomb" approach to occupations and careers, where the individual does not experience tension or the need to make the kinds of compromises most people's lives realistically require. Nor do we espouse an approach that suggests every job be available to every person. Compromise and choice should be the person's responsibility, but people should have opportunities to make choices in a free and open marketplace.

It would seem reasonable, although shortsighted, for organizations to be more concerned with organizational than occupational or vocation choice. After all, one might say, an organization should be concerned with who works for it. However, as was shown in Fig. 5.3, artificial limitations on people's occupational alternatives also limit the kinds of people the organization may ultimately attract.

The key issue in understanding attraction to an organization is "fit." Fit refers to individuals' impressions that they fit or match an organization—in terms of such diverse features as pay, job characteristics, supervision practices, organizational reputation, promotion opportunities, convenience (easy to get to), and so on. Unfortunately, we do not have a good theory of the ways in which people view the various attributes of organizations, but we show some more recent attempts in this regard in the later section on current issues.

What we do know is that different people seem to focus on different attributes of organizations in estimating the degree to which they fit them, which suggests that what is attractive to one person may be completely unattractive to another. At the time of this writing, there were about a dozen such laboratory and field studies that report support for the idea that people are attracted to organizations they fit. In these studies, fit has been both subjectively assessed ("How well do you think you fit the organization?") and objectively assessed (assessing applicant values and organizational values; assessing applicant personality and organizational attributes). An example of an early lab study is one by Turban and Keon (1993), who found that organizations were differentially attractive to potential applicants as a function of the interaction of their personality (e.g., need for achievement, locus of control) and hypothetical organizations presented to them (e.g., organizations that varied in terms of centralization, size, and reward structures). They showed, for instance, that people with high needs for achievement were more attracted to organizations that paid on an incentive system.

Cable and Judge (1996) and Judge and Cable (1997) revealed similar findings in field studies. They found that applicants' perceptions of fit significantly predicted their job choice intentions—even after the attractiveness of job attributes was controlled for. Indeed, they showed that these perceptions of fit predicted later commitment, satisfaction, willingness to recommend the organization to others, and intentions to remain.

A study by Cable and Parsons (2001) found results similar to these earlier studies. Cable and Parsons studied 129 people pre- and postentry into jobs over a 2-year period. They found that

1. Preentry values congruence with the organization at time 1 predicted subjective fit perceptions on the part of newcomers at time 2.
2. Preentry values congruence was strongly correlated (0.65) with postentry values congruence. This means that people's values did not change much over time despite the socialization tactics employed by the organizations studied. Indeed, they showed

that the more offers a person had prior to entry, the stronger was pre- and postvalues congruence. This means that the more choice people have, the better they fit the organization they join—that is, you do not need to know much about socialization tactics when looking at values congruence, especially when people have choices about where they go to work.
3. They showed that values congruence at time 2 and perceptions of fit at time 2 predict turnover at time 3. This means that poor congruence and poor perceived fit yields attrition.

This should provide a good flavor of the research on attraction to organizations. It should be noted that, as Schneider (1987) hypothesized, this kind of research suggests that similar people are attracted to similar kinds of organizations. So, if we can predict based on a person's personality or values the kinds of organizational attributes to which they are likely attracted, it follows that those who end up applying for jobs there and those eventually hired will be more like each other than they are like people in other organizations. There is now some intriguing evidence to indicate this is true (Schneider, Smith, Taylor, & Fleenor, 1998).

The important point here is to know that people are not randomly assigned to the organizations to which they make application; there is information about the organization no matter what its validity is that leads people to be attracted to some but not other potential employers. Here is where the way an organization deals with incumbent employees becomes doubly important—people are attracted to organizations based on what they know about them, and they are likely to have more positive knowledge about an organization that deals well with the job satisfaction and commitment of its employees as a key part of their overall management philosophy. Such organizations can be expected to have larger pools of applicants and, as we learn in chapter 6, large pools of job applicants is a key to ensuring the best people are available from which to make a choice.

Next, we turn to what we know about actively finding applicants for jobs in organizations. As we will see, there is also evidence from the standpoint of organizations that they also are looking for a fit or match.

External Recruitment Research

Barber (1998) prepared an excellent summary of what we know about the external recruitment of candidates for jobs. The book focuses on the recruitment activities of primarily large firms, so it is important for us to state here that our belief is that most recruitment for organizations happens informally rather than formally through active recruitment tactics. Thus, 80% of the people in the United States work in firms with 500 or fewer employees (National Alliance of Business, 1997), so these firms are not visiting college campuses across the country trying to interest them in a job. Most of the recruitment for most of the firms in fact is logically based on personal contacts, and this is why we have so often emphasized the internal recruitment issues of job satisfaction and organizational commitment—because this is likely to yield a larger and perhaps superior pool of applicants.

That having been said, Barber (1998) discussed what we know about recruitment in terms of the three stages of recruitment: (a) outreach to potential applicants, (b) maintaining applicant interest in applying for a job, and (c) persuading applicants to accept a job offer. There are different actors who are important in each phase and different activities that characterize those actors in the different phases. Here, we summarize Barber's extensive treatment of these issues in a few pages.

Outreach to Potential Applicants

This topic is what we think of as "recruitment" when it is discussed generically (e.g., where do you go to find applicants). The most important issue here is making decisions about the kinds of applicants the firm requires. Such determinations emerge out of job analyses and projections about labor requirements and the availability of labor pools in the short and long term. The most important decision here concerns whether to grow or buy talent—to hire inexperienced people and produce through training and development experienced talent, or to hire already competent people. Thus, as we learned in chapter 2, an important issue to assess concerns the degree to which the KSAOs required by the job are typically learned before or after hire. With these data in hand, one can then pursue applicants and the issue becomes one of identifying the most appropriate sources for applicants. The research in this arena reveals that informal sources of applicants (e.g., referrals from existing employees and also self-applications) outperform (higher job performance, less turnover and absenteeism) those that emerge from more formal sources (campus visits, newspaper ads; Breaugh, 1992). There is some good research to indicate that this phenomenon is related to such informal sources of applicants having superior knowledge about organizations, especially the nontask attributes of working in them (e.g., the shifts that must be worked, the stress on the job). Incidentally, the single best source of applicants from a performance standpoint are rehires, suggesting that the prior knowledge of the organization is a key ingredient in superior results for these applicants.

This issue of the realism of information that applicants have for jobs has received considerable research attention under the notion of the realistic job preview (RJP). As noted earlier, recruitment should be concerned with the dual problem of securing people the company needs and ensuring they will remain. The RJP "tells it like it is." One argument against "telling it like it is" can be that it is so bad that no one would be interested. The solution here, of course, is to change the job/organization matrix of rewards offered, not to lie about it. The research evidence (cf. Phillips, 1998) suggests that when people are provided realistic information about the job, there is a marked improvement in turnover rates. Obviously, job analyses and organizational diagnoses provide one source of information for sketching out the organization/job for potential employees.

However, the emphasis should not only be to describe the job, but also to describe the organization. It must be assumed that people desire to work for a particular organization and on a particular job. Further, the person–organization relationship will probably last longer than the person–job relationship. This factor reemphasizes the importance of giving people adequate information on which to base their organizational choices.

Research on RJPs has been done primarily, although not exclusively in field settings. In the typical effort, a random half of applicants for a job are provided "fly paper" and the other half is provided realistic information. Research suggests that those who receive the realistic information have somewhat lower expectations of how satisfied they will be on the job (Meglino, DeNisi, Youngblood, & Williams, 1988) and that commitment of those who received the RJP tended to be significantly higher (Premack & Wanous, 1985). These findings help provide some explanation for why RJPs seem to have positive outcomes with regard to eventual turnover.

Barber (1998) noted that there may be a slight effect on the attractions of a job in an organization when potential employees are provided an RJP. She went on to note, however, that this negative effect may not be of negative consequence for organizations: "If those who withdrew from the hiring process were more likely to have quit shortly after being hired, the organization is probably better off without them" (p. 92). In contrast, some research suggests that the most qualified applicants tend to weight negative information the strongest, meaning the best applicants may be the most "turned off" by realistic job information (Bretz & Judge, 1998). So if RJPs are to be used, they must be carefully developed and administered.

In sum, it seems true there are potentially important long-term effects associated with having applicants who are knowledgeable about the organization, whether that knowledge comes from the organization at recruitment time via the RJP or comes with applicants to jobs because of information about the organization either they have gathered themselves or have obtained from job incumbents. When we combine this summary of the research on source effects with the extensive literature we presented earlier on attraction to organizations that stimulates self-applications, it becomes clear that the more people are aware of the true image of an organization the more likely it is for organizations to make wise decisions about the people they hire. Such wise decisions will be made because the pool of applicants the organization gets to pursue will be knowledgeable about what it might be like to work in the organization and be attracted to the organization because they think they will fit it. It is important to note once again that there are individual differences in what is attractive to people so organizations need not be of a similar sort to be attractive to potential candidates; what they need to be is attractive to the kinds of candidates they require in their talent pool. Once they can find such applicants, the key issue is to begin the active recruitment process and maintain applicant interest in the firm.

Maintaining Applicant Interest

The big issue studied here concerns the effects of the recruiter on potential applicants. One can think of recruiter effects as an initial face-to-face contact between an organizational representative and a potential applicant. So, "recruiter" effects also address any screening interviews done by firms as they decide whether to further pursue an applicant, as well as interviews done in campus visits, at job fairs, and so forth (we discuss recruitment interviews more fully in chapter 10). The bottom line on this research is that there appear to be recruiter main effects on applicants such that when applicants perceive recruiters to be warm, competent, and informative, they as applicants have increased

EXTERNAL RECRUITMENT 273

intentions to pursue employment with the company. This result appears to be a function of applicants reading into recruiter behavior attributes of the organization as a whole (i.e., applicants believe the recruiter portrays what it is like to work in the organization). This effect is called a "signaling effect" because it is believed that recruiters "signal" the organization through their behavior (Rynes, Bretz, & Gehrhart, 1991). Although there appear to be some effects for how recruiters handle themselves, Barber (1998) reported essentially no important effects on applicants with regard to the focus of the interview with applicants, the content of the interview, and whether the interview is structured or unstructured. In other words, what gets talked about (interview content) is less important than how the recruiter behaves. However, these results only address the initial contact between applicant and recruiter with the next stage of contact that between the applicant and the larger organization, through site visits, follow-up interviews, and so forth.

The research on later phases of the attempt to maintain applicant interest reveals that there are two significant turn-offs for applicants (Rynes et. al., 1991): (a) time delays—delays between the initial contact and later interest shown by the organization; and (b) site visits where applicants believed they were not treated as special people (e.g., by the level of people with whom they interacted) and where the scheduling of the visit appeared to lack flexibility. As in the research described earlier, these kinds of activities appear to be serving some signaling function for applicants—either interest in the applicant or lack of interest in the applicant. With regard to formal testing and assessment, these appear to be positive characteristics of the contact with organizations, apparently signaling to applicants that the company is interested in them (Smither, Reilly, Millsap, Pearlman, & Stoffey, 1993).

In summary, applicants can either feel they are special and be inclined to pursue the organization further, or they can be turned off by the way the organization treats them. The bottom line is that applicants are desperately seeking information to use as a basis for understanding what the organization might be like as a place to work, and anything the organization does (or fails to do) is interpreted by them as a "signal" of the interest the organization might have. However, the deal has not been closed, and this is the topic of the next stage of the recruitment process. Box 5.1 provides some examples of "minor" things organizations do that contribute to people rejecting their potential job offer.

> **Box 5.1 It's Always the Little Things**
>
> One of the authors has conducted several studies examining what contributes to people's perceptions of jobs and organizations and how they react to selection procedures. Here is a list of relatively minor things that people have used to form rather strong negative impressions about the job and organization:
>
> 1. *Rude secretary.* In several studies, we have found that one rude secretary can cause a lot of damage. Many people will state something along the lines of, "The secretary was rude and frustrating. Why would I want to work there?" A strong signal from a single person.
> 2. *Confusing application process.* When the application procedures are confusing, applicants will often infer the organization must be poorly managed or staffed with incompetent people.
> 3. *Complicated application process.* Likewise, complicated application procedures give the impression of a bureaucratic nightmare.
> 4. *Web sites that do not work or are out of date.* Much is inferred from Web sites—those that look cutting edge give impressions of innovation, those that are youth oriented give impressions of being modern, and those that are outdated (or contain broken links, etc.) give impressions of a poorly run organization. See also Box 5.2.
> 5. *Appearance of recruiters.* A recruiter who looks disinterested, yawns during interviews, is not dressed appropriately, and so on gives the impression of disinterest.
>
> Fortunately, many or most of these things can be addressed easily. The difficulty is ensuring that point-of-contact employees are respectful of all applicants at all times. Therefore, whenever possible, strive to ensure
>
> 1. The applicants have been provided with a complete explanation of the recruitment and staffing process—why it is being used and how it is being implemented.
> 2. The recruitment and selection procedures are either job related or are explained to show why they are job related.

EXTERNAL RECRUITMENT

3. The hiring decision is clearly explained and conveyed to applicants—even those who are rejected. This goes a long way toward minimizing resentment and hurt feelings.
4. The recruitment and selection process is as short and convenient as possible. Inform applicants when they will learn of the hiring decision and provide an adequate explanation to rejected applicants immediately after the hiring decision.
5. There is a friendly and informed contact person who can provide information and assistance in a friendly and respectful manner.

Closing the Deal

Readers to this point must be asking themselves, "How difficult is it anyway to recruit a few people for a couple of jobs?" Of course, the answer is that if you want to do it right, there are hundreds of things that have to be taken into account. There is the internal recruitment itself to focus on and then there are the different issues surrounding attraction and the early phases of the external recruitment process. The present section addresses the issue of influencing the candidate decision to choose your organization once you make an offer. So, you have gone through the steps to attract candidates, to meet with them initially, to invite them back for a second round of interviews and/or tests, and now you really want them to accept the offer you made. As Barber (1998) noted, the decision to accept a job offer is a difficult one for many candidates because it seems like such a final thing to do because accepting one offer closes all other possibilities (at least for a while). Until this point in the process, the offering organization is usually in the driver seat, but now the candidate, with an offer in their pocket, becomes the dictator of the process. What are the issues the candidate takes into account?

A critical issue for candidates in general concerns the various attributes of the job and the employment contract; issues like job attributes, hours, pay, benefits, and so forth. There is research where candidates are asked to rank these various attributes, but direct ranking of attributes has resulted in considerable inconsistency across studies (Barber, 1998). In addition to monetary rewards, people seeking employment consider these factors in their decisions, but there is little consistency across studies in the rank orders assigned to them:

1. The *intrinsic rewards* available in an environment. That is, the job/organization potential for satisfying many of the kinds of feelings an individual wants to experience (e.g., social, esteem, self-actualization).
2. The status *image* of an organization. That is, the impact that the fact of working in a particular organization has in the larger life of the individual (recall the job embeddedness discussion; Mitchell et al., 2001).
3. The *supervisory style* typical of an organization. That is, the dominant mode—supportive or democratic versus task oriented or authoritarian—the individual perceives to exist.
4. The dominant *values* of an organization. That is, whether the organization's goals stress innovation, service, or entrepreneurship.

As a side note, it is worth summarizing research on the question of the job preferences of men and women; the data show they are very similar (Lacy, Bokemeier, & Shepard, 1983). Lacy et al. studied about 7,000 people, equally split by gender, and discovered that there was essential agreement that important and meaningful work is their most preferred characteristic. This was followed by income and promotion and, last, by job security or working hours. Although other studies indicate that pay levels are most directly associated with the acceptance of job offers, it may be likely that job applicants get accurate information only about their starting pay level. Unfortunately, Taylor and Sniezek (1984) found that recruiters tend to ignore applicants' needs for job-relevant information.

If there is little consistency across studies in the attribute rankings, what is an organization to do to discover what attributes its own candidates prefer in jobs? An organization can do what is called policy capturing research (discussed briefly in chapter 4). In this kind of research, the objective is to capture the policy or reasoning behind a choice. In the present case, if we wanted to understand the policy or reasoning behind the choice of candidates for jobs, we would present each candidate with a series of 20 job descriptions in which we systematically varied such issues as hours, pay, benefits, and job attributes.

After reading each description, candidates would rate what their intentions are with regard to employment in the job described. After rating all job descriptions, we would be able to determine statistically which systematically varied job attributes across the job descriptions were most strongly correlated with intentions. Although not much research has been done using this procedure, any company could do this kind of research to discover for its own candidates the combination of attributes that makes jobs most attractive to its candidates. Thus, because the relevant issues vary so much across organizations and jobs, it is probably not feasible to identify the rankings of job attributes that cut across jobs and candidates. There is just too much variability on both.

More research has been done on the "fit" approach to understanding candidate job choice decisions. This kind of research follows the same logic and approach as the research described earlier concerning attraction to jobs. In brief, it turns out that people are not only attracted to jobs they think they fit, but they also tend to accept offers from companies with job attributes they think they fit (Cable & Judge, 1996; Judge & Cable, 1997). "Fit" turns out to be a central issue in career and job and organizational choice (and job satisfaction and turnover, too; Kristof, 1996), so it is important for organizations to know the image they portray to the outside world because this image will be an important factor in the job choice decisions potential candidates make. Obviously, the contacts candidates have with organizations during the external recruitment process influences the image candidates have so those contacts, as we noted earlier, need to be carefully managed for what they signal about the organization.

What candidates seem to do is collate the information they have or think they have and then make some judgments about the degree to which they will fit the organization. These judgments themselves have been the subject of some debate, with some (Vroom, 1966) arguing that people make decisions in quite conscious ways by weighing alternatives, and Soelberg (1967) presenting the alternative perspective that people accept the first job offer that meets their minimal requirements (see Power & Aldag, 1985, for a review). As usual, there is evidence for both perspectives, and we think the determining factor is the degree to which there is an abundance of alternatives perceived by an applicant to be available. So, for an MBA graduate there may be many possible alternatives available so those alternatives are weighed. In contrast, for an unskilled laborer who believes there are not many alternatives, any job offer is the one to accept.

There has been some interesting research on things organizations can do to help those with a job offer make up their mind and choose the organization. For example, Ivancevich and Donnelly (1971) took as the basis for their study one of the major problems in recruitment, the number of people who accept job offers and then back out of their acceptance. The rate of such behavior is between 10% and 15% for management jobs. Ivancevich and Donnelly hypothesized that if people who accept a job offer are given some social reinforcement of their acceptance, they are less likely to back out than those whose acceptance is treated with a typical impersonal acknowledgment. One half of a group of prospective managers ($N = 196$) were telephoned immediately on their acceptance of a job offer. In the following weeks, they were telephoned twice more. The other half of the group ($n = 196$) were sent some literature and a confirmation of receipt of their acceptance (incidentally, the average number of offers each person received was three). Of those receiving the personalized treatment, 2.6% backed out; 12.5% of those receiving the usual treatment backed out. Similarly, a study by Gilliland et al. (2001) demonstrated how writing selection decision letters in a sensitive and respectful manner had a positive influence on subsequent reapplications. This fits with the notion we presented earlier of people wanting to feel "special."

So, once it is realized that recruits like MBAs may be choosing among various alternatives, special corporate attention to them may pay important dividends. MBAs who are seeking a job consider a mix of alternatives in making their decisions, not one alternative in isolation from all other possibilities. If this is true, then as Power and Aldag (1985) noted, the nature of recruit desires and goals is of great importance to organizations if they are going to be able to recruit effectively. Because recruit self-esteem is also related to recruit desirability to organizations (Ellis & Taylor, 1983), the importance of paying attention to the particular characteristics of attractive recruits is clear.

An interesting study supporting this idea concerns Type A people. Type A people are characterized by such factors as high ambition, competitiveness, high need for achievement, time urgency, and working frenetically at more than one thing at a time (Chesney & Rosenman, 1980; Friedman & Rosenman, 1974). Research on Type A people reveals they tend to experience some significant physiological disorders (heart disease, ulcers) but, at the same time, they tend to be effective and successful. The question Burke and Deszca (1982) asked was would Type A persons prefer (be more satisfied with, successful, and more likely to join) some organizations more than others?

To answer this question, they gave 188 people a measure of preferred organizational climate and a measure of their Type A behavior (the Jenkins Activity Survey; Jenkins, Rosenman, & Friedman, 1967). The data showed that people with higher Type A scores preferred climates characterized by high performance standards, spontaneity, ambiguity, and toughness—again showing the importance of fit in job choice decisions.

Methods of External Recruitment

Our attention now turns to a review of some of the practices actually used by organizations to do their external recruitment. Table 5.2 shows a listing of the 14 most frequent sources of recruits for different kinds of occupations. The figures in Table 5.2 refer to the percentage of companies who report that they use the procedure. Table 5.3 shows some

TABLE 5.2
Recruit Source By Occupation

	Occupation				
Source	Office/ Clerical	Plant/ Service	Sales	Professional/ Technical	Management
Employee Referrals	92	94	74	68	65
Walk-ins	87	92	46	46	40
Newspaper Advertising	68	88	75	89	82
Local High Schools or Trade Schools	66	61	6	27	7
U.S. Employment Service (USES)	63	72	34	41	27
Community Agencies	55	57	22	34	28
Private Employment Agencies	44	11	63	71	75
(company pays fee)	(31)	(5)	(49)	(48)	(65)
Career Conferences/ Job Fairs	19	16	19	37	17
Colleges/Universities	17	9	48	74	50
Advertising in Special Publications	12	6	43	75	57
Professional Societies	5	19	17	52	36
Radio-TV Advertising	5	8	2	7	4
Search Firms	1	2	2	31	54
Unions	1	12	0	3	0

Figures are percentages of companies providing data for each employee group.
Reprinted by special permission from *Personnel Policies Forum*, Copyright © 1979, by the Bureau of National Affairs, Inc., Washington, D.C.

TABLE 5.3
Effectiveness of Recruiting Source By Occupation

Source	Office/ Clerical	Plant/ Service	Sales	Professional/ Technical	Management
Newspaper Advertising	39	30	30	38	35
Walk-ins	24	37	5	7	2
Employee Referrals	20	5	17	7	7
Private Employment Agencies	10	2	23	25	27
U.S. Employment Service (USES)	5	6	0	1	1
Local High Schools or Trade Schools	2	2	0	0	0
Colleges/Universities	1	1	8	15	2
Community Agencies	1	3	0	1	2
Unions	0	2	0	0	0
Career Conferences/ Job Fairs	0	1	2	2	1
Professional Societies	0	1	1	0	2
Search Firms	0	0	2	5	17
Radio-TV Advertising	0	1	0	0	1
Advertising in Special Publications	0	0	3	5	8

Figures are percentages of companies providing data for each employee group. Columns may add to more than 100 percent because of multiple responses or less than 100 percent because of nonresponses. These percentages are votes for effectiveness.
Reprinted by special permission from *Personnel Policies Forum*, Survey No. 126. Recruiting Policies & Practice, pp 4–5. Copyright © 1979, by the Bureau of National Affairs, Inc., Washington, D.C.

differences in source of recruits as a function of the type of job being filled. For example,

1. At the extremes, office/clerical and plant/service people are recruited predominantly via employee referrals, walk-ins, and newspaper advertisements. Management persons, however, are located from private employment agencies rather than walk-ins, as well as through newspaper ads and employee referrals.

2. Professional/technical persons tend to be located through ads in specialized publications far more frequently than are people for other occupations.

3. Search firms, a topic we spend some time on later, are used primarily for management occupations. However, they are also used by some firms for finding professional/technical candidates.

EXTERNAL RECRUITMENT 281

Most of the strategies listed in Table 5.2 are fairly obvious to anyone who has looked for a job. However, a few of them warrant some additional attention. Community agencies are formal sources of job information other than the U.S. Employment Service and affiliated state employment services. The latter two agencies have access to all manner of jobs for all kinds of interests and ability levels. Community agencies tend to specialize in particular kinds of persons based on such characteristics as gender, race, religion, disability (e.g., physically or emotionally disabled), or occupational history (e.g., some agencies specialize in working with the chronically unemployed). Some examples of these agencies would be the Catholic Youth Organization, the Urban League, and Alcoholics Anonymous.

Career conferences/job fairs are programs set up by community groups, schools (high schools and colleges/universities) to which many companies, governments (federal, state, and local), and the military (for armed services) may come to present information about themselves. These large job marketplaces are used by organizations to publicize what they do, what they have to offer, and the kinds of persons in which they may have interest. The analogy to a fair is appropriate in that typically these events have a wide variety of opportunities—and a wide variety of risks. In other words, such fairs can be slick in the extreme—the fly-paper approach to recruitment typified. This is not always true, but it is worth remembering when seeking a job.

In Table 5.2, colleges/universities refers to what is also called *campus recruitment*. Numerous large companies and large colleges/universities have an annual recruiting season (spring) that occupies graduating seniors for a number of months. It works something like this: The university has a staff of people, usually in a counseling, placement, or job center. These people maintain two lists, one of candidates and the other of companies that will be visiting. Companies place notices about the kinds of people for whom they are looking, and candidates file a document summarizing their interests and accomplishments. A match is made, sometimes by the placement center, but more often by the companies looking over the paperwork filed by interested candidates. It is important to note here that the quality of the paperwork filed may have a bearing on whether an interview is obtained.

The fact that some companies that do campus recruitment have teams of people traveling from campus to campus during "the season" reveals the magnitude of the investment companies make in attempting to attract the kinds of people they want. Some companies will have numerous teams comprised of professional personnel/recruiting persons, as well as people who actually do the kind of work for which the team is seeking recruits. These teams may travel for 6 or 8 weeks to campuses, conducting interviews and living as a group.

Search firms are specialized employment agencies that work only from requests made by companies. That is, in contrast to employment agencies, candidates do not contact search firms; search firms contact people in whom their client may have an interest. For example, imagine that a company wants to hire a new comptroller. Comptrollers are typically responsible for all accounting/financial issues in a corporation. As such they are very well paid, critical to the company, and difficult to hire. So the company might call a search firm (sometimes called an executive search firm or a head hunter). For example, the head hunter may be asked to find a comptroller with 5 year's experience in a real estate firm to be paid $225,000 in salary, $50,000 in other compensation (e.g., moving expenses, purchase of the current home, reduced mortgage rates on a move, stock options and bonuses), plus additional help as required (e.g., finding a job for a spouse, a school for the children.)

Executive search firms are a post—World War II phenomenon, the first being founded in 1946 (Wareham, 1980); now there are thousands of such firms, ranging from one-person operations to firms with billings in the hundreds of millions of dollars. For finding a comptroller as described previously, the head hunter would be paid about one third of the comptroller's first year salary. In return, the search firm would agree to the following:

1. *Confidentiality*: In exchange for the opportunity to search for a comptroller, the search firm agrees to not divulge its mission. Search firms are frequently used to find likely successors for people who do not know they are about to be succeeded.
2. *Off limits*: Search firms generally agree to leave placed recruits in the new job at least 2 years before trying to recruit them again for another job.
3. *Retention*: Search firms also usually agree to help the company locate another recruit if one that is placed fails to last a year—and they do this for no additional fee. So, recruiters have a stake in making placements that last. Indeed, some search firms have extensive questionnaires for both the corporate client and the recruit to try and arrange the most effective match (Zippo, 1980).

EXTERNAL RECRUITMENT

Most search firms belong to the Association of Executive Search Consultants (AERC). The AERC has a code of ethics, does some research on the industry, and publishes statistics of use to companies seeking executive search information. The most important facet of executive or technical search, however, is names. Search firms must maintain elaborate systems replete with thousands of biographies because when a client needs someone, time is usually important. Clients want more than one name—they want to make a choice. To keep the file up to date, search firms retain or develop "clipping" services, which track news articles about particular people and/or particular industries. Biographical files in the best firms are continuously updated—not a bad model for organizations themselves.

Prior to leaving this topic, the issue of employee relocation deserves brief comment. Regardless of whether internal or external recruitment is involved, employees frequently have to be relocated. Just as there are search firms, there are also relocation firms. These firms, most of which are members of the Employee Relocation Council, handle everything connected with a move: mortgages, storage of household goods, temporary living arrangements, house hunting trips (with the whole family), redecorating of new home, repurchase of old home, and so on. There is not much research on relocation. One study showed that, contrary to the popular press image of resistance and hardship, executives and their families are generally willing to move for career enhancement and job challenge, and adjustment to the new location does not present great hardship (Brett, 1980).

Finally, the Internet has so quickly transformed our way of life that it is hard to imagine recruiting without it. In fact, there are so many companies that advertise and recruit on the Internet that one would think they all know it is effective. Just think of massive Web sites such as www.monster.com and a plethora of online job postings run by local newspapers, and you quickly appreciate how much the Internet has transformed recruiting. However, the reality is that research is just starting to identify the benefits—and potential limitations—of Internet recruiting. It is for this reason that Tables 5.2 and 5.3 do not discuss Internet recruiting; the data have not yet been collected! Yet, research is starting to examine these issues.

An interesting report by Karr (2000) suggested approximately one fourth of job seekers in their sample rejected organizations from further consideration due to the poor quality of their Web site. Dineen, Ash, and Noe (2002) found perceptions of person–environment fit were affected by organizational Web sites, and this fit information contributed to perceptions of organizational attractiveness. Williamson, Lepak, and King (2003) found Web sites designed for recruitment (rather than screening) purposes contributed to enhanced perceptions of organizational attractiveness. However, the relationship was not direct but mediated through perceptions of the usefulness of the Web site. Cober and colleagues (Cober, Brown, Keeping, & Levy, 2004; Cober, Brown, Levy, Cober, & Keeping, 2003) helped show empirically and conceptually how Web sites influence organizational attraction. Their research suggests both the content (i.e., technical information about the job and organization) and style (i.e., aesthetic appeal) of the Web site can influence organizational attraction. Like the recruitment interview, it is not only what the Web site says, but also how it says it, that influences organizational attraction.

It is worth noting that Internet recruiting may prove to be a great equalizer for smaller business competing against larger business with larger recruiting budgets. For example, the Web site can be used to convey the values and benefits of working for the organization. It might reach applicants who otherwise would not have been recognized. A simple Web site that conveys the important information and is reasonably visually appealing is actually quite affordable.

However, preliminary research suggests organizations cannot simply put information about their company on a Web site without it having some effect on applicant perceptions. Rather, Web sites need to be carefully thought out, developed, and kept current for them to prove a useful recruiting tool. Box 5.2 provides some suggestions for how organizations can improve their Web site effectiveness.

Applicant Reactions to Recruiting Practices

Related to the topic of external recruitment is the topic of how applicants perceive and react to recruiting and staffing practices. This focus is slightly different from recruitment in that the orientation is from the applicant and not the organization. That is, here we want to understand how and why applicants form impressions of not only recruiting practices, but also selection tests and procedures. This information contributes to a better understanding of applicant behavior and decision processes, and thus provides HR personnel with an understanding of how to create staffing practices in the most acceptable manner.

EXTERNAL RECRUITMENT

> **Box 5.2 Some Suggestions for Improving Web Site Recruitment Effectiveness**
>
> Although we are not aware of any comprehensive guidelines for developing Web sites to enhance recruiting, the following suggestions seem reasonable and tend to work well in our experience:
>
> 1. The visual display of the Web site should be aesthetically appropriate (e.g., not too busy).
> 2. The information on the Web site should be current.
> 3. According to Cober et al. (2003), appealing Web sites are those that repeat key visual cues (e.g., borders, colors, links), align links and icons consistently within and across pages, and group similar topics in close proximity.
> 4. A Web page should contain no more information than can be viewed without requiring more than one click to scroll down.
> 5. Many argue that good Web sites follow a rule of "three clicks," meaning that the information one wants to obtain should be found within three mouse clicks of the home page. However, increasingly, organizations use a single click to take a person from the home page to the recruitment page.
> 6. Ensure the Web site conveys the appropriate image the organization wants to convey. For example, strategic use of video employee testimonials, pictures, and text can reaffirm the organization's core values and attract applicants who hold similar values.
> 7. Ensure the Web site is compliant with the U.S. federal government's regulations for designing Web sites to not disadvantage those with disabilities (e.g., violate Section 508).
> 8. Do not be quite at the cutting edge. Requiring a Web site to have the newest version of Flash, Java, or related software will probably mean the Web site will not work for many (if not most) applicants. It is better to use a slightly older—but more common—version of the Web software.
> 9. If you use the Internet to take applications, send applicants an e-mail acknowledging receipt of the application and a timeline for next steps.
> 10. Offer a link where applicants can provide feedback and suggestions.

This research only started becoming popular during the early 1990s. We suspect the interest in applicant reactions came from three sources. First, the world of staffing was becoming increasingly litigious and some hoped such litigation could be reduced by paying attention to what applicants thought and wanted. Second, the economy was strong and unemployment was at record lows, contributing to it being an "applicant's market," with fierce competition for qualified individuals. Finally, the search for selection tests that minimized racial and gender subgroup differences led some researchers to examine whether test differences came from the perceptions of applicants themselves (we discuss subgroup differences more in chapter 7).

These societal influences created the perfect environment for Gilliland (1993) to publish a paper that energized, and continues to direct, applicant reactions research. Gilliland integrated the staffing process with principles derived from organizational justice theories to identify to which staffing factors applicants attend, why they attend to them, and what the consequences are for such perceptions. In brief, his model predicted that applicants are fundamentally concerned about issues of fairness, and these fairness perceptions drive their subsequent choices and behavior. There are two kinds of fairness: *process fairness* is a perception of the fairness of the staffing procedures, and *outcome fairness* is a perception of the fairness of the hiring decision. As the theory goes, these two fairness perceptions influence a host of perceptions (e.g., organizational attractiveness), job choices, and behaviors (e.g., withdrawal, test scores). Gilliland further identified numerous "justice rules" that if satisfied contribute to process fairness:

- *Job relatedness*: Does the test or staffing procedure appear related to the job?
- *Opportunity to perform*: Did applicants get an adequate opportunity to demonstrate their skills?
- *Opportunity to be reconsidered*: Do applicants have some way to retake the exams (or have their exams rescored)?
- *Consistency*: Are the selection procedures administered consistently across applicants and time?
- *Feedback*: Was feedback provided about test scores and performance in the system?
- *Selection information*: Was an adequate explanation provided for the use of the staffing procedures?

- *Honesty*: Were HR personnel honest in their communications with applicants?
- *Interpersonal effectiveness*: Were HR personnel respectful of applicants?
- *Two-way communication*: Do applicants have an opportunity to speak to and receive answers from HR personnel?
- *Question propriety*: Did the selection tests ask questions that were too invasive or personal?

Three rules also contribute to outcome fairness:

- *Equity*: Hiring decisions are based on test performance.
- *Equality*: Hiring decisions are made to be equal across racial and gender subgroups.
- *Needs*: Hiring decisions are based on who needs the job the most.

This model stimulated quite a bit of research, much of which is summarized in Ryan and Ployhart (2000). Some of the highlights suggest that attending to these justice rules does in fact enhance fairness perceptions and lead to positive outcomes such as increased likelihood of recommending the organization to others and organizational attractiveness (e.g., Bauer, Maertz, Dolen, & Campion, 1998). Giving applicants detailed information about how and why they were selected, and why the particular selection procedures are being used the way they are, contributes to favorable perceptions (e.g., Ployhart, Ryan, & Bennett, 1999). However, Ryan and Ployhart (2000) noted few of the effects on fairness perceptions relate to actual test performance or such behaviors as withdrawal from the selection system. Instead, it appears these perceptions relate most strongly to attitudinal consequences such as organizational attractiveness, job attitudes, and self-perceptions.

In a different vein of research, the idea was to link perceptions to actual test performance. The goal was to understand whether the frequently observed differences between Whites and Blacks on cognitively oriented tests were due to differences in perceptions and motivation (these subgroup differences are discussed in chapter 7). Arvey, Strickland, Drauden, and Martin (1990) demonstrated that White–Black differences on work sample tests were due in part to differences in test-taking motivation. Chan, Schmitt, DeShon, Clause, and Delbridge (1997) showed a quite interesting effect where part of the White–Black test score difference was due to subgroup differences in test-taking motivation (such that Blacks exhibited lower motivation). Ryan (2001) documented several test perceptions that are held more negatively by Blacks, hypothesized to be caused by a more general distrust of testing than Whites. However, again we do not see large effects of these perceptions on test scores or subgroup differences.

Overall, a few conclusions seem appropriate. First, perceptions of applicants do not tend to produce dramatic effects on test scores or job choice variables. However, there is some relationship, and for this reason, it is worthwhile to consider such effects. Second, applicant perceptions appear to produce stronger effects on more attitudinal outcomes such as organizational attraction. So, from a public relations perspective, such perceptions are important. Third, although the effects of fairness or test perceptions may not always be strong, the cost of even a single person suing due to mistreatment is quite expensive, and anything that can be done to reduce such litigation is money well spent. Finally, most of the research on applicant reactions suggests little things, such as treating applicants with respect or telling them why the tests are being used, go a long way toward keeping them more satisfied with the process. Organizations should be doing these things anyway, and this line of research identifies some potentially serious negative consequences when they do not (see Box 5.1).

CURRENT CHALLENGES FOR INTERNAL AND EXTERNAL RECRUITMENT

In this final section, we consider the many current challenges and issues facing recruitment research and practice. We consider several such issues, including levels of fit, theories of recruiting, selection system withdrawal, targeted recruiting, and using recruitment as a source of competitive advantage.

The Meaning of Fit

We have seen throughout this chapter how the concept of person–environment (or person–job) fit has important consequences on both applicant and organizational decision processes. Much of this research has focused on the consequences of person–job fit or person–organization fit, but more research is examining other consequences and also trying to determine what contributes to fit perceptions. For example, Kristof-Brown, Jansen, and Colbert (2002) used a policy capturing design to find person–job, person–group, and person–organization fit; each had independent effects on job satisfaction. Perhaps more important, they showed that people use rather complex ways of integrating various pieces of information on fit. This poses some challenges—and opportunities—for recruiters because it means several forms of fit information may need to be conveyed in the recruiting message.

Other research by Cable and Parsons (2001) is described previously and helps understand what contributes to perceptions of fit (and how applicants use fit perceptions). For example, they show that those with more options perceive better fit with the organization they choose to join. However, an issue that has plagued the study of fit is how one should measure and statistically analyze fit. Some argue for objective fit measurement, where fit is defined in terms of congruence or statistical interaction; others argue for subjective fit measurement in which applicants are simply asked how well they fit with the environment. These issues are intertwined with how to analyze fit, and there have been several debates about various approaches.

Research is starting to better articulate what contributes to fit perceptions, how applicants use these perceptions, and how malleable these perceptions are over time. However, until there is more consensus about how to measure and analyze fit, such research will be difficult to conduct. Given fit is so central to many concepts in staffing and organizational behavior, it is imperative these issues get resolved.

Theories of Recruiting

It may seem odd that after reading so many pages of recruitment research, we have not discussed any theories or models of recruitment. We briefly noted Gilliland's (1993) model of justice in staffing contexts, but this model was not intended to capture the breadth of recruiting. Barber (1998) and Breaugh and Starke (2000) lamented this state of affairs, arguing for a need for theories of the recruitment process. In particular, there is a continuing need to consider recruiting processes longitudinally because applicant and organizational decisions evolve over time. For example, a job may seem attractive to an applicant until another job offer is obtained; studying job choice at different times can lead to different conclusions.

One important step in this direction was provided by Lievens and Highhouse (2003). They applied a theory of branding to recruitment. Branding is a concept borrowed from marketing. The idea is that different brands evoke different cognitions, feelings, and emotions. For example, think of Coke or Pepsi, Honda or Chevrolet, and Yamaha or Harley-Davidson; each produces different reactions. The same concept was hypothesized to apply to recruiting, such that organizations are essentially brands, and therefore organizations must manage and promote their brand. Thus, a brand is something less tangible than job and organizational characteristics (e.g., pay). Lievens and Highhouse argued that this branding perspective is important because many organizations cannot be differentiated in terms of objective characteristics such as salary or location. However, and consistent with their results, organizations can enhance their recruiting effectiveness by emphasizing their brand. For example, one organization may seem innovative, whereas another may seem conservative. As we note in the section on career choices, different people will be attracted to these different organizational brands.

The branding approach seems provocative because there is considerable research on it in the marketing literature, and hence, may contribute to quickly building a theory of recruiting. It may signal a shift in recruiting from an emphasis on salary and working conditions to an emphasis on the "personality" of the organization. However, whether the branding strategy is sufficient to capture the complexities of recruiting practice remains to be seen.

Selection System Withdrawal

In many civil service testing programs, such as those for police and firefighters, there are often several thousand applicants, and the selection ratio is frequently less than 5%. These testing programs often last several weeks or months from the time of application, through testing and background checks, to the time of actual employment. One curious finding in these testing programs is the excessive withdrawal that occurs in these processes. For example, a study of police applicants found 39% (1,106 applicants) applied for the job but did not show up to take the first set of exams (Ployhart, McFarland, & Ryan, 2002). This is not uncommon, and the problem is exacerbated by the fact that most of those who withdraw are minority applicants. This may contribute to less opportunity to hire qualified minority candidates (if those who withdraw are qualified), and possibly less diversity. It is also true that when qualified applicants who would be hired withdraw or turn down job offers, the utility of the staffing system will suffer (Murphy, 1986).

Ryan and colleagues studied the factors that contribute to withdrawal. In their research, they have identified the reasons why applicants withdraw (Schmit & Ryan, 1997) and what predicts withdrawal (Ryan, Sacco, McFarland, & Kriska, 2000). For example, Ryan et al. (2000) showed those who withdrew from a police officer selection process were more likely to hold negative images of the organization. Further, those who applied to many jobs were more likely to withdraw at late stages in the process, emphasizing the importance of studying withdrawal longitudinally. Given so many people withdraw and that many of these are minority applicants, it is important for future research to understand what can be done to either reduce the number of those who apply or ensure individuals remain within the application process.

Targeted Recruiting

One topic that is important, but for which there is essentially no research, is the topic of targeted recruiting. Targeted recruiting is particularly important in light of increasing diversity and the sometimes high withdrawal rates from minority applicants discussed previously. Many organizations strive to hire highly qualified minority candidates, but if these same candidates do not apply or drop out of the selection process, the organization has no opportunity to hire them. Therefore, it is not uncommon for organizations to seek places where those qualified applicants may be present. A general example of this we discussed earlier is when an organization recruits at select college campuses. The idea is that these colleges have such a good standard of education that it becomes a safer bet to hire an applicant from a particular college than anywhere else.

However, a more important question is whether recruiting efforts targeting minority candidates are successful and actually result in hiring more minority applicants. Civil service agencies use this approach frequently. For example, in an attempt to increase the number of Hispanics in the police force, the District of Columbia actually recruits in other states with large numbers of Hispanics, and even Puerto Rico. In fact, they will advertise in Puerto Rico and then travel there to administer the exams. Applicants who pass the exams are then required to move to Washington, DC, of course, but once they arrive there are numerous resources to help them adjust to living in the United States (Dvorak, 2002). Clearly, research must examine whether such practices are cost effective, but the criterion for such research must be broadly defined. Rather than just looking at job performance, one must also look at other outcomes—in the case of the DC Police Department, enhanced community relations by having a police force representative of the local citizenry. Unfortunately, data demonstrating the effectiveness of such efforts is nearly nonexistent. A simulation study by Tam, Murphy, and Lyall (2004) suggests that such efforts are unlikely to do much to increase minority representation and hiring rates.

Recruitment as Competitive Advantage

Projections from many sources predict there will be severe shortages of qualified labor in many areas of the economy, including medical, technical, and science professions. Consequently, organizations that are better able to attract, select, and retain qualified individuals will be organizations that are more competitive and successful. We believe management will increasingly see recruiting as a source of competitive advantage rather than solely an administrative function.

If so, organizations must learn how to use recruitment as a source of competitive advantage and more carefully craft and tailor their recruitment message to attract top talent. Research is just starting to illuminate ways this might be done, but we believe the key will be to present an organizational image that appeals to applicants. Highhouse and colleagues (Highhouse, Zickar, Thorsteinson, Stierwalt, & Slaughter, 1999; Lievens & Highhouse, 2003) conducted several studies helpful in this regard. Building from their work and incorporating a few suggestions of our own, the process could go something like the steps presented in Table 5.4.

One benefit of an approach such as this is that is makes the organization unique and allows it to stand out from competitors. This may be particularly important when the organization is unable to offer higher compensation, a better location, or working conditions than competitors. Another benefit is that these steps require few resources to complete (in fact, the marketing department may already have this information). Research will be needed to understand whether such an approach leads to better success in the acquisition and retention of targeted applicants, but until that time these suggestions seem reasonable.

TABLE 5.4
Suggestions for Using Recruitment as a Means to
Differentiate the Organization for Competitive Advantage

1. Identify the desired target applicant group.
2. Survey this group to learn how they perceive the image and attributes of your organization and those of competitor organizations (see Highhouse et al., 1999, for a discussion of how to do this).
3. Identify which attributes most differentiate your company in a positive way from competitors (see Highhouse et al., 1999, for a discussion of how to do this).
4. Emphasize those differentiating attributes that are valued by your target applicant group in your recruitment message (e.g., Web site, recruitment interviews).

SUMMARY

In this chapter, we present the argument that recruiting has two facets—internal recruitment and external recruitment. The reason for this somewhat artificial but important distinction is to focus on organizational conditions that will keep competent employees so when future needs emerge, already socialized, trained, and experienced persons will be available. As became clear, organizations must create conditions that will satisfy employees and gain employee commitment to the organization if competent persons are to be retained. The importance of this issue cannot be overstressed because many organizations view turnover as an employee problem when, in fact, turnover can be attributed to inappropriate recruitment processes, poor selection processes, and a poor work environment—all of which are organizational responsibilities.

The second part of the chapter, external recruitment, emphasizes the dual ideas that people are looking for certain organizational characteristics at the same time organizations try to find people. The RJP is presented as a possible mechanism for helping organizations and individuals create a match. Some support for the RJP approach is summarized, emphasizing the idea that the more knowledge candidates have about the jobs they eventually take, the more likely it is that positive consequences will follow for both the individual and the organization. In support of this idea, research on the effectiveness of recruiting practices is summarized, showing that employee referrals and, perhaps, rehires of former employees may yield more effective recruits. It is too easy to overlook two other important conclusions we reached: (a) candidates for jobs like to believe they are "special," so it is important both in the early visits to companies and in the follow-up to offers that this comes through for candidates; and (b) candidates likely make their choices among alternatives by weighing a variety of job attribute issues and then choosing the one they believe they best fit. The former issue requires that considerable attention be paid to the way candidates are treated, and the latter issue reflects the importance of the image of the organization—in all its complexity and variety—be carefully managed so the choices made by candidates on the basis of fit perceptions are accurate.

Finally, we examine several current issues facing recruitment. In that section, we discuss how recruitment research is slowly becoming more theoretical, as we search for a better understanding of what recruitment practices actually do to applicants. We suspect recruitment will become an even more important research topic in coming years.

REFERENCES

Arvey, R. D., Strickland, W., Drauden, G., & Martin, C. (1990). Motivational components of test taking. *Personnel Psychology, 43*, 695–716.

Barber, A. E. (1998). *Recruiting employees: Individual and organizational perspectives.* Thousand Oaks, CA: Sage.

Bateman, T. S., & Organ, D. W. (1983). Job satisfaction and the good soldier: The relationship between effort and employee "citizenship." *Academy of Management Journal, 26*, 887–895.

Bauer, T. N., Maertz, C. P., Dolen, M. R., & Campion, M. A. (1998). Longitudinal assessment of applicant reactions to employment testing and test outcome feedback. *Journal of Applied Psychology, 83*, 892–903.

Baysinger, B. D., & Mobley, W. H. (1983). Turnover. In K. M. Rowland & G. R. Ferris (Eds.), *Research in personnel and human resources* (Vol. 2, pp. 269–319). Greenwich, CT: JAI Press.

Blau, P. M., Gustad, J. W., Jesson, R., Fames, H. R., & Wilcox, R. C. (1956). Occupational choice: A conceptual framework. *Industrial and Labor Relations Review, 9*, 531–543.

Borman, W. C., & Motowidlo, S. J. (1997). Task performance and contextual performance: The meaning for personnel selection research. *Human Performance, 10*, 99–110.

Boudreau, J. W. (1991). Utility analysis for decisions in human resource management. In M. D. Dunnette & L. M. Hough (Eds.), *Handbook of industrial and organizational psychology* (2nd ed., Vol. 2, pp. 621–745). Palo Alto, CA: Consulting Psychologists Press.

Brayfield, A. H., & Crockett, W. H. (1955). Employee attitudes and employee performance. *Psychological Bulletin, 52*, 415–422.

Breaugh, J., & Starke, M. (2000). Research on employee recruiting: So many studies, so many remaining questions. *Journal of Management, 26*, 405–434.

Breaugh, J. A. (1992). *Recruitment: Science and practice.* Boston: PWS-Kent.

REFERENCES

Brett, J. M. (1980). The effects of job transfer on employees and their families. In C. L. Cooper & R. Payne (Eds.), *Current concerns in occupational stress* (pp. 99–136). New York: Wiley.

Brett, J. M., Stroh, L. K., & Reilly, A. H. (1992). What is it like being a dual career manager in the 1990s? In S. Zedeck (Ed.), *Work, families, and organizations* (pp. 138–167). San Francisco: Jossey-Bass.

Bretz, R. D., & Judge, T. A. (1998). Realistic job previews: A test of the adverse self-selection hypothesis. *Journal of Applied Psychology, 83*, 330–337.

Brief, A. P. (1998). *Attitudes in and around organizations*. Thousand Oaks, CA: Sage.

Burke, R. J., & Deszca, E. (1982). Preferred organizational climates of Type A individuals. *Journal of Vocational Behavior, 21*, 50–59.

Cable, D. M., & Judge, T. A. (1996). Person-organization fit, job choice decisions, and organizational entry. *Organizational Behavior and Human Decision Processes, 67*, 294–311.

Cable, D. M., & Parsons, C. K. (2001). Socialization tactics and person–organization fit. *Personnel Psychology, 54*, 1–24.

Chan, D., Schmitt, N., DeShon, R. P., Clause, C. S., & Delbridge, K. (1997). Reactions to cognitive ability tests: The relationships between race, test performance, face validity perceptions, and test-taking motivation. *Journal of Applied Psychology, 82*, 300–310.

Chesney, M. A., & Rosenman, R. H. (1980). Type A behavior in the work setting. In C. L. Cooper & R. Payne (Eds.), *Current concerns in occupational stress* (pp. 187–212). New York: Wiley.

Cober, R. T., Brown, D. J., Keeping, L. M., & Levy, P. E. (2004). Recruitment on the net: How do organizational web site characteristics influence applicant attraction? *Journal of Management, 30*, 623–646.

Cober, R. T., Brown, D. J., Levy, P. E., Cober, A. B., & Keeping, L. M. (2003). Organizational web sites: Web site content and style as determinants of organizational attraction. *International Journal of Selection and Assessment, 11*, 158–169.

Crites, J. O. (1978).*Career maturity inventory* (rev. ed.). Monterey, CA: MTB/McGraw-Hill.

Dineen, B. R., Ash, S. R., & Noe, R. A. (2002). A web of applicant attraction: Person–organization fit in the context of web-based recruitment. *Journal of Applied Psychology, 87*, 723–734.

Dvorak, P. (2002). DC police hire 60 from Puerto Rico. *The Washington Post, June 24*, B1.

Ellis, R. A., & Taylor, M. S. (1983). Role of self-esteem within the job search process. *Journal of Applied Psychology, 68*, 632–640.

Feldman, D. C. (Ed.). (2002). *Work careers: A developmental perspective*. San Francisco: Jossey-Bass.

Friedman, M., & Rosenman, R. H. (1974).*Type A behavior and your heart*. New York: Knopf.

Gatewood, R. D., & Field, H. S. (1994). *Human resource selection* (3rd ed.). Orlando, FL: Dryden.

Gilliland, S. W. (1993). The perceived fairness of selection systems: An organizational justice perspective. *Academy of Management Review, 18*, 694–734.

Gilliland, S. W., Groth, M., Baker, R. C., IV, Dew, A. F., Polly, L. M., & Langdon, J. C. (2001). Improving applicants' reactions to rejection letters: An application of fairness theory. *Personnel Psychology, 54*, 669–703.

Griffith, R. W., Hom, P. W., & Gaertner, S. (2000). A meta-analysis of antecedents and correlates of employee turnover: Update, moderator tests, and research implications for the next millennium. *Journal of Management, 26*, 463–488.

Hall, D. T. (2002). *Careers in and out of organizations*. Thousand Oaks, CA: Sage.

Hall, D. T., & Kahn, W. A. (2001). Developmental relationship at work: A learning perpective. In C. Copper & R. J. Burke(Eds.), *The new world of work* (pp. 49–74). London:Blackwell.

Harter, J. K., Schmidt, F. L., & Hayes, T. L. (2002). Business-unit-level relationships between employee satisfaction, employee engagement, and business outcomes: A meta-analysis. *Journal of Applied Psychology, 87,* 268–279.

Heneman, R. L., & Judge, T. A. (2004). *Staffing organizations* (4th ed.). Middleton, WI: Mendota House.

Highhouse, S., Zickar, M., Thorsteinson, T., Stierwalt, S., & Slaughter, J. (1999). Assessing company employment image: An example in the fast food industry. *Personnel Psychology, 52,* 151–172.

Holland, J. L. (1997). *Making vocational choices: A theory of vocational personalities and work environments* (3rd ed.). Odessa, FL: PAR.

Hulin, C. L. (1991). Adaptation, persistence, and commitment in organizations. In M. D. Dunnette & L. M. Hough (Eds.), *Handbook of industrial and organizational psychology* (Vol. 2, pp. 445–506). Palo Alto, CA: Consulting Psychologists Press.

Hulin, C. L., Roznowski, M., & Hachiya, D. (1985). Alternative opportunities and withdrawal decisions: Empirical and theoretical discrepancies and an integration. *Psychological Bulletin, 97,* 233–250.

Iaffaldano, M. T., & Muchinsky, P. M. (1985). Job satisfaction and job performance: A meta-analysis. *Psychological Bulletin, 97,* 193–200.

Ivancevich, J. M., & Donnelly, J. M. (1971). Job offers acceptance behavior and reinforcement. *Journal of Applied Psychology, 55,* 119–122.

Jackson, S. E., & Schuler, R. S. (2003). *Managing human resources: Through strategic partnerships* (8th ed.). Mason, OH: South-Western.

Jenkins, C. D., Rosenman, R. H., & Friedman, M. (1967). Development of objective psychological tests for the determination of the coronary-prone behavior pattern in employed men. *Journal of Chronic Diseases, 20,* 371–379.

Judge, T. A. (1992). The dispositional perspective in human resources research. In G. R. Ferris & K. M. Rowland (Eds.), *Research in personnel and human resource management* (Vol. 10, pp. 31–72). Greenwich, CT: JAI Press.

Judge, T. A., & Cable, D. M. (1997). Applicant personality, organizational culture, and organization attraction. *Personnel Psychology, 50,* 359–394.

Judge, T. A., Heller, D., & Mount, M. K. (2002). Five-factor model of personality and job satisfaction: A meta-analysis. *Journal of Applied Psychology, 87,* 530–541.

Judge, T. A., Thoresen, C. J., & Bono, J. E. (2001). The job satisfaction—job performance relationship: A qualitative and quantitative review. *Psychological Bulletin, 127,* 376–407.

Judge, T. A., Thoresen, C. J., Bono, J. E., & Patton, G. K. (2001). The job satisfaction-job performance relationship: A qualitative and quantitative review. *Psychological Bulletin, 127,* 376–407.

Karr, A. R. (2000). A special report about life on the job and trends taking shape there. *Wall Street Journal, 4,* A1.

Katz, D., & Kahn, R. L. (1978). *The social psychology of organizations* (2nd ed.). New York: Wiley.

Korman, A. K. (1999). Motivation, commitment, and the "new contracts" between employers and employees. In A. I. Kraut & A. K. Korman (Eds.), *Evolving practices in human resource management: Responses to a changing world of work* (pp. 23–40). San Francisco: Jossey-Bass.

Kraut, A. I. (1992). Organizational research on work and family issues. In S. Zedeck (Ed.), *Work, families, and organizations* (pp. 208–235). San Francisco: Jossey-Bass.

REFERENCES

Kraut, A. I. (Ed.). (1996). *Organizational surveys: Tools for assessment and change.* San Francisco: Jossey-Bass.
Kristof, A. L., (1996). Person-organization fit: An integrative review of its conceptualizations, measurement and implications. *Personnel Psychology, 49,* 1–50.
Kristof-Brown, A. L., Jansen, K. J., & Colbert, A. E. (2002). A policy-capturing study of the simultaneous effects of fit with jobs, groups, and organizations. *Journal of Applied Psychology, 87,* 985–993.
Lacy, W. B., Bokemeier, J. L., & Shepard, J. M. (1983). Job attribute preferences and work commitment of men and women in the United States. *Personnel Psychology, 36,* 315–330.
Lievens, F., & Highhouse, S. (2003). The relation of instrumental and symbolic attributes to a company's attractiveness as an employer. *Personnel Psychology, 56,* 75–102.
Locke, E. A. (1976). The nature and causes of job satisfaction. In M. D. Dunnette (Ed.), *Handbook of industrial and organizational psychology* (pp. 1297–1350). Chicago: Rand McNally.
Louis, M. (1990). Acculturation in the workplace: Newcomers as lay ethnographers. In B. Schneider (Ed.), *Organizational climate and culture* (pp. 40–84). San Francisco: Jossey-Bass.
Maertz, C. P., & Campion, M. A. (1998). 25 years of voluntary turnover research: A review and critique. In C. L. Cooper & I. T. Robertson (Eds.), *International review of industrial and organizational psychology* (Vol. 13, pp. 49–83). Chichester, UK: Wiley.
March, J. G., & Simon, H. A. (1958). *Organizations.* New York: Wiley.
Meglino, B. M., DeNisi, A. S., Youngblood, S. A., & Williams, K. J. (1988). Effects of realistic job previews: A comparison using an enhancement and reduction preview. *Journal of Applied Psychology, 75,* 259–266.
Meyer, J. P., & Allen, N. J. (1997). *Commitment in the workplace: Theory, research, and application.* Thousand Oaks, CA: Sage.
Mitchell, T. R., Holtom, B. C., Lee, T. W., Sablynski, C. J., & Erez, M. (2001). Why people stay: Using job embeddedness to predict voluntary turnover. *Academy of Management Journal, 44,* 1102–1121.
Mobley, W. H. (1982). *Employee turnover—Causes, consequences and control.* Reading, MA: Addison-Wesley.
Morrow, P. C. (1983). Concept redundancy in organizational research: The case of work commitment. *Academy of Management Review, 8,* 486–500.
Motowidlo, S. J., Borman, W. C., & Schmit, M. J. (1997). A theory of individual differences in task and contextual performance. *Human Performance, 10,* 71–84.
Murphy, K. R. (1986). When your top choice turns you down: Effect of rejected job offers on the utility of selection tests. *Psychological Bulletin, 99,* 133–138.
National Alliance of Business. (1997). *Workforce development trends.* New York: Author.
Organ, D. W. (1988). *Organizational citizenship behavior.* Lexington, MA: Lexington Books.
Organ, D. W. (1997). Organizational citizenship behavior: It's construct clean-up time. *Human Performance, 10,* 85–98.
Phillips, J. M. (1998). Effects of realistic job previews on multiple organizational outcomes: A meta-analysis. *Academy of Management Journal, 41,* 673–690.
Ployhart, R. E., McFarland, L. A., & Ryan, A. M. (2002). Examining applicants' attributions for withdrawal from a selection procedure. *Journal of Applied Social Psychology, 32,* 2228–2252.
Ployhart, R. E., Ryan, A. M., & Bennett, M. (1999). Explanations for selection decisions: Applicants' reactions to informational and sensitivity features of explanations. *Journal of Applied Psychology, 84,* 87–106.

Podsakoff, P. M., & MacKenzie, S. B. (1997). The impact of organizational citizenship behavior on organizational performance: A review and suggestions for further research. *Human Performance, 10,* 133–151.

Podsakoff, P. M., MacKenzie, S. B., Paine, J. B., & Bachrach, D. G. (2000). Organizational citizenship behaviors: A critical review of the theoretical and empirical literature and suggestions for future research. *Journal of Management, 26,* 513–563.

Power, D. J., & Aldag, R. J. (1985). Soelberg's job search and choice model: A clarification, review and critique. *Academy of Management Review, 10,* 48–58.

Premack, S. L., & Wanous, J. P. (1985). A meta-analysis of realistic job preview experiments. *Journal of Applied Psychology, 70,* 706–719.

Price, J. L. (1977). *The study of turnover.* Ames: Iowa State University Press.

Roe, A. (1957). Early determinants of vocational choice. *Journal of Counseling Psychology, 4,* 212–217.

Rousseau, D. M. (1995). *Psychological contracts in organizations: Understanding written and unwritten agreements.* Thousand Oaks, CA: Sage.

Ryan, A. M. (2001). Explaining the Black/White test score gap: The role of test perceptions. *Human Performance, 14,* 45–75.

Ryan, A. M., & Ployhart, R. (2000). Applicants' perceptions of selection procedures and decisions: A critical review and agenda for the future. *Journal of Management, 26,* 565–606.

Ryan, A. M., Sacco, J., McFarland, L. A., & Kriska, S. D. (2000). Applicant self-selection: Correlates of withdrawal from a multiple hurdle process. *Journal of Applied Psychology, 85,* 163–179.

Rynes, S. L., Bretz, R. D., Jr., & Gerhart, B. (1991). The importance of recruitment in job choice: A different way of looking. *Personnel Psychology, 44,* 487–521.

Scarpello, V., & Campbell, J. P. (1983). Job satisfaction: Are all the parts there? *Personnel Psychology, 36,* 577–600.

Schmit, M. J., & Ryan, A. M. (1997). Applicant withdrawal: The role of test-taking attitudes and racial differences. *Personnel Psychology, 50,* 855–876.

Schneider, B. (1987). The people make the place. *Personnel Psychology, 40,* 437–453.

Schneider, B., Gunnarson, S. K., & Wheeler, J. K. (1992). The role of opportunity in the conceptualization and measurement of job satisfaction. In C. J. Cranny, P. C. Smith, & E. F. Stone (Eds.), *Job satisfaction: How people feel about their jobs and how it affects their performance* (pp. 53–63). Lexington, MA: Lexington Books.

Schneider, B., Hanges, P. J., Smith, D. B., & Salvaggio, A. N. (2003). Which comes first: Employee attitudes or organizational financial and market performance? *Journal of Applied Psychology, 88,* 836–851.

Schneider, B., Smith, D. B., Taylor, S., & Fleenor, J. (1998). Personality and organization: A test of the homogeneity of personality hypothesis. *Journal of Applied Psychology, 83,* 462–470.

Schuler, R. A. (1984). *Personnel and human resource management* (2nd ed.). St. Paul, MN: West.

Smith, P. C., Kendall, L. M., & Hulin, C. L. (1969). *The measurement of satisfaction in work and retirement.* Chicago: Rand McNally.

Smither, J. W., Reilly, R. R., Millsap, R. E., Pearlman, K., & Stoffey, R. W. (1993). Applicant reactions to selection procedures. *Personnel Psychology, 46,* 49–76.

Soelberg, P. O. (1967). Unprogrammed decision-making. *Industrial Management Review, 8,* 19–29.

Spector, P. E. (1997). *Job satisfaction: Application, assessment, and consequences.* San Francisco: Jossey-Bass.

REFERENCES

Staw, B. (2004). The dispositional approach to job attitudes: An empirical and conceptual review. In B. Schneider & D. B. Smith (Eds.), *Personality and organizations* (pp. 163–191). Mahwah, NJ: Erlbaum.
Steele, R. P. (2002). Turnover theory at the empirical interface: Problems of fit and function. *Academy of Management Review, 27,* 346–360.
Super, D. E. (1953). A theory of vocational development. *American Psychologist, 8,* 185–190.
Tam, A. P., Murphy, K. R., & Lyall, J. T. (2004). Can changes in differential drop out rates reduce adverse impact? A computer simulation study of a multi-wave selection system. *Personnel Psychology, 57,* 905–934.
Taylor, M. S., & Sniezek, J. A. (1984). The college recruitment interview: Topical content and applicant reactions. *Journal of Occupational Psychology, 57,* 157–168.
Turban, D. B., & Keon, T. L. (1993). Organizational attractiveness: An interactionist perspective. *Journal of Applied Psychology, 78,* 184–193.
Viteles, M. S. (1932). *Industrial psychology.* New York: Norton.
Vroom, V. H. (1964). *Work and motivation.* New York: Wiley.
Vroom, V. H. (1966). Organizational choice: A study of pre- and post-decision processes. *Organizational Behavior and Human Performance, 1,* 212–225.
Wareham, J. (1980). *Secrets of an executive headhunter.* New York: Athenum.
Williamson, I. O., Lepak, D. P., & King, J. (2003). The effect of company recruitment web site orientation on individuals' perceptions of organizational attractiveness. *Journal of Vocational Behavior, 63,* 242–263.
Zippo, M. (1980). Getting the most out of an executive search firm. *Personnel Psychology, 57,* 47–48.

6

VALIDATION STRATEGIES AND UTILITY

AIMS OF THE CHAPTER

This chapter is about how staffing researchers demonstrate the potential meaningfulness and usefulness of the procedures we can develop to assist in making staffing decisions. We cover the following topics: (a) the ways people who do staffing research and practice have come to understand the relationship between performance on a selection procedure and performance on the job—generically called validity; (b) ways to summarize the vast number of attempts to show the relationship between selection procedures and job performance; and (c) approaches for demonstrating the monetary utility of selection procedures.

As we state in chapter 2, validity refers to the degree to which inferences made from predictor scores or other selection procedures are correct or accurate. For example, knowing the content of the GRE and the demands of graduate school, one might look at students' scores on the GRE and infer that they would or would not do well in graduate school. If such inferences are correct (relatively speaking) more often than expected by chance alone, we would say that the GRE yields valid inferences about subsequent performance in graduate school.

Books in measurement and I/O psychology and the *Standards for Educational and Psychological Testing* (American Psychological Association, 1974) used to indicate that there are three "types" of validity: criterion-related, content, and construct validity. *Criterion-related validity* refers to the extent that scores on measures used to make staffing

AIMS OF THE CHAPTER

decisions are correlated with job performance measures. *Content validity* is the degree to which the responses required of job candidates by the potential selection measures are representative of the behaviors to be exhibited in some area of job performance about which we want to make inferences. *Construct validity* represents the degree to which test scores are consistent with our theoretical notions about what the selection measure assesses.

For several reasons, this categorization of the different kinds of validity produced considerable problems when psychologists attempted to explain validity to people not trained in test construction and validation. It has also been the source of considerable confusion among professionals themselves. In the most recent version of the *Standards for Educational and Psychological Testing* (American Educational Research Association (AERA), American Psychological Association, & National Council on Measurement in Education, 1999), it has been recognized that the distinctions themselves are artificial and that all forms of validity relate to the appropriateness of the conceptual arguments that are made about test scores. This makes construct validity the core idea with other forms of validity evidence varieties of construct validity. The 1999 version of the *Standards* describes eight types of evidence for validity and the need to integrate this evidence in a meaningful way as the bases on which to support the inferences we draw from test scores.

I/O psychologists have acted as though the presentation of a significant criterion-related validity coefficient was proof of the test's usefulness and should convince an enlightened management of the great importance of our work. Occasionally, we assert that our procedures are content valid and leave the management decision makers the chore of comparing our assertions with those of others who may document their claims and requests in terms more meaningful to management. Those meaningful terms are usually monetary and are almost always directly related to the company's continued survival, at least as far as management is concerned. For example, when a production manager comes with a request for new machinery, this request is supported with projected increases in productivity and resultant decreases in unit costs of production. The sales manager supports the request for a centralized computer system to process orders with figures concerning the amount of salesperson time saved and estimates of the increased amount of sales that could be generated during this time. The maintenance manager supports the request to hire three new people with figures concerning the amount of downtime due to equipment malfunction that could be reduced and the resulting savings that could be achieved by the addition of three new persons working at the same rate as the current maintenance people. The point is if HR managers do not compete for scarce organizational resources in the same way as their colleagues, the chances of successfully competing are small. This chapter includes a brief introduction to the manner in which staffing personnel can provide a convincing monetary justification for the use of staffing measures. Interestingly, as we show, the monetary utility of a selection procedure is related to the criterion-related validity of that procedure.

Finally, significant advances in our understanding of the construct validity and the generalizability of our research on validity have come as the result of meta-analysis and validity generalization work initiated by Schmidt and Hunter (1977, 1998). This work takes the results of many criterion-related validity studies and summarizes it, yielding inferences about the usefulness of a particular selection procedure across settings and samples. We provide a brief explanation of the approach, the results of some meta-analyses, and some thoughts on how these kinds of results might inform staffing procedures and decisions in the future.

APPROACHES TO VALIDATION

As mentioned previously, the 1999 version of the *Standards* describes eight sources of validity evidence. Evidence based on test content and test–criterion relationships reflects issues that were once discussed under the labels of content and criterion-related validity. Evidence based on response processes, internal structure, relationships to other variables, validity generalization, and convergent and discriminant evidence reflects issues that were once considered to be aspects of construct validity. Also discussed is evidence about the intended and unintended consequences of testing, termed consequential validity by Messick (1998). We discuss each type of evidence and provide examples that illustrate each. As you will see, there can be considerable overlap when discussing these issues with regard to validity, but each of the eight forms of validity (plus consequential validity) addresses a facet of the general idea of validity that requires explication and attention.

APPROACHES TO VALIDATION

Test–Criterion Relationships

Criterion-related validity applies when we want to infer from some measure of individuals' attributes their standing on a criterion. The criterion in industrial applications is usually some measure of job performance, including sales, raw productivity per hour, helping others, customer service, or turnover—whatever is important for the organization to be successful. In staffing, we are interested in predicting individuals' future levels on the criterion from their current standing on a test. Even in the case of "job samples," the reason for using the job sample as a basis for making hiring decisions is not that we are interested in their scores on the job sample, but that we believe those scores permit us to make valid inferences about future job performance. It follows that when we intend to measure ability or aptitude, we are hypothesizing a relationship of ability to job performance, and in a criterion-related validity study, we are investigating the validity of that hypothesis, not the validity of the test (Guion, 1976). This criterion-related validity would seem simple enough to establish: (a) Collect scores on the predictors to be evaluated, (b) measure the criterion job performance, and (c) correlate the two sets of measures. However, there are significant problems associated with criterion-related validity studies that make their feasibility problematic in many situations. Most complications arise from various practical constraints on the type of research design we are able to employ in criterion-related validation studies.

Concurrent and Predictive Validation Designs

There are two basic designs by which I/O psychologists have sought to establish the criterion-related validity of their instruments: concurrent and predictive. In a *concurrent validity study*, predictor and criterion data are collected at the same time from current employees. In a *predictive validity study*, selection data are collected on a group of applicants but not used for selection decisions. When it is possible to measure their subsequent job performance, the job performance data are correlated with the original scores on the selection procedures being evaluated. The difference between concurrent and predictive criterion-related validity studies is shown in Fig. 6.1.

I/O psychologists have nearly always preferred the predictive design if it is feasible. Concurrent studies—using people already on the

	TIME₁	TIME₂
Concurrent validity	–Collect test data –Collect criterion data –Correlate test and criterion data (Study participants are job incumbents.)	
Predictive validity	–Collect test data from job applicants but do not use it to make hiring decisions.*	–Collect criterion data –Correlate test and criterion data

FIG. 6.1. The difference between concurrent and predictive criterion-related validity studies.
*Frequently, organizations employ a predictive design to evaluate criterion-related validity, but employ the test in making decisions, thereby producing a restriction in the range of predictor scores and, to the degree that the test is valid, in the criterion as well. The correlation computed between predictor and criterion in these instances will be an underestimate of criterion-related validity.

job—were criticized because experience on the job might affect test scores and because of the idea that incumbent employees might not approach the test with the same motivation or anxiety as would a job applicant. There are other obvious potential problems with the concurrent design: (a) Low-performing employees may have left the firm, and (b) the best performing employees may have been promoted to new jobs, resulting in the collection of data on a group of employees whose job performance—and test scores—are much less variable than they would be for an unselected group. The latter problem scientists assumed should produce underestimates of the validity of tests through what is called a restriction of range problem (see Fig. 6.2).

However, actual empirical comparisons of criterion-related validity coefficients derived from concurrent and predictive studies indicate that they yield virtually identical estimates (Barrett, Phillips, & Alexander, 1981; Schmitt, Gooding, Noe, & Kirsch, 1984), but there remain good conceptual reasons to maintain reservations about concurrent studies (Sussmann & Robertson, 1986). These concerns are really about whether inferences made from test scores on incumbents are identical to test scores obtained with applicants.

Feasibility of Criterion-Related Validity Studies

I/O psychologists have long preferred the use of criterion-related studies (whether concurrent or predictive) over other more judgmental or theory-based approaches we discuss. However, since the

APPROACHES TO VALIDATION 305

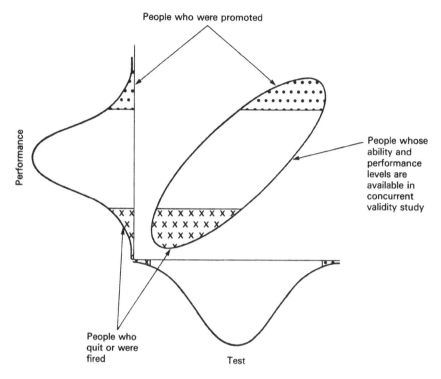

FIG. 6.2. Restriction of range caused by low performing workers being fired or leaving (X X X X) and high performing workers being promoted (• • •) in a concurrent criterion-related validity study.

mid-1980s, we recognize that criterion-related studies are infeasible in many situations. Furthermore, we recognize that judgment is also basic to criterion-related validity in that someone must decide what the criterion data will be. If the criteria with which tests are correlated are not relevant (i.e., contaminated or deficient in some manner as discussed in chapter 4), then our estimates of the validity of the tests will also not be appropriate.

In an excellent analysis of the feasibility of criterion-related validation studies, Schmidt, Hunter, and Urry (1976) examined the implications of both lack of criterion reliability and restriction of range in test validation research. Their basic point was that, given (a) the sample size available in most criterion-related studies, (b) the modest usual level of validity and criterion reliability, and (c) the usual range restriction in those studies, then the probability of finding a statistically significant criterion-related validity coefficient is very low. Their analyses were based on two formulae that quantify the problems of range restriction and criterion unreliability.

Previously, we note that the use of the concurrent criterion-related validity design can reduce the range of individual differences in the sample in terms of both the criterion of interest (poor performers leave; good performers get promoted) and, of course, the predictors—because those who are no longer there cannot be tested (see Fig. 6.2). As shown in Fig. 6.2, the portion of the scatterplot that is available for study with those missing is more nearly circular (recall from chapter 2 that a circular scatterplot represents a correlation of zero); hence, the estimate of validity in this situation will be lower than is actually the case. The diagram is intended to depict a concurrent study, but similar restriction likely occurs in predictive studies as well because organizations frequently may use a predictive validation design but do not hire very low-scoring persons or hire based on other criteria (e.g., friends of the owner, other attractive characteristics), which are correlated with the predictor being studied.

As an example of the effect of restriction of range, consider a situation in which the true validity of a prediction for a population is 0.60, the standard deviation of scores of the entire applicant group on some test is 20, and the standard deviation of the test score of those actually selected is 10 (which would be true if the top 30% of all applicants were selected). The expected observed validity in this case can be estimated by modification of a formula provided by Thorndike (1949) for correction of range restriction:

$$r_{xy} = \frac{(s_r/s_u)R_{xy}}{\sqrt{(s_r/s_u)^2 R_{xy}^2 - R_{xy}^2 + 1}} = \frac{(10/20)0.60}{\sqrt{(10/20)^2 0.60^2 - 0.60^2 + 1}} = 0.35$$

where s_r = standard deviation on the test for those selected,
s_u = standard deviation on the test for all applicants,
r_{xy} = observed validity of the test, and
R_{xy} = true validity of the test if calculated on the basis of hiring all applicants.

Essentially, this formula shows how much of a reduction in validity would occur if the distribution of scores was reduced or restricted. The formula and computations indicate that even though the true validity of the test is 0.60, a study conducted in this situation of restriction of range would yield an estimate of only 0.35.

APPROACHES TO VALIDATION

In addition, the validity of the test could be underestimated because the criterion against which the test is evaluated is not measured with perfect reliability. Assume a criterion reliability of 0.81. The correction for attenuation due to lack of reliability (recall our discussion of this issue in chapter 2) and the calculation for our hypothetical situation is as follows:

$$r_{xy} = (\sqrt{r_{yy}})(r_{xyc}) = (\sqrt{0.81})(0.35) = 0.315$$

where r_{xy} = observed validity,
r_{yy} = criterion reliability, and
r_{xyc} = validity after the correction for restriction of range.

These calculations, then, suggest that the actual observed validity given this relatively realistic set of data would be only 0.31. The problem with the expectation that our observed validity coefficient is going to be 0.31 as opposed to 0.60 is that a much larger sample size will be required for us to conclude that the criterion-related validity coefficient is statistically significant. In tests of criterion-related validity, we are testing the hypothesis that the validity is greater than chance ($r = 0.00$). If we fail to find a significant coefficient (see chapter 2), as is more likely with a small sample size, we falsely conclude that the test is not predictive of job performance.

It is possible to calculate the sample sizes required to be reasonably confident of concluding that a validity coefficient of typical magnitude would be statistically significant (Schmidt et al., 1976). The results are often quite discouraging. For example, if we estimate the true validity coefficient to be 0.35, we want a 90% probability of finding a significant result, our criterion measure has a reliability of 0.70, and we are selecting the top 30% of the applicants, the required sample size would be 428. It is also important to note that, aside from the issue of finding statistical significance, the underestimates of validity that result from range restriction and lack of criterion reliability lower our estimates of the utility of selection instruments, as is obvious when we discuss utility estimation at the end of this chapter.

Also, when we discuss validity generalization later in the chapter, it becomes clear why the use of small samples for estimating criterion-related validity has resulted in many erroneous conclusions about the actual effectiveness in terms of criterion-related validity that many of our selection procedures actually reveal.

Issues Requiring Attention in Conducting a Criterion-Related Validity Study

Given the idea that there are many potential problems one can encounter in doing a criterion-related validity study, we recommend that careful consideration be given to each of the following issues prior to actually beginning the study:

1. Decide on the alpha level. Alpha level (sometimes called the p value) is the probability that an observed correlation occurred by chance. An alpha level of .05 is the level most frequently used, and it tells us that the chances are only 5 in 100 that an observed value occurred by chance. However, if we are willing to increase this probability, say to .10, we will be able to reduce the necessary sample size. Thus, a correlation with a sample of 50 people may not reach the p value required for .05 when the same sample and correlation would reach the p value of .10, obviating the necessity to gather additional samples.

2. Read the literature to decide what level of criterion-related validity you can expect to find in your situation, or make a determination of the level of correlation you need to make the whole selection procedure worthwhile (see the utility discussion at the end of this chapter).

3. Decide on an acceptable level of probability that you will conclude the test has no validity. That is, what is a reasonable probability that you will decide not to use a potentially worthwhile selection instrument?

4. Apply the following formula to determine the necessary sample size:

$$N = \left(((Z_1 - Z_2)^2)/E(F_z)^2\right) + 3$$

where N = sample size,
 Z_1 = standard normal deviate corresponding to the alpha probability,
 Z_2 = standard normal deviate corresponding to the probability that we will not find a significant validity coefficient, and
 $E(F_z)$ = Fisher z equivalent of the expected validity coefficient after correction as shown for restriction of range and criterion unreliability.

Fisher z equivalents of correlations can be found in tables in most measurement books. For example, assume the estimate of observed validity calculated previously; namely, 0.35. Its Fisher z equivalent is 0.326.

APPROACHES TO VALIDATION 309

Also assume the alpha level is .05; z is then 1.96. Finally, assume we must be 90% certain that we will detect a significant correlation (validity); z is then -1.28. Solving for N in the previous equation yields the following:

$$N = (((1.96 - (-1.28))^2)/(0.326^2)) + 3 = 102$$

So, we must have a sample of 102 or more to conclude that the criterion-related validity study is feasible within the levels of probability set previously. Even when feasible, it may be more desirable to rely on the results of validity generalization studies outlined later to get the most accurate assessment of criterion-related validity.

Given the implications of the Schmidt et al. (1976) analysis, different strategies of test validation have become increasingly popular. Indeed, work on validity generalization summarized later in this chapter should make it less obligatory to conduct criterion-related validation studies of tests that have been frequently validated for the same job for which some new test usage is proposed. Where existing validity evidence is substantial, it should be sufficient to show that use of a test to select people for a particular job is similar to the jobs and the criteria for which extensive validity data already exist. Box 6.1 discusses some options for small businesses who want to show the validity of their staffing procedures, but do not have the sample size to perform a criterion-related study.

The Use of Consortia for Validity Studies

There is another approach that solves the problem of small sample sizes that any one company might encounter. Some companies within a single industry (e.g., banking, electric utilities, life insurance) cooperate in validation studies for jobs that are similar across the companies. Such consortium studies substantially increase the sample size available for any given job. We believe these studies can be valuable even when there is meta-analytic support for predictor–criterion relationships. Thus, although we have sufficient evidence for validity generalization of an increasingly large number of tests, there are still always reporting deficiencies in the primary database and in theory-based questions (Bobko, Roth, & Potosky, 1999; Borman, White, Pulakos, & Oppler, 1991; Schmitt, Rogers, Chan, Sheppard, & Jennings, 1997) that make large sample studies useful. Our position is that continued criterion-related validation of novel testing strategies with larger sample sizes, perhaps multiorganizational cooperative validation efforts, are needed and are currently being conducted.

> **Box 6.1 Validity Options for Small Businesses**
>
> Throughout this chapter, we discuss various ways to demonstrate validity for our staffing procedures. Small business owners frequently do not have the sample size or resources available to conduct a criterion-related validity study. What then can they do?
>
> 1. *Join a consortium and use synthetic validity.* In this strategy, the small business owner joins a consortium of similar businesses with similar jobs. Although any one of these businesses does not have the sample size to conduct a criterion-related study, together they might and the criterion-related study would be based on employees from multiple organizations. This type of validity evidence would be known as synthetic validity.
>
> 2. *Use job analysis and content validity.* We illustrate in chapter 3 (Box 3.1) how a small business could use the O*NET system to conduct a job analysis. This same information can then be used to demonstrate the content validity of the selection procedure. For example, interview questions may be based on the KSAOs identified by O*NET. Alternatively, one may use the PAQ as a means to argue the job is sufficiently similar to other jobs that have already been extensively studied.
>
> 3. *Validity generalization.* Examination of validity generalization studies in the academic literature may identify the key predictors of performance for a given job, or the validity of a particular test or construct across jobs. Justifying the use for a test may be better served by basing it on validity generalization evidence when sample sizes are small.
>
> 4. *Use an existing test battery.* Good test publishers follow stringent development methods and ensure the validity of their measures for the appropriate purposes. Test publishers frequently provide information similar to validity generalization evidence supporting the use of a test for certain types of jobs.

Why Are Predictors That Should Be Valid Not Valid?

Another issue requiring attention concerns attempts to understand why predictors that should work fail to work, even when sample size, criterion reliability, and range restriction problems are minimal

APPROACHES TO VALIDATION 311

or nonexistent. Contextual factors may (a) permit a correlation between a predictor and a criterion but suppress performance levels, or (b) prevent a predictor from revealing a strong relationship to a criterion. Researchers (e.g., Klein & Kozlowski, 2000) have recognized the fact that units or even organizations may be characterized by constraints that do not allow the expression of individual differences. We described a data analytic technique (i.e., hierarchical linear modeling) in chapter 2 that allows us to investigate such level effects, and in chapter 12, we discuss the implications of such issues for staffing research. Here, we provide an example of how such constraints might operate.

In the situation indicated in Fig. 6.3, we have two departments in an organization doing the same work. If we consider the relationship between ability and performance in these two departments, we find the same relationship. We also see the same relative levels of ability in the groups as a whole (represented by their position on the horizontal axis). However, performance levels in the two units differ, and the overall ability–performance relationship for the two units combined is lower (more nearly circular) than the relationship in the individual units. A number of factors may provide an explanation for this phenomenon. For example, norms in one department may dictate that

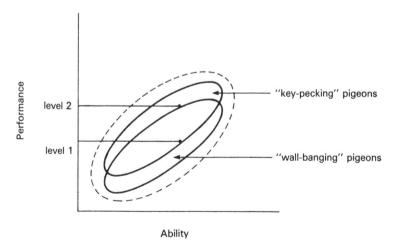

FIG. 6.3. How relationships between predictors and criteria can be the same in two or more groups with differing levels of performance between groups. Note the dotted line around the scatterplots for both key-pecking pigeons and wall-banging pigeons is representative of what we would obtain if the two groups were undifferentiated. This dotted scatterplot is more circular, indicating a lower validity coefficient.

workers limit their production (e.g., like in the Hawthorne studies described in chapter 1), or the equipment in one group may be obsolescent compared with the equipment in the other group, or one department has responsibilities not considered part of the performance index used to judge employees. The point is that there can be artificial limitations on performance in one group such that the average level of performance is lower. Although in appropriate subgroups (e.g., shown in Fig. 6.3) validities may be equal, studies conducted across the groups (departments, organizations) may yield lower (or higher) validities.

When something in the context has an effect on the validity coefficient, we say that the context serves as a moderator of the predictor–criterion relationship (see chapter 2). Some efforts designed to identify contextual factors that moderate predictor–criterion relationships have been a failure (Schneider, 1978, 1983). However, one reason these factors were not found is that it was assumed the effect of contextual issues was to suppress correlations. That is, true moderators were sought wherein it was expected that the predictor–criterion relationship in one context would be shown to be significantly different from the relationship in another context. In fact, the more likely outcome was that the relationships (slopes) were the same in the two contexts, but the levels of performance on the criteria differed (intercepts), as shown in Fig. 6.3. This harkens back to our discussion of HLM in chapter 2, and using HLM would be the appropriate way to model these questions. So, when we remove barriers to employee performance, we can expect all employees to perform better, not just some subgroup or organization. We are only beginning to research these kinds of contextual effects (called "levels effects"; see Klein & Kozlowski, 2000) associated with predictor–criterion relationships. For example, Ployhart, Weekley, and Baughman (in press) examined relationships between personality, job satisfaction, and performance across 85 jobs and 12 organizations. The data were analyzed with HLM. Consistent with prior meta-analyses, they found no significant variability in validities (slopes) across jobs and organizations. However, they found between-job and between-organization differences in job satisfaction and performance that was explained by *aggregate personality*. Ployhart and Schneider (2005) provided theoretical and analytical frameworks that serve as the basis for a more thorough look at the manner in which organizational and group constraints or enhancers may influence the outcomes staffing researchers hope to achieve and the relationships they observe when they conduct criterion-related validity studies.

Summary

We began this section by saying that it should be quite straightforward to establish the criterion-related validity of a given staffing procedure: Administer the measure to applicants, and then later collect the criterion data on them and correlate the two sets of data. As with most things in life, what looks straightforward is complex beneath the surface. We showed here that such issues as restriction in range, criterion unreliability, small sample size, and contextual effects can all lead us astray in the conclusions we reach about the criterion-related validity of a given measure or staffing procedure. These possibilities each require attention in evaluating the conduct and outcome of such a study, and they suggest the potential usefulness of meta-analysis and the use of consortia for the conduct of new studies.

VALIDITY BASED ON TEST CONTENT

Both the APA *Standards* (AERA et al., 1999) and the recent version of the Society for Industrial and Organizational Psychology's *Principles* (2003) recognize the importance of an examination of the degree to which the content of a selection procedure or measure reflects the construct(s) it is intended to measure. Both documents also recognize that test content includes the questions and tasks included on the test, as well as format (computer versus face to face) and wording (language level of required reading) of questions, the response format (multiple choice versus behavioral responding), and the administration and scoring of the test. With regard to administration, for example, if one develops a selection procedure or a performance measure and wants to estimate how a person performs on it, the measure (test or criterion) must represent a sampling of those situations that are believed to represent a particular domain. This universe of situations in the employment context must include job behaviors. Test content has always been of primary concern to people constructing knowledge or achievement tests. However, a similar concern should guide the construction of personality and aptitude tests, as well as the criteria of job performance that staffing researchers use as standards to evaluate tests in criterion-related validity studies. In all cases, when the nature of the test content is the only evidence of validity, there must be evidence that the item content of the measure is a representative sample of behaviors to be exhibited in some performance domain on the job. Consequently,

- The performance domain must be carefully specified.
- The objectives of the test user must be clearly formulated.
- The method of sampling from the possible performance domain must be adequate.

In applications of content validation to personnel selection, this approach demands a careful specification of the job domain or content, as well as careful test construction and examination of resultant test content to ensure the content of the job is accurately reflected in the test or criterion. This specification usually comes from a careful job analysis of the type described in chapter 3. Evidence of validity based on test content should include the following activities and/or considerations:

1. *Thorough and reliable job analysis.* This job analysis is conducted with individuals who have considerable knowledge about the position under study and is designed to identify the KSAOs and/or tasks required for satisfactory job performance and their relative importance within the position. After lists of tasks and KSAOs are identified, it is often the case that a representative sample of experienced job incumbents are asked to rate these items on a variety of scales (see chapter 2; Brannick & Levine, 2002; Schmitt & Chan, 1998, for examples) to determine what SMEs consider important or relevant to successful job performance.

2. *Construction of the examination content.* The results of the job analysis exercise outlined previously determine the content areas around which the measures should be developed. One such test specification is provided in Fig. 6.4, where a test plan has been developed for a 100-item test to assess the various objectives associated with a course in measurement. So, Fig. 6.4 shows that there were six objectives of the course, and these objectives were associated with three key constructs that helped define the item content. For a job, there might be five performance objectives (e.g., raw productivity, absenteeism, safety, helping others, scrappage), and a test would be designed to assess three content domains (job aptitude, personality conscientiousness, and personality extroversion).

3. *Concerns for exam format.* In constructing the test, we must also be concerned with whether the examination format conforms as closely as possible to the job task. The more closely the actual performance required on the job is to the behavior actually tested, the more likely

VALIDITY BASED ON TEST CONTENT

Educational Objective[a]	TEST CONTENT			
	Reliability	Validity	Item Analysis	Total
Knowledge	10	10	5	25
Comprehension	10	10	2	22
Application	15	20	12	47
Analysis	2	0	1	3
Synthesis	2	0	0	2
Evaluation	1	0	0	1
Total	40	40	20	

[a]The objectives listed are from the *Taxonomy of Educational Objectives* (Bloom, 1956). These objectives are universal in that they cover all potential objectives of achievement tests.

FIG. 6.4. Test plan for a course in measurement. [Provided as an example only. An actual plan may involve more detail in both content and objectives. The numbers in the blocks of the table represent the number of test items and, assuming equal item variability, are indices of the importance of particular content areas.]

it is that the required KSAOs are being measured. An example of the presumed content validity of several possible test formats to measure the "ability to relate to members of varied socioeconomic levels" is shown in Fig. 6.5. There we see that a personality measure assessing dogmatism would not be believed to be content relevant, whereas a careful documentation of the degree to which an applicant has had these kinds of experiences in their previous jobs would likely be seen to have content validity.

4. *SME judgments of the content validity of a measure.* After content selection and examination construction, content validation of tests requires informed judgments from people who know about the job, its knowledge and ability requirements, and the test content. These SMEs

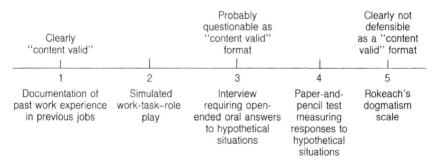

FIG. 6.5. The "content validity" of various test formats as possible measures of the ability to relate to members of varied socioeconomic levels.
Source: From Campbell, (1973).

must be carefully chosen and instructed and, for legal reasons, their credentials must be documented. Judges are most often asked to evaluate the job relevancy, accuracy (of the situation presented and the alternative answers if a multiple-choice question), and the fairness of the item (manner of presentation). Of most importance, of course, is job relevance. Often, the judges are asked to link each test item or exercise back to an important KSAO as determined in the job analysis making the link between the examination and the job analysis explicit.

5. *Setting "cut scores" for examination performance.* Cut scores for important decisions (selection/rejection or promotion/no promotion) must be considered. The issue is where one draws the line for the score or scores that indicate acceptance versus rejection. This may be the most difficult task. The reason for this difficulty lies in the fact that most individual differences are continuously, not dichotomously, distributed. The ability to distinguish colors may be a minimal requirement for an interior decorator, but it would be less easy to say what minimal level of interpersonal skill is essential. The solution is again judgmental. A panel of judges is asked to categorize the test tasks or content into those that require skills or knowledge that all employees must possess and those test tasks that discriminate between average and superior employees. These data are cumulated and a specific minimum score for being hired is established (Livingston & Zieky, 1982).

It should also be pointed out that establishment of minimal cutoff scores does not imply that those above the minimum cut score should be chosen randomly. Selecting the most qualified applicants from the top down in rank order almost always results in higher average expected criterion performance than other strategies do. Cutoff scores for tests for which we have evidence of a relationship to important outcomes can be developed by reference to a particular level of job performance (Fig. 6.6). Thus, when criterion-related validity exists, the cut score for exam performance is a direct function of the establishment of the standard for criterion performance. Then, the exam score that produces that level of criterion performance is set as the standard and becomes the cut score.

As well as matching the format of a predictor to the performance requirements of the job, the test constructor must also use communication devices or techniques that are consistent with the requirements of the job. For example, when written tests are used, the reading level

1. Regress the outcome variable on the predictor variable or a weighted composite of a set of predictors.
2. Determine what level of performance on the outcome is acceptable. If the outcome variable is a rating scale and a "3" on this scale is labeled "satisfactory," that might be the target or acceptable performance level.
3. Use the regression equation to determine what predictor level would be associated with the acceptable outcome (e.g., acceptable outcome = 3).
Let's say the regression equation is as follows:

$$Y = 0.4(x) + 2.2$$

Then an acceptable level of performance would be associated with a predictor value of 2:

$$3 = 0.4(x) + 2.2$$
$$-0.4(x) = -0.8$$
$$x = 2$$

4. Then, the cutoff might be moved up or down using the standard error of estimate (see chapter 2) to determine with some level of confidence that persons with a specified predictor score would have a certain probability of achieving a performance outcome of 3 or "satisfactory" better.

FIG. 6.6. Demonstration of setting a test cutoff score when the relationship between the test and a criterion is known.

required by the test should match that necessary to read job-related manuals or instructions. For example, even when the intent of a test is to measure the ability to perform physical aspects of a job, the instructions should be clearly understood by all examinees. Reading-ease formulas are available to make appropriate assessments of reading difficulty (Flesch, 1949; McLaughlin, 1969) and are now part of most commonly used word processing packages (Fig. 6.7).

Generally speaking, this discussion of the construction of predictors with the special attention directed to the content of the predictor has emphasized two issues: (a) the criticality of the job analysis, and (b) the relationship between the actual job behaviors and the behaviors measured by the proposed predictor of job performance. In general, we can state that the more the predictor is psychologically similar to the criterion, the more the predictor is content valid. By "psychologically

Text

After its purchase, the corporate management of the Weldon Company was taken over by Harwood, but the plant manager and the managerial and supervisory staffs in the Weldon plant were retained. A few additional supervisors were appointed subsequently. The Weldon plant is located in Williamsport, Pennsylvania and employs approximately 800 employees.

1. Count the word length in syllables and compute an average word length, then multiply by 100.
   Total Number of Syllables = 102
   Total Number of Words = 51
   Average Word Length = 2.0
   Multiply by 100 = 200.0
2. Compute the average number of words in each sentence.
   Total Number of Words = 51
   Total Number of Sentences = 3
   Average Sentence Length = 17.0
3. Compute the Reading Ease Formula as Follows:
   R.E. = 206.835 − .846 (100 × Average Word Length) − 1.05 (Average Sentence Length)
   = 206.835 − 169.2 − 17.85
   = 19.785
4. Refer to the table below to determine the level of reading difficulty represented by this text.

PATTERN OF "READING EASE" SCORES

"Reading Ease" Score	Description of Style	Typical Magazine	Syllables per 100 Words	Average Sentence Length in Words
0 to 30	Very difficult	Scientific	192 or more	29 or more
30 to 50	Difficult	Academic	167	25
50 to 60	Fairly difficult	Quality	155	21
60 to 70	Standard	Digests	147	17
70 to 80	Fairly easy	Slick-fiction	139	14
80 to 90	Easy	Pulp-fiction	131	11
90 to 100	Very easy	Comics	123 or fewer	8 or fewer

FIG. 6.7. Application of Flesch Reading Ease formula to a short text.

similar," we mean that the predictor makes the same kinds of psychological demands on an applicant as the job will make on a worker. Trainers refer to psychological similarity as psychological fidelity—the overlap of the training with the requirements of on-the-job performance (Goldstein & Ford, 2002). We can use the same analogy for the overlap of a predictor with on-the-job requirements. Lawshe (1975) presented a way to quantify this overlap. He proposed that a content validity ratio (CVR) be calculated from the independent relevancy ratings of expert judges. Each judge is asked to indicate whether the knowledge, skill, or ability measured by the item is essential, useful but not essential, or not necessary to the performance of the job. The CVR for each item is then computed using the following formula:

$$CVR = (n_e - N/2)$$

where n_e equals the number of judges indicating the item was essential, and N is the total number of judges. Items are then eliminated if most of the judges do not indicate the item is essential (CVR = 0, when half the judges do not believe the item is essential). A content validity index can also be computed for the test as a whole by averaging the CVRs for the items remaining after all individual items (exercises, role plays, or whatever) are rated. The content validity index, then, represents the extent to which the judges perceive overlap between the KSAOs relevant to the job performance domain and the test performance domain.

Administering and scoring an exam developed in this manner is no different than administering or scoring any other exam. Examinees should be informed of the test content, perhaps employing sample items; efforts should be made to minimize extraneous variables such as test anxiety, lighting and noise problems, and so on. Above all, standardization of the testing situation for all applicants must be assured. Finally, item analyses and reliability estimates (appropriate estimates may include test–retest, intercoder, or internal consistency) of the test should be made following test administration. Given this information, the tests may be further revised for future administration.

VALIDITY BASED ON THEORETICAL MEANINGFULNESS OF TESTS

Several types of evidence identified in the *Standards* (AERA et al., 1999) relate to the degree to which test responses are consistent with conceptual or theoretical arguments. Validity in this case is evaluated by investigating the qualities a test measures to determine the degree to which certain explanatory psychological concepts or constructs account for performance on a test. Using the terminology from chapter 2, we are interested in whether variance in manifest variables is caused by latent constructs. A psychological construct is an idea that is used to organize or integrate existing knowledge about some phenomenon. For example, a researcher might observe that some people move about more frequently than others while waiting for an examination to begin. Because of the circumstances and the researcher's previous experience with examinations, it may be inferred that individuals who move about more are experiencing more anxiety. Anxiety itself is not observable, but the movement that is a hypothesized correlate of anxiety is observable and is the basis for the inference (this harkens back to the distinction between latent constructs and manifest measures in chapter 2).

Anxiety might also be associated with certain body changes and oral reports of being worried. In fact, investigators interested in studying anxiety have developed measures of it, such as paper-and-pencil questionnaires or the galvanic skin response (which assesses minute changes in the electrical conductivity of the skin).

Information about the constructs measured takes many forms. First, the different measures themselves (e.g., the questionnaires, the galvanic skin response) should yield the same information about people's anxiety states; that is, they should correlate highly. Second, the measures should be sensitive to situations in which the experience of anxiety is logically high or low. That is, people engaged in a competitive sporting event, taking an important licensing exam, getting married, or caught in a severe storm should get high scores on the anxiety measure. People relaxing at home should get low scores. Third, individuals who typically appear excitable or nervous to others should get high scores, whereas those seen by others as always calm should receive low scores. The point here is that development of measures to assess psychological constructs must yield evidence of support for the construct and its measurement in numerous ways—for example, observation of people, reports by those people, documentation that the measures behave as predicted under different circumstances, and high intercorrelation of different kinds of measures of the same construct.

Evidence Based on Response Processes That Underlie Test Performance

In criterion-related validity and content validity designs, responses to the test content are the focus of interest: (a) for criterion-related validity, how responses to the test items correlate with the performance criterion, and (b) for content validity, how judgments of the overlap of test content with what the job analysis has shown to be important for job performance. It may be useful, however, to look behind the responses to item content and ask what psychological processes might be the likely correlates (or even causes) of performance on test items. Such an exercise would provide insight into why some people score differently than others on the measure of interest.

In short, examination of the response processes of test takers—what lies behind their responses to test content—can provide evidence about the meaning of test scores. We provide a detailed description of the predictor response process in chapter 7, but let us provide a simple example of this principle here. Suppose we have a test that requires a job applicant to read some directions for use of a piece of equipment used on the job and then demonstrate the use of the piece of equipment. Because the equipment is job relevant, this test could be seen as displaying content validity, but suppose we find that test scores are highly related to examinees' vocabulary level. An important question would then become whether mastery of the vocabulary is necessary to learn how to use this equipment. Vocabulary now becomes a precursor to performance on the piece of equipment.

As another example, a common problem in the use of biodata measures (see chapter 9) concerns evidence that some respondents try to figure out what the "correct" responses are to the questions about their life history experiences and choose those responses rather than those that represent what they have actually experienced. In fact, McDaniel and Timms (1990) reported that respondents to biographical data questions took longer to respond to the biodata items when there was evidence that they were lying.

Frederiksen (1986) detailed the development of a test for admission of medical students to medical school. Frederiksen began by constructing a criterion measure to assess a doctor's ability to solve "patient management problems." The test simulates, on paper, a situation that might be encountered by a resident on duty in the emergency ward of a hospital. At first, the examinees are given a small amount of information about a patient and are asked to indicate what diagnosis comes to mind. Then, they are allowed to seek further information, which is provided if they seek it, and they start a new cycle of information gathering and hypothesis testing concerning diagnosis. Assuming the best way to predict performance on this criterion test was to use a test similar to it, the selection tests were constructed using problems requiring similar information gathering strategies and hypothesis generation components, but not requiring medical knowledge (which a medical school applicant would not have). A third set of variables, which Frederiksen termed process variables, was included to investigate the construct validity of both the selection and criterion measures. These were called process variables because they represent hypotheses concerning what might be happening of an internal or process nature that would account for success on either or both the criterion and predictor variables. These tests included tests of medical knowledge, medical school grades, and ratings, as well as cognitive and personality tests. All three sets of measures—the criterion tests, the selection measures, and the process measures—were administered to fourth-year medical students in a concurrent validity strategy.

Frederiksen then assessed validity in several different ways. First, the correlations between criterion and selection tests represent concurrent criterion-related validity assuming the job relevance of the criterion measures. Second, future evidence of predictive validity can be gathered by correlating scores of incoming students on the selection tests with their subsequent performance in medical school. By correlating criterion measures with the process variables mentioned previously and examining the correlations in light of reasonable hypotheses about diagnostic problem solving, information about the construct validity of the criterion measures will accumulate. Similarly, we can examine the relationship between the new selection tests and other more traditional medical school entrance exams. The pattern of correlations between the selection tests and process variables on the one hand, and the criterion tests and process measures on the other hand, should be similar if the criteria and predictors involve similar constructs. Finally, comparisons of fourth- and first-year medical students with respect to the relationships between selection tests and process measures would provide evidence of the effect of medical training on the constructs measured. In brief, what Frederiksen did was to hypothesize two process variables, information gathering and hypothesis generation, and then showed that assessment of these variables related to performance on other predictors of medical student performance and to other indicators of medical student performance in medical school. This suggests that these constructs are more basic or more fundamental process variables that underlie more surface—albeit important—test and criterion performance.

Evidence Based on Internal Structure of the Test

As is obvious from the attention we paid to measurement issues in chapter 2, another source of evidence about tests or measures concerns statistical analysis of the internal nature or structure of the measure. In these analyses, we search for (a) the degree to which different items in a test (or different exercises in an assessment center) yield correlated results, and (b) the degree to which different items designed to measure

THEORETICAL MEANINGFULNESS OF TESTS 323

one construct can be differentiated from another set of items designed to measure another construct. Thus, analyses of the interrelationships among test items or test components can tell us whether responses to a test are consistent with the conceptual or theoretical bases of the test. These analyses include item means, standard deviations, and item intercorrelations; intercorrelations among subcomponents of the test; analysis of coefficient alpha; and exploratory or confirmatory factor analyses.

For example, analyses of 25 items—Goldberg (1999) thought to measure the Big Five personality dimensions (Conscientiousness, Emotional Stability, Openness, Extraversion, and Agreeableness; see chapter 9)—are presented in Table 6.1A. The first two columns of Table 6.1A present the means and standard deviations of the items. In this case, responses were made on 5-point scales ranging from "Very accurate" to "Very inaccurate." Because items were scored in a direction that represents a positive appraisal of oneself, it is reasonable to expect item means above 3.00, and with 5-point scales, a standard deviation of one is reasonable (3 ± 2 SDs equals responses of 1 to 5 on the 5-point scale). The third column of numbers is the correlation between each item and the average of the other four items in the scale measuring each of the five constructs. This is sometimes called an *item-total correlation* because it represents how one item relates to the total of the other items. If all five items are measuring the same construct, these item-total correlations should be positive and at least moderate (0.3 to 0.6). This column shows the degree to which items designed to measure a specific construct yield correlated results. The last five columns represent factor loadings resulting from a principal components analysis of the 25 items. These columns show not only whether items designed to measure a construct relate to each other, but also the degree to which items designed to measure one construct are relatively unrelated to items designed to measure another construct. The numbers in these columns should display an interpretable pattern. For example, the first five items were intended to be measures of Extraversion; therefore, loadings of these items should be high on a single factor (F_1, in this case) and low on the other four factors. The second set of five items was written to measure Openness. They display the intended pattern of factor loadings (all five load highest on F_5 and low on the other four factors). However, in this case, two items ("Am not interested in abstract ideas" and "Have excellent ideas") perhaps should be loaded

TABLE 6.1
Internal Analysis of Big Five Personality Measure

Item[d]	Mean	SD	Item-Total Correlation[a]	Exploratory Factor Analysis[b]				
				F_1	F_2	F_3	F_4	F_5
A. Item-Level Data								
Don't like to draw attention to myself	2.86	1.13	0.43	0.62	−0.05	0.08	−0.09	0.01
Talk to a lot of different people at parties	3.63	1.11	0.43	0.63	0.06	0.06	0.16	−0.01
Don't mind being the center of attention	3.64	1.13	0.55	0.76	−0.03	0.07	−0.06	0.10
Have little to say[c]	3.72	0.99	0.47	0.64	0.05	−0.06	0.22	0.07
Don't talk a lot[c]	3.63	1.07	0.63	0.77	−0.01	0.03	0.18	0.03
Am not interested in abstract ideas[c]	3.40	1.02	0.24	−0.06	−0.17	0.12	0.32	0.42
Have excellent ideas	3.79	0.79	0.29	0.32	0.38	0.04	0.09	0.38
Use difficult words	2.93	0.91	0.41	−0.01	0.04	−0.10	−0.08	0.77
Have difficulty understanding abstract ideas[c]	3.64	0.85	0.41	0.08	0.04	0.22	0.27	0.55
Have a rich vocabulary	3.31	0.98	0.49	0.11	0.13	−0.02	−0.06	0.78
Make people feel at ease	4.11	0.79	0.37	0.21	0.12	0.11	0.54	−0.01
Insult people[c]	3.72	1.03	0.27	−0.19	0.23	0.26	0.51	−0.07
Feel little concern for others[c]	4.11	1.01	0.38	0.08	0.03	0.03	0.64	0.04
Take time out for others	4.00	0.75	0.45	0.18	0.19	−0.07	0.61	0.14
Have a soft heart	4.07	0.89	0.43	0.05	0.02	−0.15	0.68	0.04
Follow a schedule	3.69	0.99	0.53	0.04	0.75	−0.10	0.08	−0.12
Am exacting in my work	3.52	0.87	0.52	0.01	0.74	−0.02	−0.01	0.02
Make a mess of things[c]	3.72	0.96	0.35	−0.07	0.48	0.19	0.17	0.05

Item	Mean	SD	r[a]	1	2	3	4	5
Pay attention to details	3.77	0.81	0.43	0.04	0.58	-0.00	0.21	0.27
Am always prepared	3.46	0.89	0.56	-0.02	0.75	-0.09	0.08	-0.12
Change my mood a lot[c]	3.01	1.14	0.43	-0.01	0.07	0.62	0.26	0.02
Get stressed out easily[c]	2.90	1.10	0.48	-0.02	-0.10	0.70	-0.09	0.19
Seldom feel blue	3.04	1.02	0.45	0.06	0.11	0.63	0.00	-0.06
Worry about things[c]	2.45	1.06	0.49	-0.03	-0.12	0.71	-0.19	0.08
Often feel blue[c]	3.44	1.08	0.52	0.20	0.13	0.68	0.13	-0.09

B. Subscale Data

Scale	1	2	3	4	5
Extraversion (1)	(0.74)[e]	0.32	0.10	0.12	0.30
Agreeableness (2)	0.22	(0.62)	0.43	0.18	0.34
Conscientiousness (3)	0.07	0.29	(0.72)	0.08	0.29
Emotional Stability (4)	0.09	0.12	0.06	(0.72)	0.18
Openness (5)	0.20	0.21	0.19	0.12	(0.61)

[a] This is the correlation of each item with the total of the other four items that comprise the a priori measure to which the item belonged conceptually.

[b] These are factor loadings resulting from varimax rotation of five principal components.

[c] Responses to these items were reversed so all items were scored positively.

[d] The first five items were intended to measure Extraversions; the second group of five, Openness; the third, Agreeableness; the fourth set, Conscientiousness; and the final five, Emotional Stability.

[e] Values in parentheses on the diagonal are coefficient alpha. Values below the diagonal are observed correlations. Correlations above the diagonal have been corrected for unreliability (see chapter 2).

more highly on Factor 5, and the loadings of the latter of the two on Factors 1 and 2 indicate that the items' meaning may be more ambiguous than we might like. As a whole, the factor loadings and item-total correlations are quite interpretable and represent a reasonable confirmation that items grouped themselves as intended. Because each set of five items was defined a priori as a measure of a different construct, we could have also used confirmatory factor analyses to test the validity of our hypotheses about the internal structure of these items (see chapter 2).

In Table 6.1B, we present another way to look at the data regarding the intercorrelations and internal consistency of the five hypothesized dimensions of personality. On the diagonal of Table 6.1B are the coefficient alpha measures of internal reliability. Below the diagonal, we present the observed intercorrelations among the dimensions. These observed intercorrelations should be low if the five subscales measure different aspects of personality as was intended. However, we know from chapter 2 that these correlations may be low if neither of the two scales being correlated is measured with adequate reliability. So, we also present the correlations corrected for unreliability above the diagonal. Both observed correlations and corrected correlations indicate positive, but relatively low, correlations among these personality dimensions. This should be the case if these five scales truly represent different constructs.

Table 6.1A and B indicates that the internal structure of these 25 items is consistent with a Big Five conceptualization of personality. This same process for looking at the internal structure of measures can be applied to the development and validation of any measure: measures designed to assess various facets or dimensions of aptitudes, interests, skills, and indeed other conceptualizations of personality. However, positive results regarding internal structure are minimal evidence of validity. We also expect correlations with other variables to be consistent with our notions about each construct.

Evidence Based on Relationships With Other Variables

Relationships with other variables should be consistent with the conceptual or theoretical meaning of a measure. In staffing, relationships between the measure and other important psychological constructs are additional evidence of construct validity. Interestingly, when we examine relationships with other variables, we occasionally hope to also find an absence of relationships. For example, noncognitive measures such as personality and biodata may be influenced by applicants' desires to present themselves in the best possible light rather than to present an accurate report of their experience or background. In this case, evidence of low correlations with measures of social desirability would be considered positive evidence regarding the construct validity of a new measure. Likewise, correlations with measures of different constructs should be low—at least low enough so the new measure will provide some practical level of nonredundant information about candidates' attributes.

In Table 6.2, we present a part of the results of a study like this by Stokes and Cooper (2001), who were interested in further understanding the constructs underlying biodata measures they had constructed for use in selection. The correlation of their biodata measures with measures of the five well-known personality constructs described earlier and a measure of impression management are presented in Table 6.2. These authors had specific a priori hypotheses about the relationships between biodata and personality that should be the strongest (see the underlined items in Table 6.2). As can be seen in Table 6.2, the underlined values are generally the highest values, reflecting some support for the authors' hypotheses and the presumed construct validity of the biodata measures. The fact that all correlations in the table are positive, with the exception of correlations with Neuroticism, may be a result of the fact that respondents are answering in a way that reflects positively on themselves. The latter hypothesis is supported by the uniformly positive correlations with a measure of impression management (Paulhus, 1991). Stronger support for construct validity would be evidenced by lower correlations with the values in Table 6.2 that are not underlined and the absence of correlations between impression management and the biodata measures.

Another study similar in approach but involving different variables was conducted to evaluate measures designed to assess the physical abilities of workers in a steel plant. In this study, Arnold, Rauschenberger, Soubel, and Guion (1982) conducted what they termed a construct validation of a set of physical ability measures. These tests were relatively simple measures of the ability to grip or lift material. Evidence of the construct validity of the physical tests consisted of the correlation between scores on these simple, inexpensive, and easy to administer tests and scores on relatively elaborate job simulations.

TABLE 6.2
Correlations of Biodata Measures with Measures of the Big Five Personality Constructs and Impression Management[a]

	Impression Management	Neuroticism	Extraversion	Openness	Agreeable	Conscientious
Cultural adaptability	0.21	−0.30	0.28	0.<u>49</u>	0.19	0.16
Conflict resolution	0.20	−0.30	0.43	0.<u>39</u>	0.12	0.26
Dependability	0.23	−0.23	0.43	0.35	0.18	0.<u>30</u>
External contacts	0.25	−0.28	0.<u>51</u>	0.21	0.16	0.<u>14</u>
Internal contacts	0.19	−0.32	0.<u>52</u>	0.33	0.21	0.<u>35</u>
Information processing	0.18	−0.30	0.40	0.<u>33</u>	0.09	0.34
Initiative/persistence	0.27	−0.31	0.39	0.<u>36</u>	0.16	0.<u>43</u>
Leadership	0.14	−0.28	0.<u>50</u>	0.20	0.03	0.<u>31</u>
Oral communication	0.18	−0.27	0.<u>43</u>	0.30	0.02	0.31
Planning	0.22	−0.27	0.32	0.11	0.12	0.<u>55</u>
Problem solving	0.20	−0.26	0.24	0.<u>42</u>	0.00	0.<u>25</u>
Stress tolerance	0.23	−0.<u>64</u>	0.34	0.27	0.17	0.28

[a] Underlined values represent relationships predicted by the authors (Stokes & Cooper, 2001), based on the nature of the biographical items.

The job simulations were difficult to administer in various locations to many job applicants, but correlations of substantial magnitude between the simple tests and the complex simulations indicated that the simpler tests were measuring the same underlying abilities (constructs) required for successful job performance that the more elaborate job simulations measured.

Convergent and Discriminant Validation

The most frequently employed approach to construct validity involves comparing correlations in a multitrait-multimethod (MTMM) matrix (Campbell & Fiske, 1959). This method is most frequently used to examine the meaning of scores on new measures in light of people's scores on other measures about which more is known. It has been usefully employed to help understand the dimensionality of multiple performance measures in a criterion-related study (e.g., Lawler, 1967; Schmidt & Johnson, 1973), different ways of framing interview questions (e.g., Huffcutt, Weekley, Wiesner, DeGroot, & Jones, 2001), different methods of collecting trait information in an assessment center (Haaland & Christiansen, 2002), and many other applications. Basic to the use of the MTMM matrix is the idea that a measure of some phenomenon should correlate highly with other measures of the same phenomenon and that it should not correlate with measures of different phenomena. When a measure of a phenomenon correlates highly with other measures of the same phenomenon as it theoretically should, this indicates *convergent validity*; when the test predictably does not correlate with measures of different phenomena, this indicates *discriminant validity*.

A hypothetical MTMM matrix is illustrated in the top half of Fig. 6.8. The triangles on the diagonal are referred to as heterotrait-monomethod (many traits, one method) triangles. The three heterotrait-monomethod triangles represent the correlations among Traits A, B, and C when each is measured by Methods 1, 2, and 3, respectively. The off-diagonal boxes, called heterotrait-heteromethod blocks (many traits, many methods), portray the correlations among traits measured by different methods. The diagonal values in these heterotrait-heteromethod blocks (labeled as V in Fig. 6.8) are the correlations between different measures of the same trait and are termed validities. If you recall the earlier discussion about how internal analyses of a Big Five measure of personality can provide evidence for construct validity, the present topic has similar logic. There we asked if items designed to

330 6. VALIDATION STRATEGIES AND UTILITY

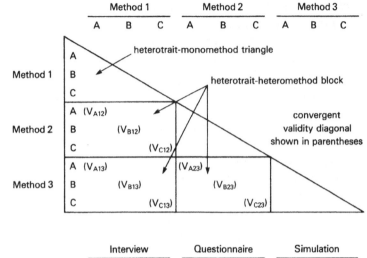

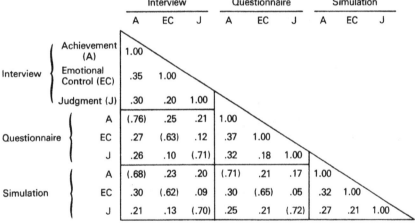

FIG. 6.8. Multitrait-multimethod matrix [A, B, and C represent different traits, V_{A12}, V_{B12}, etc. are the validities or the correlations between different measures of the same trait].

assess different dimensions of personality correlate with each other but not with items designed to measure other facets of personality. Here, we ask if measures of different constructs measured in different ways yield results like those we hypothesize to be so.

Four criteria have been suggested by Campbell and Fiske (1959) for evaluating the MTMM matrix. First, the correlations between similar traits measured by different methods (i.e., the convergent validities) should be both statistically significant and high enough to warrant

THEORETICAL MEANINGFULNESS OF TESTS

further consideration. Second, the convergent validities should be higher than the correlations between different traits measured by different methods (i.e., the off-diagonal elements in the heterotrait-heteromethod blocks). Third, the convergent validities should be higher than the correlations between different traits measured by the same method (i.e., the validity diagonals should be higher than the heterotrait-monomethod triangles). Finally, a similar pattern of trait intercorrelations should be apparent in the heterotrait-monomethod triangles and the heterotrait-heteromethod submatrices.

We can see that when a new test or measure is developed to assess a particular construct called a trait in MTMM language, determining how it correlates with other known measures of the same construct is useful information. That is, if convergent and discriminant validity are evident, this leads to confidence in the construct validity of the new measure. In the bottom half of Table 6.8, we present a set of correlations that meet the Campbell-Fiske criteria reasonably well.

The bottom half of Fig. 6.8 shows three methods (Interview, Questionnaire, and Simulation) for assessing three constructs (Achievement, A; Emotional Control, EC; and Judgment, J). If we take the four issues suggested for examining MTMM, it becomes clear that

1. The convergent validities should be strong enough to warrant further consideration. We note that the convergent validities in the three heterotrait-heteromethod blocks are all above 0.60 (these are shown in parentheses in Fig. 6.8).
2. The convergent validities should be higher than the other correlations in the heterotrait-heteromethod blocks. We note that the convergent validities all exceed 0.60, whereas no other correlation in these blocks exceeds 0.30.
3. The convergent validities should be higher than the correlations in the heterotrait-heteromethod triangles. Again, the convergent validities all exceed 0.60, yet the strongest correlation in the heterotrait-monomethod triangle is 0.37.
4. The heterotrait-monomethod triangles should have a similar pattern to the off-diagonal entries in the heterotrait-heteromethod blocks. This means, for example, that the interview assessments of A, EC, and J should be intercorrelated similarly to the way the questionnaire and simulation measures of these three variables are intercorrelated. As can be seen in the bottom half of Fig. 6.8, the patterns are similar.

Today, researchers use much more sophisticated methods (see Lance, Noble, & Scullen, 2002) than is implied by the Campbell and Fiske (1959) criteria cited previously to analyze the degree to which data in a MTMM matrix support the construct validity of our measures. Because there are specific hypotheses about the manner in which the variables in a MTMM should correlate with each other, CFA outlined in chapter 2 is frequently used to analyze MTMM matrices. Although CFA and other multivariate procedures provide more quantitative and detailed analyses of convergent and discriminant validity, the underlying logic is identical to the original Campbell and Fiske formulations.

Validity Generalization

Early work on test–criterion relationships indicated that many of the instruments used by staffing researchers exhibited practically useful validities (Ghiselli, 1966, 1973). However, this work often indicated that observed validity coefficients varied considerably from study to study across different settings, even when the jobs and tests involved appeared similar or even identical. The conclusion was that tests must be validated by staffing researchers whenever they are used in a different situation or company or with a different set of employees. Aside from the tremendous effort expended in conducting validation studies each time a test was used in a new situation, this variability across situations also made it difficult to draw conclusions about the individual difference constructs that underlie these tests.

Schmidt and Hunter (1977) attacked this situational specificity hypothesis. Their view was that validity did generalize across situations and that observed differences in validity coefficients were due to various defects in the validity studies themselves, rather than differences in the situations. They alleged that much of the variability in study results could be attributed to small sample sizes and the resulting sampling error, differences in range restriction, unreliability of measurement, and other artifacts whose influence on study results varied across studies. They began to cumulate results across the very large number of criterion-related studies that had been conducted. They also developed meta-analytic methods (Hunter & Schmidt, 1990) that assessed the impact of these presumed artifacts on observed variation across studies, as well as provided estimates of population effect sizes (i.e., the true correlation between individual difference constructs and

THEORETICAL MEANINGFULNESS OF TESTS

performance outcomes). Their work on validity generalization (as opposed to situational specificity) has fundamentally changed the thinking of staffing researchers about the validity and utility of tests. In the next few pages, we describe the basics of meta-analysis, what Hunter and Schmidt called a "bare bones" analysis. We also briefly describe the complexities that characterize a more thorough meta-analysis and provide a brief summary of some meta-analytic results that are most central to staffing researchers. Examples of meta-analytic research are also described elsewhere in this book (see chapters 7–10).

The Bare Bones Approach to Validity Generalization

Meta-analyses of correlational data begin with the computation of the average correlation ($\bar{r}$) across studies weighted by the sample size in the study that produced the correlation:

$$\bar{r} = \sum (N_i r_i) \Big/ \sum N_i \qquad (6.1)$$

where r_i is the correlation recorded from an individual study, N_i is the sample size in study i, and $\sum$ indicates that the product, $N_i r_i$, is summed across studies. In this formula, r_i is weighted by the sample size, giving more weight to studies performed with large sample sizes. This will always provide a better estimate of the population correlation, except in those instances in which a single or small number of large sample size studies comes from a population in which the correlation is different for some reason from small sample studies. We can check for this possibility by performing separate analyses for the two sets of studies or by correlating sample size with the correlations or effect sizes. In the event of differences between groups of studies or a large correlation between study sample size and the effect size, we will need to report the difference and pursue potential reasons for this difference.

The variance (σ_r^2) of the observed correlations is estimated across studies as follows:

$$\sigma_r^2 = \sum \left(N_i (r_i - \bar{r})^2 \right) \Big/ \sum N_i \qquad (6.2)$$

An Example of a Bare Bones Study With Extensions

An example of a set of 10 studies is presented in Table 6.3, along with calculations of $\bar{r}$ and σ_r^2. For these hypothetical data, half the studies were conducted in academic settings, and half were conducted in

TABLE 6.3
An Example of a Bare Bones Meta-Analysis

Study	Year	Setting	N	r	Nr	$r_i - \bar{r}$	$N_i(r_i - \bar{r})^2$
1	1992	Work	57	0.38	21.66	0.00	0.00
2	1998	Academic	72	0.19	13.68	-0.19	2.60
3	2002	Academic	98	0.23	22.54	-0.15	2.21
4	2001	Work	212	0.41	86.92	0.03	0.19
5	1991	Academic	70	0.23	16.10	-0.15	1.58
6	1988	Work	580	0.39	226.20	0.01	0.06
7	2002	Work	320	0.50	160.00	0.12	4.61
8	1997	Academic	79	0.26	20.54	-0.12	1.14
9	1979	Academic	412	0.31	129.58	-0.07	2.02
10	1982	Work	291	0.47	136.77	0.09	2.36
			2191		833.99		16.77

A. An Application of Meta-Analytic Formulas to the Achievement Motivation–Performance Relationship

$\bar{r} = (833.99)/2{,}191 = 0.38$
$\sigma_r^2 = 16.77/2{,}191 = 0.0076$
$\sigma_e^2 = 7.3/2{,}191 = 0.0033$

B. Summary of Moderator Analysis for the Achievement Motivation–Performance Relationship

Work Setting
$\bar{r} = 0.28$
$\sigma_r^2 = 0.0019$
$\sigma_e^2 = 0.0063$
$\sigma_p^2 = -0.0044$

Academic Setting
$\bar{r} = 0.43$
$\sigma_r^2 = 0.0022$
$\sigma_e^2 = 0.0028$
$\sigma_p^2 = -0.0008$

a work setting. The correlations represent the relationship between achievement motivation and performance (in academic settings, the outcome was grade point average; in work settings, the outcome was a performance rating). The experimenter, in this instance, might believe achievement motivation would be differentially related to performance across these two settings. The average r was 0.38, and the variance in correlations was 0.0076. However, Hunter and Schmidt (1990) pointed out that some of that variability may be due to actual differences in population correlations, but that some, or even most, is due to differences in study sample sizes or sampling error. They maintained that there are also other sources of error, but a bare bones analysis provides only an estimate of variability due to sampling error. To estimate the actual variance in correlations, the value of σ_r^2 must be corrected for variance due to sampling error. The variance due to sampling error is provided by the following formula:

$$\sigma_e^2 = \left((1 - \bar{r}^2)^2 K\right) \Big/ \sum N_i \qquad (6.3)$$

where K is the number of studies. This formula uses $\bar{r}$ as the estimate of the population correlation. In this hypothetical set of motivation–performance relationships, the estimate of sampling error is 0.0033. The difference between the observed variability (0.0076) and the variance explained by sampling error is 0.0043.

This remaining variance (0.0043) can now be used to compute a *credibility interval* regarding our expectations of the magnitude of the motivation–performance relationship in other situations. Taking the square root of the remaining variance leaves 0.066. The 95% credibility interval then would be 0.38 ± 1.96 (0.066) or 0.25 to 0.51. This credibility interval means that we would expect to find a nonzero relationship between achievement motivation and performance when conditions similar to those in the studies reviewed prevail. In terms of our concern about construct validity, it means that the achievement motivation construct is reliably correlated with performance and that we can expect this relationship to generalize to similar situations using similar measures.

Computation of credibility intervals is more complicated when corrections are made for other potential artifacts, such as range restriction, unreliability, and differences in the factor structure of the measures used to operationalize the variables (in this case, achievement motivation and performance) in different studies. Range restriction and unreliability of measurement serve to lower estimates of effect size as we saw in chapter 2. They may also serve to increase the variability of our effect size estimates, so corrections are made to our estimate of σ_r^2. Corrections for these various artifacts add considerable complexity to the estimation of population effect sizes, and the reader should consult Hunter and Schmidt or similar texts for help in these computations. Several software packages are also available to provide these computations (e.g., see www.powerandprecision.com).

When the variability remaining after correction for various artifacts is large and the credibility interval is relatively broad, the remaining variance in correlations between predictors and criteria may be due to *moderator variables*. That is, we may have two or more groups of studies across which the effect size varies. If the meta-analyst has information about study characteristics, the impact of such moderator variables on the magnitude and variability of the observed effect sizes can be examined. A good meta-analyst will understand the primary literature well enough to hypothesize what might moderate effect sizes and code the primary studies to include information on the moderator variable, assuming the original researcher recorded the required information. In our hypothetical example, it is reasonable to hypothesize that the academic and workplace studies might provide different estimates of the motivation–performance relationship because these studies differ in terms of the performance outcome used, the age of the study participants, the manner in which achievement motivation is measured, and perhaps in other ways as well.

If the setting of the study does moderate this relationship, we would expect the correlations in the two sets of studies to differ and that the variance of the effect size estimates within each setting to be close to zero. That is, by taking into account setting, we would remove the between-setting variance. A summary of the analyses relevant to this question is provided at the bottom of Table 6.3. As we can see, the correlations of achievement motivation and performance are higher (0.43) in academic settings than similar correlations in work settings (0.28). Within each set of correlations, sampling error actually accounts for more variance than is observed. This frequently occurs when the number of studies is small and is referred to as second-order sampling error. The data at the bottom of Table 6.3 are consistent with the hypothesis that setting moderates the relationship studied and that within each of the two settings there is a nonzero relationship between achievement motivation and performance.

This bare bones meta-analysis provides a good illustration of the manner in which meta-analyses are conducted. Moreover, Hunter and Schmidt (1990) documented that for many examples of relationships relevant to staffing researchers the correction for sampling error alone removes a large portion of the variability in observed coefficients. However, as indicated at the outset of this section of our chapter, meta-analyses that correct for other artifacts and that consider multiple and sometimes interacting moderators are very complex. In addition, the primary studies have not reported information that makes it possible to correct for certain artifacts or examine potential moderators. Efforts to use existing data on such artifacts as a means of estimating their impact on a set of studies are also provided in Hunter and Schmidt. A list of the information that should always be reported by staffing researchers so such "guesses" about the influence of these artifacts and moderators are no longer necessary is provided in Table 6.4.

Schmidt and Hunter (1998) summarized the results of a great deal of primary research and meta-analytic data on the validity of various

TABLE 6.4

Information That Should Be Reported in Validation Studies to Facilitate Data Aggregation

1. **Firm**: The sponsor of the study and the organization or firm with which the study was done (or the type of firm when this information is proprietary).
2. **Problem and setting**: The problem to which the study was addressed and the social, economic, and organizational elements of the setting.
3. **Job title and code**: The title and code of the job performed as taken from the O*NET.
4. **Job description**: A description supplementing the O*NET description when necessary.
5. **Sample**: The sample size and the characteristics of the people studied (i.e., their gender, age, education, ethnic status, job level, job experience, proportion of the total population represented in the sample and applicant versus study participant characteristics).
6. **Predictors**: The kinds of data being investigated for their usefulness in guiding personnel actions are described to include appropriate reliability estimates and the intercorrelations among predictors.
7. **Criteria**: Detailed description of the criterion data collected (including their reliability) and a discussion of their relevance. If ratings were used, some estimate of the amount of contact the rater had with the employee should be reported; if production records were used, the duration of the data collection period and whether any unusual events occurred during that period should be reported.
8. **Data reported**: The means, variances, and intercorrelations of variables for applicant and employee groups should be reported as fully as possible. The methods used to analyze the data (i.e., regression, analysis of variance, contingency tables) should be reported in detail.

predictor constructs. Their summary indicates that validities of many of the constructs and methods used by staffing researchers display sizable and generalizable validity. In chapters 7 to 10, we draw on their studies and those of other meta-analytic researchers to characterize the nature of the constructs we measure and the relationships of these constructs to various performance criteria.

Necessary Evidence When Relying on Validity Generalization Results

If validity generalization studies become the major or only support for the use of a selection procedure, we believe the following evidence must be presented.

1. There should be evidence that the position for which we want to use a test to select employees requires knowledge, skills, and abilities similar to those represented in the studies that are part of the validity generalization database.

2. The new or proposed tests should have the same factor structure as the tests that are part of the validity generalization database. We can certainly provide job analyses and research data relevant to both these requirements, but some professional(s) will be required to use their knowledge of the research literature and the job in question to make these judgments.

Summary

In this section, we indicate numerous ways by which staffing researchers can show evidence for the theoretical meaningfulness of the procedures they use or propose to use as a basis for making personnel selection decisions. These approaches range from studying the underlying processes believed to be related to test performance, to studies of the internal nature of the measures and their constructs with regard to the way those measures should behave theoretically, to examination of the way different measures of a construct relate to other measures of the same constructs (or fail to relate to constructs they should not relate to), and to studies across settings to defend the notion that selection procedures generalize in their validity across those settings due mostly to differences in sample sizes. You can see that when these methods for exploring the way proposed selection procedures are combined, they yield important evidence for the constructs they assess, the way they likely link with other similar constructs, the ways in which they can be distinguished from other measures of other constructs, and the likely reason why they relate as they do to performance criteria of interest across settings. Armed with this kind of conceptual insight, staffing researchers are in a strong conceptual position to understand why their tests and measures behave as they do with regard to criterion performance outcomes. However, there is more to evaluating selection than test validity; we must also be concerned with other outcomes from testing, the topic we turn to next.

VALIDITY BASED ON THE CONSEQUENCES OF TESTING

Evidence about the intended and unintended consequences of the use of tests has been termed *consequential validity* by Messick (1998). Primarily concerned with educational applications of tests, his contention was that researchers must consider the unintended consequences of the use of tests, such as the disproportionate selection or rejection of members of different demographic groups for access to desirable or undesirable educational opportunities or, in our case, jobs. Thus, similar notions have been espoused in the use of tests in work organizations. Some authors have maintained that such issues should not concern staffing researchers. Their position is that the primary obligation of staffing researchers should be documentation that the constructs measured by our selection procedures are relevant to the job performance outcomes we want to assess. We believe there are two reasons why this position may be untenable. First, it is often the case that staffing researchers are quite narrow in what performance outcomes they consider, restricting themselves to task performance (see chapter 4) and, occasionally, contextual performance. Considering the individual's role or the organization's responsibility in the attainment of broader societal goals (i.e., social responsibility, employment of disadvantaged applicants, cultural diversity) is quite rare. Considering this broader set of goals will almost certainly result in the examination of what has been termed the consequences of test use. The second reason we do not believe staffing researchers can ignore test use consequences is purely practical. The organization that does not consider the impact of its selection procedures on the selection of minorities and women risks social, economic, and legal consequences and often also risks the loss of important human talent with diverse competencies, styles, and perspectives.

The practical consequences of the use of selection procedures that exhibit adverse impact (i.e., lower selection rates of minority or female groups relative to White males) have long been of concern to staffing researchers. In fact, they have often generated the research data, methods of analysis, and definitions that have become standard in various professional documents (e.g., the *Standards*, AERA et al., 1999). Staffing researchers must continue to generate the information that allows corporate decision makers to proceed having as complete information as possible about the impact of hiring decisions on how sometimes conflicting goals will be served by the recommended selection procedures.

Since the passage of the Civil Rights Act of 1964 and the increased frequency with which hiring decisions are challenged in the courts, there has been a rapid accumulation of research on possible ethnic differences in the predictive meaning of test scores. Racial group differences are repeatedly observed on standardized knowledge, skill, ability, and achievement tests (e.g., Bobko et al., 1999; Hartigan & Wigdor, 1989; Neisser et al., 1996; Schmitt, Clause, & Pulakos, 1996). These studies, which are actually summaries of many hundreds of primary studies, usually indicate that African Americans score up to about 1 *SD* lower than Caucasian groups, Hispanics score about 0.75 *SD* lower than Causasians, and Asians often score higher than Caucasians on mathematical tests and lower on verbal measures. Gender differences usually indicate that women score slightly higher on verbal tests and somewhat lower on mathematics tests than men. Gender differences favoring men on physical ability tests are often very large (>1 *SD*) especially when upper body strength is assessed (for a summary of these studies, see Hogan, 1991). These mean differences can translate into rather large differences in the proportion of a lower-scoring subgroup being selected, especially when the organization has a favorable selection ratio (many applicants for each job opening) and selects from candidates in rank order from the top down using their scores on these tests.

One display of the implications of the impact for selection of lower-scoring groups is provided in Table 6.5. In this table, we can see how the size of the mean difference between groups and the proportion of the applicants an organization selects relates to the proportion of a minority group that is selected if the organization chooses from the top down on the predictor. Subgroup differences and low selection ratios result in very low rates of selection among lower-scoring groups. Similarly,

TABLE 6.5
What Standardized Group Differences Mean for Minority Selection

	Majority Group Selection Ratio		
Standardized Group Difference (d)	10%	50%	90%
0.0	0.100	0.500	0.900
0.1	0.084	0.460	0.881
0.3	0.057	0.382	0.836
0.5	0.038	0.309	0.782
0.7	0.024	0.242	0.719
0.9	0.015	0.184	0.648
1.0	0.013	0.159	0.610
1.1	0.009	0.136	0.571
1.3	0.005	0.097	0.492
1.5	0.003	0.070	0.413

Note: Adapted from Sackett & Wilk, 1994.
Minority group selection ratio when cutoff for majority group is set at 10%, 50%, and 90%.

more detailed consideration of the impact of tests with varying degrees of subgroup mean differences under different circumstances are provided by Sackett and Wilk (1994). As one might expect, the use of even appropriately validated procedures with high adverse impact has generated legal scrutiny of these measures. The legal status of staffing procedures will be addressed in greater detail in chapter 11 and in other chapters discussing particular selection procedures. Chapter 7 provides an overview of validity and utility for various predictor constructs and measures. For now, we discuss some definitional issues and provide a description of the typical analysis of relevant data in these situations.

The problems investigated are generally subsumed under the heading of test bias. The main questions raised regarding test bias pertain to validity coefficients (referred to as slope bias) and to the relationship between group means on the test and the criterion (often referred to as intercept bias). Current examination of these questions treats them as part of a general "differential prediction" problem (Linn, 1978). The scientific literature and *Standards* (AERA et al., 1999) have accepted a definition first proposed by Cleary (1968):

> A test is biased for members of a subgroup of the population if, in the prediction of a criterion for which the test was designed, consistent non-zero errors of prediction are made for members of the subgroup. In other words, the test is biased if the criterion score predicted from the common regression

line is consistently too high or too low for members of the subgroup. With this definition of bias, there may be a connotation of "unfair," particularly if the use of the test produces a prediction that is too low. If the test is used for selection, members of a subgroup may be rejected when they were capable of adequate performance. (p. 115)

Studies of differential prediction most often employ moderated multiple regression (Bartlett, Bobko, Mosier, & Hannan, 1978) of the type described in chapter 2. We can also subgroup study participants on the basis of sex, race, or ethnic group, and then compute and compare regression slopes (reflected in the regression coefficients associated with the test or predictor), intercepts (reflected in the constants associated with the regression equations), and standard errors of estimate. If we find evidence for some form of differential prediction, the implication is that a single prediction formula will result in over- or underprediction of the performance of one or more subgroups.

When prediction systems are compared in this fashion, the most frequently occurring difference is in intercepts, not slopes (i.e., the performance of one group is predicted to be superior to the performance of another group). In one review, researchers found significant intercept differences in about 18% of 1,190 racial group comparisons and slope differences in 5% of the comparisons (Bartlett et al., 1978). In the few instances in which differences are observed, we commonly find that the prediction system for the nonminority group slightly overpredicts minority group performance. That is, minorities would tend to do less well on the job than their test scores indicate.

In Fig. 6.9, we illustrate the regression approach to examining the possibility of differential prediction given the usual research findings, namely, overprediction of minority job performance using a common regression line (in which a single regression equation is computed for all study participants). Use of the minority regression equation to predict minority job performance for the one test score level illustrated would yield a predicted value of P_{MIN}. If we used a common regression line or the majority regression equation to make predictions for the same test score, we would obtain higher predicted performance values for minority applicants with this test score; in this case, P_C or P_{MAJ}. The differences between P_C and P_{MIN} and P_{MAJ} and P_{MIN} represent overprediction. Similar analyses for other possibilities of subgroup differential prediction are available in Anastasi and Urbina (1997), and an excellent analysis of actual subgroup differences in prediction is presented by Gael, Grant, and Ritchie (1975).

VALIDITY BASED ON THE CONSEQUENCES OF TESTING 343

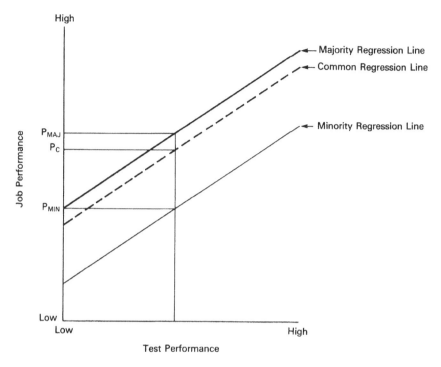

FIG. 6.9. Illustration of most common case of differential prediction.

It should also be noted that examinations such as that depicted in Fig. 6.9 provide evidence of psychometric or predictive bias. Psychological reactions or perceptions of the tests used or the results of the use of tests are often unrelated to the conclusions of tests of predictive bias. These reactions have become the focus of staffing researchers' attention in the past decade and are summarized in chapter 5. The legal treatment and definition of the consequences of testing are considered in chapter 11.

Prior to leaving this section, it is important to note that the differences in racial, ethnic, and gender performance on tests we have described exist primarily for paper-and-pencil, multiple-choice tests of cognitive ability—and for physical performance when considering men and women. Evidence presented later shows that on personality inventories these differences tend to disappear, and this also tends to be true with regard to performance on interviews for jobs and interpersonal role-play exercises in assessment centers.

Summary

In this section, we learn that there are multiple lines of evidence that lead to the conclusion that the inferences we draw from scores on a selection procedure are appropriate. Beginning with test construction, there should be evidence that the test items or parts represent some hypothetical construct or job-relevant domain of interest as defined by the test constructor. In workplace situations, it is hoped that we have evidence that test scores correlate with relevant measures of criterion performance. We consider various criterion-related validity research designs to assess relationships with criteria and the methods of correcting for some potential limitations (e.g., range restriction and unreliability in the criterion). We also considered several broader and more theoretical approaches to the development of conceptual and empirical bases that provide confidence in the proposed use of tests through the understanding gained in studying the bases for performance on selection tests. Of these latter approaches, validity generalization is likely to be the most useful in staffing applications because of the large database available on various test–criterion relationships and because the database usually directly addresses the degree to which a test predicts job outcomes. Finally, because of the social, political, and legal context within which staffing researchers work, they must be sensitive to the consequences of the use of the instruments they develop and propose and the idea that just because a test predicts performance does not mean the end of the pursuit of alternative procedures with fewer unintended negative consequences.

TEST UTILITY

Fortunately, there are not only negative unintended consequences of test use, but there are both intended and unintended positive consequences that accompany the use of carefully developed selection procedures. For example, one positive social and societal consequence of the use of such procedures is the opportunity that they may offer for people to demonstrate competencies and occupational interests that they were unaware of prior to the testing. So, broad-based assessments of job candidates can sometimes identify in people abilities that have not been otherwise developed and/or interests that might benefit from additional pursuit (Holland, 1997). A more obvious positive consequence would be the resultant increase in the average level of performance for job incumbents as a function of hiring more competent people. Or, a decrease in turnover rates for people hired into a job because people with the appropriate interests are hired for a job they are more likely to enjoy and find satisfying.

Staffing researchers have also considered the impact of their procedures on other organizational outcomes, including economic returns. This research is often referred to as research on the economic utility of the use of tests that have been shown to reveal criterion-related validity. As indicated at the beginning of this chapter, we must provide management decision makers with information that allows them to compare the worth of HR interventions with other demands for organizational resources as they attempt to maximize organizational effectiveness and survival, and we do this through utility analysis.

Early work on test utility (Brogden, 1946; Cronbach & Gleser, 1965; Taylor & Russell, 1939) provided the basic theoretical or mathematical formulations for modern test utility theory. All these early approaches depended on some practical monetary unit to provide economic meaning to correlations, means, and standard deviations that have been the language of staffing researchers. Beginning in the 1980s, several researchers provided methods and discussions that have helped staffing researchers translate their research outcomes into the language of economists (Boudreau, 1991; Cascio, 2000; Schmidt, Hunter, McKenzie, & Muldrow, 1979). Unless we have the means for measuring and tracking productivity and showing how our personnel techniques relate to productivity, we cannot know the short- or long-term economic effects of our contributions. In short, we must be able to translate our validity coefficients into dollar and cents values, however crude those translations may be. This translation is called utility analysis, and the 1980s produced a large number of studies on test utility spurred by the development of an index of the standard deviation of employee economic worth (SD_y). Experience with these translations has often led to conclusions about the utility of various personnel programs that have exceeded staffing researchers' fondest hopes and expectations.

There are at least three reasons why utility analyses are important. The first we have alluded to above—namely, as a means to strengthen and support the HR practices we espouse vis-à-vis other claims for organizational resources. The second reason is certainly implicit in the first: We should have better information ourselves concerning the relative cost–benefit of various HR efforts, such as training, selection, recruitment, and career development programs. For example, is it more cost beneficial to invest in sophisticated selection procedures or training programs when preparing to open a fast-food restaurant? Or, more specifically relevant to selection, how much money should be invested in the selection of the manager of a nuclear facility, the hiring of an airplane pilot, or the security personnel at airports after September 11? What are the costs of a failure when making these hiring decisions? The third reason is that, on a societal level, it is more important to provide information concerning the relative costs and benefits of making different kinds of choices when confronted with conflicting goals regarding HR allocation. A good example is the apparent conflict between productivity and affirmative action goals. Hiring those persons whose predicted performance is greatest based on paper-and-pencil tests of cognitive ability may result in the nearly total exclusion of minority individuals, given the mean differences in subgroup test performance described previously. These conflicts have been made explicit in formulations by Hunter, Schmidt, and Rauschenberger (1977). These analyses indicate that selection of applicants proportional to their representation in the applicant pool with selection of members of each group from the top down in terms of their estimated qualifications would be the best solution if we want to maximize both expected productivity and the employment of a diverse workforce. Negative effects on productivity are minimized relative to rank order selection without identification of ethnic status (i.e., purely on the basis of qualifications) and are far superior to a strategy that selects individuals randomly with proportional representation above some minimal cutoff. This solution, though, is based on ethnic group representation and has been socially and legally unacceptable, at least in the United States.

A final point concerning utility estimation should be mentioned. Many psychologists object to the estimation of human outcomes in dollar terms; such calculations may be in conflict with the set of values that brought them to careers in psychology. However, refusal to express our contributions in these terms ignores the reality of the language of business and ensures maintenance of the status quo. That is, by failing to consider utility issues, we say that the bases for decisions that are currently being made and the values that guide those decisions are fine. By refusing to speak the language of business, we can lack influence on decision makers and, thus, ensure potentially valuable HR programs are not fully used. The result of wearing such blinders is a potential loss to both individuals and organizations.

In the next section of this chapter, we trace the development of ideas and methodology in utility analyses and present techniques that have been found useful in various contexts. We end the chapter describing the need to link investments in human talent to organizational strategic success.

History of the Development of Selection Utility Models

The question of the utility of selection procedures has been of long-term interest to some industrial psychologists, but most early attempts focused on psychometric considerations rather than dollars. Most early procedures yielded evidence that (a) the use of valid selection procedures did not improve things much, and (b) only tests with high validity would yield important improvements. For example, one procedure examined the relative size of the errors of prediction (Hull, 1928) or the percent of variance in the job performance measure that is accounted for by knowledge of predictor scores. This estimate, called the *index of forecasting efficiency* is equal to $1 - \sqrt{1 - r_{xy}^2}$ where r_{xy} is the criterion-related validity coefficient. The index of forecasting efficiency compares the standard deviation of the errors we make in predicting job performance scores using test information with the standard deviation of the errors that would result from the use of random selection or nonvalid information. For example, the index of forecasting efficiency for a test of 0.60 validity would show that the predictor does only 20% better than chance:

$$\text{Efficiency} = 1 - \sqrt{1 - 0.36} = 1 - \sqrt{0.64} = 1 - 0.80 = 0.20$$

If a validity of 0.30 is placed into the formula, the result shows it would only do about 5% better than chance. These are not very positive interpretations of validity.

Another indicator that yielded not very positive interpretations of validity is called the *coefficient of determination*, r_{xy}^2. It became popular in the 1930s and 1940s and continues to be cited by some staffing researchers as a measure of utility. This coefficient is the amount of variance in the job performance measure that is accounted for by the selection procedure. A test with validity equal to 0.60 would be described as a measure that accounts for 36% of the criterion variance. Likewise, a test with validity of 0.3 would account for 9% of the variance.

PROPORTION OF EMPLOYEES CONSIDERED SATISFACTORY = .50
SELECTION RATIO

r	.05	.10	.20	.30	.40	.50	.60	.70	.80	.90	.95
.00	.50	.50	.50	.50	.50	.50	.50	.50	.50	.50	.50
.05	.54	.54	.53	.52	.52	.52	.51	.51	.51	.50	.50
.10	.58	.57	.56	.55	.54	.53	.53	.52	.51	.51	.50
.15	.63	.61	.58	.57	.56	.55	.54	.53	.52	.51	.51
.20	.67	.64	.61	.59	.58	.56	.55	.54	.53	.52	.51
.25	.70	.67	.64	.62	.60	.58	.56	.55	.54	.52	.51
.30	.74	.71	.67	.64	.62	.60	.58	.56	.54	.52	.51
.35	.78	.74	.70	.66	.64	.61	.59	.57	.55	.53	.51
.40	.82	.78	.73	.69	.66	.63	.61	.58	.56	.53	.52
.45	.85	.81	.75	.71	.68	.65	.62	.59	.56	.53	.52
.50	.88	.84	.78	.74	.70	.67	.63	.60	.57	.54	.52
.55	.91	.87	.81	.76	.72	.69	.65	.61	.58	.54	.52
.60	.94	.90	.84	.79	.75	.70	.66	.62	.59	.54	.52
.65	.96	.92	.87	.82	.77	.73	.68	.64	.59	.55	.52
.70	.98	.95	.90	.85	.80	.75	.70	.65	.60	.55	.53
.75	.99	.97	.92	.87	.82	.77	.72	.66	.61	.55	.53
.80	1.00	.99	.95	.90	.85	.80	.73	.67	.61	.55	.53
.85	1.00	.99	.97	.94	.88	.82	.76	.69	.62	.55	.53
.90	1.00	1.00	.99	.97	.92	.86	.78	.70	.62	.56	.53
.95	1.00	1.00	1.00	.99	.96	.90	.81	.71	.63	.56	.53
1.00	1.00	1.00	1.00	1.00	1.00	1.00	.83	.71	.63	.56	.53

FIG. 6.10.

Neither the index of forecasting efficiency nor the coefficient of determination bear any direct relationship to the actual economic value of a selection instrument. In addition, neither recognizes that the value of a test varies as a function of various situational parameters (e.g., the current level of job incumbent performance). Both lead to the conclusion that validities must be very high before a test has much economic value. As we see in this section, this is false. Various authors have shown that the validity coefficient itself is an inappropriate index of the utility of a selection device (Brogden, 1946; Cronbach & Gleser, 1965; Taylor & Russell, 1939).

A more widely used and well-known approach to utility is that described by Taylor and Russell (1939). Their approach took into account not only the validity coefficient, but also the selection ratio (the proportion of applicants who actually get hired) and the base rate (the percentage of employees considered successful prior to the introduction of a new selection procedure) to make estimates of the value of the selection instrument. Estimates of the utility of a procedure are made in terms of the proportion of successful candidates hired using the new procedure. One portion of the numerous tables they developed as aids in utility estimation is presented in Fig. 6.10. This table indicates the

TEST UTILITY

PROPORTION OF EMPLOYEES CONSIDERED SATISFACTORY = .60
SELECTION RATIO

r	.05	.10	.20	.30	.40	.50	.60	.70	.80	.90	.95
.00	.60	.60	.60	.60	.60	.60	.60	.60	.60	.60	.60
.05	.64	.63	.63	.62	.62	.62	.61	.61	.61	.60	.60
.10	.68	.67	.65	.64	.64	.63	.63	.62	.61	.61	.60
.15	.71	.70	.68	.67	.66	.65	.64	.63	.62	.61	.61
.20	.75	.73	.71	.69	.67	.66	.65	.64	.63	.62	.61
.25	.78	.76	.73	.71	.69	.68	.66	.65	.63	.62	.61
.30	.82	.79	.76	.73	.71	.69	.68	.66	.64	.62	.61
.35	.85	.82	.78	.75	.73	.71	.69	.67	.65	.63	.62
.40	.88	.85	.81	.78	.75	.73	.70	.68	.66	.63	.62
.45	.90	.87	.83	.80	.77	.74	.72	.69	.66	.64	.62
.50	.93	.90	.86	.82	.79	.76	.73	.70	.67	.64	.62
.55	.95	.92	.88	.84	.81	.78	.75	.71	.68	.64	.62
.60	.96	.94	.90	.87	.83	.80	.76	.73	.69	.65	.63
.65	.98	.96	.92	.89	.85	.82	.78	.74	.70	.65	.63
.70	.99	.97	.94	.91	.87	.84	.80	.75	.71	.66	.63
.75	.99	.99	.96	.93	.90	.86	.81	.77	.71	.66	.63
.80	1.00	.99	.98	.95	.92	.88	.83	.78	.72	.66	.63
.85	1.00	1.00	.99	.97	.95	.91	.86	.80	.73	.66	.63
.90	1.00	1.00	1.00	.99	.97	.94	.88	.82	.74	.67	.63
.95	1.00	1.00	1.00	1.00	.99	.97	.92	.84	.75	.67	.63
1.00	1.00	1.00	1.00	1.00	1.00	1.00	1.00	.86	.75	.67	.63

FIG. 6.10. (*continued*)

percent of all employees who will be successful after introducing a test with varying levels of validity and various selection ratios when half of the current workforce is already considered successful (i.e., without using the new procedure, 50% are already performing successfully). Taylor and Russell provided similar tables for other base rates as well. Thus, using the first half of Fig. 6.10, a procedure with validity of 0.50 used in a situation in which the selection ratio is 0.50 produces a workforce that is now 67% successful as opposed to 50% successful. This represents a 34% (67–50/50) increase in the percent of successful employees or an absolute increase of 17%. Similarly, a selection ratio of 0.20 and validity of 0.50 would yield 78% successful employees, an increase of 56% in the proportion of successful employees or an absolute increase of 28% successful employees. An examination of the rows in the first half of Fig. 6.10 will indicate the "percent successful" is (a) always greatest when selection ratios are low, and (b) tests of quite low validity can be useful if the selection ratio is low. We can see how the organization's attractiveness to prospective employees can pay the organization back in unexpected ways: Increased attractiveness should increase the number of applicants, lowering the selection ratio, and raising the value associated with the use of any selection device of any validity.

We can also see by examining figures for various base rates (percent of current employees considered successful) that we achieve maximal increases in absolute percent successful when the base rate is close to 0.50, given both constant validity and constant selection ratio. At low base rates, there will be small changes in absolute percent successful but large relative changes. For example, when the base rate equals 0.10 (the second half of Fig. 6.10), the selection ratio is 0.50, the validity of the test is 0.40, and the percent successful when employing the test is 16. This represents a 6% increase in successful employees (16–10) but a 60% improvement (16–10/10). An examination for a similar situation (same validity and selection ratio) in which the base rate is 0.50 indicates that 63% of the selected persons would be successful. This is a 13% increase (63–50) in successful employees but only a 26% (63–50/50) improvement. The principle is that the poorer the present base rate, the more the potential relative gain.

Although the Taylor-Russell tables represented a significant improvement in utility estimation by showing the benefits of using selection procedures under different circumstances, there are two distinct disadvantages associated with their use:

1. The Taylor-Russell approach neglects any consideration of (a) the cost of gathering information concerning applicants (i.e., testing, interviewing, checking references), and (b) the cost of recruiting an applicant pool that is of similar quality but much larger. Recall that the utility of a test is greatly influenced by the selection ratio and that to maintain a low selection ratio one must recruit and test many applicants.

2. Perhaps the more important limitation is that workers are either put in a "successful" or a "not successful" group. Splitting employees into two performance groups is called dichotomizing the criterion. When we dichotomize the criterion, we are forced to treat all employees in the "successful" group as equally successful; similarly, in the "unsuccessful" group, all employees are treated as equally unsuccessful. Obviously, this does not reflect reality because employees perform at more than two different levels of effectiveness, and the contributions of those different levels are lost when we dichotomize the criterion (Cronbach & Gleser, 1965).

Subsequent treatments of utility have solved both of these problems and revealed the more likely true advantages accruing to companies when they use validated tests. For example, Brogden (1946) showed that the validity coefficient is a direct index of utility. His formulation reveals that, assuming the predictor and criterion are continuous and have identical distributions and that the predictor and criterion are linearly related, a test with validity of 0.50 could be expected to produce 50% of the gain in utility that would result if we had a test of perfect validity. So, if an organization has turnover costs of $200,000 annually, a test battery that predicts turnover with a validity of 0.50 would save the company half of the $200,000, or $100,000 per year.

Brogden also showed how the selection ratio and the standard deviation of job performance in dollars (SD_y) influence the utility of a selection procedure. Mathematically,

$$\Delta \text{Utility/Person selected} = r_{xy}(\bar{z}_x)(SD_y)$$

where Δ Utility = the average per person gain in utility in dollars,
$\bar{z}_x$ = the mean standard score on the test of those persons selected (standard scores are based on the applicant distribution).

Note that the lower the selection ratio, the higher $\bar{z}_x$ will be, assuming the highest scoring people are selected. SD_y is the standard deviation of job performance in dollars among employees hired without use of the test. Determination of SD_y in dollar terms is discussed as follows.

The product of the validity and the mean standard score $r_{xy}(\bar{z}_x)$ is equal to the mean standard score of the criterion $\bar{z}_y$ for those selected. So, the Brogden formula reduces to an important formula: ΔUtility/Person selected $= \bar{z}_y(SD_y)$. As stated later in this section, figures for $\bar{z}_y$ given various levels of validity and selection ratio are given in Fig. 6.11. Utility per selectee is the mean standard score on the criterion for those selected times the standard deviation of the performance criterion in dollar form. Multiplying this figure by the number selected gives the total dollar gain of the selection procedure.

The implications of these mathematics are worked out for various levels of validity and selection ratios by Brown and Ghiselli (1953) and are reproduced here as Fig. 6.11. The values in the body of Fig. 6.11 are the mean standard criterion score of those selected ($\bar{z}_y$). Using the table for the situation in which predictor validity is 0.50 and the selection ratio is 0.20 yields the information that the average standard criterion score of those selected is 0.70. By comparison, the average standard criterion score of a randomly selected group or one selected with a nonvalid test would be 0.00. If we also have information that indicates

Selection Ratio	.00	.05	.10	.15	.20	.25	.30	.35	.40	.45	.50	.55	.60	.65	.70	.75	.80	.85	.90	.95	1.00
.05	.00	.10	.21	.31	.42	.52	.62	.73	.83	.94	1.04	1.14	1.25	1.35	1.46	1.56	1.66	1.77	1.87	1.98	2.08
.10	.00	.09	.18	.26	.35	.44	.53	.62	.70	.79	.88	.97	1.05	1.14	1.23	1.32	1.41	1.49	1.58	1.67	1.76
.15	.00	.08	.15	.23	.31	.39	.46	.54	.62	.70	.77	.85	.93	1.01	1.08	1.16	1.24	1.32	1.39	1.47	1.55
.20	.00	.07	.14	.21	.28	.35	.42	.49	.56	.63	.70	.77	.84	.91	.98	1.05	1.12	1.19	1.26	1.33	1.40
.25	.00	.06	.13	.19	.25	.32	.38	.44	.51	.57	.63	.70	.76	.82	.89	.95	1.01	1.08	1.14	1.20	1.27
.30	.00	.06	.12	.17	.23	.29	.35	.40	.46	.52	.58	.64	.69	.75	.81	.87	.92	.98	1.04	1.10	1.16
.35	.00	.05	.11	.16	.21	.26	.32	.37	.42	.48	.53	.58	.63	.69	.74	.79	.84	.90	.95	1.00	1.06
.40	.00	.05	.10	.15	.19	.24	.29	.34	.39	.44	.48	.53	.58	.63	.68	.73	.77	.82	.87	.92	.97
.45	.00	.04	.09	.13	.18	.22	.26	.31	.35	.40	.44	.48	.53	.57	.62	.66	.70	.75	.79	.84	.88
.50	.00	.04	.08	.12	.16	.20	.24	.28	.32	.36	.40	.44	.48	.52	.56	.60	.64	.68	.72	.76	.80
.55	.00	.04	.07	.11	.14	.18	.22	.25	.29	.32	.36	.40	.43	.47	.50	.54	.58	.61	.65	.68	.72
.60	.00	.03	.06	.10	.13	.16	.19	.23	.26	.29	.32	.35	.39	.42	.45	.48	.52	.55	.58	.61	.64
.65	.00	.03	.06	.09	.11	.14	.17	.20	.23	.26	.28	.31	.34	.37	.40	.43	.46	.48	.51	.54	.57
.70	.00	.02	.05	.07	.10	.12	.15	.17	.20	.22	.25	.27	.30	.32	.35	.37	.40	.42	.45	.47	.50
.75	.00	.02	.04	.06	.08	.11	.13	.15	.17	.19	.21	.23	.25	.27	.30	.32	.33	.36	.38	.40	.42
.80	.00	.02	.04	.05	.07	.09	.11	.12	.14	.16	.18	.19	.21	.22	.25	.26	.28	.30	.32	.33	.35
.85	.00	.01	.03	.04	.05	.07	.08	.10	.11	.12	.14	.15	.16	.18	.19	.20	.22	.23	.25	.26	.27
.90	.00	.01	.02	.03	.04	.05	.06	.07	.08	.09	.10	.11	.12	.13	.14	.15	.16	.17	.18	.19	.20
.95	.00	.01	.01	.02	.02	.03	.03	.04	.04	.05	.05	.06	.07	.07	.08	.08	.09	.09	.10	.10	.11

FIG. 6.11. Mean standard criterion score ($\bar{z}_y$) of selected cases in relation to test validity and selection ratio.
Source: From Brown and Ghiselli, 1953, p. 342.

the standard deviation of job performance in dollar terms per year is $1,000, then the average gain per person selected with a predictor with validity (r_{xy}) equal to 0.50 is $700 (0.7 × $1,000). Further, if 10 are selected, the test procedure would result in a net gain of $7,000 (10 × $700) in the first year these people would be employed.

Both Figs. 6.10 and 6.11 indicate that utility increases without limit as the selection ratio decreases. However, Brogden (1949) showed that, at the extremes, selection ratio and utility are curvilinearly related. This is because for very low selection ratios we incur relatively large testing costs because we test many applicants to make few selections. Hence, the cost of testing per selectee can become large. To make up for the cost of testing, the SD_y must be large or the net gain in utility becomes negative (i.e., the costs outweigh the benefits). Brogden successfully addressed the two major problems with the Taylor-Russell work: (a) the necessity to dichotomize the criterion, and (b) the failure to consider the costs in recruiting and testing.

Cronbach and Gleser (1965) published a book titled *Psychological Tests and Personnel Decisions* that presented detailed and sophisticated formulations about utility when employing various placement, classification, and sequential selection strategies, as well as strategies involving simple hire–reject decisions. In classification, an employer has multiple jobs, and the task involves assigning individuals to jobs in such a way as to maximize productivity while ensuring each job receives the required number of workers. Classification usually involves multiple predictors and a separate prediction equation for each job. The case in which there is only a single predictor and multiple jobs is called placement.

We do not discuss the classification and placement models, except to point out that classification can produce gains in utility over selection because classification systems automatically decrease the selection ratio. That is, when we can consider each applicant as an applicant for all jobs, the selection ratio becomes smaller for all jobs. The gain in utility that may result from classification as opposed to selection also depends on the differences in prediction equations and the standard deviation of the contribution of persons (SD_y) in different jobs to the organization. If our prediction equations are similar, then we predict that the same set of persons be hired for all jobs. If we predict the same value for the performance of a given person on two jobs, and if the jobs do not differ in SD_y, then the expected gain in utility resulting from the separate consideration of these two jobs would be zero. In contrast, if there are large differences in SD_y for the two jobs, the most appropriate assignment of a person with identical predicted job performance for the two jobs would be to the job that has the greatest SD_y.

With some simplifying assumptions, Brogden (1951, 1959) provided examples and tables expressing the gain that could be realized by the use of classification over selection. Because of the complexity of the mathematics and the relatively rare applicability of classification strategies (the military being a notable exception), little attention has been directed to classification problems.

SD_y Estimates Key to Utility Analyses

Obviously, a critical feature of this analysis is the standard deviation in dollar terms of the performance criterion (SD_y). The rule is that for jobs in which the contribution of individual employees to the organization differs greatly, valid testing will result in large dollar gains. In those situations in which individual contributions are relatively similar (e.g., for jobs that are essentially machine paced), a valid testing procedure and a low selection ratio will not result in large dollar gains.

Brogden's work in the 1940s and Cronbach and Gleser's (1965) formal introduction of the cost of testing in utility formulations provided the mathematical solution to problems of estimation of utility. However, applications of their formulations were not frequent until industrial psychologists developed methods to calculate SD_y, the standard deviation of individuals' dollar contributions to the organization. Prior to this development, only complex cost accounting procedures could be used to provide such estimates, and in most cases, human performance differences would not lend themselves to such analyses. Schmidt et al. (1979) reported a pilot study in which they derived estimates of SD_y for budget analysts. Supervisors were used as judges of the worth of their budget analysts because they were believed to have had the best opportunities to observe actual performance and output differences among employees. Schmidt et al.'s method was based on the following reasoning: If job performance in dollar terms is normally distributed (see chapter 2 for a description of the normal curve), then the difference between the value to the organization of the products and services produced by the average employee and those produced by an employee at the 85th percentile (one whose performance is as good as or better than 85% of the employees) is equal to SD_y. Similarly, the estimated difference between the average performer and the person who performs at the 15th percentile (as good as or better than only 15% of the employees) ought to be equal to SD_y. Furthermore, those two separate estimates ought to be equal to each other and also to the difference between the 85th and 97th percentile values if the normality assumption is correct (see Bobko, Karren, & Parkington, 1983, for an evaluation of this assumption).

Budget analyst supervisors in the pilot study (Schmidt et al., 1979) were asked to estimate the contribution of employees at both the 50th (average) and 85th percentile, and these values were averaged over 62 supervisors. The average SD_y was $11,327 for the budget analyst position. Because the instructions to the judges who make SD_y estimates are a critical element of the procedure, we reproduce the Schmidt et al. instructions in Fig. 6.12. Subsequent research has suggested that a reasonable estimate of SD_y is often 40% of the salary of people in the position in question. Obviously, the higher the average salary, the higher the SD_y and given equivalent validity and the greater the utility of using a selection procedure. Researchers have investigated various methods of estimating SD_y, the accuracy of the estimates, how to provide directions to the judges who provide the estimates, the extent to which such estimates actually influence organizational decision makers, and ways in which that influence can be enhanced. This research is reviewed in Boudreau and Ramstad (2003) and Cabrera and Raju (2001).

Suppose we as personnel administrators or consultants are interested in the utility that would accrue to our organization if we institute a new selection procedure. The job is that of secretary. Job incumbents perform a variety of duties: answering telephone calls, doing word processing, and answering client or customer questions and complaints. Thirty supervisory personnel, following the Schmidt et al. (1979) procedure, provide a mean estimate of SD_y equal to $10,000 (average earnings in this position are $25,000 annually). The true validity (corrected for attenuation due to unreliability in the criterion—see chapter 2) of the selection procedure is found to be 0.40. The cost of testing associated with the new procedure is shown to total $100 per applicant. We want to estimate what the utility of this selection procedure would be for various selection ratios. Currently, you have 200 applicants for the 100 positions you must fill annually, so the selection ratio is 0.50. An estimate of current recruiting and job advertising costs is approximately $50 per applicant. Using the Brogden formula and Fig. 6.11, which produces $\bar{z}_y$, the product $r_{xy}\bar{z}_x$, we can estimate utility:

$$\text{Average } \Delta\text{Utility/Person selected} = \bar{z}_y(SD_y)$$

The instructions to the supervisors were as follows:

The dollar utility estimates we are asking you to make are critical in estimating the relative dollar value to the government of different selection methods. In answering these questions, you will have to make some very *difficult judgments*. We realize they are difficult and that they are judgments or estimates. You will have to ponder for some time before giving each estimate, and there is probably no way you can be absolutely certain your estimate is accurate when you do reach a decision. But keep in mind three things:

1. The alternative to estimates of this kind is application of cost accounting procedures to the evaluation of job performance. Such applications are *usually* prohibitively expensive. And in the end, they produce only imperfect estimates, like this estimation procedure.
2. Your estimates will be averaged in with those of other supervisors of computer programmers. Thus errors produced by too high and too low estimates will tend to be averaged out, providing more accurate final estimates.
3. The decisions that must be made about selection methods do not require that all estimates be accurate down to the last dollar. Substantially accurate estimates will lead to the same decisions as perfectly accurate estimates.

Based on your experience with agency programmers, we would like for you to estimate the yearly value to your agency of the products and services produced by the average GS 9–11 computer programmer. Consider the quality and quantity of output typical of the *average programmer* and the value of this output. In placing an overall dollar value on this output, it may help to consider what the cost would be of having an outside firm provide these products and services.

Based on my experience, I estimate the value to my agency of the average GS 9–11 computer programmer at _____ dollars per year.

We would now like for you to consider the *"superior" programmer*. Let us define a superior performer as a programmer who is at the 85th percentile. That is, his or her performance is better than that of 85 percent of his or her fellow GS 9–11 programmers, and only 15 percent turn in better performances. Consider the quality and quantity of the output typical of the superior programmer. Then estimate the value of these products and services. In placing an overall dollar value on this output, it may again help to consider what the cost would be of having an outside firm provide these products and services.

Based on my experience, I estimate the value to my agency of a superior GS 9–11 computer programmer to be _____ dollars per year.

Finally, we would like you to consider the *"low-performing" computer programmer*. Let us define a low-performing programmer as one who is at the 15th percentile. That is, 85 percent of all GS 9–11 computer programmers turn in performances better than the low-performing programmer, and only 15 percent turn in worse performances. Consider the quality and quantity of the output typical of the low-performing programmer. Then estimate the value of these products and services. In placing an overall dollar value on this output, it may again help to consider what the cost would be of having an outside firm provide these products and services.

Based on my experience, I estimate the value to my agency of the low-performing GS 9–11 computer programmer at _____ dollars per year.

FIG. 6.12. Instructions for the estimation of SD_y.
Source: Reproduced with permission from Schmidt, Hunter, McKenzie, and Muldrow, 1979, page 621.

$\bar{z}_y$ from Fig. 6.11 with $r_{xy} = 0.40$ and a selection ratio of 0.50 is equal to 0.32. SD_y is $10,000.00 as stated. So, Average ΔUtility/Person selected = 0.32 × $10,000, or $3,200 per person selected. Multiplying this figure by 100, the number of persons selected in a given year, yields a value of $320,000. However, the use of the selection tests costs the company $200 per applicant, or $40,000 (200 × $200) for the total applicant pool, and recruiting costs are equal to $100 × 200, or $20,000. So, the net gain for the year would be equal to $260,000.

Now suppose we can, with an increased recruiting effort, reduce the selection ratio to 0.40. This increased emphasis on recruiting increases our per applicant recruiting cost to $150. For a selection ratio of 0.40, we have the following figures. First, the benefit in dollar terms would be equal to 0.39 (mean standard criterion score of selected persons from Fig. 6.11) × $10,000 ($SD_y$) × 100 (the number selected), or $390,000. Costs of testing would be equal to $200 × 250 (the number of applicants needed to achieve SR = 0.40), or $50,000. Cost of recruiting would be $150 × 250, or $37,500. Net gain for the year would be $390,000 − $50,000 − $37,500, or $302,500.

If further emphasis on recruiting produces a selection ratio of 0.20 and per applicant recruiting costs of $400, we have the following computations. Our dollar benefit would be 0.56 (mean standard criterion score from Fig. 6.11) × $10,000 ($SD_y$) × 100 (the number selected), or $560,000. Costs of testing would be $200 × 500 (the number needed to achieve SR = 0.20), or $100,000 and the cost of recruiting would be 500 × $400, or $200,000. So, the net gain for one year would be $560,000 − $200,000 − $100,000, or $260,000.

In this particular example, we can see that all three approaches to the use of the selection battery indicate that it has substantial predicted economic utility. The best strategy in terms of overall utility ($302,500) would be to increase recruiting efforts nominally to achieve a selection ratio of 0.40. At a selection ratio of 0.20, the added costs of recruiting offset the gain realized by the opportunity to raise the average standard score performance of the people selected. However, it is important to point out that if the gains associated with a more competent workforce are projected beyond the first year of employment, the analysis would certainly favor a selection ratio of 0.20 because recruiting and testing costs occur only at the time of hire. The more competent group of employees hired with a selection ratio of 0.20 will likely add to organizational productivity for more than a single year. Also, our analysis assumes an organization can recruit equally competent persons to lower the selection ratio. This may not be the case as Murphy (1986) has noted, and the utility gains will not be as great as would be estimated by the procedures outlined previously. Boudreau (1983a, 1983b) detailed some of the complexities of projecting gains beyond the first year of employment. These data gathering and analysis refinements associated with recruitment, multiyear accounting, employee flows, labor market effects, and others certainly made utility estimates more accurate (although this has been debated; Sturman, 2000), but they also made them more complex and may have contributed to a reluctance on the part of some staffing researchers and managers to use these methods or attend to the results of such analyses (Cronshaw, 1997; Latham & Whyte, 1994; Rauschenberger & Schmidt, 1987; Whyte & Latham, 1997). More recent discussion of utility analysis has focused on the impact of utility analysis on the actual decisions made by organizations (Boudreau & Ramstad, 2003; Cascio, 2000) and the connection of utility analyses to the actual strategic goals of an organization.

Direct Estimates of the Importance of Human Resources in Organizational Effectiveness

Beginning in the 1990s, some researchers have tried to link the use of various HR practices, including selection, to total organizational outcomes, including financial. Terpstra and Rozell (1993), for example, showed that five HR practices related to staffing were significantly related to firm profit. Huselid (1995), in a study of 968 firms, found that 13 HR practices that were considered high-performance work practices were related with turnover, employee productivity, and company financial performance. Delaney and Huselid (1996) found that selection, training, compensation, grievance procedures, and internal promotional practices were related to two measures of organizational performance based on perceptions others had of the firms. Becker and Huselid (1998) provided a review of the literature indicating that increasingly sophisticated HR practices can have quite significant impacts on firm financial outcomes.

Boudreau and Ramstad (2003) suggested that it is important for staffing researchers to stop focusing on single jobs in companies and to begin to focus on *talent pools*, groups of people who collectively determine important organizational outcomes. For example, suppose we have a university that has as a goal of student satisfaction. Given student satisfaction as a university goal, then faculty members would be part of the talent pool, but so would housing administrators, advisors, teaching assistants, and secretarial staff. The potential impact of staffing on customer contact for this talent pool would then be examined, rather than the SD_y associated with people in a single job. An important key issue is how to identify talent pools that are most critical to organizational success. Viewing the utility problem in this manner generates some counterintuitive results. We often assume key talent is represented by positions that are critical to the success of an organization, and in some utility formulations their average salary. However, the key question for Boudreau and Ramstad is not which talent has the greatest average value (e.g., faculty), but rather the talent that has the greatest potential performance variation in their impact on key strategic goals. The average value of a pilot to an airline is certainly high, but because of the high levels of certification and training required by law, there is little variability in their performance. However, individuals' performance variability in lower-level jobs, that is, the baggage handlers and ticket agents may be very high, so their impact on some aspects of the airline's organizational performance should be much greater. Focusing HR interventions on these groups may have much greater impact than focusing additional resources on the pilots who have low variability. The link between employee variability and organizational value is the essential part of SD_y. The talent pool logic for strategic HR management has been linked to a reduction in employee turnover and increased overall market performance (Richard & Johnson, 2001). The bottom line is that staffing researchers must continue in their attempts to make this link, but research to date certainly supports the conclusion that substantial organizational advantage can result from the use of valid selection procedures.

SUMMARY

In the latter section of this chapter, we present the position that utility analyses are essential even using imperfect estimates. We trace the history of utility analyses, beginning with the development of purely psychometric definitions of utility. Taylor-Russell tables represent an advance in that they focused attention on the organizational consequences of a selection and the situation (base rate and selection ratio) in which the selections are made. Brogden developed a utility formulation that eliminated the necessity to dichotomize the criterion (a problem with Taylor-Russell tables) and began work that incorporated the cost of testing. Cronbach and Gleser extended the Brogden formulations to placement and classification problems. The Brogden and Cronbach-Gleser formulations were used rarely until methodology was developed to estimate SD_y. The Schmidt et al. (1979) work and related formulations (Cascio, 2000) have generated renewed interest, use, and research on utility. More recently, researchers have sought to link HR interventions, including staffing directly to organizational outcomes. Boudreau and Ramstad (2003) believed that progress in utility research is most likely when HR interventions are linked to specific business processes that in turn have impact on organizational strategic goals.

Finally, we present validity and utility in the same chapter because of the intimate relationship between them. Validity has numerous overlapping subcomponents and processes that yield improved understanding of why tests and measures behave as they do when used in making staffing decisions. A key issue in the matrix of validity subcomponents concerns criterion-related validity that reveals a direct estimate of the statistical correlation between the staffing procedure and criterion outcomes. It is this specific form of validity that is required for estimating economic utility, which suggests an excellent reason for striving for this approach to validity when feasible.

REFERENCES

American Educational Research Association (AERA), American Psychological Association, & National Council on Measurement in Education. (1999). *Standards for educational and psychological testing*. Washington, DC: AERA.

American Psychological Association. (1974). *Standards for educational and psychological testing*. Washington, DC: Author.

Anastasi, A., & Urbina, S. (1997). *Psychological testing*. Upper Saddle River, NJ: Prentice Hall.

Arnold, J. D., Rauschenberger, J. M., Soubel, W. G., & Guion, R. M. (1982). Validation and utility of a strength test for selecting steelworkers. *Journal of Applied Psychology, 67*, 588–604.

Barrett, G. V., Phillips, J. S., & Alexander, R. A. (1981). Concurrent and predictive validity designs: A critical reanalysis. *Journal of Applied Psychology, 66*, 1–6.

Bartlett, C. J., Bobko, P., Mosier, S. B., & Hannan, R. (1978). Testing for fairness with a moderated multiple regression strategy: An alternative to differential analysis. *Personnel Psychology, 31*, 233–241.

Becker, B. E., & Huselid, M. (1998). High performance work systems and firm performance: A synthesis of research and managerial implications. *Research in Personnel and Human Resource Management, 16*, 53–101.

Bobko, P., Karren, R., & Parkington, J. J. (1983). Estimation of standard deviations in utility analyses: An empirical test. *Journal of Applied Psychology, 68*, 170–176.

REFERENCES

Bobko, P., Roth, P. L., & Potosky, D. (1999). Derivation and implications of a meta-analytic matrix incorporating cognitive ability, alternative predictors, and job performance. *Personnel Psychology, 52*, 561–589.
Borman, W. C., White, L. A., Pulakos, E. D., & Oppler, S. H. (1991). Models of supervisory job performance ratings. *Journal of Applied Psychology, 76*, 863–872.
Boudreau, J. (1983a). Effects of employee flows on utility analysis of human resource productivity improvement programs. *Journal of Applied Psychology, 68*, 396–406.
Boudreau, J. W. (1983b). Economic considerations in estimating the utility of human resource productivity improvement programs. *Personnel Psychology, 36*, 551–576.
Boudreau, J. W. (1991). Utility analysis for decisions in human resource management. In M. D. Dunnette & L. M. Hough (Eds.), *Handbook of industrial and organizational psychology* (Vol. 2, pp. 621–745). Palo Alto, CA: Consulting Psychologists Press.
Boudreau, J. W., & Ramstad, P. M. (2003). Strategic industrial and organizational psychology and the role of utility analysis models. In W. C. Borman, D. R. Ilgen, & R. J. Klimoski (Eds.), *Handbook of psychology* (Vol. 12, pp. 193–224). Hoboken, NJ: Wiley.
Brannick, M. T., & Levine, E. L. (2002). *Job analysis*. Thousand Oaks, CA: Sage.
Brogden, H. E. (1946). On the interpretation of the correlation coefficient as a measure of predictive efficiency. *Journal of Educational Psychology, 37*, 65–76.
Brogden, H. E. (1949). When testing pays off. *Personnel Psychology, 2*, 171–183.
Brogden, H. E. (1951). Increased efficiency of selection resulting from replacement of a single predictor with several differential predictors. *Educational and Psychological Measurement, 11*, 173–196.
Brogden, H. E. (1959). Efficiency of classification as a function of number of jobs, percent rejected, and the validity and intercorrelation of job performance estimates. *Educational and Psychological Measurement, 19*, 181–190.
Brown, C. W., & Ghiselli, E. E. (1953). Percent increase in proficiency resulting from use of selective devices. *Journal of Applied Psychology, 37*, 341–345.
Cabrera, E. F., & Raju, N. (2001). Utility analysis: Current trends and future directions. *International Journal of Selection and Assessment, 9*, 92–102.
Campbell, D. T., & Fiske, D. W. (1959). Convergent and discriminant validation by the multitrait-multimethod matrix. *Psychological Bulletin, 56*, 81–105.
Cascio, W. F. (2000). *Costing human resources. The financial impact of behavior in organizations*. Cincinnati, OH: Southwestern.
Cleary, T. A. (1968). Test bias: Prediction of grades of Negro and White students in integrated colleges. *Journal of Educational Measurement, 5*, 115–124.
Cronbach, L. J., & Gleser, G. C. (1965). *Psychological tests and personnel decisions*. Urbana: University of Illinois Press.
Cronshaw, S. F. (1997). Lo! The stimulus speaks: The insider's view on Whyte and Latham's "The Futility of Utility Analysis." *Personnel Psychology, 50*, 611–615.
Delaney, J. T., & Huselid, M. (1996). The impact of human resource practices on perceptions of organizational performance. *Academy of Management Journal, 39*, 949–969.
Flesch, R. (1949). *The art of readable writing*. New York: Collier.
Frederiksen, N. (1986). Construct validity and construct similarity: Methods for use in test development and test validation. *Multivariate Behavioral Research, 21*, 3–28.
Gael, S., Grant, D. L., & Ritchie, R. L. (1975). Employment test validation for minority and nonminority clerks with work sample criteria. *Journal of Applied Psychology, 60*, 420–426.
Ghiselli, E. E. (1966). *The validity of occupational aptitude tests*. New York: Wiley.
Ghiselli, E. E. (1973). The validity of aptitude tests in personnel selection. *Personnel Psychology, 26*, 461–478.

Goldberg, L. R. (1999). A broad-bandwidth, public domain, personality inventory measuring the lower-level facets of several five-factor models. In I. Mervielde, I. Deary, F. DeFruyt, & F. Ostendorf (Eds.), *Personality psychology in Europe* (Vol. 7, pp. 7–28). Tilburg, The Netherlands: Tilburg University Press.
Goldstein, I. L., & Ford, J. K. (2002). *Training in organizations.* Belmont, CA: Wadsworth.
Guion, R. M. (1976). Recruiting, selection, and job placement. In M. D. Dunnette (Ed.), *Handbook of industrial and organizational psychology* (pp. 777–828). Chicago: Rand-McNally.
Haaland, S., & Christiansen, N. D. (2002). Implications of trait-activation theory for evaluating the construct validity of assessment center ratings. *Personnel Psychology, 55,* 137–163.
Hartigan, J. A., & Wigdor, A. K. (1989). *Fairness in employment testing.* Washington, DC: National Academy Press.
Hogan, J. C. (1991). Physical abilities. In M. D. Dunnette & L. M. Hough (Eds.), *Handbook of industrial and organizational psychology* (Vol. 2, pp. 753–832). Palo Alto, CA: Consulting Psychologists Press.
Holland, J. L. (1997). *Making vocational choices: A theory of vocational personalities and work environments* (3rd ed.). Odessa, FL: PAR.
Huffcutt, A. I., Weekley, J. A., Wiesner, W. H., DeGroot, T. G., & Jones, C. (2001). Comparison of situational and behavior description interview questions for higher-level positions. *Personnel Psychology, 54,* 619–644.
Hull, C. L. (1928). *Aptitude testing.* Yonkers, NY: World Book Co.
Hunter, J. E., & Schmidt, F. L. (1990). *Methods of meta-analysis.* Newbury Park, CA: Sage.
Hunter, J. E., Schmidt, F. L., & Rauschenberger, J. M. (1977). Fairness of psychological tests: Implication of four definitions for selection utility and minority hiring. *Journal of Applied Psychology, 62,* 245–260.
Huselid, M. (1995). The impact of human resource management practices on turnover, productivity, and corporate financial performance. *Academy of Management Journal, 38,* 635–672.
Klein, K. J., & Kozlowski, S. W. J. (Eds.). (2000). *Multilevel theory, research, and methods in organizations.* San Francisco: Jossey-Bass.
Lance, C. E., Noble, C. L., & Scullen, S. E. (2002). A critique of the correlated trait-correlated uniqueness models for multitrait-multimethod data. *Psychological Methods, 7,* 228–244.
Latham, G. P., & Whyte, G. (1994). The futility of utility analysis. *Personnel Psychology, 47,* 31–47.
Lawler, E. E., III. (1967). The multitrait-multirater approach to measuring managerial job performance. *Journal of Applied Psychology, 51,* 369–381.
Lawshe, C. H. (1975). A quantitative approach to content validity. *Personnel Psychology, 28,* 563–575.
Linn, R. L. (1978). Single-group validity, differential validity, and differential prediction. *Journal of Applied Psychology, 63,* 507–512.
Livingston, S. A., & Zieky, M. J. (1982). *Passing scores.* Princeton, NJ: Educational Testing Service.
McDaniel, M. A., & Timms, H. (1990, August). *Lying takes time: Predicting deception in biodata using response latency.* Paper presented at the annual meeting of the American Psychological Association, Boston.
McLaughlin, G. H. (1969). SMOG grading a new readability formula. *Journal of Reading, 12,* 639–646.
Messick, S. (1998). Test validity: A matter of consequence. *Social Indicators Research, 45,* 35–44.

Murphy, K. A. (1986). When your top choice turns you down: Effect of rejected job offers on the utility of selection tests. *Psychological Bulletin, 99,* 133–138.
Neisser, U., Boodoo, G., Bouchard, T. J., Jr., Boykin, A. W., Brody, N., Ceci, S. J., Halpern, D. F., Loehlin, J. C., Perloff, R., Steinberg, R. J., & Urbina, S. (1996). Intelligence: Known and unknowns. *American Psychologist, 51,* 77–101.
Paulhus, D. L. (1991). Measurement and control of response bias. In J. Robinson, P. Shavear, & L. Wightsman (Eds.), *Measures of personality and social psychological attitudes* (pp. 17–59). San Diego: Academic Press.
Ployhart, R. E., & Schneider, B. (2005). Multilevel selection and prediction: Theories, methods, and models. In A. Evers, O. Smit-Voskuyl, & N. Anderson (Eds.), *Handbook of personnel selection* (pp. 495–516). Chichester/London: Wiley.
Ployhart, R. E., Weekley, J. A., & Baughman, K. (in press). The structure and function of human capital emergence: A multilevel investigation of the ASA model. *Academy of Management Journal.*
Rauschenberger, J. M., & Schmidt, F. L. (1987). Measuring the economic impact of human resource programs. *Journal of Business and Psychology, 2,* 50–59.
Richard, O. C., & Johnson, N. B. (2001). Strategic human resource management effectiveness and firm performance. *International Journal of Human Resource Management, 12,* 299–301.
Sackett, P. R., & Wilk, S. L. (1994). Within-group norming and other forms of score adjustment in preemployment testing. *American Psychologist, 49,* 929–954.
Schmidt, F. L., & Hunter, J. E. (1977). Development of a general solution to the problem of validity generalization. *Journal of Applied Psychology, 62,* 529–540.
Schmidt, F. L., & Hunter, J. E. (1998). The validity and utility of selection methods in personnel psychology: Practical and theoretical implications of 85 years of research findings. *Psychological Bulletin, 124,* 262–274.
Schmidt, F. L., Hunter, J. E., McKenzie, R., & Muldrow, T. (1979). Impact of valid selection procedures on workforce productivity. *Journal of Applied Psychology, 64,* 609–626.
Schmidt, F. L., Hunter, J. E., & Urry, V. W. (1976). Statistical power in criterion-related validation studies. *Journal of Applied Psychology, 61,* 473–485.
Schmidt, F. L., & Johnson, R. H. (1973). Effect of race on peer ratings in an industrial situation. *Journal of Applied Psychology, 37,* 237–241.
Schmitt, N., & Chan, D. (1998). *Personnel selection: A theoretical approach.* Thousand Oaks, CA: Sage.
Schmitt, N., Clause, C. S., & Pulakos, E. D. (1996). Subgroup differences associated with different measures of some job-relevant constructs. In C. R. Cooper & I. T. Robertson (Eds.), *International review of industrial and organizational psychology* (Vol. 11, pp. 115–140). New York: Wiley.
Schmitt, N., Gooding, R., Noe, R. A., & Kirsch, M. (1984). Meta-analyses of validity studies published between 1964 and 1982, and the investigation of study characteristics. *Personnel Psychology, 37,* 407–422.
Schmitt, N., Rogers, W., Chan, D., Sheppard, L., & Jennings, D. (1997). Adverse impact and predictive efficiency of various predictor combinations. *Journal of Applied Psychology, 82,* 717–730.
Schneider, B. (1978). Person–situation selection: A review of some ability–situation interaction research. *Personnel Psychology, 31,* 281–298.
Schneider, B. (1983). Interactional psychology and organizational behavior. In L. L. Cummings & B. M. Staw (Eds.), *Research in organizational behavior* (pp. 1–31). Greenwich, CT: JAI Press.
Society for Industrial and Organizational Psychology. (2003). *Principles for the validation and use of personnel selection procedures.* Bowling Green, OH: Author.

Stokes, G. S., & Cooper, L. A. (2001). Content/construct approaches in life history form development for selection. *International Journal of Selection and Assessment, 9,* 138–151.
Sturman, M. C. (2000). Implications of utility analysis adjustments for estimates of human resource intervention value. *Journal of Management, 26,* 281–299.
Sussmann, M., & Robertson, D. U. (1986). The validity of validity: An analysis of validation study designs. *Journal of Applied Psychology, 71,* 461–468.
Taylor, H. C., & Russell, J. T. (1939). The relationship of validity coefficients to the practical effectiveness of tests in selection. *Journal of Applied Psychology, 23,* 565–578.
Terpstra, D. E., & Rozell, E. J. (1993). The relationship of staffing practices to organizational level measures of performance. *Personnel Psychology, 46,* 27–48.
Thorndike, R. L. (1949). *Personnel selection.* New York: Wiley.
Whyte, G., & Latham, G. P. (1997). The futility of utility analysis revisited: When even an expert fails. *Personnel Psychology, 50,* 601–610.

7

Hiring Procedures: An Overview

AIMS OF THE CHAPTER

This chapter provides an organizing framework for thinking about the many predictors of job performance. We have seen in previous chapters, particularly chapters 4 and 6, that predictors derive their importance from criteria. However, identifying the criterion domain is but one step in the process. Even when the KSAO characteristics that will be used as the basis for selection are identified (i.e., the latent nature of individual differences that contribute to effective job performance), there is much work to do. The task facing the staffing specialist is to best develop manifest measures of these latent KSAOs, and then determine how to administer, score, and combine those manifest measures. As we describe in the next four chapters, there are a number of predictors available, including cognitive ability tests, personality tests, interviews, simulations, assessment centers, and so on. At first glance one might think the task of choosing a predictor is an easy one, and sometimes staffing experts have treated this task in an unthinking manner. However, in reality, there are a host of issues that must be considered. For example, "What format will be used to measure the latent construct—face-to-face interaction, paper-and-pencil tests, a video-based test, or so on?", "What is the criterion-related validity of the construct and/or measure?", "Will applicants find the test acceptable, or will its use evoke negative reactions and potential litigation?", "Will the predictor disadvantage protected minority groups?", and "Will the benefits accrued from the test be minimal relative to the costs of administration (low utility)?"

Answering these questions is greatly facilitated by considering predictors within an organizing framework. Such predictor frameworks have been scarce in staffing psychology, although a few noteworthy attempts have influenced this chapter (e.g., Klimoski, 1993; Schmitt, Cortina, Ingerick, & Wiechmann, 2003). Let us now turn to what we mean by the term "predictor."

PREDICTORS: A GENERAL DEFINITION

For our purposes, we may define a *predictor* as follows:

Any manifest measure whose scores represent individual differences on latent knowledge, skills, abilities, or other characteristics linked to effective criterion performance, and which is used (at least in part) as the basis for selection or promotion.

There are several key features to this definition. First, a predictor is a manifest measure of a latent construct or constructs. Such manifest measures may be a test, an interview, a knowledge test, how many sit-ups can be done in 1 minute, or almost any measure. As manifest measures, we know predictors will always have some degree of unreliability, and we should always assess the construct validity of the measure to ensure it adequately assesses those latent KSAOs of interest. Second, the latent KSAOs are linked to criterion performance, meaning that lacking those KSAOs, criterion performance would suffer or simply be impossible. Third, we assume there are individual differences on the latent construct, and thus, we should find individual differences in the manifest measure. The implication of this fact is that we can draw from the extensive literature on cognitive ability, physical ability, personality, and so on, to conceptualize the latent nature of our predictors. Fourth, distinguishing between the latent construct(s) and the manifest measure leaves open the possibility for using a variety of manifest measures for the same construct, a point we return to shortly. Finally, we define a predictor as comprising a single score, but multiple predictors may be used that together comprise a *predictor battery*.

A MODEL OF THE PREDICTOR-RESPONSE PROCESS

Now that we have a working definition of a predictor, it is possible to describe the predictor-response process (i.e., the ways in which people approach and respond to the predictors). So, it is one thing to focus on the content of the predictor (and we will) and another to focus on the factors influencing the ways in which people approach responding to predictors. Ployhart (in press) described a model of the predictor-response process that we summarize in this section. The model states that when performing on any predictor, whether it be an interview, knowledge test, or interactive simulation, performance on that predictor will require (at the most general level) two cognitive operations: comprehending the test stimulus, and providing a response to the stimulus. For example, when taking a multiple-choice test of aptitude such as the SAT or the ACT, one must comprehend (perceive and interpret) the question, and then pick out of four options the correct response. Similarly, an applicant being interviewed must comprehend the interview question and then respond orally with an answer. Wrong answers may come from incorrectly interpreting the question or from providing an incorrect or incomplete response. Therefore, individual differences in predictor performance come from differences in comprehending and responding to questions/items on the predictor measure. Of interest, obviously, are the latent individual differences that influence comprehending and responding.

Figure 7.1 shows a general model that can be used to illuminate the factors influencing comprehension of a test item, on the one hand, and responding to an item, on the other hand. Recall in looking at Fig. 7.1 that boxes indicate manifest measures, and circles represent the latent constructs (see chapter 2 if you need a refresher). In Fig. 7.1, there is only one manifest variable—the response—because it is the response to the collection of items (interview questions, assessment center exercises) that is of ultimate interest to us. Notice in Fig. 7.1 that not only latent individual differences in KSAOs of immediate content relevance to the item are of concern, but also of importance are potential contaminants and motivations that can influence comprehending and responding to an item. With regard to the contaminants identified in Fig. 7.1, it is interesting to note that this is the same construct we introduced in discussing criteria—criterion contamination—but here we apply it

7. HIRING PROCEDURES

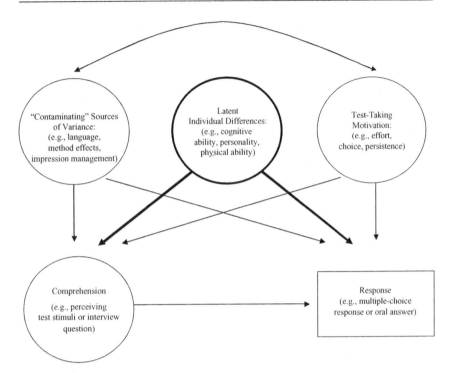

FIG. 7.1. Heuristic model of the predictor response process. Circles represent latent constructs, boxes represent manifest measures. Bold circles represent "true score" variance on the test response process. *Note.* Adapted from Ployhart (in press).

to the idea that predictor comprehension and responding can also be contaminated by seemingly irrelevant factors.

First and most important, we expect the predictor responses are primarily affected by "true score" variance, the *latent individual differences* we hope to assess and use as the basis for selection. This "true score" variance is denoted by the bold circle and arrows in Fig. 7.1. For example, if we develop a measure of cognitive ability to be used as a basis for hiring decisions, we hope the comprehension and responses to the cognitive ability test are primarily influenced by latent cognitive ability.

Second, Fig. 7.1 acknowledges two potentially confounding sources of variance. The first confounding source of variance is *motivation* and reflects how much applicants choose to focus their attention on performing on the predictor, how much effort they will expend in answering the questions, and how persistent they will be in solving difficult questions. Unless one were trying to select based on motivation, one

A MODEL OF THE PREDICTOR-RESPONSE PROCESS 369

hopes motivation will be constant across most applicants and hence contribute little to test scores (a condition of maximum performance tests). However, this is not always the case, and sometimes differences in motivation influence how hard people try on a test or how frightened they are by the testing situation and thus their test performance irrespective of latent true score variance (a condition of typical performance tests). For example, if you incorrectly believe a test is very difficult, you may not expend as much effort and persist on difficult questions like you would if you believe the test is within your ability. Or, imagine being interviewed for a promotion, but believing that the managers who will make the promotion decision are biased against your gender—you may not try as hard in the interview because you believe that no matter how well you perform, you will not get the promotion.

The second source of confounding variance is *contaminating sources* of variance that reflect latent constructs that influence the comprehension-response process for reasons unrelated to the latent individual difference of interest. For example, suppose the predictor is a personality test written in English, yet some applicants speak Spanish as their first language. Performance on this written test for Spanish-speaking applicants may be more affected by language than by latent personality, thus the language of the predictor is a contaminating source of variance on test scores.

Finally, notice that latent motivation and contamination sources of variance are related to each other (denoted by the curved, two-sided arrow). This reflects the fact that motivation may be influenced by contaminants, and vice versa. For example, applicants participating in an interview may misrepresent themselves and use impression management tactics to convey them as more qualified than they really are. Yet, such applicants may answer an identical set of questions completely honestly if responding to the questions for research-only purposes. Thus, the value or valence of effective performance on the predictor may influence the use of impression management tactics. These mutual contaminant-motivation influences contribute to affecting the predictor-response process irrespective of the true score variance of interest.

This simple predictor-response model can be used to understand a variety of influences on predictor scores. As we see in later chapters, issues such as racial subgroup differences in test scores, applicant impression management, faking on noncognitive measures, and differences between incumbent and applicant validation studies may be explained easily by consideration of the various latent factors identified in this heuristic model.

THE STRUCTURE AND FUNCTION OF THE PREDICTOR DOMAIN

One may compare and contrast predictors on multiple dimensions. Here, we focus on three ways of conceptualizing predictors: constructs versus measurement methods, psychological versus physical fidelity, and distal versus proximal predictors. As we explore these different ways to think about predictors, Fig. 7.1 is a useful frame of reference.

Constructs Versus Measurement Methods

Although the distinction between latent constructs and manifest measurement methods has often been ignored in the past, our scientific understanding of staffing is greatly enhanced by acknowledging this distinction exists. For example, we may want to use cognitive ability (the latent construct) as a basis for staffing decisions. However, we have many choices in what types of manifest measures we will use to assess cognitive ability: a written multiple-choice test; a computerized multiple-choice test; a Web-based, multiple-choice test; and so on. This distinction becomes important because different measurement methods can have different consequences for assessing the same latent construct. For example, we might find that using a Web-based cognitive ability test saves considerable money (and hence shows greater utility) than a written test. However, if minority applicants or applicants from less developed countries are less familiar with computers (thus, computer familiarity is a contaminating factor), there may be greater subgroup differences with the Web-based test and thus the written test is preferable. We discuss these consequences more fully in later sections.

Table 7.1 shows some common types of predictor constructs and the measurements frequently used to assess them. Notice that some constructs such as job knowledge may be measured by multiple measurement methods. For example, consider knowledge tests that must be passed to obtain a driver's license, they involve both written and performance-based (driving) assessments. In contrast, some predictor constructs such as physical ability can only be assessed with more hands-on, performance-based methods. You might ask a person how

STRUCTURE AND FUNCTION OF THE PREDICTOR DOMAIN

TABLE 7.1
Overview of Predictor Construct Domains and Measurement Methods

Predictor Construct Domains	Measurement Methods				
	Written	Computer/ Web	Visual and Aural/Oral	Interview	Performance Based
Cognitive					
Cognitive ability	X	X			
Job knowledge	X	X	X	X	X
Noncognitive					
Experience	X	X		X	
Biographical information	X	X	X	X	
Personality	X	X	X	X	X
Motivation				X	X
Interests/values	X	X	X	X	X
Performance Based					
Physical ability					X

Xs indicate common ways of measuring each construct.

fast they can run a mile, but the best way to obtain this information is to time the person actually running the mile. However, Table 7.1 is a simple overview, and with some effort most any predictor construct could theoretically be assessed by most any measurement method. Table 7.1 simply illustrates the most commonly used measures for different predictor constructs. Let us now briefly discuss the main predictor construct domains and predictor measurement domains.

Predictor Construct Domains

Predictor domains represent the types of individual differences found in differential psychology (i.e., the study of individual differences). There are many such domains (see Ackerman & Humphreys, 1990), as well as fine distinctions within domains, but here we have focused only on the three major domains. The first is the *cognitive domain* and refers to individual differences reflecting intellectual functioning, overall cognitive ability, academic aptitudes, knowledge, and related cognitive processes. We discuss these in more detail in chapter 8. The second is the *noncognitive domain* and refers to individual differences reflecting dispositions, traits, motivation, choice, and interest. Common examples include personality, biographical information (background data such as biographies and accomplishments), experience, motivation, interests, and values. We discuss these in more detail in chapter 9.

The third is the *performance-based* domain and refers to test methods such as interviews and simulations, as well as physical ability testing. We discuss these in more detail in chapter 10. Although these domains are very general, they tend to capture the majority of individual differences of interest to staffing specialists. That is, they cover individual differences in what people can do (see chapter 8), will do (see chapter 9), and are physically capable of doing (see chapter 10; Cronbach, 1960; Lubinski, 2000; Schmitt et al., 2003). Because we discuss these domains in more detail in the following three chapters, we focus here on measurement methods.

Predictor Measurement Methods

Measurement methods are simply the means we use to assess the latent construct; they are the manifest indicators on which selection decisions are ultimately based. As such, one can measure any given construct with multiple methods. Although there have been decades of research examining the structure of individual differences, there has been considerably less research designed to understand the structure and function of various measurement methods. This occurs despite the long-standing knowledge that measurement methods can influence the assessment of the construct and construct validity (Campbell & Fiske, 1959). Part of this neglect comes from the inherent difficulty in trying to develop a taxonomy or structure for method effects. For example, why would a written measure of personality differ from an interview measure of personality? Only recently has there been systematic investigation into the psychological nature of method effects (Schmitt, 1994; Schmitt, Pulakos, Nason, & Whitney, 1997; Williams & Anderson, 1994).

The model shown in Fig. 7.1 helps understand these method effects across predictors (e.g., contamination due to reading or language, viewing a computer screen). Specifically, we can use Fig. 7.1 to compare and contrast the different measurement method contaminants that may influence the comprehension-response process. That is, we can compare and contrast measurement methods by the nature of the contaminants they produce on test comprehension and response. Table 7.2 provides an overview of these potential contaminants.

Written (Paper-and-Pencil) Methods. Written methods require both the comprehension and response functions to occur in a paper-and-pencil format. A good example is the typical college in-class exam.

TABLE 7.2
Common Measurement Methods and Their Potential Contaminants

Measurement Methods	Primary Contaminants (Method Effects)	
	Comprehension	Response
Written	Reading ability, language	Reading ability, language, writing
Computer and/or Internet	Familiarity and comfort with computers/Internet, reading ability, language, ability to backtrack, hardware/software sophistication	Familiarity and comfort with computers/Internet, familiarity with mouse/typing, ability to backtrack, hardware/software sophistication
Visual and aural/oral	Eyesight, hearing, ability to ask questions	Hearing, vocabulary, ability to ask questions, impression management
Interview	Hearing, social anxiousness	Social anxiousness, vocabulary, impression management, appearance
Performance based	Presentation of stimuli, distractions, hardware/equipment	Social anxiousness, hardware/equipment

The test is in a written format, and you also answer the questions in writing. Written tests may have many types of item presentation and response formats, including multiple choice, reading comprehension, and open-ended (essay) responses. The primary potential contaminants of written methods are reading ability and language, unless the test was specifically designed to assess these KSAOs.

Computer and Web-Based Methods. Computer and Web-based methods present the items and response options on the computer screen. The items may consist of written items in which case the same kinds of contaminants and issues evident with written tests may be present. An example of this kind of predictor is the GRE, which is now administered entirely via computer. Computers and Web-based testing allow much greater flexibility than traditional written formats because such measures might present video information, aural information, or any combination of these. The response modes for computer/Web-based methods may involve simply using the mouse or keyboard to click on an appropriate answer (e.g., the GRE), or they may be more interactive, such as assessing reaction times to stimuli and monitoring mouse movements for moving targets.

Note that even if the computer/Web-based predictors involve the simple presentation of written information, additional contaminants may be present. For example, not everyone will be equally comfortable using a computer, and there may be differences in how well people can use a mouse. For example, this will clearly influence the scores that are based on mouse movements. In fact, one of the biggest contaminants on computer/Web-based responses is the inability to backtrack, or difficulty with navigating, through the software (Richman, Kiesler, Weisband, & Drasgow, 1999; Spray, Ackerman, Reckase, & Carlson, 1989). This contaminant can be removed simply by making the software user friendly, that is, by pilot-testing the procedure until the procedure itself is no longer a contaminant.

In general, research suggests that computerized testing is similar to paper-and-pencil testing (unless they are speeded tests) (see Mead & Drasgow, 1993; Potosky & Bobko, 1997; and Richman et al., 1999). However, little of this research has examined noncognitive constructs with applicant samples, so it is not certain whether different contaminants may be present and thus influence test scores. Likewise, there have been few studies comparing Web-based tests to paper tests. Some studies have found substantial similarity (Salgado & Moscoso, 2003), but other studies based on applicant samples find some differences (Ployhart, Weekley, Holtz, & Kemp, 2003), although the psychometric properties tend to favor the Web-based format (e.g., test scores yield more normal distributions). Future research needs to carefully consider not only the measurement method, but also the conditions under which the testing occurs (e.g., applicants versus incumbents).

We should also be careful to distinguish between computerized and Web-based testing. They are similar measurement methods, but they are not identical and may contain different contaminants (Lievens & Harris, 2003). For example, having test scores transmitted over the Internet to a remote server may evoke different concerns about test security than simply having the test administered and scored on a computer in one's plain sight. Another important issue more relevant to Web-based testing is whether the test is proctored. A proctored test means an organizational representative is present during the testing session, to ensure test security and to ensure the applicants are in fact the people completing the test. Compare this to an unproctored testing situation where applicants complete the test anywhere at any time (e.g., at home, at work). Unproctored testing offers many advantages, including reduced costs and flexibility. The major potential disadvantage is that one does not know who actually takes the test, whether persons cheat by having colleagues help them, and so on. Understanding the similarities and differences between Web-based, computerized, and paper tests is currently an active area of research.

Visual and Aural/Oral Methods. These methods refer to various ways of either presenting the test/predictor questions (thus influencing comprehension) or the ways examinees are to respond to the test. Visual methods most often involve the presentation of predictor stimuli in a video-based format. For example, one might administer a test of situational judgment via a VCR where test takers watch the video situations, and then respond in writing (e.g., Chan & Schmitt, 1997). Thus, contaminants on the response process will be similar to written tests, but contaminants on the comprehension process might additionally involve eyesight, the quality of the video, and related factors—even color-blindness, for example.

Aural-based methods present stimuli only in an aural format so test takers can only hear the questions—there are no reading or visual cues to aid in the comprehension process. Oral-based methods require responses to only occur in an oral format so test takers must speak their responses. Example contaminants may involve hearing ability or vocabulary (if vocabulary was not the purpose for the test). Perhaps the best example of an aural/oral measurement system is interactive voice recording (IVR) assessments. With IVR, applicants dial a phone number and are administered a set of prearranged questions over the phone (hence, the aural presentation of stimuli). After each question, they speak their response (hence the oral response), which is recorded by a computer and scored automatically. Of course, one may mix and match visual/aural/oral responses with most any stimuli/response, (e.g., having a written test be answered orally, or having applicants watch a short movie and then answer a set of written questions). Note that such measurement methods are different from an interview in that they do not involve another person or persons being physically present in the assessment exercise.

Interview Methods. The hallmark of the interview is the face-to-face interaction between the interviewer (or interview panel) and applicant. We see in chapter 10 how there are many variations in the format of interviews, and thus, in the comprehension-response processes. However, these various methods all share a common element of face-to-face interaction. Because of this social interaction component, the primary potential contaminants of scores relate to social-cognitive processes. For example, impression management may occur when applicants tend to make themselves appear more favorable than they really are. We are all familiar with popular press recommendations for how to "ace" an interview by answering the questions in a particular way—even if the responses are not necessarily accurate. Likewise, there are many recommendations for how to dress and act in an interview that may have nothing to do with the latent construct. Alternatively, applicants may suffer from social anxiousness that inhibits their comprehension and responses. Thus, to understand the interview, one must understand these social-cognitive processes.

Performance-Based Methods. Performance-based methods refer to a diverse variety of measurement methods. They tend to not be computerized (when they are, they fall under the computerized testing methods), but instead involve more interactive, hands-on kinds of assessments. Role plays, assessment centers, work sample tests, simulations, or performance tests are all examples of performance-based methods (these predictors are discussed more in chapter 10). The defining feature of performance-based methods is a need for the applicant to act out or perform the action.

Thus, performance-based methods represent a variety of measurement systems, and as such have many potential sources of contamination. Beyond the sources discussed earlier for written, computer/Web-based, or interview methods, additional contaminants may include the nature of the task (e.g., group interactions may constrain the applicant's behavior), the types of equipment/hardware used to perform the task, or even how the exercises are structured.

A Note on Measurement Methods as Predictors

We have thus far described constructs and methods as though the distinction was very straightforward. Often it is not, and the lines between the two get blurred. Interviews and simulations are perhaps the best example of this. Interviews, for example, may be used to assess a knowledge construct, but other times they are simply a measurement method that assesses a variety of interpersonal/social constructs. The important point to take away from this discussion of predictor measurement methods is the following: The measurement method chosen for the assessment of the latent individual differences (KSAOs) of interest can have unintended consequences on comprehension and responding such that actual scores represent more than the KSAO constructs of interest.

Psychological and Physical Fidelity of the Method

A second way to conceptualize predictors is to examine their psychological and physical fidelity to the performance task. *Fidelity* refers to how closely related the predictor is to the performance domain; *psychological fidelity* refers to the extent the predictor taps the same KSAOs required for successful performance on the criterion, and *physical fidelity* refers to the extent the predictor is physically similar to the criterion. Of the two types of fidelity, psychological fidelity is more important than physical fidelity, but one should strive to enhance both (e.g., Goldstein, Zedeck, & Schneider, 1993).

Notice that psychological fidelity roughly corresponds to the nature of the latent construct, whereas physical fidelity roughly corresponds to the nature of the measurement method. For example, suppose we want to develop a predictor assessing knowledge of flight procedures for airline pilots. If we use a written job knowledge test as our predictor, it will have moderate psychological fidelity but low physical fidelity. If we instead ask pilots to play a flight video game, it may have more physical fidelity but may lack psychological fidelity. In the first instance, the predictor taps the critical KSAOs necessary for the job, which is ultimately most important. The best of both worlds would come from an actual flight simulator, which would be high on both psychological and physical fidelity. In general, performance tests have higher levels of physical fidelity and, when carefully constructed also have psychological fidelity, making such tests high on content validity. An attractive feature of content validity is that this measurement method appears to applicants to be job relevant—perhaps because it is.

Distal Versus Proximal Predictors

A final way to conceptualize predictors is to place them on a continuum from distal to proximal influences on criterion performance. As we have noted, predictors derive their importance from criteria, so if one starts with the criterion, one can ask which predictors are conceptually most related to—that most determine—criterion performance? Those predictors that assess KSAOs most directly related to performance are known as proximal predictors (Schmitt et al., 2003). As one moves farther from

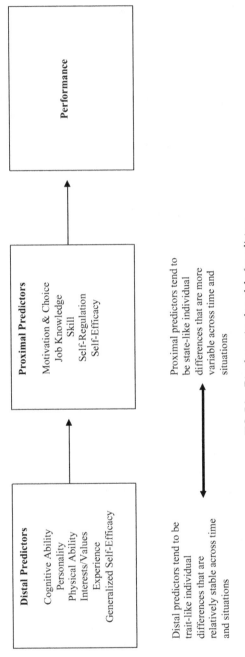

FIG. 7.2. Distal-proximal model of predictors.

EVALUATING THE ADEQUACY OF PREDICTORS

the criterion, the predictors are known as distal—they are related to criterion performance, but not as immediately strongly as the proximal predictors.

Figure 7.2 illustrates this distal-proximal continuum for the prediction of overall job performance (all relationships are individual-level relationships). Proximal predictors tend to assess more state-like constructs such as motivation, job knowledge, and self-regulation of behavior. These constructs are the immediate precursors to performance, and change across situations and time as performance changes. These tend to be the constructs often examined in organizational behavior and training because they are more malleable. In contrast, distal predictors tend to reflect trait-like, stable individual differences. Constructs such as general intelligence and personality do not change dramatically over short periods of time, but rather develop slowly over many decades. Most staffing research is focused on these stable individual differences because when a person is hired, the person is expected to have the characteristics that will result in effective performance across time and situations. However, proximal constructs, particularly job knowledge and job-specific skills, do play an important place in staffing.

Note that equating distal-proximal predictors with trait-like and state-like constructs only works when talking about predictor *constructs*. When talking about predictor measurement methods such as interviews, situational judgment, and assessment centers, they tend to predict performance strongly because they capture a variety of distal-proximal constructs.

EVALUATING THE ADEQUACY OF PREDICTORS

Having discussed the structure and function of the predictor domain, we may now consider the dimensions that can be used to evaluate the adequacy of predictors. Obviously, "good" predictors are those that tap the KSAOs most related to job performance (see chapter 6), but there is still considerable flexibility in the types of constructs and measures that may be used in practice. In this section, we briefly consider the primary dimensions of predictor adequacy: criterion-related validity, subgroup differences, user acceptability, and overall utility (see Austin, Klimoski, & Hunt, 1996, for similar set of dimensions). Table 7.3 summarizes these across predictor constructs and measures. We only briefly discuss these dimensions here; subsequent chapters describe the research on these dimensions for each predictor in much greater detail.

TABLE 7.3
Adequacy of Predictor Constructs and Measurement Methods

Predictors	Criterion-Related Validity	Subgroup Differences (Race/Gender)	User Acceptability	Utility
	Correlation	*d value*		
Effect sizes	0.10 = Low 0.20 = Moderate 0.30 = High	0.20 = Small 0.50 = Moderate 0.80 = Large	N/A	N/A
		Predictor Constructs		
Cognitive				
Cognitive ability	High	Large/small	Moderate	High
Job knowledge	High	Moderate/small	Favorable	High
Noncognitive				
Experience	Moderate	Small/small	Moderate	Low
Biographical information	Moderate	Small/small	Unfavorable	Moderate
Personality	Low/moderate	Small/small	Unfavorable	Moderate
Motivation	Low	Small/small	Moderate	Low
Interests/values	Low	Small/small	Unfavorable	Low
Performance Based				
Physical ability	High	Small/large	Moderate	Moderate
		Predictor Measures		
Interview—structured	High	Small/small	Moderate	Moderate
Interview—unstructured	Low	Small/small	Moderate	Low
Situational judgment	Moderate	Moderate/small	Favorable	High
Performance based	Moderate/High	Small to moderate/small	Favorable	Moderate

Estimates for validity and subgroup differences are based on uncorrected values.

Criterion-Related Validity

As we saw in chapters 4 and 6, criterion-related validity is of critical importance for staffing because this correlation describes the relationship between the predictor and the criterion. As such, criterion-related validity indicates the likelihood that basing hiring decisions on the predictor will lead to enhanced performance. It is perhaps the primary consideration in evaluating the adequacy of a predictor, but the difficulty comes in determining just what magnitude of validity is "important." This is a difficult question to answer because it depends on the cost of testing, selection ratio, number of qualified applicants in the population, the importance of the job, and related features as discussed in chapter 6. Popular conventions would indicate small, medium, and large effect sizes (correlations) of 0.10, 0.30, and 0.50, respectively (Cohen, 1988). These do not work well in staffing research and practice because uncorrected effect sizes are rarely above 0.30. Thus, we consider uncorrected validities of 0.10, 0.20, and 0.30 to be small, medium, and large, respectively. [Do not be confused when validities are corrected for unreliability, range restriction, sampling error, and related artifacts (discussed in chapter 6). When making these corrections, validities will often be much higher than 0.30. Although these corrected validities are useful for estimating population values, they do not characterize the kinds of validities observed in a given selection context.]

Table 7.3 shows that cognitive ability, job knowledge, physical ability, structured interviews, and performance-based tests all show relatively high criterion-related validity. Experience, biographical information, and situational judgment tests are more moderate, whereas personality, motivation, interests/values, and unstructured interviews tend to show relatively low to moderate validity. Note, however, that even a test of low validity may have some value as a basis for making hiring decisions. For example, personality provides an important complement to cognitive ability tests, as we see later. At the extreme, any test that has a validity greater than zero does a better job of hiring employees than random hiring, so even predictors with low validity will result in better hiring decisions.

Subgroup Differences

Popular convention treats demographic groups as majority and minority, where minority groups are those historically disadvantaged in society (e.g., women, Blacks, Hispanics, disabled, elderly). These groups are defined by employment litigation and such acts as the 1964 and 1991 Civil Rights Acts and Title VII of Executive Order 11246 (these are discussed in more detail in chapter 11). Typically, these protected groups are defined by gender, race, ethnicity, age, disability status, and related demographic categories. These tend to be called *subgroups* because they are categories of a more general group; thus, men and women comprise subgroups of gender. This is why the examination of mean differences on predictors are called *subgroup differences*.

The amount of subgroup difference represents how much of a mean difference on the predictor scores exists between two or more demographic groups. The effect size used to estimate the mean subgroup difference on predictors is the d statistic (Cohen, 1988), which expresses any subgroup difference in terms of standard deviation units (more specifically, d = ([majority group mean − minority group mean]/pooled standard deviation)). Thus, a d value of 0.75 says that one group scores 0.75 standard deviation units higher than the other group. For the sake of argument, we follow convention (Cohen, 1988) and define small, medium, and large subgroup differences as having d values of 0.20, 0.50, and 0.80.

To the extent there is a mean difference between demographic groups on test scores, members from the demographic group that scores higher on the test will be more likely to be selected (this assumes applicants are hired in a top-down manner based on their test scores, such that those with the highest scores are hired first). Thus, if men score higher than women on a test of physical ability, men are more likely to be hired than women. The greater the subgroup difference, the greater the likelihood of there being an adverse impact (different hiring rates) against the lower scoring group. For example, Fig. 7.3 shows the consequences of different subgroup differences on a predictor with test scores ranging from 0 to 100 (higher numbers indicate better performance). In this example, all applicants who score 60 and above are hired. Notice that when the d value is 0.25, many more minority applicants would be hired (indicated by the shaded areas) than when the d value is 1.00. Thus, the greater the mean difference, the less likely one would hire members from the lower-scoring group.

Table 7.3 summarizes the relative size of the subgroup difference for the various predictors. In general, cognitively oriented predictors tend to show large racial subgroup differences, around the order of 0.60 to more than 1.00 (Jensen, 1998; Roth, Bevier, Bobko, Switzer, &

EVALUATING THE ADEQUACY OF PREDICTORS

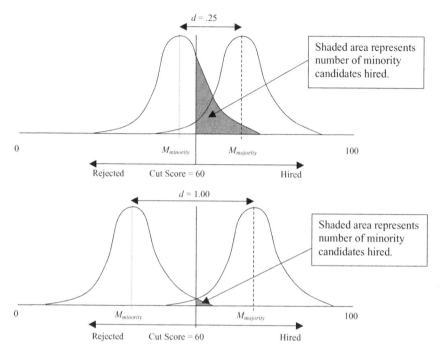

FIG. 7.3. Example of subgroup differences. Notice how fewer members from the lower-scoring group are hired when $d = 1.00$ as opposed to when $d = 0.25$. M, mean.

Tyler, 2001; Sackett & Wilk, 1994). Noncognitive constructs tend to show small to moderate subgroup differences, with values ranging from 0.00 to about 0.30 (Hough, Oswald, & Ployhart, 2001; Schmitt, Clause, & Pulakos, 1996). Physical ability tests produce large gender differences, ranging from about 0.20 to more than 2 *SDs* (Hough et al., 2001; Schmitt et al., 1996). Measurement methods such as interviews and performance-based tests show smaller subgroup differences, although this is sometimes due to the lower reliability of these measures. Research by Bobko, Roth, and Buster (2005) suggests work sample subgroup differences may be larger than previously thought. Other things being equal (which they rarely are), it is obviously wise to use measures of constructs with smaller subgroup differences. A challenge for staffing specialists is to combine predictor measures in ways to optimize both criterion-related validity and lack of subgroup differences. This is one reason why it is useful to study a number of predictor methods and constructs simultaneously so exploration of possible trade-offs can be made based on evidence.

User Acceptability

Validity and subgroup differences are objective characteristics of the predictor scores; user acceptability refers to whether people involved with using the test (e.g., applicants, staffing personnel, managers) *perceive* the predictors as being an appropriate basis for making hiring and promotion decisions. For example, despite validity evidence that supports making admissions decisions based on SAT, ACT, GRE, GMAT, or LSAT scores, many test takers refuse to believe these tests predict performance. Relative to research on predictor validity, research on user acceptability is still fairly new, but it is important because if the predictors are to be used appropriately, the users of the tests must believe in them (Arvey & Sackett, 1993). Indeed, in our experience, it is critical that a predictor appear "credible" or the organizational personnel expected to administer the predictor will never use it appropriately; hence, the validity and utility of the predictor will suffer. We cover much of the material relating to applicants' perceptions of predictors and staffing practices in chapter 5. In Table 7.3, we summarize most of this research by noting whether the various predictors tend to be perceived as unfavorable, moderate, or favorable (these estimates are drawn primarily from Kluger & Rothstein, 1993; Ryan & Ployhart, 2000; Rynes & Connerley, 1993; and Smither, Reilly, Millsap, Pearlman, & Stoffey, 1993). In general, tests that show greater physical fidelity (or content validity) are perceived more favorably.

Utility

As noted in chapter 6, utility refers to the (usually monetary) value of using the predictor versus nothing. Utility is first and foremost affected by the validity of the predictor, as well as by the cost of testing. Classic utility models (see chapter 6) have ignored issues relating to subgroup differences and user acceptability. This can present a bit of a misleading picture because a test with high utility (high validity, low cost) could produce large subgroup differences and/or be unacceptable to test users (and thus never implemented). Again, it is difficult to put a specific dollar value or utility estimate on various predictors, but in Table 7.3 we have simply put low, moderate, and high as summary descriptions. In general, tests that are inexpensive to develop, that show high validity, and are reasonably acceptable to users show the greatest utility. However, as utility is so highly affected by other characteristics of the staffing situation (e.g., selection ratio), these estimates should not be taken literally or overinterpreted.

Implications

Table 7.3 provides several implications about the adequacy of predictors. As shown in Table 7.3, there is no one perfect predictor that maximizes all dimensions. Cognitively oriented predictors maintain special status among all predictors because they generally show high validity, utility, and reasonably favorable user reactions. Their major drawback, particularly with measures of overall cognitive ability, is their large racial subgroup differences. Noncognitive predictors tend to show moderate to low levels of validity and utility and have moderate user reactions, but importantly, they show few subgroup differences. This makes noncognitive predictors attractive supplements to a predictor battery containing cognitively oriented measures (discussed shortly). With the exception of unstructured interviews, performance-based predictor measures show moderate to high validity and user reactions, and smaller subgroup differences, but their high cost of development and administration tends to produce only moderate utility (the exception is situational judgment tests). Thus, there is no perfect predictor (if there was, we would have little need for a book on staffing).

USE OF PREDICTORS IN STAFFING PRACTICE

In this section, we consider a variety of choices facing the staffing practitioner, including the use of predictor batteries, methods for combining predictors, the use of multiple hurdle systems, and compound predictor constructs.

Use of Predictor Batteries

A predictor battery consists of multiple individual predictors. Jobs are rarely so homogeneous in their KSAO determinants that a single predictor adequately captures all the relevant variance in job performance (Murphy, 1996). Therefore, many organizations will use a battery of cognitive, noncognitive, and performance-based predictors. If well developed, this overall battery of tests nearly always outpredicts any one predictor (although sometimes the incremental validity and utility of adding additional predictors may be small).

Today, one of the most common forms of predictor batteries supplements overall measures of cognitive ability with noncognitive measures of personality, biographical information, or related predictors. There have been many studies demonstrating that predictor batteries combining cognitive and noncognitive predictors tend to show the highest validity, smallest subgroup differences, and highest utility (Bobko, Roth, & Potosky, 1999; Sackett & Ellingson, 1997; Sackett & Roth, 1996; Schmidt & Hunter, 1998; Schmitt, Rogers, Chan, Sheppard, & Jennings, 1997). The use of predictor batteries is not perfect, however, because a sizeable amount of racial subgroup difference is still present even with the inclusion of noncognitive measures (which tend to show smaller subgroup differences). The addition of these other predictors will also not always add in an incremental way to the utility of prediction. There are also no simple rules of thumb to describe when a battery of certain tests will demonstrate a given pattern of validity, subgroup differences, and utility. The reason is because these characteristics are a function of the criterion-related validities, subgroup differences, and intercorrelations among the predictors in the battery (see Sackett & Ellingson, 1997; Sackett & Roth, 1996).

Methods for Combining Predictors

One issue that must be considered when using predictor batteries is how the predictors will be combined into an overall score. There are multiple ways one may combine various predictors, including using empirical versus rational weights (we touched on this issue in chapter 2). As noted earlier, the best human judge can only combine predictors as good as a regression-based prediction equation; this has to be true because the regression model provides a statistically optimal combination. So, whenever the predictors are combined in a manner different from the regression-based weights (and to a lesser extent, unit weights), validity will suffer. Note, however, the regression model defines "best" in terms of prediction and hence validity. The regression model does not consider subgroup differences, user acceptability, or utility.

To better balance validity with subgroup differences, some may choose to weight different predictors according to their amount of subgroup difference. For example, in a predictor battery consisting of cognitive ability and personality, one might weight personality twice as much as cognitive ability because it demonstrates a smaller subgroup difference. Doing so could result in a loss of validity (and utility because validity is a primary determinant of utility). The question becomes whether the reductions in subgroup differences are worth the reductions in validity and utility. There is currently a great deal of debate among applied psychologists, the legal system, and society in general as to what the correct answer is in this situation. Some would argue that small trade-offs in validity are worth the increased number of minority applicants who are hired; others would claim such an approach mixes politics and values with science. It is the courts that ultimately will decide the answer to this question, but I/O psychologists play a pivotal role by providing expert opinion and advice (noted in chapter 11).

A different issue concerns whether the predictor battery will be compensatory or noncompensatory. In a *compensatory model*, low scores on one predictor can be made up with high scores on another. Regression models are inherently compensatory models. For example, a person with high ability but only average motivation will have the same predicted performance as will someone with average ability but high motivation. In *noncompensatory models*, low scores on one predictor cannot be made up by high scores on a different predictor. For example, a pilot with high ability but extremely poor eyesight will not be a very good pilot (this is one reason why there is a limit on how old a commercial pilot can be and have a license). Use of cut scores, where only applicants who score above the cut score are considered for employment, is an example of a noncompensatory system. For an excellent overview of other ways of combining predictors, see Gatewood and Field (1998).

Multiple Hurdle Systems

When implementing a staffing system, it is often not the case that all the predictors are administered to all applicants. Particularly in staffing procedures with large numbers of applicants, organizations may use a multiple hurdle system to reduce the costs of testing and the burden on applicants who will stand little chance of ultimately being selected. Thus, multiple hurdle systems use several stages of testing. In the early stages, one usually uses predictors that are relatively cost effective and easy to administer to large numbers of applicants. For example, written or Web-based cognitive ability and personality tests might be used because they are easily administered to large numbers of people. However, a cut score will be placed on the tests in this first stage, such that only those applicants who score above a certain point on the ability and personality tests will continue in the staffing process. For those applicants, the second stage of testing might involve more expensive and time-consuming predictors, such as interviews or simulations. As another example, many police departments require that applicants do not have any outstanding tickets or warrants (this is used as a first hurdle before testing). Although this might seem obvious, one of the authors worked on a project where the police staff helping with a large-scale test administration arrested several applicants who showed up for the test because they had warrants for their arrest!

Thus, multiple hurdle strategies are noncompensatory staffing systems that progressively narrow the number of applicants at each stage of testing. Your authors are familiar with numerous organizations that in practice use a telephone "screen" to assess applicants for jobs and make a "reject" or "continue" judgment based on responses to the screen. These are efficient ways of reducing the applicant pool to more manageable numbers, but it is important to recognize that each cut score is essentially a hiring decision that must be demonstrated to have reasonable validity. As such, multiple hurdle systems must be able to show that those who are rejected at early stages truly are unable to perform the job.

"Compound" Predictors

Beyond the distinction between predictor constructs and predictor measures, one may also combine multiple predictor constructs into an overall multidimensional composite measure. For example, combining the specific and relatively homogenous personality traits of neuroticism, agreeableness, and conscientiousness into an overall heterogeneous "customer service orientation" score often results in better prediction (Frei & McDaniel, 1998). In this example, the specific traits of neuroticism, agreeableness, and conscientiousness can be thought of as "elements," whereas the composite measure of service orientation can be thought of as a "compound" (Hough & Schneider, 1996). Another example is the concept of core self-evaluations, which are comprised of self-esteem, neuroticism, locus of control, and generalized self-efficacy (Judge, Erez, Bono, & Thoresen, 2002). Sometimes these compound traits are referred to as "criterion-focused occupational personality scales" because only those test items that are empirically related to the job performance criterion remain in the scale to be scored (e.g., Ones & Viswesvaran, 2001).

Examples of compound measures include customer service orientation, integrity (honesty) tests, managerial potential, sales potential, violence scales, and stress tolerance (e.g., Hogan & Hogan, 1992; Hough & Schneider, 1996; Ones & Viswesvaran, 2001). Each compound measure contains items from multiple homogenous personality scales; the exact combination of items is based on their empirical relationship to the criterion. Thus, they are maximally predictive of a certain type of criterion measure.

The main reason for using such compound measures of personality is because they predict the relevant performance criterion more strongly than the more abstract and homogeneous measures. For example, Ones and Viswesvaran (2001) demonstrated how compared with homogeneous personality scales, compound measures predict overall ratings of supervisory performance on a magnitude nearly twice as large. For example, in their study conscientiousness predicts performance 0.23, whereas stress tolerance predicts performance 0.41. This should come as no surprise because the compound measure contains only those items related to the criterion; it is thus a predictor capturing only that variance related to the criterion of interest. In contrast, "element" traits (e.g., those based on the FFM of personality) are used to describe normal adult personality across a range of everyday situations.

Thus, although the criterion-related validities of personality tend to be modest, the validities can be quite impressive if the personality traits are combined in ways to maximally predict performance. However, although prediction may be enhanced, scientific understanding of predictor–criterion relationships may be slowed by the use of such measures. We revisit this topic again in chapter 9.

CURRENT ISSUES IN PREDICTOR DEVELOPMENT

Because predictor development occurs in a dynamic, changing world, a variety of new issues present challenges for the staffing practitioner. We have highlighted three such challenges here: use of predictors in multilevel contexts, cultural influences on predictor development, and technological developments for predictor measurement and application. As you read the next three chapters describing the various predictors in more detail, you may want to consider how the following concerns impact modern predictor development.

Predictors in Multilevel Contexts

In this chapter, we talk about predictors primarily in terms of individual-level constructs and measures, with linkages to individual criteria. Yet, predictors still have importance in multilevel contexts such as those involving team-level staffing and performance. However, the nature of the predictor constructs and measures likely changes when thinking about predicting higher-level criteria such as team performance (Chen & Bliese, 2002; Gully, Incalcaterra, Joshi, & Beaubien, 2002; Klimoski & Jones, 1995). In chapters 1 and 4, we note how different criteria may exist at different levels of analysis. As a result, the types of predictor constructs most proximal to performance may change across the two levels. Figure 7.4 shows this potential disconnect across two levels of analysis. At the individual level, notice that predictors are expected to directly relate to individual performance (even though there are both distal and proximal predictors). At the team level, the composition or aggregate of individual KSAOs is expected to have an indirect (mediated) relationship with team performance. Mediating team process constructs, such as team cohesion, shared mental models, and coordination, are the direct determinants of performance. Note that individual-level KSAOs determine team composition and may also predict team mediating processes (e.g., teams composed of more agreeable members may get along better).

There have been few staffing studies that have examined these relationships across levels, but the results are supportive of the model in Fig. 7.4. Barrick, Stewart, Neubert, and Mount (1998) demonstrated teams with more cognitive ability, and certain personality traits received more favorable ratings of team performance and team viability (a measure of how well the team could maintain itself over time). However, the Extraversion and Neuroticism personality traits were only related to team viability through the team mediating processes of cohesion. They also demonstrated how a low-performing team member could influence the effectiveness of the overall team. Neuman and Wright (1999) similarly demonstrated different KSAOs would show different relations to performance across individual and group levels. Together, the Barrick et al. (1998) and Neuman and Wright (1999) studies suggest that some personality constructs may actually be stronger predictors of performance for teams than for individuals; this is most likely attributable to the key role of team member interactions that comprise the proximal determinant of team performance. However,

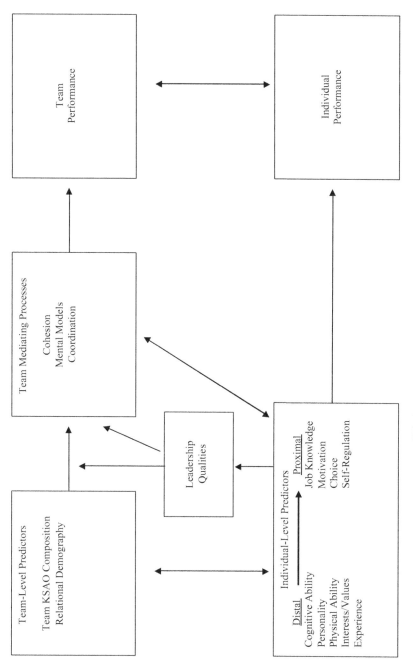

FIG. 7.4. Model of staffing in team contexts.

whether the team composition of member KSAOs should be homogeneous, heterogeneous, or some combination of the two is highly dependent on the task, as both homogeneity and heterogeneity can predict different parts of the team criterion space (Barrick et al., 1998).

Also of importance in Fig. 7.4 is the key role played by the team leader. Notice that leadership qualities are individual KSAOs, but the leader has both direct and interactive effects on team mediating processes and hence team performance. For example, LePine, Hollenbeck, Ilgen, and Hedlund (1997) found team performance was best when both the leader and team were high on cognitive ability and conscientiousness. Thus, the characteristics of the leader interact with the characteristics of the team to influence team performance. Similarly, Lim and Ployhart (2004) found ratings of transformational leadership partially mediated the effects of leader personality on typical team performance and fully mediated the effects of leader personality on maximum team performance.

This disconnect between the determinants of performance across levels of analysis has obvious implications for staffing research and practice. Ployhart and Schneider (Ployhart, 2004, in press; Ployhart & Schneider, 2002) discussed these issues and how to model staffing within a multilevel context. There are some general conclusions from this work. First, the comparison of predictors shown in Table 7.3 is largely true only for the individual level of analysis. Second, the kinds of KSAOs used for staffing at the individual level may not necessarily be the most appropriate at the team level, and entirely different predictor constructs and measures may be required to best predict and explain performance at different levels. Third, the types of staffing interventions may need to change across levels of analysis, such that selection may need to occur for team composition as the criterion, rather than solely selecting for individual-level performance. Or, staffing for the team leader may comprise a more important influence on team performance than changing any one team member. Fourth, whether homogeneity or heterogeneity in team composition impacts performance needs to be considered, and depending on the team task, staffing practice may search for individuals with skills similar to or different from the existing team members. Finally, with the vast majority of predictor research having been conducted at the individual level of analysis, a major task facing future staffing experts is to understand the cross-level and multilevel linkages between predictors and criteria.

Cultural Influences on Predictor Development and Use

Staffing research is just starting to examine the extent to which predictors might function differently across cultures. Such cultural variation may be caused by language and translation difficulties (where the meaning of the items and questions differ across cultures), actual cultural differences in the latent structure of the predictor construct, or even cultural differences in the acceptability of different predictor constructs and measures.

At a macro level, Ryan, McFarland, Baron, and Page (1999) demonstrated that across 20 countries, the cultural dimensions of uncertainty avoidance and power distance can influence the types of staffing systems that are used. However, it is instructive to compare the simple frequency of test use across cultures. For virtually every standardized testing procedure they examined, it was used less frequently in the United States than was the average across companies in all 20 countries. Salgado and Anderson (2002) likewise pointed out that cognitive ability test use was greater in 10 of 12 European countries than it was in the United States and speculated that some of this difference was a function of the legal differences regarding test use in the United States and Europe. Thus, cultural differences in laws can influence choice of predictors.

Several studies have also examined whether the factor structure of various predictors are equivalent across cultures, generally finding that with appropriate translation procedures cognitive ability and personality measures can demonstrate cross-cultural equivalence (e.g., Jensen, 1998; McCrae & Costa, 1997; Schmit, Kihm, & Robie, 2000). Criterion-related validity also appears to be largely similar across cultures, as meta-analyses in the United Kingdom and Europe find validities for both cognitive ability and personality similar to those in the United States and Canada (Bertua, Anderson, & Salgado, in press; Salgado, 1997, Salgado & Anderson, 2002; Salgado, Anderson, Moscoso, Bertua, & Fruyt, 2003). In terms of user reactions, Steiner and Gilliland (1996) compared applicant reactions to multiple predictors across the United States and France. Although participants from both countries perceived face validity to be the key determinant of acceptability, other fairness judgments of the various predictors differed between the two countries. Salgado and Anderson (2002) also showed user acceptance of cognitive ability tests are very similar in Europe to those in the United States.

Yet, it is important to recognize that culture is an inherently multidimensional construct that varies across countries (e.g., Hofstede, 1980, 2001; Schwartz, 1994). Future research must directly measure these cultural dimensions to determine if they influence validity, subgroup differences, reactions, and utility.

Technological Advances and Predictors

Although we have already discussed the application of computers, the Internet, and related predictor measurement methods, it is worth emphasizing again that technology is having a fundamental impact on predictor measurement. This impact is felt most directly by those who practice staffing, where many consulting companies and private organizations either use or specialize in the development and administration of Web-based testing. The Web-based application of predictors challenges many of the current professional, legal, and scientific guidelines practitioners must follow. Indeed, a special task force of the American Psychological Association more recently prepared a document describing the key issues to consider when using Web-based testing (Naglieri et al., 2004).

The flexibility of Web-based testing is also what creates the greatest opportunities for abuse. For example, there are now private companies that offer tests entirely online, such that an organization need only pay a fee and point applicants to the Web-site to complete the predictors. An obvious concern, beyond the issue of validity of course, is whether the tests are being administered, scored, and interpreted correctly; an important part of good professional practice ensures testing procedures are supervised by a qualified professional (see chapter 11). Likewise, research has not comprehensively examined the consequences of various administration features of Web-based test administration, such as the consequences of administering tests in unproctored settings. Many of those in practice have embraced Web-based testing because of the benefits it provides in terms of cost savings and administrative efficiency (and hence utility). Unfortunately, what we know about Web-based tests and the possible testing conditions provides an incomplete foundation on which to base this practice. This is an issue of great research interest, and we suspect much more will be known about Web-based testing within the next few years. Indeed, an entire issue of the *International Journal of Selection and Assessment* was devoted to technology and assessment (Viswesvaran, 2003).

Box 7.1 Development of an Interactive, Internet-Based Assessment of Leadership Processes

One of the authors more recently developed an interactive, Internet-based measure of leadership processes. Our goals with this assessment were to measure the more proximal, cognitive processes that good leadership requires for U.S Army captains. The measure needed to be administered over the Internet, take no more than 15 minutes to complete, and be as objectively scored as possible. To ensure it was realistic and engaging, it was designed with the help of numerous captains.

We ultimately developed a scenario-based, interactive measure. The scenario consisted of two parts: one was an offensive action, and the other was a defensive action. The assessment was interactive such that what a captain did influenced subsequent outcomes in the program. In the simulation, captains were presented on the monitor with a map, troop icons, radio buttons, and unit status reports. The radio buttons lit up when a soldier from a different unit had information to report (some of this information was relevant, some of it irrelevant). The captains clicked on a button to hear a sound bite of a radio message. They had to keep track of their troop movements (dragging icons around the map) by listening to these messages. Then, there was an important change in the simulation, and the captains had to be able to identify and correctly respond to the change.

A slide from the simulation is shown in Fig. 7.5. In Fig. 7.5, the buttons on the right indicate the radio buttons that would turn green when there was a transmission. The bottom left corner of the map shows various troop icons that can be dragged to particular sections on the map to update troop movements. The bottom middle of Fig. 7.5 provides information about the status of different units (represented by different colors). As the captains navigate the Web site, we measure time, reaction time, mouse clicks, the placement of icons, and the distance between where the soldiers put the icons and where they objectively exist.

This simulation measured such constructs as situational awareness and environmental monitoring, strategy formulation and implementation, and motivating change. The assessments were objective and required no judgment (other than that used to develop the simulation). The data are collected on a central server and scored automatically. Although it was extremely challenging to develop the simulation, it has been well received by soldiers and is fairly realistic.

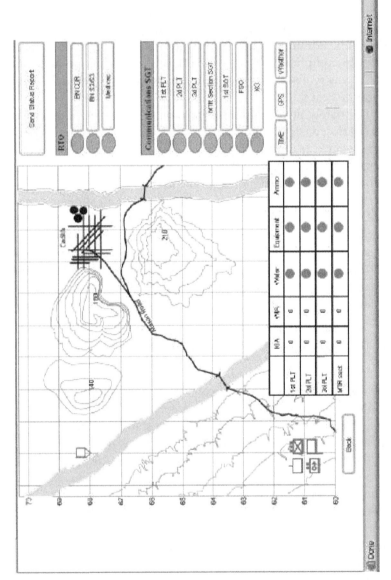

FIG. 7.5. Slide form interactive Internet-based assessment of leadership.

Other than the enthusiasm for computer or Web-based testing, staffing specialists have been slow to adopt other forms of technology into their research and practice. For example, many of the measurement systems frequently employed by cognitive psychologists, such as reaction time measures and measures of implicit cognitions, have been largely ignored. Just as one example, some theories argue working memory capacity plays an important role in cognitive ability (see Drasgow, 2003), yet few applied psychologists have considered the role of working memory either as a component of cognitive ability or as a factor influencing comprehension or measures of constructs. Likewise, there are more cognitively oriented theories of personality that have direct implications on personality predictor measurement, but have not been explored (e.g., Matthews, 1997). Also lacking is a good understanding of the cognitive processes involved with completing predictors; a kind of cognitive task analysis for predictor measures. The predictor response model in Fig. 7.1 is only a very basic first step in this regard. Adoption of these technologies from other areas of science offers potential benefits to applied psychologists. If for no other reason, measures such as reaction times and implicit cognitions are relatively objective and unlikely to be affected by conscious attempts to distort, cheat, and lie. The promise of objective predictor measurement clearly deserves more attention from staffing researchers. Box 7.1 gives an example of such an attempt by one of your authors.

SUMMARY

Predictor development perhaps best captures the very essence of staffing, yet an organizing structure to compare and contrast predictor constructs and measures has been lacking. We try to provide such a heuristic organizing structure in this chapter. We provide a simple, but generalizable, model of the predictor response process to facilitate later discussions on how various factors (e.g., applicant versus incumbent validation strategies) may influence test scores. We have then provided dimensions on which to identify the similarities and differences of predictors, followed by comparing the adequacy of predictors in terms of their multifaceted consequences. The chapter concludes by discussing the use of predictors in staffing practice, followed by current challenges to predictor development. Readers should keep the issues developed in this chapter in mind as they read the subsequent chapters detailing cognitive, noncognitive, and simulations/performance-based predictors.

REFERENCES

Ackerman, P. L., & Humphreys, L. G. (1990). Individual differences theory in industrial and organizational psychology. In M. D. Dunnette & L. M. Hough (Eds.)., *Handbook of industrial and organizational psychology* (pp. 223–282). Palo Alto, CA: Consulting Psychologists Press.

American Educational Research Association (AERA), American Psychological Association, & National Council of Measurement in Education. (1999). *Standards for educational and psychological testing.* Washington, DC: AERA.

Arvey, R. D., & Sackett, P. R. (1993). Fairness in selection: Current developments and perspectives. In N. Schmitt & W. C. Borman (Eds.), *Personnel selection in organizations* (pp. 171–202). San Francisco: Jossey-Bass.

Austin, J. T., Klimoski, R. J., & Hunt, S. T. (1996). Dilemmatics in public sector assessment: A framework for developing and evaluating selection systems. *Human Performance, 9,* 177–198.

Barrick, M. R., Stewart, G. L., Neubert, M. J., & Mount, M. K. (1998). Relating member ability and personality to work-team processes and team effectiveness. *Journal of Applied Psychology, 83,* 377–391.

Bertua, C., Anderson, N., & Salgado, J. F. (in press). The predictive validity of cognitive ability tests: A UK meta-analysis. *Journal of Organizational and Occupational Psychology.*

Bobko, P., Roth, P. L., & Buster, M. A. (2005). Work sample selection tests and expected reduction in adverse impact: A cautionary note. *International Journal of Selection and Assessment, 13,* 1–10.

Bobko, P., Roth, P. L., & Potosky, D. (1999). Derivation and implications of a meta-analytic matrix incorporating cognitive ability, alternative predictors, and job performance. *Personnel Psychology, 52,* 561–589.

Campbell, D. T., & Fiske, D. W. (1959). Convergent and discriminant validation by the multitrait-multimethod matrix. *Psychological Bulletin, 56,* 81–105.

Chan, D., & Schmitt, N. (1997). Video-based versus paper-and-pencil method of assessment in situational judgment tests: Subgroup differences in test performance and face validity perceptions. *Journal of Applied Psychology, 82,* 143–159.

Chen, G., & Bliese, P. D. (2002). The role of different levels of leadership in predicting self- and collective efficacy: Evidence for discontinuity. *Journal of Applied Psychology, 87,* 549–556.

Cohen, J. (1988). *Statistical power analysis for the behavioral sciences* (2nd ed.). Hillsdale, NJ: Erlbaum.

Cronbach, L. J. (1960). *Essentials of psychological testing* (2nd ed.). New York: Harper.

Drasgow, F. (2003). Intelligence in the workplace. In W. C. Borman, D. R. Ilgen, & R. J. Klimoski (Eds.), *Handbook of psychology: Industrial and organizational psychology* (Vol. 12, pp. 107–130). Hoboken, NJ: Wiley.

Frei, R. L., & McDaniel, M. A. (1998). Validity of customer service measures in personnel selection: A review of criterion and construct evidence. *Human Performance, 11,* 1–27.

Gatewood, R. D., & Field, H. S. (1998). *Human resource selection.* Fort Worth, TX: Dryden/Harcourt Brace College.

Goldstein, I. L., Zedeck, S., & Schneider, B. (1993). An exploration of the job analysis-content validity process. In N. Schmitt & W. C. Borman (Eds.), *Personnel selection in organizations* (pp. 3–34). San Francisco: Jossey-Bass.

Gully, S. M., Incalcaterra, K. A., Joshi, A., Beaubien, J. M. (2002). A meta-analysis of team-efficacy, potency, and performance: Interdependence and level of analysis as moderators of observed relationships. *Journal of Applied Psychology, 87,* 819–832.

REFERENCES

Hofstede, G. (1980). *Culture's consequences: International differences in work-related values.* Beverly Hills, CA: Sage.

Hofstede, G. (2001). *Culture's consequences: Comparing values, behaviors, institutions and organizations across nations.* Thousand Oaks, CA: Sage.

Hogan, R., & Hogan, J. (1992). *Manual for the Hogan Personality Inventory.* Tulsa, OK: Hogan Assessment Systems.

Hough, L. M., Oswald, F. L., & Ployhart, R. E. (2001). Determinants, detection, and amelioration of adverse impact in personnel selection procedures: Issues, evidence, and lessons learned. *International Journal of Selection and Assessment, 9,* 152–194.

Hough, L. M., & Schneider, R. J. (1996). Personality traits, taxonomies, and applications in organizations. In K. R. Murphy (Ed.), *Individual differences and behavior in organizations* (pp. 3–30). San Francisco: Jossey-Bass.

Jensen, A. R. (1998). *The g factor.* Westport, CT: Praeger.

Judge, T. A., Erez, A., Bono, J. E., & Thoresen, C. J. (2002). Are measures of self-esteem, neuroticism, locus of control, and generalized self-efficacy indicators of a common core construct? *Journal of Personality and Social Psychology, 83,* 693–710.

Klimoski, R., & Jones, R. G. (1995). Staffing for effective group decision making: Key issues in matching people to teams. In R. Guzzo, E. Salas, & Associates (Eds.), *Team effectiveness and decision making in organizations* (pp. 291–332). San Francisco: Jossey-Bass.

Klimoski, R. J. (1993). Predictor constructs and their measurement. In N. Schmitt & W. C. Borman (Eds.), *Personnel selection in organizations* (pp. 99–134). San Francisco: Jossey-Bass.

Kluger, A. N., & Rothstein, H. R. (1993). The influence of selection test type on applicant reactions to employment testing. *Journal of Business and Psychology, 8,* 3–25.

LePine, J. A., Hollenbeck, J. R., Ilgen, D. R., & Hedlund, J. (1997). Effects of individual differences on the performance of hierarchical decision-making teams: Much more than g. *Journal of Applied Psychology, 82,* 803–811.

Lievens, F., & Harris, M. M. (2003). Web-based recruitment and testing. In C. L. Cooper & I. T. Robertson (Eds.), *International review of industrial and organizational psychology* (Vol. 18, pp. 131–165). Chicester: Wiley.

Lim, B. C., & Ployhart, R. E. (2004). Transformational leadership: Relations to the Five Factor Model and team performance in typical and maximum contexts. *Journal of Applied Psychology, 89,* 610–621.

Lubinski, D. (2000). Scientific and social significance of assessing individual differences: Sinking shafts at a few critical points. *Annual Review of Psychology, 51,* 405–444.

Matthews, G. (1997). Extraversion, emotion, and performance: A cognitive-adaptive model. In G. Matthews (Ed.), *Cognitive science perspectives on personality and emotion* (pp. 399–442). Amsterdam: Elsevier Science.

McCrae, R. R., & Costa, P. T., Jr. (1997). Personality trait structure as a human universal. *American Psychologist, 52,* 509–516.

Mead, A. D., & Drasgow, F. (1993). Equivalence of computerized and paper-and-pencil cognitive ability tests: A meta-analysis. *Psychological Bulletin, 114,* 449–458.

Murphy, K. R. (1996). Individual differences and behavior in organizations: Much more than g. In K. R. Murphy (Ed.), *Individual differences and behavior in organizations* (pp. 3–30). San Francisco: Jossey-Bass.

Naglieri, J. A., Drasgow, F., Schmit, M., Handler, L., Prifitera, A., Margolis, A., & Velasquez, R. (2004). Psychological testing on the Internet: New problems, old issues. *American Psychologist, 59,* 150–162.

Neuman, G. A., & Wright, J. (1999). Team effectiveness: Beyond skills and cognitive ability. *Journal of Applied Psychology, 84,* 376–389.

Ones, D. S., & Viswesvaran, C. (2001). Integrity tests and other criterion-focused occupational personality scales (COPS) used in personnel selection. *International Journal of Selection and Assessment, 9*, 31–39.

Ployhart, R. E. (2004). Organizational staffing: A multilevel review, synthesis, and model. In J. Martocchio (Ed.), *Research in personnel and human resource management* (Vol. 23, pp. 121–176). Oxford, UK: Elsevier.

Ployhart, R. E. (in press). The predictor response model. In J. A. Weekley & R. E. Ployhart (Eds.), *Situational judgment tests*. Mahwah, NJ: Erlbaum.

Ployhart, R. E., & Schneider, B. (2002). A multilevel perspective on personnel selection: Implications for selection system design, assessment, and construct validation. In F. J. Dansereau & F. Yammarino (Eds.), *Research in multi-level issues. Volume 1: The many faces of multi-level issues* (pp. 95–140). Oxford, UK: Elsevier Science.

Ployhart, R. E., & Schneider, B. (2005). Multilevel selection and prediction: Theories, methods, and models. In A. Evers, O. Smit-Voskuyl, & N. R. Anderson (Eds.), *Handbook of personnel selection* (pp. 495–516). Chichester/London: Wiley.

Ployhart, R. E., Weekley, J. A., Holtz, B. C., & Kemp, C. F. (2003). Web-based and paper-and-pencil testing of applicants in a proctored setting: Are personality, biodata, and situational judgment tests comparable? *Personnel Psychology, 56*, 733–752.

Potosky, E., & Bobko, P. (1997). Computer versus paper-and-pencil administration mode and response distortion in noncognitive selection tests. *Journal of Applied Psychology, 82*, 293–299.

Richman, W. L., Kiesler, S., Weisband, S., & Drasgow, F. (1999). A meta-analytic study of social desirability distortion in computer-administered questionnaires, traditional questionnaires, and interviews. *Journal of Applied Psychology, 84*, 754–775.

Roth, P. L., Bevier, C. A., Bobko, P., Switzer, F. S., III, & Tyler, P. (2001). Ethnic subgroup differences in cognitive ability in employment and educational settings. A meta-analysis. *Personnel Psychology, 54*, 297–330.

Ryan, A. M., McFarland, L., Baron, H., & Page, R. (1999). An international look at selection practices: Nation and culture as explanations for variability in practice. *Personnel Psychology, 52*, 359–391.

Ryan, A. M., & Ployhart, R. E. (2000). Applicants' perceptions of selection procedures and decisions: A critical review and agenda for the future. *Journal of Management, 26*, 565–606.

Rynes, S. L., & Connerley, M. L. (1993). Applicant reactions to alternative selection procedures. *Journal of Business and Psychology, 7*, 261–277.

Sackett, P. R., & Ellingson, J. E. (1997). The effects of forming multi-predictor composites on group differences and adverse impact. *Personnel Psychology, 50*, 707–721.

Sackett, P. R., & Roth, L. (1996). Multi-stage selection strategies: A monte carlo investigation of effects on performance and minority hiring. *Personnel Psychology, 49*, 1–18.

Sackett, P. R., & Wilk, S. L. (1994). Within-group norming and other forms of score adjustment in preemployment testing. *American Psychologist, 49*, 929–954.

Salgado, J. F. (1997). The five factor model of personality and job performance in the European community. *Journal of Applied Psychology, 82*, 30–43.

Salgado, J. F., & Anderson, N. (2002). Cognitive and GMA testing in the European community: Issues and evidence. *Human Performance, 15*, 75–96.

Salgado, J. F., Anderson, N., Moscoso, S., Bertua, C., & Fruyt, F. D. (2003). International validity generalization of GMA and cognitive abilities: A European community meta-analysis. *Personnel Psychology, 56*, 573–605.

Salgado, J. F., & Moscoso, S. (2003). Internet-based personality testing: Equivalence of measures and assessors' perceptions and reactions. *International Journal of Selection and Assessment, 11*, 194–205.

REFERENCES

Schmidt, F. L., & Hunter, J. E. (1998). The validity and utility of selection methods in personnel psychology: Practical and theoretical implications of 85 years of research findings. *Psychological Bulletin, 124,* 262–274.

Schmit, M. J., Kihm, J. A., & Robie, C. (2000). Development of a global measure of personality. *Personnel Psychology, 53,* 153–193.

Schmitt, N. (1994). Method bias: The importance of theory and measurement. *Journal of Organizational Behavior, 15,* 393–398.

Schmitt, N., Clause, C. S., & Pulakos, E. D. (1996). Subgroup differences associated with different measures of some common job-relevant constructs. In C. L. Cooper & I. T. Robertson (Eds.), *International journal of industrial and organizational psychology* (Vol. 11, pp. 115–140). New York: Wiley.

Schmitt, N., Cortina, J. M., Ingerick, M. J., & Wiechmann, D. (2003). Personnel selection and employee performance. In W. C. Borman, D. R. Ilgen, & R. J. Klimoski (Eds.), *Handbook of psychology: Volume 12: Industrial and organizational psychology* (pp. 77–105). Hoboken, NJ: Wiley.

Schmitt, N., Pulakos, E. D., Nason, E., & Whitney, D. J. (1997). Likability and similarity as potential sources of predictor-related criterion bias in validation research. *Organizational Behavior and Human Decision Processes, 68,* 272–286.

Schmitt, N., Rogers, W., Chan, D., Sheppard, L., & Jennings, D. (1997). Adverse impact and predictive efficiency of various predictor combinations. *Journal of Applied Psychology, 82,* 719–730.

Schwartz, S. H. (1994). Beyond individualism and collectivism: New cultural dimensions of values. In U. Kim, H. C. Triandis, C. Kagitcibasi, S. C. Choi, & G. Yoon (Eds.), *Individualism and collectivism: Theory, method, and applications* (pp. 85–119). Thousand Oaks, CA: Sage.

Smither, J. W., Reilly, R. R., Millsap, R. E., Pearlman, K., & Stoffey, R. W. (1993). Applicant reactions to selection procedures. *Personnel Psychology, 46,* 49–76.

Society for Industrial and Organizational Psychology. (2003). *Principles for the validation and use of personnel selection procedures.* Bowling Green, OH: Author.

Spray, J. A., Ackerman, T. A., Reckase, M. D., & Carlson, J. E. (1989). Effect of the medium of item presentation on examinee performance and item characteristics. *Journal of Educational Measurement, 26,* 261–271.

Steiner, D., & Gilliland, S. W. (1996). Fairness reactions to personnel selection techniques in France and the United States. *Journal of Applied Psychology, 81,* 134–141.

Viswesvaran, C. (2003). Introduction to special issue: Role of technology in shaping the future of staffing and assessment. *International Journal of Selection and Assessment, 11,* 107–112.

Williams, L. J., & Anderson, S. E. (1994). An alternative approach to the estimation of method effects by using latent variable models: Applications in organizational behavior research. *Journal of Applied Psychology, 79,* 323–331.

# 8

Hiring Procedures I: Cognitive Ability and Certification Exams

AIMS OF THE CHAPTER

This chapter has three major sections. The first section focuses on cognitive ability as a predictor of performance, including a brief discussion of the structure of cognitive ability, as well as its measurement and validity. In the second section, we provide a discussion of the issue of subgroup differences in cognitive ability. The principles associated with the use of measures that are associated with subgroup differences apply to any predictor, but they are especially relevant in the use of cognitive ability tests because differences between Caucasian groups and some minority groups are relatively large. The final section discusses certification exams that are frequently used to ensure professional competence and public safety.

PAPER-AND-PENCIL TESTS OF GENERAL ABILITY

In chapter 1, we place considerable emphasis on the concept of individual differences and on the late nineteenth-century attempts of Binet and Simon to diagnose children who were most likely to benefit from the typical school curriculum. The items for their tests were developed on the basis of observation of the kinds of knowledge, skills, and abilities required for success at school and validated against teacher judgments of whether the students were fast or slow learners.

PAPER-AND-PENCIL TESTS OF GENERAL ABILITY

Note that because Binet and Simon selected the actual items for their test based on the item's validity, a person's score on the test would be a simple sum of the items answered correctly. That is, the score would be a single number. Note also that the situation for which the test would be valid was the classroom, a fairly well-standardized situation—a single teacher in the front of the room, rows of desks, about 20 children, and so forth. In addition, because the items selected for the test were those that correlated with school performance as judged by teachers, adaptations and translations of the test into many languages (in the United States by Lewis Terman at Stanford University, resulting in what is called the Stanford-Binet) created a tidal wave of testing in Western countries. Subsequently, adult forms of the Stanford-Binet were developed and a competing test, the Wechsler-Bellevue (Wechsler, 1944), subsequently the Wechsler Adult Intelligence Scale (WAIS), was produced for assessing adult intelligence. Although the Stanford-Binet and WAIS required one tester for each examinee (individually administered tests), other tests were produced (e.g., Otis Quick Scoring Mental Abilities) that were suited to group administration (group-administered tests).

Today, there are many off-the-shelf tests of cognitive ability available. Tests of general cognitive ability have been shown to be valid in many studies, especially in the selection of first-line supervisors (Ghiselli, 1966, 1973; Schmidt & Hunter, 1998; Schmidt, Hunter, Pearlman, & Shane, 1979). As mentioned previously, tests of cognitive ability often show majority–minority differences in average scores, resulting in unequal hiring rates across groups. This finding has produced a great deal of legal and social scrutiny despite the considerable validity evidence that supports the use of these measures (Gottfredson, 2002). We return to the evidence for the validity of cognitive ability tests and the minority–majority group difference, but first we look at the body of literature that examines the nature of the cognitive ability construct itself and its validity in the prediction of job performance. The debate about whether cognitive ability is a single dimension (g) or comprised of multiple related dimensions has occupied researchers interested in individual differences in ability for at least a century (Ree & Caretta, 2002; Spearman, 1904). This debate is important for both theoretical and practical reasons. On the theoretical side, evidence for multiple dimensions of ability would suggest that different mental and learning processes are involved in the acquisition of knowledge. On the practical side, the underlying structure of mental ability determines the number of dimensions one needs to predict and understand performance in various domains, including work performance.

Multidimensional Concepts of Cognitive Ability

Perhaps the earliest work on the structure of ability was that conducted by Spearman (1923, 1927) and summarized in two books. Using factor analysis, Spearman believed a general factor (g) was responsible for performance on a large number of test tasks and correlations between these tests. In addition, each test seemed to also be a function of a specific factor (s) that was unique to that test. Although it may be the case that Spearman investigated the relationships among the specific components in his tests (Carroll, 1993), his notion of ability is usually characterized as one in which there is an all-encompassing g factor and many specific factors that are unrelated.

Thurstone (1941) believed the idea of general intelligence was an oversimplification. Spearman had used factor analysis to show that whenever a set of tests was given to a group of people, the intercorrelations of the tests always revealed something in common—that is, there was always some overlap or relationship between the tests. He interpreted this as support for his concept of g or general ability. Thurstone was interested in how many and what kinds of factors are needed to account for the observed correlations among tests of ability. His factor analyses led him to believe that g did not exist, but that the intercorrelations among tests could be explained by the following primary mental abilities (Thurstone, 1941):

1. *Verbal Comprehension*—found in tests of vocabulary, antonyms and synonyms, completion tests, and various reading comprehension tests
2. *Word Fluency*—found in tests of anagrams, rhyming, and producing words with a given initial letter, prefix, or suffix
3. *Space*—any test in which the task of the subject is to visualize and mentally manipulate an object in a two- or three-dimensional space
4. *Number*—tests that require the subject to rapidly and accurately do relatively simple calculations, not problem solving
5. *Memory*—test of the ability to memorize quickly
6. *Reasoning*—tests requiring the subject to discover rules or principles covering the material of a test (Two such factors seemed

to emerge; one for deductive reasoning, and another for inductive reasoning.)

Guilford (1967) made another major effort to identify factors of intellect. He attempted not only to identify ability factors, but also to conceptualize the structure of intellect. That is, given existing factors he asked the question: How can I organize what already exists so I can know what is still to be discovered? In reviewing the research of other scholars, and by conducting his own research over a number of years, Guilford presented a three-factor, or the three faces of, intellect model:

1. The kind of process or operation the abilities represent such as memory, cognition, convergent thinking, evaluating
2. The content of the ability; that is, the nature of the kinds of information the individual is required to operate on such as figures, symbols, words, or behavior
3. The product or outcome of the operation of the ability, such as forming relations, systems, or transformations

The most recent attempt to provide a theory of the structure of cognitive ability is provided by Carroll (1993). After an encyclopedic review and reanalyses of the relationships between many different tests of ability used with different groups, Carroll hypothesized the structure displayed in Fig. 8.1. At the top of this hierarchical structure is a general factor that is believed to be responsible for some positive correlation among all ability tests. At Stratum II in Carroll's hierarchy are group factors such as those of Thurstone (1941) that represent measures whose intercorrelations are higher than those of tests that represent the other group factors. The relative degree to which these secondary factors are a function of the general factor is represented by their distance from the general factor in Fig. 8.1. At the third level are the specific factors that are representative of each group factor. Note that in some groups at this stratum, Carroll differentiates between the level of performance on this factor and the speed of performance in this area. In addition, two of the Level II factors are measures of speed. Given the number and variety of studies and data that Carroll has considered in developing this structure, it is not surprising that a number of assumptions and judgments were made as to the nature of the measure involved. These assumptions and ambiguities are discussed in the last section of Carroll's book. Even given these limitations, Carroll's effort is bound to serve as the reference point for future research on the structure of cognitive ability.

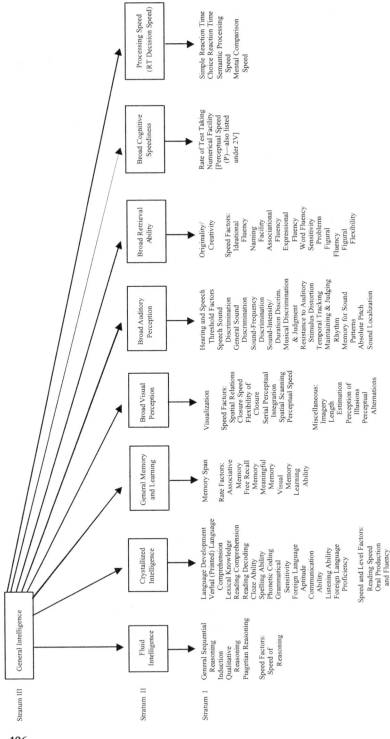

FIG. 8.1. The structure of cognitive abilities. *Source:* Adapted from Carroll, 1993.

The previous models of cognitive ability are usually referred to as the psychometric approach; that is, they employ factor analyses of the relationships on many different measures in an effort to understand the underlying structure of cognitive ability. More recent theories of intelligence have also arisen from information processing and neuropsychological research (see Drasgow, 2003). Sternberg's (1977) componential model seeks to describe the elements (e.g., encoding information; processing) involved in cognitive tasks and then how these various components get combined in dealing with information. This componential aspect of intelligence describes how problems get resolved. A contextual intelligence involves an understanding of how to modify or adapt a situation or select a new environment, and an experiential aspect of intelligence relates to the ability of a person to use their past experience in solving problems (Sternberg, 2000). Perhaps Sternberg's greatest influence on staffing has derived from the use of his notion of practical intelligence (Wagner, 2000), which stimulated the more recent upsurge in the use of situational judgment measures (Motowidlo, Dunnette, & Carter, 1990).

Naglieri and Das (1997) provided a theory of intelligence they claim is based in neuropsychological research. They posit three major functional areas of intelligence: planning, attention, and simultaneous or successive information processing. Tests based on their hypothesized areas of functioning have been published, but applications in work situations are not yet available to our knowledge.

The ideas presented in this section of the chapter are representative theories of the structure of cognitive ability across various applications of psychology, including those involving work performance. In the next section of this chapter, we present a brief description of some actual test batteries that have been developed to represent some of the factors hypothesized in these various models.

Multidimensional Measures of Job-Related Aptitudes

There are single measures of aptitude for specific occupations (e.g., clerical, sales) and multiaptitude batteries. The latter are sets of tests, administered as a package, that yield aptitude scores on anywhere from 8 to 20 potentially job-relevant dimensions.

Of course, because these batteries are commercially available, the tests are not specific to a particular job. The tests are designed to be applicable to a wide variety of jobs and settings. In contrast, the care taken in developing the measures to ensure their generality, reliability, and potential validity make them useful in approximating the contribution more specific tests can make to predicting effective job performance.

Multiaptitude batteries provide the most information on job candidates in the least amount of time. Profiles of aptitude scores can be used to evaluate individuals from a multi-dimensional viewpoint rather than a unidimensional viewpoint. Such a view may be a step toward conceptualizing personnel selection decisions in terms of complex humans. However, several additional points are important. Multiaptitude batteries do not assess all the important job aptitudes. In each situation, job analyses should reveal the salient aspects of the job situation, thereby pinpointing which characteristics of potential employees should be assessed. Further, we should not expect that all aptitudes are best assessed by "paper-and-pencil" tests; this is a point we consider in the next two chapters. Finally, as we see later in this chapter, the degree to which information on multiple dimensions affords better prediction of job performance outcomes than does a general measure of cognitive ability such as the Wonderlic is debatable (Ree & Caretta, 2002). Three of the better known multiaptitude batteries are as follows:

- Flanagan Aptitude Classification Tests (FACT)
- General Aptitude Test Battery (GATB)
- Employee Aptitude Survey (EAS)

Table 8.1 presents the subtests offered in each. It can be noted that there is considerable overlap in content among the tests and between multiaptitude batteries and tests of more general intellectual factors. Although there is overlap in content between EAS and the other batteries, the EAS (besides having considerable validity evidence) takes about 1 hour to complete, and the FACT takes about 4 hours. The GATB was formerly used by individuals applying for jobs at State Employment Services; it is the source of a great deal of data on the validity and generalizability of cognitive ability test batteries (Hartigan & Wigdor, 1989; Hunter & Hunter, 1984).

Another point regarding Table 8.1 is that two of the batteries, GATB and EAS, include tests of fine motor abilities, whereas the FACT does not. These "eye–hand" kinds of abilities, along with the "pure" sensory tests of vision (including color blindness) and hearing, represent some of the earliest and most continually valid measures we have for tasks involving a high sensorimotor component. Measurement of physical abilities is so important that we discuss the topic in a special section in

TABLE 8.1
Some Multiaptitude Batteries and What They Measure

FACT	GATB	EAS
1. Inspection	1. General intelligence	1. Verbal Comprehension
2. Coding	2. Verbal	2. Numerical ability
3. Memory	3. Numerical	3. Visual pursuit
4. Precision	4. Spatial	4. Visual speed and accuracy
5. Assembly	5. Form perception	5. Space visualization
6. Scales	6. Clerical perception	6. Numerical reasoning
7. Coordination	7. Motor coordination	7. Verbal reasoning
8. Judgment and comprehension	8. Finger dexterity	8. Word fluency
9. Arithmetic	9. Manual dexterity	9. Manual speed and accuracy
10. Patterns		10. Symbolic reasoning
11. Components		
12. Tables		
13. Mechanics		
14. Expression		

FACT, Flanagan Aptitude Classification Tests; GATB, General Aptitude Test Battery; EAS, Employee Aptitude Survey.

chapter 10. You may believe the aptitudes listed in Table 8.1 are very general in that they appear somewhat abstract and somewhat removed from the kinds of activities people engage in at work. This perception is accurate because the general kinds of aptitude batteries discussed so far emphasize basic or primary abilities, and it may require expert judgment to make the inference about which of these aptitudes is most likely to be valid for a particular job.

Some test publishers have also seen the potential for problems here, and they have presented batteries of tests more clearly related to the kinds of abilities and skills required at work. Thus, they have reasoned that the basic or primary aptitude tests can be useful for making general assessments of people's basic abilities, but that more specificity in assessment could be valuable for the prediction of effectiveness at a specific job.

Table 8.2, for example, presents the 20 subtests available in one battery of specific aptitude tests (basic skills tests; Psychological Services Inc., PSI 1982). The titles and descriptions suggest that these tests may have more work relevance. However, a factor analysis of the 20 tests yields clusters of the tests with familiar names such as:

- verbal ability
- quantitative ability
- reasoning ability
- perceptual speed

TABLE 8.2

Specific Aptitudes Tested by the Psychological Services; Inc., Basic Skills Tests

1. Language Skills—ability to recognize correct spelling, punctuation, capitalization, grammar, and usage
2. Reading Comprehension—ability to read a passage and answer literal and inferential questions about it
3. Vocabulary—ability to identify the correct meanings of words
4. Computation—ability to solve arithmetic problems involving operations with whole numbers, decimals, percents, and simple fractions
5. Problem Solving—ability to solve "story" problems requiring the application of arithmetic operations
6. Decision Making—ability to read a set of procedures and apply them to new situations
7. Following Oral Directions—ability to listen to information and instructions presented orally, taking notes if desired, and to answer questions about the content
8. Following Written Directions—ability to read and follow a set of rules
9. Forms Checking—ability to verify the accuracy of completed forms by comparison to written information
10. Reasoning—ability to analyze facts and to make valid judgments on the basis of the logical implications of such facts
11. Classifying—ability to place information into predetermined categories
12. Coding—ability to code information according to prescribed system
13. Filing Names—ability to insert names in a list in alphabetical order
14. Filing Numbers—ability to insert numbers in a list in numerical order
15. Visual Speed and Accuracy—ability to see differences in small detail
16. Memory—ability to recall information after having a chance to study it
17. Typing—Practice Copy—a preparatory typing test to tests 18, 19, and 20; test 18 assesses typing of straight copy; test 19 the typing from copy that has been revised in handwriting; and test 20 measures the ability to set up and type tables according to specific directions

It is always likely that similar analyses of any set of tests will yield factors similar to those labeled primary mental abilities by Thurstone, but tests can be developed whose content appears more directly job relevant.

The Importance of g

Earlier in this chapter, we refer to literature that indicated that from a practical predictive standpoint, inclusion of measures of specific cognitive ability measures does not seem to provide additional predictive value beyond g. This view is strongly held by Ree and colleagues, and convincing support for this view from a variety of studies is provided in Ree and Caretta (2002). In a large-scale study of the performance of military personnel, McHenry, Hough, Toquam, Hanson, and Ashworth

(1990) found that g predicted core technical proficiency and general soldiering proficiency with correlations of 0.63 and 0.65 when range restriction corrections were applied. Adding interest and personality predictors, as well as some specific ability measures, failed to increase these correlations by more than 0.02. The role of noncognitive predictors and g was reversed, however, when other contextual criteria such as citizenship behavior (see chapter 4) were considered. Ree, Earles, and Teachout (1994) reported a similar outcome for members of various enlisted U.S. Air Force jobs.

Similar findings have been reported in studies that have used training performance as an outcome. Thorndike (1986) studied the relative effectiveness of g and specific ability measures in the training success of U.S. Army enlistees in technical schools and found that measures of specific abilities incremented the predictive validity of g by no more than 0.03. Olea and Ree (1994) studied the prediction of the training grades and work sample performance of U.S. Air Force pilots and navigators. They found that g predicted a composite of these outcomes at 0.40 and 0.49 for pilots and navigators, respectively. Specific aptitudes did increment the prediction by 0.08 for pilots, but only 0.02 for navigators.

Schmidt and Hunter (1998) provided a *corrected* meta-analytic estimate of the validity of cognitive ability of 0.51 for performance outcomes and 0.56 for training criteria. They also examined the degree of validity for a variety of other predictors described in chapters 9 and 10 and the degree to which these other predictors added to the predictability afforded by cognitive ability. Their analyses showed impressive validities for a wide variety of predictors, but only the validity of work sample predictors rivaled the validity of cognitive ability. Increments in validity ranged from 0.01 to 0.14. Measures of specific ability were not included; personality predictors and interviews provided the best incremental validity over that of cognitive ability alone.

The procedure employed by Schmidt, Hunter, and their colleagues is detailed in their various papers, as well as in a book (Hunter & Schmidt, 1990). The basics of the meta-analyses procedures they use are described in chapter 6. In a wide variety of jobs, the major cognitive and perceptual tests always exhibited potentially useful validities (Pearlman, 1980; Schmidt, Hunter, & Pearlman, 1981). Hunter and Hunter (1984) extended this work by broadening the job base to include virtually the entire job spectrum and to include psychomotor abilities (see our discussion of measures of psychomotor abilities in chapter 10).

A meta-analysis by Hunter and Hunter (1984) of 515 validation studies conducted by the U.S. Employment Service for jobs encompassing the entire job spectrum covered by the *Dictionary of Occupational Titles* (DOT) indicated that cognitive ability as represented by the GATB (see Table 8.1 for a list of the subtests in this battery) was a valid predictor of performance on all jobs. Of the 515 validation studies, 425 used a criterion of job performance, whereas 90 used a criterion of training success. The average sample size in the 515 studies was 75.

Table 8.3 presents the results of the 515 validity studies in a number of different ways. Section A in Table 8.3 is a summary of what Hunter and Hunter (1984) found when they looked at the 515 validity studies. That is, they found that (a) on average, the validity of the different subtests of the GATB was 0.25 (mean observed correlation in the table); (b) the standard deviation of the distribution of validity coefficients was about 0.15; (c) about 90% of the observed validities exceeded 0.05; and (d) about 10% exceeded 0.45. Section A in Table 8.3, then, is a summary

TABLE 8.3

Distribution of Observed and True Validity for Cognitive Ability, Perceptual Ability, and Psychomotor Ability for All Jobs

	Cognitive Ability	Perceptual Ability	Psychomotor Ability
A. Distribution of Observed Validity Coefficients Across All Jobs			
Mean observed correlation	0.25	0.25	0.25
Observed standard deviation	0.15	0.15	0.17
Observed 10th percentile	0.05	0.05	0.03
Observed 90th percentile	0.45	0.45	0.47
B. Distribution of Observed Validity Coefficients Corrected for Sampling Error			
Mean observed correlation	0.25	0.25	0.25
Corrected standard deviation	0.08	0.07	0.11
Corrected 10th percentile	0.15	0.16	0.11
Corrected 90th percentile	0.35	0.34	0.39
C. Distribution of True Validity Across All Jobs (Validity Corrected for Unreliability in the Job Performance Measure and Restriction of Range of Test Scores for Study Participants)			
Mean true validity	0.47	0.38	0.35
Standard deviation of true validity	0.12	0.09	0.14
10th Percentile of true validity	0.31	0.26	0.17
90th Percentile of true validity	0.63	0.50	0.53

Source: Adapted from Hunter & Hunter (1984).

of the actual distribution of validity coefficients. Like any distribution, the mean and standard deviation can be calculated, even when the "scores" being distributed are 515 validity coefficients.

By displaying not only the mean of the distribution, but also the standard deviations and 10th and 90th percentiles, Hunter was showing us that there is a great deal of variability in the distribution. However, Hunter and Hunter (1984) statistically remove the variability due to sample size differences in Section B of Table 8.3. We can now see that much of the variability in validity coefficients was due to differences in the sample sizes in the various validity studies. That is, the 10th and 90th percentiles indicate a much smaller range of validity coefficients around the mean once the validities are corrected for sample size. Moreover, subtracting 2 SDs from the mean observed validity coefficients for the psychomotor tests $(0.25 - (2 \times 0.11))$ indicates that there is little likelihood that these tests will have zero validity for any job. Recall from our discussion of the normal curve in chapter 2 that plus or minus 2 SDs from the mean includes more than 96% of the cases.

Finally in Section C of Table 8.3, Hunter and Hunter (1984) corrected the mean observed validity coefficients for lack of reliability in the criterion and for restriction of range of test scores in the study population (only those applicants who were selected were study participants). With those corrections, the observed validity coefficients shown in Section A turned out to be considerable underestimates of the true validity; the logic and nature of both corrections is explained in chapter 2. These corrections produce substantial changes in mean validity. Also, the 10th percentiles of the true validity distributions are all substantially above 0.00, indicating that the GATB ability composites are valid in predicting performance in virtually all jobs.

Although Hunter and his colleagues have repeatedly presented evidence that cognitive ability tests are valid for virtually all jobs, the primary purpose of the Hunter and Hunter (1984) paper was to show that the type of test that is most valid for a given job depends on the complexity of the job. In Table 8.4, we present some of the data from their analysis of the effect of job complexity based on indices available in the DOT (now replaced by the O*NET; see chapter 3). Cognitive ability tests increase in validity as job complexity increases, whereas psychomotor ability increases in validity as job complexity decreases. The pattern of validities for perceptual ability tests is similar to that of the cognitive ability tests.

TABLE 8.4

Mean Observed Validity for Various Levels of Job Complexity
Using a Modified Form of the Data and Things Categories
for Cognitive, Perceptual, and Psychomotor Ability

	Observed Validity			
Complexity Level	Cognitive Ability	Perceptual Ability	Psychomotor Ability	Number of Jobs
1. High	0.34	0.35	0.19	21
2.	0.30	0.21	0.13	60
3.	0.28	0.27	0.24	205
4.	0.22	0.24	0.30	209
5. Low	0.13	0.16	0.35	20

The primary conclusion that should be drawn from the Schmidt-Hunter work is that paper-and-pencil cognitive ability tests are indeed valid and that there is no empirical basis for requiring separate validity studies for each job. Tests can be validated at the level of job families. Further, although these tests can be used in many, if not all, contexts, the data in Table 8.4 indicate that there are substantial differences among validities for jobs that vary in complexity. Although the data presented in Hunter and Hunter (1984) were controversial when first published, the notions underlying their work and the conclusion that cognitive ability measures are valid predictors of job performance is widely accepted. Also accepted is the notion that the complexity of the job moderates the level of validity (see Schmidt, 2002), and these results generalize to other countries and cultures (Salgado, Anderson, Moscoso, Bertua, & de Fruyt, 2003).

Since the mid-1980s, meta-analysis has been used to evaluate the generalizability of the validity of a wide variety of selection procedures (Schmidt & Hunter, 1998). Some of this additional research relevant to the use of other predictors is described in chapters 9 and 10.

Interestingly, it seems that cognitive ability is a correlate of a wide variety of life outcomes, in addition to job performance (Hernstein & Murray, 1994; Jensen, 1998; Lubinski & Humphries, 1997). Occupational level, divorce, delinquency and criminal behavior, income, adjustment indices, and accidents appear to be correlated with measures of general cognitive ability. The causal mechanisms that mediate the role of cognitive ability in the determination of these outcomes have been rarely studied, even in the case of job performance (Borman, White, Pulakos, & Oppler, 1991; Hunter, 1986).

One potential downside of the overwhelming evidence that cognitive ability is an important predictor of job performance is the fact that we see few efforts to update measures or conduct new studies of validity. The currently available database that is the source of most meta-analytic reviews of the validity of cognitive ability (e.g., Schmidt, 2002; Schmidt & Hunter, 1998) was collected at least 30 years ago. Project A (see the Summer 1990 issue of *Personnel Psychology*) is an exception in that a variety of measures were used to predict the performance of military personnel on a range of criteria. Beyond this major study, however, not many primary studies have added to the database that support the validity of cognitive ability since the 1970s. The one exception to this has been more recent meta-analyses on the relationship between cognitive ability and job performance and training performance in the European community. Salgado and colleagues (Salgado, Anderson, Moscoso, Bertua, & de Fruyt, 2003; Salgado et al. 2003) found criterion-related validities similar to or slightly larger than those found in Hunter and Hunter (1984), and also showed the increasing validity of cognitive ability as job complexity increased, supporting the robustness of the cognitive–performance relationship internationally.

Given the impressive evidence reviewed briefly here and in the sources we cited, it may be surprising to learn that cognitive ability tests are not used more frequently. For example, Ryan, McFarland, Baron, and Page (1999) examined the use of selection procedures in 20 different countries. They surveyed 959 companies who responded to questions about their use of different selection procedures using the following response options: 1 = never, 2 = rarely (1%–20%), 3 = occasionally (21%–50%), 4 = often (51%–80%), and 5 = almost always or always (81%–100%). The average response of U.S. companies regarding the use of cognitive ability tests on this scale was 2.09, equivalent to slightly more than 20%. The average across all 20 countries was 2.98. Salgado and Anderson (2002) similarly pointed out that reported cognitive ability test use was greater in 10 of 12 European countries than it was in the United States.

Why are cognitive ability tests not used more frequently in the United States? One reason may be that the general public does not believe cognitive ability tests are sufficiently valid for predicting job performance (i.e., they lack face validity; Jensen, 2000). Salgado and Anderson (2002) and others (see Gottfredson, 2002) have suggested that the major reason is that there are large differences in majority and minority subgroup mean scores on these measures. The resultant social, political, and legal problems associated with the use of measures with these differences have discouraged organizations from using them in making staffing decisions. In the next section of this chapter, we review the research on this complex issue and describe the nature of the problem.

RACIAL SUBGROUP DIFFERENCES AND COGNITIVE ABILITY

Racial subgroup differences are repeatedly observed when researchers use cognitive ability measures in an employment setting. African Americans score about 0.75 to 1 SDs lower than Caucasians, and Hispanic Americans score approximately 0.75 SDs lower than Caucasians. Asians usually score higher than Whites on quantitative measures and lower on verbal measures (Bobko, Roth, & Potosky, 1999; Niesser et al., 1996; Roth, Bevier, Bobko, Switzer, & Tyler, 2001; Schmidt, 1988; Schmitt, Clause, & Pulakos, 1996). These mean differences can translate into substantial differences in hiring rates across groups (see Table 6.5 in chapter 6 and Figure 7.3 in chapter 7). As subgroup mean differences increase and an organization is increasingly selective (selection ratios are decreased), smaller proportions of a lower-scoring subgroup will be selected. This adverse impact on minority groups has occupied the research attention of scholars in several related fields for at least the last four decades and has resulted in federal guidelines and considerable case law, which is described in chapter 11.

These continuously observed subgroup differences have led to concern that they result in biased predictions about the potential performance of lower-scoring minorities. However, it does not appear that the performance of minority groups is underpredicted using the regression approach described in chapter 6. This lack of predictive bias is widely recognized in the scientific literature (Niesser et al., 1996; Sackett & Wilk, 1994) and is recognized in the most recent edition of the APA *Standards* (American Educational Research Association, American Psychological Association, & National Council of Measurement in Education, 1999). It may also be believed that the cognitive tests themselves are biased against minority groups. However, a wealth of research does not find support for this finding (Jensen, 1998; Schmidt, 1988; see also chapter 7 on cultural influences on testing).

The finding that cognitive ability is consistently valid across jobs, but produces substantial differences in subgroup hiring rates when used in a selective fashion, produces a quandary for organizations that want to increase the skill levels of their workforce and workforce diversity (Murphy, 2002). The trade-offs associated with these two goals have been known for some time (see Schmidt, Mack, & Hunter, 1984, for an early example) but have become increasingly salient as the courts and federal law appear to be banning any form of minority group preference (Sackett & Wilk, 1994). However, the more recent *Grutter v. Bollinger* (2003) court case involving law school admissions at University of Michigan seems to represent a legal recognition that racial group membership may still be an appropriate hiring consideration in some situations (we discuss this case in detail in chapter 11).

In the next section of this chapter, we consider attempts to reduce or eliminate subgroup mean differences in ability and the corresponding differential hiring rates when these tests are used to help make selection decisions.

Ways to Reduce Subgroup Differences When Using Cognitive Ability Tests

Various strategies designed to reduce or understand the apparent conflict between productivity and diversity goals have been investigated by researchers and practitioners and are described in Sackett, Schmitt, Ellingson, and Kabin (2001). In this section, we consider these strategies. Box 8.1 describes research conducted by one of the authors and the attempt to increase minority admission rates to universities.

Use a Predictor Battery of Cognitive and Noncognitive Predictors

One approach is to use noncognitive measures that usually display smaller or nonexistent differences in subgroup means, along with cognitive measures. When such measures are valid indices of job-relevant constructs, this would certainly be an appropriate strategy (Outtz, 2002) in that it may increase validity and decrease the magnitude of subgroup differences. However, the magnitude of the subgroup differences that occur when a cognitive ability test is combined with another measure that displays no subgroup difference is still relatively large (Sackett & Ellingson, 1997), unless the alternative measure actually results in lower majority group performance.

Box 8.1 Attempts to Minimize Irrelevant Subgroup Differences

Over the last several years, one of the authors has been involved in a project in which one objective is to minimize subgroup differences on college admissions exams and improve prediction of college student success. At the beginning of this project, we realized that high school GPA and ACT or SAT tests were excellent predictors of college GPA and that both produced substantial differences between racial subgroups. We also realized that most educational institutions hoped to influence students along dimensions that were not represented in the usual college GPA measures that were used in most studies validating admissions procedures. An examination of college mission statements indicated that most universities were hoping to develop a graduate with a strong sense of ethics/integrity, social responsibility, leadership, multicultural appreciation, and other noncognitively loaded dimensions.

The research team then developed biodata and situational judgment instruments targeted to these dimensions. They also attempted to measure students' performance on these dimensions using self- and peer ratings on BARS and a set of BOS that reflected the frequency at which students engaged in behavior relevant to each dimension. Currently, the researchers also hope to broaden the scope of their outcome measures to include aspects of OCB, institutional commitment or loyalty, and perseverance or withdrawal.

The results indicate that there are minimal subgroup differences on the biodata and situational judgment measures and that they are highly correlated with performance measures of the same dimensions. Interestingly, they also contribute to the incremental validity of a battery of prediction instruments that include traditional college admissions tests (i.e., SAT/ACT) and high school GPA when those instruments are used to predict college grades. As indicated in the text, this combination of predictors still results in sizable subgroup differences because of the large differences typically found on the ACT/SAT measures and frequently for high school GPA as well.

Whether to implement the use of biodata or situational judgment measures for admissions purposes is still being debated. If they are implemented, two important and related concerns must be addressed. These instruments are certainly coachable and fakable if used in a high-stakes situation such as college admission.

> One possibility the researchers are exploring is to provide information on these instruments to all students, hence providing an "equal playing field." In this case, the concern would be what happens to the validity of the instruments and the magnitude of the subgroup differences. Another possibility is to use them only for developmental purposes. In any event, this project represents three of the ways in which we can reduce subgroup differences and enhance prediction of performance: Consideration of a broader set of relevant performance dimensions, developing noncognitive predictors of those dimensions, and the provision of information or coaching to all applicants to ensure that all have equal opportunity to do well.

In one examination of various combinations of predictors, Pulakos and Schmitt (1996) used a verbal ability measure that displayed a Black–White mean difference of 1.03 SDs. When combined with a biodata measure, a situational judgment test, and a structured interview, the mean difference between these two groups dropped to 0.63 SD. A combination of the three alternative tests alone produced a mean difference of 0.23 SDs. This three-test composite correlated 0.41 with the performance measure in that study; adding verbal ability increased the multiple correlation to 0.43. So, a mean difference of 0.40 (0.63–0.23) SDs was associated with the addition of the verbal ability measure with a corresponding increase of only 0.02 (0.43–0.41) in the prediction of performance. A simulation (Schmitt, Rogers, Chan, Sheppard, & Jennings, 1997) involving various predictors with a range of realistic intercorrelation values produced similar conclusions. That is, mean differences can be reduced by the combination of cognitive ability measures with other noncognitive predictors, but there remain rather large subgroup differences in most circumstances if the cognitive ability measure is used, and the use of these combinations of predictors will still result in differential hiring rates in many circumstances. Finally, a similar analysis using intercorrelation values based on a meta-analysis (Bobko et al., 1999) produced results that were almost identical to those reported by Schmitt et al. (1997).

Reconceptualizing the Performance Criterion

A second approach to the reduction of subgroup differences in the measures used to select employees involves a consideration of the performance criterion against which tests are validated. This solution is based on the fact that assessments of predictive bias using regression as mentioned previously and outlined in chapter 6 are predicated on the assumption that the criterion is unbiased. If the criterion is biased, deficient, or contaminated in some way, then predictors associated with the bias (in this case, potentially, racial group membership) may be more heavily weighted in any composite of predictors than should be the case. Given more current views of the performance construct that include contextual performance and task performance, this concern may be important because many noncognitive predictors are more highly correlated with contextual performance than task performance. If contextual performance is not part of the criterion against which tests are validated, then the role of noncognitive predictors that typically display smaller subgroup mean differences will be underestimated, and the weighting associated with cognitive ability will be overestimated. The magnitude of the difference this makes in hiring rates has been investigated in two studies (DeCorte, 1999; Hattrup, Rock, & Scalia, 1997). Although both studies confirm that the weighting of criterion elements can change the weighting associated with cognitive and noncognitive predictors, the weight associated with task performance (relative to contextual performance) would have to be minimal to make any appreciable difference in minority hiring because task performance is so strongly related to cognitive ability.

Mode of Stimuli Presentation and Required Responses

Another attempt to reduce subgroup differences on written paper-and-pencil tests is to change the mode in which questions are asked and the type of response required. The idea underlying this approach is that we often overemphasize reading or cognitive abilities in a selection battery when we are really interested in measuring other constructs. In terms of the predictor response model shown in Fig. 7.1 (chapter 7), the idea is that the construct we are interested in (let's say fighting fires) is not largely dependent on cognitive ability, but due to the nature of the assessment, verbal ability or cognitive ability is contaminating the measure. It is, however, often difficult to separate the construct measured from the methods used to measure them (Schmitt et al., 1996). For example, it is easier to measure interpersonal or oral communication skills in an interview or a role-play simulation than in a paper-and-pencil test. It may also be difficult to measure leadership in a paper-and-pencil format, but many have attempted to do so (e.g., Blake & Mouton, 1964). In one study, Chan and Schmitt (1997) compared subgroup differences on paper-and-pencil and video-based situational judgment tests. The latter was based on the same script as was the paper-and-pencil version. Although the written version of this test produced a mean difference of 0.95 SDs favoring Whites over Blacks, the mean difference on the video version was only 0.21. However, Sackett (1998) reported on the development of a video-based version of the Multistate Bar Examination, a multiple-choice test of legal knowledge and reasoning. Black–White mean differences on the two versions of the test were identical: 0.89 SDs in favor of the White group. The difference in the two studies may be due to the fact that different constructs were targeted. Chan and Schmitt (1997) were interested in assessing interpersonal skills, whereas the Sackett research was focused on cognitive constructs. The large subgroup mean difference for the verbal version of the Chan and Schmitt situational judgment test may actually have been the result of the inappropriate inclusion of cognitive components in a test of interpersonal skills. It would be useful to have more research on the nature of stimuli and responses required in examinations, but these two studies do point to the importance of carefully separating construct and method in assessing both subgroup differences and validity. If the staffing researcher is interested in assessing noncognitive attributes, the measure of those attributes should require minimal cognitive ability.

A wide variety of other assessment methods have been developed to assess KSAOs. Assessment centers and situational judgment tests (see chapter 10) in which applicants are required to perform a variety of tasks designed to closely replicate actual job tasks and situations are often used to measure a variety of KSAOs, including cognitive ability. Although some of these methods produce smaller subgroup differences with good validity, it is often the case that these methods are less reliable than paper-and-pencil measures, which will result in underestimation of both validity and subgroup mean differences (see Sackett et al., 2001, for a critical review).

Another test that shows low racial subgroup differences is the accomplishment record. Accomplishment records require that examinees describe major past accomplishments that are illustrative of their level of competence on targeted KSAO dimensions. These accomplishments are then rated by staffing personnel using BARS. Hough (1984) used accomplishment records to evaluate attorneys and validated the accomplishment record scores against performance ratings. Validity coefficients ranged from 0.17 to 0.25 and Black–White mean differences were only 0.33 SDs. However, accomplishment records correlated near zero with scores on the bar exam, the LSAT, and grades in law school, again raising the question of what was being measured by the accomplishment record. It is likely that motivational and social constructs, as well as ability, were important determinants of accomplishment record scores. As previously, in comparing these alternative assessments of KSAOs, it is important to know that we are assessing the same constructs with equal reliability.

Test-Taking Motivation

Another explanation for subgroup differences is that members of different groups have differing test-taking motivation and anxieties and that these influence test performance in non–job-relevant ways. In the predictor response model in Fig. 7.1, the idea is that subgroup differences in ability test performance are due to minorities having less test-taking motivation or experiencing higher levels of stress/anxiety. Hence, motivation is not a constant across subgroups and contributes to lower test scores for minority applicants. Chan, Schmitt, DeShon, Clause, and Delbridge (1997) showed that the relationship between race and test performance was partially mediated by test-taking motivation, although this mediating effect accounted for a very small portion of the individual differences in test performance. In another study, Chan, Schmitt, Sacco, and DeShon (1998) found that pretest reactions to tests affected test performance and mediated the relationship between belief in tests and test performance, but this motivational effect was similar for White and Black examinees.

The most controversial studies of the impact of motivational factors on test performance have been those involving stereotype threat. Steele and Aronson (1995) hypothesized that knowledge of cultural stereotypes relevant to a group with which an examinee identifies affect test performance. When minority examinees encounter "high-stakes" tests, their awareness of the common finding of large subgroup differences in test performance leads to a concern that they will do poorly and confirm the stereotype. As a result of multiple laboratory studies, Steele and his colleagues (Steele & Davies, 2003, Steele, Spencer, & Aronson, 2002) maintained that stereotype threat has been shown to negatively affect the performance of multiple groups, including minority groups and women on a variety of important tests such as the GRE, SAT, GMAT, and LSAT. These studies involve a manipulation of threat that includes a description of the cognitive ability construct measured and a measure of the degree to which the examinees identify themselves as members of a given racial group. Steele usually equates groups in terms of their previous performance on an important test such as the SAT. Minority members in groups that are then presented with information about the test do less well than majority group members. No such differences are found in groups that are not provided with test information, presumably because no stereotype threat exists.

Only one field study has examined stereotype threat in testing contexts. Cullen, Hardison, and Sackett (2004) examined two large archival datasets, one being a student sample linking SAT scores to GPA and a second being a military sample linking cognitive ability to job performance. They found no evidence for stereotype threat effects in either sample. However, they did not manipulate threat, and field studies that directly assess stereotype threat do not exist to our knowledge for good reasons. Ethically, it would be difficult to justify a stereotype threat manipulation when real job opportunities are at stake. Practically, in most situations, the applicants know why they are being tested and what is being evaluated so any manipulation of stereotype threat would likely appear artificial or contrived. It would not have impact over and above the threat that would already be inherent in the situation, so it would also be difficult to imagine an appropriate control group. Replications of stereotype threat in laboratory situations that simulated a selection context did not find much support for the effect, but until the effect is studied in real testing circumstances, it is uncertain if it has any real-world impact (see a special issue of *Human Performance*, 2003). Sackett (2003) and Steele and Davies (2003) provided very different views of the existing body of research on stereotype threat.

Also motivational in nature are coaching and orientation programs provided to familiarize examinees with test content and procedures. Ryan, Ployhart, Greguras, and Schmit (1998) evaluated what seems to be an optimal test orientation program for firefighter job applicants. The purposes of the program were to provide applicants with knowledge of the test format and types of test questions partly to inform and partly to reduce anxiety about the test. Blacks, women, and more anxious examinees were more likely to attend, but attendance had no impact on test scores. Of course, because those who attended were not randomly chosen, it is possible that had they not attended they would have performed more poorly. In this regard, longer coaching attempts for educational tests seem to produce effect sizes between 0.10 and 0.25 *SDs* (Messick & Jungeblut, 1981), although the latter effect was likely to happen only after extensive coaching. In studies involving preparation for the GRE, Powers (1987) reported that test preparation efforts did improve examinee performance, but that there were no performance improvement differences between groups.

An approach similar to coaching is allowing applicants to retake the tests. Hausknecht, Farr, and colleagues (Hausknecht, Trevor, & Farr, 2002; Sin, Farr, Murphy, & Hausknecht, 2004) examined this question with state police officer applicants. They found small to moderate improvements in test scores when applicants retook the employment tests, but there was no change in subgroup mean differences through repeated test taking.

Banding

A final means of reducing the consequences of subgroup differences in hiring rates is known as banding. In this procedure, one tries to reduce the effects of subgroup differences after the test scores have been collected. The basic logic is as follows: All tests have a certain degree of unreliability. This unreliability prohibits making fine distinctions between individual test scores (as the standard error of difference described in chapter 2 illustrates). Therefore, in any score distribution there are "bands" of scores that, due to unreliability, are statistically indistinguishable. If these scores are statistically indistinguishable, then within a band one can hire any individual regardless of their rank order. Because minorities tend to score lower in the overall distribution of predictor scores, the odds are better that they will be hired if they fall within the top bands of the distribution.

You are already familiar with banding; classes that provide grades of A, B, C, and so on are banding systems. You might receive 91% of all the available points, and your friend might receive 89% of all the available points, but you will receive an A and your friend will receive a B. Alternatively, another student who receives 99% of the points will receive the same grade as you, an A. Let's say we were hiring students from your class. In top-down hiring, we would hire the person with a 99%, then the 91%, and finally the 89%. In some banding procedures, we would first hire all people with an A—but the person with the 99% or the

91% would stand an equal chance if we used random hiring within the band. Thus, banding makes no distinctions among scores within bands, only between bands. Since minorities tend to score lower on cognitive ability tests, creating these bands of indistinguishable scores helps increase the chances they will fall into a top band and thus be hired.

There are two main ways through which banding can increase minority representation. The first is setting the size of the bands sufficiently large so that they will include minority test scores. This is usually, but not always, based on the standard error of the difference. The second is determining how to make hiring decisions within a band. Here there are several choices, including random hiring, minority preference, or using scores on a second predictor with lower subgroup differences (e.g., personality). Unfortunately, the best way to increase minority representation is by giving minority preference, but this has been difficult to justify in the U.S. courts (Campion et al., 2001). In other within band selection procedures, banding closely resembles simple top-down hiring and does not greatly affect minority hiring rates (Sackett & Roth, 1991), and produces less utility because a strict top-down approach is not used (see chapter 6). A variation of these procedures is called a sliding band. In a sliding band, the band will drop as soon as the highest scoring person in the band is hired. In the class example noted above, let's assume the band was nine points, and the top band therefore contains two people scoring 99% and 91%. If the person with a 99% is hired, the top of the band slides down to the next highest score. In this situation, the highest score in the band will be 91% and the next highest score will be 89%. This approach may help increase minority hiring rates because the bands can move down the score distribution more quickly. There are other instances where banding may help minority hiring rates, but it is unlikely to be a major long-term solution. A book by Aguinis (2004) discusses a variety of issues with banding.

To conclude, in assessments of cognitive ability, one can expect subgroup differences. These differences have been resistant to a variety of efforts to reduce or eliminate them at least within the assessment or testing context. Organizations can, however, engage in efforts that are more likely to produce a reaction among examinees that they are fairly evaluated on measures relevant to the jobs for which they are being considered. Test items and formats can be constructed to be face valid. Coaching or orientation programs can be provided that familiarize all candidates with testing procedures and item formats. In assessing candidate's KSAOs, it is also important that the full range of

KSAOs be considered and that the validity of procedures be evaluated against job-relevant criteria that are comprehensive and neither deficient nor contaminated. Banding test scores based on the unreliability of the predictor measure may help, but it is unlikely to be a satisfactory solution in the long run.

LICENSURE AND CERTIFICATION EXAMS

One of the most frequent uses of testing in the United States is for licensing purposes. It is increasingly common for professional organizations (including the APA) to require practicing members to demonstrate they have up-to-date knowledge and skills. One of the best examples is the bar exam taken by aspiring lawyers, but professions ranging from police and firefighters to electricians and truck drivers all require periodic reassessment and demonstration of ability to maintain their job (Shimberg & Schmitt, 1996).

Licensure allows people to engage in a given occupation when they demonstrate minimal competency necessary to ensure public safety and health. In many ways, licensure is similar to obtaining a driver's license or graduating from high school and college—one only obtains the certificate if the tests are passed. Shimberg (1981) made a useful distinction between exams used for licensing and exams used for predicting job success. This difference in purpose may lead to differences in the KSAOs assessed, as well as the way decisions are made based on the tests.

For example, a friendly, empathic physician or nurse may be more instrumental in producing patient satisfaction with health care and even, perhaps, physical health. However, licensing exams exist to protect public safety and health, so a licensing exam would not likely include measures of interpersonal skills. The emphasis would be on diagnostic skills, knowledge of health problems and appropriate care, and, perhaps, performance of critical tasks.

The second major difference is the way a test score is used to make decisions. If the aim is to predict job success, the optimal use of test scores is to pick in rank order those individuals who do best. For licensing purposes, the goal is to deny access of individuals to a profession when there is a possibility that the individual's incompetence would result in client harm. So, a minimal cutoff is usually established. In contrast, in predicting job success we are interested in identifying those persons whose performance will be the best. Even though there may

LICENSURE AND CERTIFICATION EXAMS

be many people who have the minimum competency to perform the job, they may not be hired because their performance is not as good as other applicants. However, in licensure, we only want to know whether individuals possess the minimum competencies. This requires setting cutoff scores.

Setting cutoffs is always judgmental. Livingston (1980) pointed out that standards exist in people's minds but not in a manner that is easily transformed into an objective decision rule. Methods of choosing a cutoff score are actually ways of expressing people's personal standards in terms of the test score. Livingston suggested three approaches to setting a cutoff score on a test, each of which employs a different type of judgment. In the first, judgments are made about some reference group. For example, we decide that 10% of those graduating from a course on real estate are incompetent. The cutoff score is then set to deny licensure to the bottom 10%. In the second approach to setting cutoff scores, we look at those practitioners or recent trainees who are judged to be clearly competent, assess their performance on the proposed exam, and decide on the cutoff score based on test performance. This is basically a concurrent validity approach to setting cutoff scores. Finally, what seems to be the most logically defensible method is to examine the test items themselves (Ebel, 1972). Must persons answer the item correctly to certify that they possess appropriate knowledge or skill? Does lack of a correct response indicate that the persons would be dangers to society? Whatever the basis for judgment, Livingston (1980) recommended four criteria to apply in setting the passing score (see also Cascio, Alexander, & Barrett, 1988):

1. The judgments must be made by qualified SMEs.
2. The judgments must be made in a way that is meaningful to those experts.
3. The decision process must take into account the purpose for which the test is being used.
4. Both types of decision errors (denying a qualified person and accepting an unqualified individual) must be considered.

A frequently used approach to setting cutoffs of this content validity method is the Angoff method. In this approach, a representative sample of job experts (e.g., supervisors) examines each test item and determines the probability that a *minimally competent person* would answer the question correctly. Minimally competent refers to a person who has the minimum requisite KSAOs necessary for adequate performance on the job. The probabilities are summed across all items and averaged across all experts, and the result is the cutoff score.

The most frequently used type of test in the licensing situation is the multiple-choice test, although many occupations also use performance (work sample) tests and oral examinations. Problems of bias in administering and grading performance exams for licensure are similar to those encountered in the selection situation. Because examinees are observed directly in most instances, rater biases are certainly possible.

The validity and reliability of licensing exams represent somewhat different problems than selection tests. Because the major decision is a pass–fail one, we should be interested in the consistency of pass–fail decisions on repeated administration of the licensing exam. Hyland and Muchinsky (1990) demonstrated this problem by finding certification tests for wastewater experts exhibited no significant criterion-related validity with a job performance criterion. They reasoned this occurred because job performance and competence are not the same thing, and competence is the appropriate criterion for a certification exam. Reliability estimates should also be made in a manner consistent with this pass–fail criterion. For a detailed discussion of these reliability estimates for what are called criterion-referenced exams, see Livingston and Wingersky (1979).

The only validation evidence presented for most licensing exams is content in nature. The questions addressed in our discussion of content validity in chapter 6 are also relevant here. In addition, we must ask whether the KSAOs assessed are represented at a level of complexity and in proportions consistent with the goal of protecting the public health, safety, and welfare. A criterion-related validity study of licensing exams would seem especially difficult. Shimberg (1981) pointed to the fact that licensees work in widely different organizations, frequently with different job performance goals. Further, it seems the appropriate criterion ought to be a dual criterion. On the one hand, it should be demonstrated that persons who did not pass a licensing exam are incompetent and in some way would harm the clients they served. On the other hand, it should be shown that those who passed did not harm the client. Clearly, we do not have information concerning the job performance of those rejected. Indirect evidence of criterion-related validity might consist of changes in malpractice occurrences before and after the passage of licensing legislation.

SUMMARY

The legal liability of licensing agencies must be considered. As discussed in chapter 11, the *Uniform Guidelines on Employee Selection Procedures* published by the Equal Employment Opportunity Commission (1978) state:

> Whenever an employer, labor organization, or employment agency is required by law to restrict recruitment for any occupation to those applicants who have met licensing or certification requirements, the licensing or certifying authority, to the extent it may be covered by Federal Equal Employment law, will be considered the user with respect to those licensing and certification requirements. (p. 38308).

Most federal circuit courts have held that Title VII of the 1964 Civil Rights Act does *not* apply to licensing agencies. Whether because of legal threat or a desire to serve the public, there has been consistent improvement in test construction and validation efforts on the part of licensing bodies (Shimberg, 1981; Shimberg & Schmitt, 1996). Perhaps this is because licensing and credentialing exams still fall under the guidelines presented by the *Standards* (AERA et al., 1999), as well as other professional governing agencies.

SUMMARY

We report on the history and concepts that underlie paper-and-pencil aptitude testing. From early assessments of scholastic aptitude to contemporary (seemingly) job-relevant specific ability tests, a huge investment has been made in the development of reliable and efficient procedures for assessing cognitive competence. The cumulative validity evidence suggests that this investment has been worthwhile in two senses. First, from an individual's standpoint, paper-and-pencil tests of cognitive ability are a great equalizer because objective scoring ignores all attributes of persons but their marks on an answer sheet. Such objectivity in scoring, we believe, has literally opened thousands of doors that would otherwise have been closed due to prejudice, stereotypes, ignorance, and bias. Second, organizations have a potentially efficient method for making wise selection decisions based on professionally developed, job-relevant cognitive ability tests. As shown in chapter 6, the utility of these tests can be quite high indeed.

Cognitive ability tests have also been challenged as vehicles of discriminatory hiring practices. Multiple reviews of the available information do not reveal evidence that cognitive ability tests are predictively biased. Various attempts to reduce the magnitude of subgroup differences have been largely unsuccessful, and any successes in this regard are usually associated with the assessment of constructs other than cognitive ability. It is hoped that other scientific efforts to develop and evaluate educational, social, and political interventions that remedy subgroup differences in the life opportunities that affect the development of ability will be undertaken and be fruitful.

REFERENCES

American Educational Research Association (AERA), American Psychological Association, & National Council of Measurement in Education. (1999). *Standards for educational and psychological testing*. Washington, DC: American Psychological Association.
Blake, R. R., & Mouton, J. S. (1964). *Building a dynamic corporation through grid organization development*. Reading, MA: Addison-Wesley.
Bobko, P., Roth, P. L., & Potosky, D. (1999). Derivation and implications of a meta-analytic matrix incorporating cognitive ability, alternative predictors, and job performance. *Personnel Psychology, 52*, 561–590.
Borman, W. C., White, L. A., Pulakos, E. D., & Oppler, S. H. (1991). Models of supervisory job performance ratings. *Journal of Applied Psychology, 76*, 863–872.
Campion, M. A., Outtz, J. L., Zedeck, S., Schmidt, F. L., Kehoe, J. F., Murphy, K. R., Guion, R. M. (2001). The controversy over score banding in personnel selection: Answers to 10 key questions. *Personnel Psychology, 54*, 149–185
Carroll, J. B. (1993). *Human cognitive abilities*. New York: Cambridge University Press.
Cascio, W. F., Alexander, R. A., & Barrett, G. V. (1988). Setting cutoff scores: Legal, psychometric, and professional issues and guidelines. *Personnel Psychology, 41*, 1–24.
Chan, D., & Schmitt, N. (1997). Video-based versus paper-and-pencil method of assessment in situational judgment tests: Subgroup differences in test performance and face validity perceptions. *Journal of Applied Psychology, 82*, 143–159.
Chan, D., Schmitt, N., DeShon, R. P., Clause, C. S., & Delbridge, K. (1997). Reactions to cognitive ability tests: The relationships between race, test performance, face validity perceptions, and test-taking motivation. *Journal of Applied Psychology, 82*, 143–159.
Chan, D., Schmitt, N., Sacco, J. M., & DeShon, R. P. (1998). Understanding pretest and posttest reactions to cognitive ability and personality measures. *Journal of Applied Psychology, 83*, 471–485.
Cullen, M. J., Hardison, C. M., & Sackett, P. R. (2004). Using SAT–GPA and ability–job performance relationships to test predictions derived from stereotype threat theory. *Journal of Applied Psychology, 89*, 220–230.
DeCorte, W. (1999). Weighing job performance predictors to both maximize the quality of the selected workforce and control the level of adverse impact. *Journal of Applied Psychology, 84*, 695–702.
Drasgow, F. (2003). Intelligence and the workplace. In W. C. Borman, D. R. Ilgen, & R. J. Klimoski (Eds.), *Handbook of psychology* (Vol. 12, pp. 107–130). Hoboken, NJ: Wiley.
Ebel, R. L. (1972). *Essentials of educational measurement*. Englewood Cliffs, NJ: Prentice Hall.
Equal Employment Opportunity Commission, Civil Service Commission, Department of Labor and Department of Justice. (1978). Adoption by four agencies of *Uniform Guidelines on Employee Selection Procedures. Federal Register, 43*, 38290–382315.

REFERENCES

Ghiselli, E. E. (1966). *The validity of occupational aptitude tests.* New York: Wiley.
Ghiselli, E. E. (1973). The validity of aptitude tests in personnel selection. *Personnel Psychology, 26,* 461–478.
Gottfredson, L. S. (2002). Where and why g matters: Not a mystery. *Human Performance, 15,* 25–46.
Grutter v. Bollinger. 539 U.S. (2003).
Guilford, J. P. (1967). *The nature of human intelligence.* New York: McGraw-Hill.
Hartigan, J. A., & Wigdor, A. K. (Eds.). (1989). *Fairness in employment testing.* Washington, DC: National Academy Press.
Hattrup, K., Rock, J., & Scalia, C. (1997). The effects of varying conceptualizations of job performance on adverse impact, minority hiring, and predicted performance. *Journal of Applied Psychology, 82,* 656–664.
Hausknecht, J. P., Trevor, C. O., & Farr, J. L. (2002). Retaking ability tests in a selection setting: Implications for practice effects, training performance, and turnover. *Journal of Applied Psychology, 87,* 243–254.
Hernstein, R., & Murray, C. (1994). *The bell curve: Intelligence and class structure in American life.* New York: The Free Press.
Hough, L. M. (1984). Development and evaluation of the "accomplishment record" method of selecting and promoting professionals. *Journal of Applied Psychology, 69,* 135–146.
Human Performance. (2003). Volume 3.
Hunter, J. E. (1986). Cognitive ability, cognitive aptitudes, job knowledge, and job performance. *Journal of Vocational Behavior, 29,* 340–362.
Hunter, J. E., & Hunter, R. F. (1984). Validity and utility of alternative predictors of job performance. *Psychological Bulletin, 96,* 72–95.
Hunter, J. E., & Schmidt, F. L. (1990). *Methods of meta-analysis.* Newbury Park, CA: Sage.
Hyland, A. M., & Muchinsky, P. M. (1990). As examination of the predictive criterion-related validity of certification tests. *Journal of Business and Psychology, 51,* 127–142.
Jensen, A. R. (1998). *The g factor: The science of mental ability.* Westport, CT: Praeger.
Jensen, A. R. (2000). Testing: The dilemma of group differences. *Psychology, Public Policy, and Law, 6,* 121–127.
Livingston, S. A. (1980). Comments on criterion-referenced testing. *Applied Psychological Measurement, 4,* 575–581.
Livingston, S. A., & Wingersky, M. S. (1979). Assessing the reliability of tests used to make pass/fail decisions. *Journal of Educational Measurement, 16,* 247–260.
Lubinski, D., & Humphries, L. G. (1997). Incorporating general intelligence into epidemiology and the social sciences. *Intelligence, 24,* 159–201.
McHenry, J. J., Hough, L. M., Toquam, J. L., Hanson, M. A., & Ashworth, S. (1990). Project A validity results: The relationship between predictor and criterion domains. *Personnel Psychology, 43,* 335–354.
Messick, S. M., & Jungeblut, A. (1981). Time and method in coaching for the SAT. *Psychological Bulletin, 89,* 191–216.
Motowidlo, S. J., Dunnette, M. D., & Carter, G. W. (1990). An alternative selection procedure: The low-fidelity simulation. *Journal of Applied Psychology, 75,* 640–647.
Murphy, K. R. (2002). Can conflicting perspectives on the role of g in personnel selection be resolved? *Human Performance, 15,* 173–186.
Naglieri, J. A., & Das, J. P. (1997). Intelligence revisited: The planning, attention, simultaneous, successive (PASS) cognitive processing theory. In R. F. Dillon (Ed.), *Handbook on testing* (pp. 136–163). Westport, CT: Greenwood Press.
Niesser, U., Boodoo, G., Bouchard, T. J., Jr., Boykin, A. W., Brody, N. Ceci, S. J., Halpern, D. F., Loehlin, J. C., Perloff, R., Sternberg, R. J., & Urbina, S. (1996). Intelligence: Knowns and unknowns. *American Psychologist, 51,* 77 101.

Olea, M. M., & Ree, M. J. (1994). Predicting pilot and navigator criteria: Not much more than g. *Journal of Applied Psychology, 79,* 845–851.
Outtz, J. L. (2002). The role of cognitive ability tests in employment selection. *Human Performance, 15,* 161–172.
Pearlman, K. (1980). Job families: A review and discussion of their implications for personnel selection. *Psychological Bulletin, 87,* 1–28.
Powers, D. E. (1987). Who benefits most from preparing for a "coachable" admission test? *Journal of Educational Measurement, 24,* 247–262.
Psychological Services, Inc. (1982). Administrator's guide to the PSI basic skills tests. Los Angeles: Author.
Pulakos, E. D., & Schmitt, N. (1996). An evaluation of two strategies for reducing adverse impact and their effects on criterion-related validity. *Human Performance, 9,* 241–258.
Ree, M. J., & Caretta, T. R. (2002). g2K. *Human Performance, 15,* 3–23.
Ree, M. J., Earles, J. A., & Teachout, M. S. (1994). Predicting job performance: Not much more than g. *Journal of Applied Psychology, 79,* 518–524.
Roth, P.L., Bevier, C. A., Bobko, P., Switzer, F. S., III, & Tyler, P. (2001). Ethnic group differences in cognitive ability in employment and educational settings: A meta-analysis. *Personnel Psychology, 54,* 297–330.
Ryan, A. M., McFarland, L., Baron, H., & Page, R. (1999). An international look at selection practices: Nation and culture as explanations for variability in practice. *Personnel Psychology, 59,* 45–57.
Ryan, A. M., Ployhart, R. E., Greguras, G. J., & Schmit, M. J. (1998). Test preparation programs in selection contexts: Self-selection and program effectiveness. *Personnel Psychology, 51,* 599–622.
Sackett, P. R. (1998). Performance assessment in education and professional certification: Lessons for personnel selection. In M. D. Hakel (Ed.), *Beyond multiple-choice: Evaluating alternatives to traditional testing for selection* (pp. 113–129). Mahwah, NJ: Erlbaum.
Sackett, P. R. (2003). Stereotype threat in applied selection settings. *Human Performance, 16,* 295–310.
Sackett, P. R., & Ellingson, J. E. (1997). The effects of forming multi-predictor composites on group differences and adverse impact. *Personnel Psychology, 50,* 707–722.
Sackett, P. R., & Roth, L. (1991). A monte carlo examination of banding and rank order methods of rest score use in personnel selection. *Human Performance, 4,* 279–295.
Sackett, P. R., Schmitt, N., Ellingson, J. E., & Kabin, M. B. (2001). High-stakes testing in employment, credential, and higher education. *American Psychologist, 56,* 302–318.
Sackett, P. R., & Wilk, S. L. (1994). Within-group norming and other forms of score adjustment in preemployment testing. *American Psychologist, 49,* 929–954.
Salgado, J. F., & Anderson, N. (2002). Cognitive and GMA testing in the European community: Issues and evidence. *Human Performance, 15,* 75–96.
Salgado, J. F., Anderson, N., Moscoso, S., Bertua, C., & de Fruyt, F. (2003). International validity generalization of GMA and cognitive abilities: A European community meta-analysis. *Personnel Psychology, 56,* 573–605.
Salgado, J. F., Anderson, N., Moscoso, S., Bertua, C., de Fruyt, F., & Rolland, J. (2003). A meta-analytic study of general mental ability validity for different occupations in the European community. *Journal of Applied Psychology, 88,* 1068–1081.
Schmidt, F. L. (1988). The problem of group differences in ability test scores in employment selection. *Journal of Vocational Behavior, 33,* 272–292.
Schmidt, F. L. (2002). The role of general cognitive ability and job performance: Why there cannot be a debate. *Human Performance, 15,* 187–210.

REFERENCES

Schmidt, F. L., & Hunter, J. E. (1998). The validity and utility of selection methods in personnel psychology: Practical and theoretical implications of 85 years of research findings. *Psychological Bulletin, 124*, 262–274.
Schmidt, F. L., Hunter, J. E., & Pearlman, K. (1981). Task differences as moderators of aptitude test validity in selection: A red herring. *Journal of Applied Psychology, 66*, 166–185.
Schmidt, F. L., Hunter, J. E., Pearlman, K., & Shane, G. S. (1979). Further tests of the Schmidt-Hunter Bayesian validity generalization procedure. *Personnel Psychology, 32*, 259–282.
Schmidt, F. L., Mack, M. J., & Hunter, J. E. (1984). Selection utility in the occupation of U.S. park ranger for three modes of test use. *Journal of Applied Psychology, 69*, 490–497.
Schmitt, N., Clause, C. S., & Pulakos, E. D. (1996). Subgroup differences associated with different measures of some job-related constructs. In C. L. Cooper & I. T. Robertson (Eds.), *International review of industrial and organizational psychology* (Vol. 11, pp. 115–140). New York: Wiley.
Schmitt, N., Rogers, W., Chan, D., Sheppard, L., & Jennings, D. (1997). Adverse impact and predictive efficiency of various predictor combinations. *Journal of Applied Psychology, 82*, 719–730.
Shimberg, B. (1981). Testing for licensure and certification. *American Psychologist, 36*, 1138–1146.
Shimberg, B., & Schmitt, K. (1996). *Demystifying occupational and professional regulation: Answers to questions you may have been afraid to ask.* Lexington, KY: Council on Licensure, Enforcement, and Regulation.
Sin, H. P., Farr, J. L., Murphy, K. R., & Hausknecht, J. P. (2004). *An investigation of Black–White differences in self-selection and performance in repeated testing.* Paper presented at the 64th annual meeting of the Academy of Management, New Orleans, LA.
Spearman, C. (1904). General-intelligence, objectively determined and measured. *American Journal of Psychology, 15*, 201–292.
Spearman, C. (1923). *The nature of 'intelligence' and the principles of cognition.* London: Macmillan.
Spearman, C. (1927). *The abilities of man.* London: Macmillan.
Steele, C. M., & Aronson, J. (1995). Stereotype threat and the intellectual test performance of African Americans. *Journal of Personality and Social Psychology, 69*, 797–811.
Steele, C. M., & Davies, P. G. (2003). Stereotype threat and employment testing. *Human Performance, 16*, 311–326.
Steele, C. M., Spencer, S. J., & Aronson, J. (2002). Contending with group image: The psychology of stereotype and social identity threat. In M. Zana (Ed.), *Advances in experimental and social psychology* (Vol. 34, pp. 379–440). New York: Academic Press.
Sternberg, R. J. (1977). *Intelligence, information processing, and analogical reasoning: The componential analysis of human abilities.* Hillsdale, NJ: Erlbaum.
Sternberg, R. J. (2000). Intelligence and creativity: In R. J. Sternberg (Ed.), *Handbook of intelligence* (pp. 611–630). Cambridge, MA: Cambridge University Press.
Thorndike, R. L. (1986). The role of general ability in prediction. *Journal of Vocational Behavior, 39*, 322–339.
Thurstone, L. L. (1941). Primary mental abilities of children. *Educational and Psychological Measurement, 1*, 105–116.
Wagner, R. K. (2000). Practical intelligence. In R. J. Sternberg (Ed.), *Handbook of intelligence* (pp. 380–395). Cambridge, MA: Cambridge University Press.
Wechsler, D. (1944). *Measurement of adult intelligence.* Baltimore: Williams & Wilkins.
Wonderlic personnel test user's manual. (1992). Libertyville, IL: Wonderlic Personnel Test, Inc.

9

Hiring Procedures II: Personality Tests, Affective Measures, Interests, and Biodata

AIMS OF THE CHAPTER

This chapter continues our discussion of different kinds of hiring procedures. We discuss personality and interest measures, application blanks, and biographical data. The goals of the chapter are (a) to review the history of the development of personality and interest measurement and show the theory lying behind their use today as parts of a modern personnel selection battery; (b) to introduce the reader to the development and use of a form of selection procedures, the biographical information blank (BIB), that consistently reveals superior predictive validity in practice; and (c) to introduce the reader to the idea of person—environment fit as a conceptual base for understanding why personality and interest measurement reveal the validity they do.

Prior to going into the details of these kinds of personnel selection prediction procedures, it is useful to make a series of general observations:

1. *People have personalities at work—again.* There was a period of time, say between 1970 and 1990, where studies of personality at work in the academic literature were sparse if not nonexistent. This is not to

AIMS OF THE CHAPTER 435

say that there was a dearth of academic research on personality or that personality was ignored by practitioners, just that academic personality researchers interested in work behavior had relegated personality to the dust bin. Barrick and Mount (1991), however, showed that progress in basic personality theory and research had potential payoffs both conceptually and empirically for those interested in prediction in the workplace. We detail how this history emerged.

2. *Context matters in personality and interest validity.* The idea that people's personalities predict their behavior is so intuitively appealing as to almost be a nonissue. However, that is just like saying people's abilities are related to their behavior when we know the reality is that individual differences in the abilities required to do specific kinds of tasks are related to behavior at those tasks. In other words, we do job analyses to discover the relevant abilities required by tasks, and then we design and use measures relevant for assessing those abilities because those are the measures likely to reveal validity. The same principle applies to personality measurement: The facet or facets of personality likely to be reflected in individual differences in behavior at work are those that are relevant for particular tasks in particular contexts. In short, not just any or all personality facets are equally relevant for understanding behavior at work; it depends.

3. *Personality and interests predict more than behavior at work; they predict adjustment and satisfaction, too.* We show in detail that the emotional lives of people at work can be understood by knowing things about those people, not just by knowing things about the context in which those people work. In other words, how we feel about work and working is partially a function of how we feel about ourselves and partially a function of the place in which we work. In chapter 5, we focus on the situation; here, we focus on the person.

4. *The best predictor of future behavior is past behavior.* This is a maxim by which personnel selection researchers have lived for decades and is likely the reason why interviewing people about their pasts can have validity for the prediction of their future behavior. However, as with the caution we made in point 2 above, it is not just any past experiences that predict future behavior, it is past behavior that is reflected in future demands that predict future behavior. Personnel selection researchers have developed a technique for discovering what are the relevant past behaviors and experiences of people that predict their future behavior at work and we describe how this BIB process works and why.

PERSONALITY AND PERSONALITY TESTS

Stogdill's (1948) early review of the predictability of leadership effectiveness from trait measures of personality discouraged researcher attention in these measures. Reviews of personality tests used for other work roles were equally discouraging (Ghiselli, 1966, 1973; Guion & Gottier, 1965). Attacks on the usefulness of these types of tests were commonplace in the texts and literature of the 1970s and 1980s (Campbell, Dunnette, Lawler, & Weick, 1970; Korman, 1977; Landy & Trumbo, 1984; Muchinsky, 1983; Schmitt, Gooding, Noe, & Kirsch, 1984). Although there had also been some reports of support for the use of such measures at work, these tended to be relatively ignored and/or treated as aberrations. For example, the Thematic Apperception Test (TAT) measures of need for achievement (nAch) and need for power (nPow) were shown to predict leadership accession and effectiveness at work (McClelland & Boyatzis, 1982); Miner's Sentence Completion Scale revealed significant predictions of managerial success (Miner, 1978); Type A behavior was shown to be reflected in academicians' productivity (Taylor, Locke, Lee, & Gist, 1984); and there were even some sympathetic reviews of projective personality measures as correlates of work behavior (Cornelius, 1983). Also, not to be discounted, practitioners in industry did not abandon the use of personality tests (see Campbell et al., 1970; Clark and Clark, 1990, for informative reviews).

What was really important for the long term, however, was the merging of two major developments. One was advancements in basic personality and personality measurement research that led to the development of the Five Factor Model (FFM) of personality, or simply the Big Five. The second was the application of meta-analysis to understand how the FFM related to job performance.

The Five Factor Model of Personality: Historical Background[1]

If you were going to develop a measure of personality, you might begin by identifying all the words that are used in common conversations to describe the personalities of others. So, you might come up with words such as warm, strong, conscientious, assertive, curious, unhappy, and so forth. Such a process could yield thousands of words just in English,

---

[1] This section benefited greatly from a chapter by Hough and Schneider (1996).

TABLE 9.1
Dimensions of the Five Factor Model of Personality With Descriptive Terms

Factor Label	Descriptive Terms
Extraversion	Assertive, sociable, dominant, achievement oriented, upbeat
Conscientiousness	Dependable, determined, attentive to detail, responsible
Neuroticism[a]	Fearful, emotional, anxious, negative
Agreeableness	Likeable, friendly, trusting, cooperative
Openness to experience	Imaginative, curious, inquiring

Source: Adapted from Digman (1990) and McCrae & Costa (1989).
[a] Sometimes called Emotional Stability with descriptive terms such as secure, calm, and low affect.

and then you could see how well they seem to capture the ways in which people differ as far as their personalities are concerned.

In fact, Allport and Odbert (1936) did precisely this. The words they generated were then used as a basis for data collection by Cattell (1945), who used factor analysis to identify the number of conceptual dimensions that might capture the meaning of the words—he said the number was 35. Because this was too many dimensions to be useful, later analyses were made of the 35 dimensions to see if they could be reduced in number further—and the number that kept emerging (not always without controversy) was the number 5 (Hough & Schneider, 1996). Table 9.1 shows the names that these five dimensions of personality are commonly referred to in the literature, with a description of the content of those dimensions. The five factors are Extraversion, Conscientiousness, Agreeableness, Neuroticism (also called Emotional Stability), and Openness to Experience. Although the labels differ slightly in reports from various personality researchers (e.g., Costa & McCrae, 1992; Goldberg, 1990), the discovery that a relatively small number of dimensions might capture much of what we know as personality made both the conceptualization of personality and its measurement much more accessible.

So what is the controversy? For decades, persons who developed measures of personality had their own theories about what the important dimensions of personality might be. In addition, recall that the 35 dimensions uncovered by Cattell (1945) were then themselves submitted to factor analysis, with the result that the 35 dimensions were reduced to 5. However, there are different rules for when to stop the factor analysis and some have argued that 6 is the appropriate stopping point

(Hogan, 1982) or that 8 is the appropriate number (Hough, 1992). In addition to the statistical arguments, there are also theoretical arguments. That is, some have proposed that the 5 factors each subsume so many different facets of personality as to be too global in their meaning. For example, Hough and Schneider (1996) reported that the factor called Extraversion contains subfacets as diverse as sociability and ambition and that these correlated differently with important work outcomes. Hough (1992) also showed that the FFM was incomplete for the prediction of numerous work-related outcomes, and she had to develop alternative personality facets, including a miscellaneous category in which to put facets not otherwise classified. She showed that this facet had the most consistent predictive validity for work outcomes.

Nevertheless, as a working framework, the FFM has proved quite durable:

- Numerous researchers from differing conceptual and statistical vantage points consistently find five strong factors in their analyses of measures of both the self-report and peer-report variety.
- When analyses are done of personality measures developed prior to the mid-1980s and the advent of the FFM, five factors typically emerge then, too.
- Analyses by Barrick and Mount (1991) of many years of research with measures of the FFM revealed consistent criterion-related validity for the FFM for diverse kinds of jobs and in diverse settings.

In fact, it was the review of the literature by Barrick and Mount (1991) that began the present love affair of selection researchers with personality. They showed that, especially for the factor labeled Conscientiousness, there were consistent although modest correlations with job performance. Their finding opened the flood gates of pent-up desire to use personality measures in personnel selection studies. This desire was partially fueled by the 1991 Civil Rights Act, which prohibited the use of within-group norming (discussed in chapter 11). This helped create a search for predictors with high validity but less adverse impact than cognitive ability.

Like all love affairs, everything has not gone smoothly, but there have been astonishing insights shed on the role of personality at work, legitimating in our opinion the importance in each real world project of considering the use of such tests as part of any personnel selection battery. We use the word "considering" because we make no claims

PERSONALITY AND PERSONALITY TESTS 439

that personality measures are always useful, but they can be useful and should be considered. What is the evidence for such enthusiasm?

Evidence for the Importance of Personality at Work

The evidence for this enthusiasm emerges from a book edited by Schneider and Smith (2004a). The book concerns the diverse ways in which the personality construct helps shed understanding on a diverse set of topics relevant to a human-based understanding of behavior in and of organizations. In the book, various authors take common topics in organizational behavior and HR management and ask the question, "What does personality contribute to our understanding of this topic?" The topics covered are as diverse as OCB, job satisfaction, stress, leadership, team performance, and organizational culture. In addition, the book contains chapters on such conceptually important topics as person–environment fit and the person–situation debate. The latter chapter concerns the issue of whether behavior at work is determined by the situations people encounter or by their own personalities; the answer is both. However, let us spend a few moments briefly reviewing each topic so the reader can gain a "feel" for why it is important to consider personality if one is designing a personnel selection system or program.

Task Performance and Organizational Citizenship Behavior

We have already discussed the similarities and differences of task performance and OCB in chapters 4 and 5. Task performance relates to the more technical, job-specific tasks that an employee is expected to perform. Meta-analyses suggest that personality tends to predict task performance to moderate degrees. The early meta-analysis by Barrick and Mount (1991) found the following validities for task performance, averaged across all types of jobs (r = uncorrected; ρ = corrected):

- Extraversion: $r = 0.08$; $\rho = 0.13$
- Neuroticism: $r = 0.05$; $\rho = 0.08$
- Agreeableness: $r = 0.04$; $\rho = 0.07$
- Conscientiousness: $r = 0.13$; $\rho = 0.22$
- Openness to experience: $r = 0.03$; $\rho = 0.04$

These validities are modest but practically useful. It is important to recognize that when the personality trait is better matched to the criterion, such as when openness to experience is related to training proficiency, the validities tend to be stronger (in this case, $r = 0.14$; $\rho = 0.25$). We return to this issue later in the chapter.

OCBs are important to organizational effectiveness because they are the behaviors that people spontaneously display that facilitate organizational accomplishment. It turns out to be true that personality predicts the likelihood that a given person will display OCB, but this issue is not simple (Organ & McFall, 2004). Because of the way research has been conducted on OCB (typically in a single organization using self-report personality measures and supervisory reports of OCB), the relationships between personality and OCB are sometimes difficult to tease out. Furthermore, it appears that the FFM is not as useful as some other facets of personality in predicting OCBs. However, it does appear to be likely that personality predicts OCB better than it predicts task performance; that personality predicts job satisfaction, which predicts OCB; and that more studies are needed comparing groups and even organizations with regard to (a) the people in them and their personalities, and (b) the relative incidence of OCB at these aggregate levels of analysis (Organ & McFall, 2004).

Job Satisfaction

What if we told you that how satisfied people are at work is partially due to their personality? Not only that, we tell you that how satisfied people are at work as an adult is predictable based on their personality as a teenager. In fact, job satisfaction appears to be a predisposition that is established relatively early in life and persists across decades, jobs, and job changes (Staw, 2004). The specific predisposition of interest here is referred to as *affectivity*—the predisposition to see events, issues, and life in positive or negative terms (Watson & Tellegen, 1985). The personality measure of positive and negative affectivity (PANAS; Watson, Clark, & Tellegen, 1988) reveals consistent relationships with job satisfaction, with negative affectivity correlating -0.40 and positive affectivity 0.52 on average across 29 studies.(Thoresen & Judge, 1997, as cited in Staw, 2004). Judge, Heller, and Mount (2002) performed a meta-analysis of the research that has accrued on the relationship between personality and job satisfaction, with the conclusions that all but Openness to Experience has significant and reliable relationships to job satisfaction. Two of the strongest correlates, and the ones that generalized best across samples, were for Neuroticism (−0.29) and Extraversion (0.25), offering some additional support for the fact that negative and positive affectivity play significant roles in job satisfaction.

What is intriguing about these kinds of findings is that job dissatisfaction in much of the research literature from about World War II until the early 1990s was presumed to be due to situational factors such as job characteristics, pay, supervision/leadership, and so forth. Prior to World War II, it was believed by researchers that job dissatisfaction was due to worker maladjustment and that the maladjustment resulted in negative views of work and life (Viteles, 1932). With the research of Staw and others (see Staw, 2004, for a review), we have come full circle as it becomes clear that different kinds of people process information about their work and work worlds (and other facets of their lives) in characteristic ways that result in them experiencing more or less satisfaction from and in those worlds—both at work and at home (Judge & Ilies, 2004).

Stress

A field of research closely related to job satisfaction concerns work-related distress. Stress is a generic term we use to describe the experience we have as the result of some external event making demands and perceived to be outside our control. The event or events are called *stressors*, and the experience of the stressors is called *strain* or *distress*. The personality correlates of work-related distress are similar to those related to job satisfaction, especially the personality facets of positive and negative affectivity (George & Brief, 2004). George and Brief, though, emphasize the fact that these personality facets are strongly related to the FFM dimensions of Extraversion (positive affectivity) and Neuroticism (negative affectivity), suggesting that if one used a measure of the FFM (to be discussed later) in a personnel selection battery, then predictions about who might be likely to experience distress at work would be feasible. What is particularly appealing about the George and Brief approach is their conclusion that the specific role of personality in work-related distress is not an issue because personality appears to play numerous roles in how people experience life and work, the kinds of stimuli in their environments on which they focus, and how those stimuli are appraised with regard to threat. For any particular job and job setting, it would be important to isolate what the issues are with regard to distress on the job and then use such insights as foci for the potential use of personality measures for a selection battery.

Leadership

Stogdill's (1948) review of personality correlates of leadership in the workplace was one of the stimuli leading to decreased interest in such testing. Stogdill found that there was little consistency across studies with regard to which facets of personality and which personality inventories predicted leadership. The problem is that this was interpreted as meaning that personality did not predict leadership. Thus, it is well known now that if the personality attributes studied do not map conceptually to the criterion of interest, the chances are poor that a significant relationship will be found (Tett, Jackson, & Rothstein, 1991). Indeed, more recent reviews of the predictability of leadership based on personality constructs have produced more favorable results. For example, Judge, Bono, Ilies, and Gerhardt (2002) and Spangler, House, and Palrecha (2004), although taking different vantage points and using different approaches to their reviews, both found support for personality correlates of both leader emergence and leader effectiveness.

Judge, Bono et al. (2002) organized their review around the FFM and conducted a meta-analysis of 222 correlation coefficients from 73 different samples of people in leadership positions. Their results revealed that Extraversion was the strongest and most consistent significant correlate of leadership ($\rho = 0.31$ across samples) and that Conscientiousness ($\rho = 0.28$), Openness to Experience ($\rho = 0.24$), and Neuroticism ($\rho = -0.24$) all had potential useful validity. Indeed, a multiple correlation across the FFM dimensions yielded a value of $R = 0.48$, suggesting the robust contributions possible from a focus on the FFM dimensions for predicting leadership. It is interesting to note that Judge et al. had a difficult time organizing their review because of the wild disparity in the traits measured as possible correlates of leadership across the different samples they included. They note that "One of the biggest problems in past research relating personality to leadership is the lack of structure in describing personality" (p. 766). Of course this is why they organized their review around the FFM; it provided the structure necessary in which different studies of variously labeled traits could be organized around common themes.

Judge, Bono et al. (2002) also did us a big favor by not only looking at the FFM, but by also looking at the validity of subdimensions of the FFM when there were sufficient samples to do so. For example, Extraversion is a composite of more narrowly specified subdimensions such as Dominance and Sociability. It turns out to be true that these subdimensions are somewhat stronger correlates of leadership than the Extraversion scale as a whole. The conclusion once again is that the FFM is a useful conceptual mapping of personality traits, but that for specific samples and specific criteria, more microdimensional analyses may be important—and factors outside the FFM may also be important.

The latter logic is one that pervades the review by Spangler et al. (2004). Spangler et al. presented the view that the traits studied by most personality researchers, such as the FFM traits, are distally removed from the behavior studied as a correlate of those traits. They argued that traits beget motives, and it is motives that get reflected in behavior. Their perspective is summarized in the following quote: "The proposition that motives are differentially relevant to leader effectiveness, contingent on the orientation of the organizations being led, has been largely overlooked in the leadership literature" (p. 263). In essence, they demonstrated in an extensive review of the literature concerning McClelland's (1961; Winter, John, Stewart, Klohnen, & Duncan, 1998) theory of motivation that (a) need for achievement predicts leadership effectiveness in entrepreneurial organizations, and (b) need for power predicts leadership effectiveness in traditional hierarchically structured organizations and politics. In other words, they claimed, and supported the claim quite well, that the universality proposed for specific aspects of personality in predicting leadership is neither conceptually nor empirically supported—that the situation matters. We see this idea of contingencies in the prediction of leadership and other phenomena again later, especially in a discussion of Miner's (1993, 2002) role motivation theory.

A unique feature of McClelland's approach is the use of a projective measure for assessing needs. So, in contrast to the survey or questionnaire measures used for assessment of the FFM and most other personality attributes, McClelland employed the TAT in his work (McClelland, Atkinson, Clark, & Lowell, 1953). Although the research McClelland did as background to the development of his theory and the research on leadership is absolutely fascinating (see Brown, 1965, for an excellent review), we do not describe it here. However, later we further discuss the TAT in additional detail when presenting projective tests as measures of personality. For now, the important point is that the TAT asks people to write stories about what is happening in a series of relatively ambiguous pictures. The stories are then coded for the themes the respondents write about in describing the relatively ambiguous pictures. These themes are then coded for need for achievement (nAch) and need for power (nPow), as well as the need for affiliation (nAff). These measures are said to tap people's *implicit* motives, motives of which they are perhaps unaware. Research based on combinations of the nPow and nAff facets of motives in what is called the Leadership Motive Profile (LMP; McClelland & Boyatzis, 1982) shows reliable relationships between the profile and leadership effectiveness in moderate to large nontechnologically oriented organizations and for managers who are separated from the work of the organization by more than one organizational level. In contrast, nAch predicts effectiveness best in smaller technologically driven organizations where leaders interact actively with workers (Spangler et al., 2004).

Spangler et al. (2004) do us a great service by not being dogmatic about their approach by suggesting an integration of trait theory approaches and about their approach based on McClelland's ideas and data. Their proposal is that the proximal cause of leadership resides in McClelland's three needs, but that the FFM as reviewed by Judge, Bono et al. (2002) contributes over and above the McClelland needs to the prediction and understanding of leadership effectiveness. Our conclusion is that this kind of model has potential merit for approaching the prediction of leadership, a complex set of behaviors, and is deserving of further research because the two traditions on which it rests both reveal predictive validity.

Prior to leaving the relationship between personality and leadership, it is salutary to reflect on the idea proposed by Hogan, Raskin, and Fazzini (1990) that personality can also serve to be a negative factor in leadership. Hogan et al. proposed and presented evidence to support the idea that there are certain personality attributes that predispose leaders to achieve high levels in organizations but then to be very costly to the organizations they lead. Without going into detail here, suffice it to say that the prediction of aberrant leader behavior rests on the use of an entire profile of attributes rather than one attribute at a time, as with the research we have reviewed using the FFM. In other words, like the work by McClelland and his colleagues summarized in Spangler et al. (2004), it is reasonable to propose that profiles may be the wave of the future rather than dealing with one dimension or facet at a time.

Team Performance

Readers might be wondering how individual personality data could be reflected in team performance. Isn't personality something that resides in individuals? The answer is "yes, personality resides in individuals." However, for a starting point on team performance and personality, consider the ideas noted in chapter 7, that team leaders have a big impact on team performance. If team leaders have an impact on team performance, perhaps the personalities of team members in the aggregate also have an effect on team performance. We discuss these issues generally in chapter 7, but let's look at this topic specifically with respect to personality testing. A review by Moynihan and Peterson (2004) reviewed this literature and made the point that their review was much easier to perform than earlier reviews (e.g., Shaw, 1981) because of the development and the acceptance of the FFM. Like Judge, Bono et al. (2002), Moynihan and Peterson organized the earlier research that had not used the FFM explicitly into FFM-related dimensions and thus had a large sample of studies with which to work.

In brief, here is what Moynihan and Peterson were able to conclude from their review:

1. Conscientiousness, Extraversion, Agreeableness, and Neuroticism in the aggregate predict team process performance (how team members work together as a team), as well as how well the teams perform. This means that the higher the average score of the team members on these dimensions, the more smoothly the team functions and the better the team performs. They also reported that when team leaders reveal higher scores on these attributes, this also contributes to team performance.

2. Organizational culture and task attributes serve as boundary conditions (moderators) on the relationships reported in number 1. This means that the universalistic conclusions noted in number 1 are weaker or stronger as a function of the organization in which the team functions and the task at which the team works. So, just as with trait and motive predictions of leadership, there are conditions when it is more or less likely. The problem here is that there are few studies on which to draw conclusions.

3. Groups in which people share the same personality attributes (homogeneity) are more cohesive, but groups in which the personalities of members differ (heterogeneity) have superior information sharing and problem solving. Moynihan and Peterson (2004) made the interesting point that these results are not necessarily in conflict because groups can be homogeneous on Conscientiousness and heterogeneous on Extraversion and demonstrate even more superior performance.

We conclude from this review that, in an era where teams are becoming more important in the workplace, the personality composition of those teams should be considered. In other words, it is reasonable to claim that personnel selection for teams is a viable and appropriate exercise for personnel selection practitioners.

Organizational Culture

If the effects of personality composition are noticeable in teams, it is not a huge inferential leap to propose that the personality composition of whole organizations has noticeable effects. The same logic applies to the effects of leader personalities on teams; perhaps leader personalities have effects on whole organizations. Schneider and Smith (2004b) proposed that both proposals are true and review a broad literature suggesting they are correct.

Their argument rests on Schneider's (1987) attraction-selection-attrition (ASA) framework, reviewed earlier. Basically, ASA begins with the founders of organizations and argues that their personalities get played out in the ways they carry out the daily work of the organization, including who gets hired, how decisions are made, and what facets of strategy and marketplace conditions are attended to (Schein, 1992). The founder thus sets the way the organization functions and the structure of the organization (Miller & Droge, 1986). It is these to which new employees are more or less attracted. Some organizations pay on an incentive system, for example, and these organizations attract different kinds of people than do organizations that pay based on a straight salary (e.g., Turban & Keon, 1993). Organizations then make decisions about who will be hired, and these decisions seem to yield people who are more like those already there than like the people in other organizations (Schneider, Smith, Taylor, & Fleenor, 1998). Finally, those who do not fit the organization well will tend to leave it.

At the time we write this book, there is a small but growing literature that suggests organizations are in fact relatively homogeneous in the personalities of the people in them and that it is not accidental that this is true (Lievens, Decaester, Coetsier, & Geirnaert, 2001; Schneider, Goldstein, & Smith, 1995). Thus, we tend to think about organizations as having people who kind of randomly go to work in them, but this is clearly not the way real organizations obtain people—at least not where the unemployment rate is low and jobs are available. For example, Lievens et al. (2001) in a study in Belgium showed that the FFM was useful in predicting to which kinds of organizations engineering and MBA students would be attracted. They showed in a sample of 359 that (a) people high on Conscientiousness were significantly more likely to prefer large rather than small organizations compared with those low on Conscientiousness, and (2) people high on Openness to Experience were more likely to prefer international organizations compared with those low on Openness to Experience. Ployhart, Weekley, and Baughman (in press) showed that personality homogeneity can actually be conceptualized in two ways; that of mean levels of personality, and that of variance in personality. They showed that homogeneity in personality was nested such that personality at job levels was more similar than personality at organizational levels. Further, they found unit-level mean personality tended to relate positively to satisfaction and performance, but unit-level variances related negatively to these outcomes. Therefore, both the unit's average personality and its variability in personality appear to be determinants of individual-level outcomes. Aldrich (1999), in perhaps the most complete look at how organizations are founded, grow, and develop, presented evidence that the informal networks of founders and early hires are a key source of new employees and that these new employees are thus similar to those already in the organization.

Later, we review the research on vocational choice and show how different kinds of people end up in different careers; the same appears to be true for organizations. It is the people in organizations and the people in careers that make them look and feel the way they do (Holland, 1997; Schneider, 1987). This harkens back to our discussion of human capital in chapter 1, supporting the point that organizations become homogenous in their human capital over time.

Summary

We go into this detail on personality correlates of behavior in and of organizations to emphasize how important a consideration of personality can be in making hiring decisions. The evidence is so compelling that the O*NET job analysis procedure we described in chapter 3 (www.onetcenter.org) contains a taxonomy of work styles that closely patterns the FFM. For example, the taxonomy contains the seven dimensions shown in Table 9.2, and we have used the FFM labels of Table 9.1 to reclassify them. Our opinion clearly is that job analyses that fail to address the potential in personality constructs to inform testing and prediction are not capitalizing on what the research literature reveals (Guion, 1998).

TABLE 9.2
O*NET Work Styles Taxonomy and the Five Factor Model

O*NET Construct Label	O*NET Descriptive Labels	FFM Factor Label
Achievement orientation	Effort, persistence, initiative	Extraversion
Social influence	Energy, leadership	Extraversion
Interpersonal orientation	Cooperative, caring, social	Agreeableness
Adjustment	Self-control, stress tolerance	Stability
Conscientiousness	Dependability, attention to detail	Conscientiousness
Independence		Openness
Practical intelligence	Innovative, analytic	Openness

Thus, the list of issues related to personality does not end in the summaries just presented. In addition to the topics just outlined, personality has also been shown to be reliably related to integrity (honesty, drug and alcohol abuse, reliability; Sackett, Burris, & Callahan, 1986), service orientation (Frei & McDaniel, 1998), and deviant behavior (Colbert, Mount, Harter, Witt, & Barrick, 2004). This cannot be a book on personality issues at work because we have so many other issues on which to focus—but it could have been. Personality is interesting, useful, valid, and involved in all forms of the behavior we observe in and of organizations; it should not be taken lightly and it appears now to be taken "heavily." Box 9.1 gives one example why this evidence is so important for staffing experts.

Personality Testing

If personality is so important in organizations, how is it assessed? In this section of the chapter, we review three kinds of assessments: (1) personality inventories such as measures of the FFM, (2) projective instruments such as the TAT used by McClelland and his associates, and (3) sentence completion scales such as Miner's (1960, 2002).

Personality Inventories

From earlier materials presented, the reader might get the impression that personality testing began with the development of the FFM and/or the Big Five. This is, of course, not true. For example, we noted that Moynihan and Peterson (2004) and Judge et al. (2002) organized the research they reviewed *around* the FFM dimensions because so much

> **Box 9.1 The Attitudes of Managers Toward Personality Tests**
>
> Personnel selection experts are often amazed and disconcerted by the attitudes of managers toward personality tests. The same manager who is hard-nosed and skeptical in evaluating production or inventory techniques will often be a soft touch for a charlatan peddling a panacea personality test. One study demonstrated this gullibility quite dramatically. Ross Stagner (1958) administered a published personality test to a group of HR managers attending a convention. As feedback, the managers did not receive their actual score, but a list of 13 statements that Stagner told them was descriptive of their personality. The managers did not know it, but they all received the same description; all statements had been taken from horoscopes, astrology books, and similar sources. Managers were asked to rate the accuracy of their scores. Eighty percent believed their profile was either an "amazingly accurate" or a "rather good" description of them. Here are some examples of the 13 statements each manager received and agreed were at least a good description of them:
>
> - Although you have some personality weaknesses, you are generally able to compensate for them.
> - At times you are extroverted, affable, and sociable, whereas at other times you are introverted, wary, and reserved.
> - Security is one of your major goals in life.
>
> Results like these help demonstrate the gullibility of unsophisticated managers and show why it is so difficult to convince managers that they must statistically validate measures before putting them into use. It also shows why "management consultants" can sell unvalidated personality inventories to unsophisticated managers.

of the research they reviewed occurred prior to the development of a measure of the FFM.

Personality inventories (e.g., surveys, blanks, scales, profiles, schedules), regardless of the era, are multidimensional in nature. As such, they, similar to aptitude batteries, attempt to define and assess sets of specific aspects or traits of the person. Also, similar to aptitude tests, these inventories are of the paper-and-pencil test variety, although they are increasingly being used in Internet and computerized applications.

Also, similar to aptitude batteries, they are developed through internal consistency analyses such as factor or cluster analysis. In this procedure, a hypothetical trait is defined by a set of questions or items in a test. Certain items are written to provoke or "tap" the trait cluster, and if responses to the items across a large number of people are strongly correlated, the items in combination are said to assess the trait. This procedure, or one similar to it, has been used to develop the personality inventories most frequently used in the employment setting in the past and also characterizes the measurement processes used with measures of the FFM and the Big Five. It is important to note that even with the advent of the domination of measures of the FFM and the Big Five that (a) practitioners in companies continue to use older tests because they have shown validity in their setting (e.g., Bentz, 1990; Howard & Bray, 1990) and (b) as noted earlier, not everyone agrees that the FFM is the final insight into personality testing. Just as an example of point (a), recall that in the George and Brief (2004) review of the research on stress and in the Staw (2004) review of personality correlates of job satisfaction, the measurement of positive and negative affectivity (PANAS; Watson et al., 1988) played significant roles. This is so even when research suggests that positive affectivity may be related to Extraversion and negative affectivity to Neuroticism (Costa & McCrae, 1980), as can be seen in Table 9.1, presented earlier.

Readers of this book who look back into earlier eras of personality testing will find personality inventories with many more than five factors. The point is that these earlier inventories and their many factors can frequently be adequately subsumed by the FFM (e.g., Barrick & Mount, 1991). Some frequently used earlier measures in the past include

- 16 Personality Factor Test
- California Psychological Inventory (CPI)
- Gordon Personal Profile (GPP)
- Jackson Personality Inventory (JPI)
- Guilford-Zimmerman Temperament Survey (G-ZTS)

All these inventories have shown validity against a variety of criteria in specific settings, but the issue has always been the generalizability of those findings to new settings, something that meta-analyses of the FFM has resolved. However, translating the many factors of these once-popular personality inventories into meta-analyses of the FFM is not the same as showing that these earlier measures themselves reveal cross-situational validity.

With regard to point (b), as noted earlier, there continues to be debate over the inclusiveness that the FFM represents. For example, Guion (1998) wrote: "Apparent omissions [in the FFM] include locus of control (Hough, 1992; Tett et al., 1991), self-esteem (Buss, 1992), activity level (Guilford, 1959), shyness (Asendorpf, 1992), 'choosiness' (Riggio & Fleming, n.d.), and Type A personality (Tett et al., 1991)" (p. 143). Although analyses of previously used personality measures for which predictive validity had already been demonstrated in specific instances can frequently be subsumed by the FFM (as is accomplished in the meta-analyses described earlier), this does not make the developers of those previous measures sanguine. For example, the Myers-Briggs Type Indicator (MBTI) is a frequently used personality measure in business and industry that McCrae and Costa (1989) say captures four of the five factors in the FFM (all but Neuroticism). Nevertheless, there are sound reasons both conceptually and practically that can be presented for retaining the measure and the constructs as originally proposed (Fitzgerald, 1997).

In summary, personality inventories have a long history in psychology and in personnel selection research. The present orthodoxy suggests that measures of the FFM (e.g., Costa & McCrae, 1992) are adequate representations of the broad facets of personality, but that in specific instances it may be important to look beyond measures of the FFM and/or look into the subfactors comprising measures of the FFM to gather reliable prediction of specific outcomes (Smith & Schneider, 2004).

Projective Tests: Pictures

A projective personality measure is one that asks respondents to relate what they see or what is happening in the picture presented to them. The pictures are either relatively or totally ambiguous. The respondents are assumed to project their own personalities onto the stimulus and, by reporting what they experience, think, or feel, to reveal the kind of people they are. Prominent examples of this procedure are the Rorschach Ink Blot Test, the Holtzman Ink Blot Test, and the TAT. The task of the test administrator is to ascertain, on the basis of what people say and/or write in response to the pictures, the personality dynamics of the respondents. There are numerous debates about how projective test responses should be scored and whether projective tests are even useful (Hogan & Hogan, 1998). Indeed, more recent issues of the scholarly journals in which academic selection researchers publish (such as *Personnel Psychology* and the *Journal of Applied Psychology*) reveal scant attention to projective tests. It is reasonable to assume decreased interest in projective tests is attributable not only to debates over scoring and validity, but also due to legal concerns with regard to job relevance and other legal issues.

The TAT used in the assessment of need for Achievement (nAch), need for power (nPow), and need for affiliation (nAch), discussed earlier, is an exception to the scoring dilemma (see McClelland & Boyatzis, 1982; Winter et al., 1998). In the TAT assessment of these needs, a person is shown a series of relatively ambiguous pictures of a person or people and asked to write a story about each picture containing information on the following:

1. What is happening? Who is involved?
2. What has led up to this situation? (That is, what has happened in the past?)
3. What is being thought? What is wanted? By whom?
4. What will happen? What will be done?

The stories are then scored for their achievement, power, and affiliation *imagery*. Imagery refers to the content suggested by what is written (i.e., does what is written connote achievement, power, affiliation, and if so, to what degree?). For example, in coding for achievement imagery, the scorer would look for themes that suggest that a person strives for accomplishment, ego enhancement, and recognition, or toward surmounting obstacles in the face of potential failure. Given that the meaning of content can be reliably scored, it is possible to not only score the stories of people who are responding to TAT cards, but also to score the content of anything written (and in some cases, unwritten symbols, as well). For example, the folk tales of different societies and the pottery designs of different ages and societies are all amenable to scoring for achievement content (e.g., McClelland, 1961). To briefly describe a long, fruitful research project, in an impressively large number of instances, high achievement imagery is associated with high performance on both the individual and the societal levels.

We can raise the question "Are high nAch scores a social response to situations or a stable personality trait invariant over situation?"

Research evidence points to an interaction—people with higher achievement motivation will display it in achievement-oriented situations (Spangler et al., 2004). Therefore, nAch scores should be most useful as predictors of success under organizational conditions where the tasks are challenging and moderately risky, people have to persist to make things happen, and work requires initiative.

This is a useful listing of the boundary conditions mentioned earlier that determine when and under what conditions personality tests are likely to prove useful for predicting leadership. Thus, one would not expect nAch to predict the success of the assembly line worker, but would expect successful prediction of executive behavior in an entrepreneurial setting; in the latter case, the TAT measure of nAch is an effective predictor (Campbell et al., 1970).

We have not been able to find more recent reviews of the validity of projective measures in the workplace, but Cornelius (1983) pointed out that positive evidence concerning the use of projective tests was noted in several reviews. Guion (1965), for example, included 11 validity studies on projective tests, 8 of which contained significant results. Although Kinslinger (1966) decried the generally poor quality of studies on projectives, he does cite 20 "fairly rigorous studies," of which 16 report positive findings. In particular, Kinslinger (1966) mentioned a sentence completion test that predicted sales success in various jobs with validities between 0.53 and 0.73. We discuss sentence completion tests, another form of projective testing, next. Similarly, the Tomkins-Horn Picture Arrangement Test yielded validity coefficients in the high 0.50s for sales positions and from 0.58 to 0.82 for tabulating equipment operators (Cornelius, 1983). Finally, in a review of predictors of managerial performance, Korman (1968) expressed early surprise at the consistently positive predictive validity of the Miner Sentence Completion Scale (MSCS), and more recent reviews have not dampened this opinion (Miner, 2002). We turn next to the MSCS.

Projective Tests: Incomplete Sentences

The MSCS deserves special mention. Along with the TAT, it is the most frequently used projective device in the selection of managers, and validity evidence has been mounting for at least four decades and is quite impressive (Miner, 1978, 1993, 2002). Further, its scoring and the theory underlying the use of the test relate directly to managerial work; this is the purpose for which the test was developed. Respondents to the MSCS (Miner, 1960) must complete sentences which may begin with words such as

- My family doctor...
- Police officers...
- Running for political office...
- Wearing a necktie...
- Presenting a report at a staff meeting...

Theoretically, Miner (1978) hypothesized that managers must competently fill six different roles, and the motivation to act in each role is scored using the response to sentences in the MSCS. These six managerial role prescriptions are described by Miner (1978) as follows:

1. Managers must be in a position to obtain support for their actions at higher levels. This requires a good relationship with superiors. It follows that managers should have a generally positive attitude toward those holding positions of authority over them. Any tendency to generalize hatred, distaste, or anxiety in dealing with people in positions of authority will make it extremely difficult to meet job demands.

2. There is a strong competitive element built into managerial work. Managers must strive to win for themselves and their subordinates and accept such challenges that other managers may offer. To meet this role requirement, managers should be favorably disposed toward engaging in competition. If they are unwilling to compete for position, status, advancement, and their ideas, they are unlikely to succeed.

3. Although the behaviors expected of a parent and those expected of a manager are not identical, both are supposed to take charge, make decisions, take such disciplinary actions as necessary, and protect other members of a group. Thus, one of the common role requirements of the managerial job is that the incumbent behave in an active and assertive manner. Those who prefer more passive behavior patterns, regardless of their gender, and those who become upset or disturbed at the prospect of behaving in an assertive manner would not be expected to possess the type of motivation needed.

4. Managers must exercise power over subordinates and direct their behavior. They must tell others what to do and enforce their words through appropriate use of positive and negative sanctions. Individuals who find such behavior difficult and emotionally disturbing, who do

not want to impose their desires on others or believe it is wrong to do so, would not be expected to meet this particular role requirement.

5. The managerial job requires people to stand out from the group and assume a position of high visibility. They must deviate from the immediate subordinate group and do things that inevitably invite attention, discussion, and perhaps criticism from those reporting to them. When this idea of standing out from the group elicits feelings of unpleasantness, then behavior appropriate to the role will occur much less often than would otherwise be the case.

6. There are administrative requirements such as constructing budget estimates, serving on committees, talking on the telephone, and filling out forms, in all managerial work, although the specific activities will vary. To meet these prescriptions, a manager must at least be willing to face this type of routine and ideally gain some satisfaction from it. If such behavior is consistently viewed with apprehension or loathing, a person's chances of success are low.

Scores for each of these roles (two scores are computed for desire to compete), as well as a total score, are computed. In 57 studies of the MSCS reviewed by Miner (1993), 92% of the studies revealed statistical significance in the theoretically predicted way. The subscale scores having to do with attitudes toward authority and the desire to exercise power were generally superior, but the motives to be assertive, stand out, and perform administrative duties were also frequently valid.

We emphasize that the projective tests described here are not for use by untrained persons. Projective test scoring requires a thorough immersion in the constructs underlying the particular measure, rigorous training in applying existing scoring guidelines, and supervised practice. We also caution that a large majority of the studies on the MSCS have been conducted by Miner; more frequent verification of the validity of this instrument by other researchers should occur.

Conditional Reasoning: A New Approach to the Development of Personality Tests

James and Mazerolle (2002) offer a theory-based alternative to the development of personality tests, tests that in the small literature that exists for them seem to have good predictive validity. Their approach goes by the label *conditional reasoning*. Conditional reasoning theory proposes that the differences in people's personalities are understandable from knowledge of differences in the way they think, especially the way they explain events. Although this is not a new idea, application to the development of personality tests has not been much developed. Items written for conditional reasoning tests do not ask people for self-reports of their behavior or even self-reports of how they think. Rather items in such tests present scenarios to people and ask them to explain the cause or causes of those situations. In some ways, this is a projective test of sorts in that the respondents are asked not to report on themselves but to report why events occurred the way they did. In other words, people are asked to project their personalities by explaining why things likely happen the way they do.

James and Mazerolle (2002) think of these explanations for events as *justification mechanisms*, with such mechanisms being representative of people's characteristic ways of dealing with the world around them. In their words: "Justification mechanisms comprise implicit biases that, unknown to the reasoner, shape, define, and guide the perceptions, understandings, hypotheses, causal explanations, and expectancies the reasoner employs to give meaning to events and to reason about how to adjust to the environment" (p. 93).

James and Mazerolle (2002) reviewed four programs of research as evidence for the conditional reasoning approach to measurement and prediction, one of which concerns achievement motivation (AM) and fear of failure (FF). Without going into too much detail, there are numerous biases or justification mechanisms that differentiate AMs from FFs; we summarize some of these in Table 9.3.

In psychology, the term "justification mechanism" has referred to the ways people defend to themselves implicitly or explicitly their own behavior. So, when we smoke knowing that it can cause cancer, we "justify" smoking by saying to ourselves something like "well, smoking does not always cause cancer" or "most people who smoke do not get cancer and die." James and Mazerolle (2002) asked themselves, "What kinds of justification mechanisms for decisions and behavior would people use who are AMs compared to those people who are FFs."

Specification of such justification mechanisms is of course a huge research undertaking in and of itself. The design of items for the measurement of conditional reasoning based on this specification is an additional challenge. We present one example of such an item in Table 9.4, along with four alternative choices. Alternatives b and c are not reasonable, so the issue is to what justification mechanism do alternatives a and d refer? James and Mazerolle (2002) described in detail their own

TABLE 9.3
Sample Justification Mechanisms for Achievement Motivation and Fear of Failure

Achievement Motivation Sample Justification Mechanisms or Biases
Positive Connotation of Achievement Striving Bias. A tendency to associate effort (intensity, persistence) on demanding tasks with "dedication," "concentration," "commitment," and "involvement."

Personal Responsibility Bias. A tendency to favor personal factors such as initiative, intensity, and persistence as the most important causes of performance on demanding tasks.

Fear of Failure Sample Justification Mechanisms
Negative Connotation of Achievement Striving Bias. A tendency to frame effort (intensity, persistence) on demanding tasks as "overloading" or "stressful." Perseverance on demanding tasks after setbacks or obstacles are encountered is associated with "compulsiveness" and "lack of self-discipline."

Leveling Bias. A tendency to discount a culturally valent—but, for the reasoner, psychologically hazardous event (e.g., approaching a demanding situation)—by associating that event with a dysfunctional and aversive outcome (e.g., cardiovascular disease).

Source: Adapted from James & Mazerolle (2002, pp. 41, 43).

TABLE 9.4
Sample Item From a Conditional Reasoning Test for Achievement Motivation and Fear of Failure

The Item Stem
Studies of the stress-related causes of heart attacks led to the identification of the Type A personality. Type A persons are motivated to achieve, involved in their jobs, competitive to the point of being aggressive, and eager, wanting things completed quickly. Interestingly, these same characteristics are often used to describe the successful person in this country. It would appear that people who want to strive to be successful should consider that they will be increasing their risk for heart attack.

The Item Alternatives
Which one of the following would most weaken the prediction that striving for success increases the likelihood of having a heart attack:
  a. Recent research has shown that it is aggressiveness and impatience, rather than achievement motivation and job involvement, that are the primary causes of high stress and heart attacks.
  b. Studies of the Type A personality are usually based on information obtained from interviews and questionnaires.
  c. Studies have shown that some people fear being successful.
  d. A number of nonambitious people have heart attacks.

Source: From James and Mazerolle (2002, pp. 186–187). Used by permission of Sage Publications. Copyright by Sage Publications.

reasoning for these potential justification mechanisms that we can only summarize briefly here, as follows:

- Alternative a is based on the Positive Connotation of Achievement Striving Bias (see Table 9.3), and as such it would represent someone with achievement motivation.
- Alternative d is based on the Negative Connotation of Achievement Striving Bias and the Leveling Bias (see Table 9.3), and as such would represent someone with fear of failure.

So, alternative a is a choice that people will make when they reason that it is not achievement motivation, but rather aggressiveness and impatience that is the cause of cardiovascular disease. Alternative d, in contrast, connotes a tendency to ascribe stress to achievement motivation and reason that stress from trying to achieve is what causes cardiovascular disease.

Emotional Intelligence

Another more recent conceptualization of personality at work, especially with regard to the issue of leadership, is one called *emotional intelligence*. Popularized by Daniel Goleman (1998), the theory in essence says that some leaders, because of their high emotional intelligence (EI), create positive moods among their followers that yield superior performance outcomes for the groups they lead (Goleman, Boyatzis, & McKee, 2002). In the latest version, EI has two superordinate dimensions, Personal Competence and Social Competence, with each having two subdimensions. Personal Competence is comprised of Self-Awareness (reading one's own emotions, and knowing one's strengths and limits) and Self-Management (being honest, flexible, optimistic, and having the drive to improve performance). Social Competence is comprised of two dimensions, Social Awareness (empathy for others, being aware of organizational issues, and being service oriented to others) and Relationship Management (motivating, persuading, managing, cultivating networks, and teambuilding).

The importance of EI is that many managers and executives have been exposed to it—the label of the theory is very catchy—and the ideas are intuitively appealing. After all, who would not want to be such a person with these admirable behavioral attributes? The challenge has been to empirically verify the claims of the theory. To do this, there had to be one or more reliable measures of EI and some validity studies to actually show that those who have more of it exhibit superior performances to those they lead. A review of the literature suggests some progress on these fronts and some cautions, as follows (Zeidner, Matthews, & Roberts, 2004):

> Overall this review demonstrates that recent research has made some important strides toward understanding the usefulness of EI in the workplace. However, the ratio of hard evidence to hyperbole is high, with over-reliance in the literature on expert opinion, anecdote, case studies, and unpublished proprietary surveys. (p. 371)

The work with EI is another arena in which the "Man from Missouri" merges into the more hard-nosed personnel selection arena. We want criterion-related validity studies, and there is a paucity of these with regard to EI. This does not invalidate the EI construct and the model (there are numerous such models and measures; Law, Wong, & Song, 2004; Zeidner et al., 2004); it merely suggests caution in adopting measures based on EI unless empirical validity studies that have been appropriately done are presented as supporting evidence. Law et al. (2004), for example, show that a measure of EI is differentiable from a measure of the FFM with only Neuroticism and Conscientiousness being modestly related (about 0.50) to the EI index. In their analyses, they also show that the EI measure contributed over and above the FFM to a measure of satisfaction. This suggests that EI may be tapping into something not immediately measured by the FFM.

Summary

In the recent past, researchers' views of personality testing have gone from at best that it is of dubious usefulness to being of importance. Acceptance of the FFM as a reasonable taxonomy of personality constructs has resulted in a series of meta-analyses revealing validity against a variety of work-relevant criteria. That is, when older studies are organized around the FFM, sufficient power to detect significant effects is achieved—and such effects are revealed.

Another reason for the acceptance of personality testing has been the fact that, regardless of whether personality inventories or projectives are used, the racial and gender subgroup differences tend to be much smaller than those found for cognitive ability tests. Hough, Oswald, and Ployhart (2001) reported these differences to be about one third of a standard deviation or less (see also chapter 7). In addition, as

Schmitt and Chan (1998) noted, the correlation between personality measures and measures of ability are usually quite low, so combining evidence from both sorts of measures improves validity over what either form of measurement itself might produce. We do have personalities again, and they are important.

Improving the Validity of Personality Testing

Although personality tests can be useful predictors of performance, there is much room for improvement. It is interesting to note that early meta-analyses examining the validity of personality tests, such as Schmitt et al. (1984), reached skeptical conclusions about the utility of personality in employment testing. Several years later, the meta-analysis by Barrick and Mount (1991) found similar observed validities but reached a different conclusion. Other meta-analyses have been conducted and found much more support for the relationship between personality and criteria, such as leadership (Judge, Bono, et al., 2002) and job satisfaction (Judge, Heller, et al., 2002), than was true in previous research. What changed? One was the application of the FFM model, as we noted earlier. However, it has really been how meta-analysis is applied that best explains the differences in the interpretation of these studies.

In particular, modern approaches to meta-analysis use various corrections for statistical artifacts (discussed in chapter 6). One of these artifacts is unreliability in the criterion and/or predictor. Others involve range restriction and sampling error. Such corrections can be informative, but it is important to recognize that the uncorrected validities for personality traits have not changed much across meta-analyses; it has been the *corrected* estimates that have changed. We discuss the appropriateness of such corrections in chapters 4 and 6, but it is worth emphasizing that the observed validities of personality tests have not gotten larger over time, only our corrected correlations have. As Schmitt (2004) noted:

> Barrick and Mount (1991) had available nearly the same set of validity studies when they did their meta-analysis as did Guion and Gottier and both concluded that observed validities did not often exceed .10. Using corrections for unreliability in both predictor and criterion and range restriction, Barrick and Mount estimated that the population validity of conscientiousness was in the mid teens (or mid 20s for subjective criteria). The ensuing rush to use

personality in selection that has occurred over the last decade or so since the Barrick and Mount review has been as unthinking, uncritical, and perhaps misguided as was the earlier response to Guion and Gottier (1965). The observed validity of personality measures, then and now, is quite low even though they can account for incrementally useful levels of variance in work-related criteria beyond that afforded by cognitive ability measures because personality and cognitive ability measures are usually minimally correlated. (p. 348)

We do not mean this to cast a negative view on personality tests in selection. This issue is equally relevant to research on other predictors such as interviews and biographical data. Rather, we emphasize this point because we want readers to be cognizant that the validities observed in practice have much room for improvement.

There have been two major thrusts to enhance the validity of personality tests. One is to use what is called an "at work" frame of reference. The typical personality test items are written without reference to a particular situation or context. For example, a typical item might be something like, "I am a tidy person." This may be true of you at work, but not necessarily at home or on the weekends. Therefore, if we ask questions about your personality at work by changing the frame of reference on this item to something like, "I am a tidy person at work," the question is whether we would find enhanced predictive validity. Schmit, Ryan, Stierwalt, and Powell (1995) examined this question and altered items on a generic personality test to reflect a work context. They found a dramatic increase in validity by simply using the "at work" frame of reference in the items. Hunthausen, Truxillo, Bauer, and Hammer (2003) extended these findings and showed that when using generic personality items the validities were similar to those found in meta-analyses, but when using items with a work frame of reference the validities were two to three times stronger. Part of the reason for the higher validity comes from the higher reliability of "at work" measures.

A second and useful means for improving the validity of personality tests is to use what Hough and Schneider (1996) called "compound" personality measures. They reviewed considerable evidence suggesting the validity of personality tests can be significantly higher when the test is composed of a composite of a number of more basic traits. For example, service orientation measures correlate strongly with customer service performance, with an observed validity of 0.24 and a corrected validity of 0.50 (Frei & McDaniel, 1998). Service orientation is not a "pure" personality trait like one of the FFM constructs, but a multidimensional set of constructs that relate to service performance. We talked about these compound predictors in chapter 7, and they are an effective means for enhancing the validity of personality-type tests in practice.

Fakability: A Persistent Issue in Personality Testing

There is one large issue in personality testing on which just about everyone comments: fakability. Our reading of this literature indicates that people certainly can fake personality tests, but the same literature is less clear on whether people do fake, and if so, to what extent they fake, why they fake, and what the effects of faking are. Smith and Robie (2004) put the issue this way: "Critics suggest that self-report measures of personality are infused with distortions based on a person's perception of the desirability of various response options. In short, rather than being veridical reports, some (currently unknown) percentage of respondents may actively attempt to manage impressions and misrepresent their true personalities" (p. 5).

Smith and Robie's (2004) review of the literature leads to the following conclusions:

1. When instructed to do so, people can fake personality inventories.
2. When people are taking personality tests for selection or promotion purposes, there is a consistent elevation in their scores compared with people who take such tests for developmental purposes.
3. The validity of personality tests is somewhat higher for concurrent validity studies (where people are already on the job) than for predictive validity studies, but the differences are small, approximately 0.07 (Hough, 1998).
4. Regardless of whether the respondent is an applicant, the factor structure of FFM inventories appears quite robust.

Landy and Conte (2004) reached the following conclusion: "The issue of faking is not 'settled' yet, but there does seem to be some agreement that it is not a fatal flaw in personality testing" (p. 104). We concur, but that still leaves the question of how to prevent the faking that occurs to make the flaw a bit better than "not fatal."

Warnings that there will be consequences if the respondent is found to be faking appear to have modest effects (Hough, 1998). Fortunately, they do not lower applicants' reactions to taking such tests (McFarland, 2003). However, Smith and Robie (2004) questioned whether warnings will continue to have effects once respondents realize there is no certainty in detecting faking.

Another attempt to detect faking is to imbed "social desirability scales" or "lie scales" within the personality measure. Social desirability scales present items that are so unrealistically positive that it is unlikely for the majority of the test takers to respond favorably. One administers numerous such items because the odds of a person endorsing more than a few favorably is practically impossible. For example, one might ask items such as, "I never get angry," or "If I get cut off in traffic, it doesn't bother me." Although a person may favorably endorse one or a few of these items, if they answer each extremely favorably they will be "flagged" as a person potentially faking. Although the premise underlying these social desirability scales is reasonable, the evidence is mixed as to whether they detect faking, and some evidence suggests they can themselves be faked (Stark, Chernyshenko, Chan, Lee, & Drasgow, 2001).

Over the years, numerous attempts have been made to alter the form of items in personality inventories to make them less likely to be faked. For example, forced-choice response formats have been tried. In this approach, respondents are confronted by items that have equal social desirability but differ in the degree to which they have been shown to relate to a criterion of interest. When confronted by items that appear to be equally desirable, the logic goes, it will be difficult to fake. Unfortunately, research with these kinds of items shows applicants do not like them (e.g., applicants do not like to respond to an item that asks questions such as "Choose the one that is most like you: (a) I am a liar or (b) I am a thief"—and these tests are still fakable when instructed to do so.

A second small literature looks at items that are more subtle in what they ask compared with items that are more transparent in what they appear to be tapping. For example, an item that asks how conscientious you are is more transparent than an item that asks how important you think it is to pay attention to details rather than the big picture when doing a task. Research suggests that subtle items are still faked although it is more difficult to fake them (Kluger & Colella, 1993), but the differences in the validity between subtle and transparent items are not known.

Perhaps the most subtle items are those described earlier when we discussed the conditional reasoning approach to test development (James & Mazerolle, 2002). Recall that these items offer opportunities to test takers to reveal how they explain events, and the alternatives they are presented with are quite subtle indeed. Research on the fakability of these items reveals that they are fakable when people are instructed to do so, but there appear to be no mean differences in one study that compared applicants, incumbents, and college students (Lebreton, Barksdale, Burgess, & James, 2004).

Summary

People can fake personality tests of all kinds when instructed to do so, but who fakes, how much faking they do, and when they do it are questions without firm answers. It appears that some people do attempt to present themselves in a positive way when confronted with personality measures they know will be used as a basis for job decisions (applicants, people being considered for promotion), and there are individual differences in how well people can fake (McFarland & Ryan, 2000). It also appears that such faking does not have much effect, if any, on the validity of personality tests for use in such situations and that such situations do not significantly alter the factor structure underlying the conceptual interpretations of, for example, one measure of the FFM (Smith & Robie, 2004). A word of caution, however: Perhaps the reason we do not find much effect for faking is because the validity of personality tests are so modest. For example, if the uncorrected validity of our best personality predictor, Conscientiousness, is only 0.13, and on average faking reduces validity by about 0.07, then we have lost half of the validity of our predictor. What we do not need are more studies comparing applicants to incumbents; what we do need are theories of faking or intentional distortion that are tested across applicant and incumbent contexts. The predictor response model shown in Fig 7.1 may be useful in this regard.

Conceptual Issues in Understanding Personality at Work

To this point in the chapter, we have presented a comprehensive review of the current status of personality correlates of behavior at work. We now know that personality relates to many forms of behavior and affect at work, that five facets of personality seem to account fairly well for the great variety of facets that have been proposed in earlier times, and that personality inventories and at least two projective tests have demonstrable validity at least for the prediction of leadership and managerial effectiveness.

What we have not done is explore two central issues related to an understanding of (a) how personality interacts with the situation in which people behave to produce the behavior and/or affect of interest, and (b) the notion of fit (i.e., that it is not so much what the personality is but whether the personality fits a situation that is critical for prediction and understanding). Obviously, these two issues are related but we will consider them one at a time for the moment.

Person–Situation Interaction

For many years, psychologists who studied personality believed it represented a stable cross-situationally displayed predisposition to respond. We see this cross-situational influence in thinking in Staw's (2004) work on job satisfaction, which showed that, even when people changed jobs and even over many years, well-being at an early point in time predicts later job satisfaction. Although this is a true statement, it is also true that for people who changed jobs the relationship between well-being and job satisfaction was weaker than for those who did not change jobs. We also presented a number of cases of what we called boundary conditions, wherein the prediction of behavior (e.g., leadership) depended on whether the organization in which the prediction was being made was more hierarchical and large versus more entrepreneurial and small (Spangler et al., 2004).

These kinds of boundary conditions or interactions achieved conceptual salience after Mischel (1968) published a book that was interpreted as saying that personality did not predict the behavior of people across situations because it is situations, not individual personalities, that predict behavior. So, he took what has come to be called a situationist perspective on behavior—and you would think he had just shot his mother. The reaction to this extreme position was not long in coming, and the result has been that most people today ascribe to some form of what is called the *interactionist perspective* (cf. Magnusson & Endler, 1977).

The interactionist perspective in general psychology and in personality theory has been stronger than it has been in the field of personnel selection research and practice. That is, as we have noted several times, in personnel selection we seek main effects—direct relationships with no encumbrances—between the personality measure and the criterion or criteria of interest. Boundary conditions just make things difficult. Like people from Missouri, personnel selection researchers say "show me," meaning show me the validity coefficient and drop the complexities and conditions.

However, we all know we behave differently as a function of the situation, and people who know us well still know the same person and can predict our behavior in those different situations. They can do this because people have characteristic ways of behaving in different situations; we do not so much behave the same way across situations as we have consistent but different ways to behave in different situations. In the bar on Friday night, we are "different" from who we are when teaching Sunday school on Sunday morning—but we tend to behave similarly on Friday nights and tend to behave similarly on Sunday morning, and differently from others, too. Personality theorists call this *coherence*. Magnusson and Endler (1977) define coherence this way: "*Coherence* means that the individual's pattern of stable and changing behavior across situations of different kinds is characteristic for him or her and may be interpreted in a meaningful way" (p. 7). This means that we accept the fact that people are not consistent in their behavior across situations, but that situations interact with personal attributes (personality) in predictable ways, indicating that taking into account both personality and situation can yield accurate prediction and understanding. That is partly why we find stronger validities for personality tests that use an "at work" frame of reference.

What is interesting in this person × situation research is that it is mostly treated as the situation moderating or being a boundary condition for the personality–criterion relationship. However, the algebraic interaction used to reveal this also says that the person attributes moderate the relationship between the situation and behavior. In other words, the situation can be viewed as the boundary condition or personality can be the boundary condition, and both are likely true. To return to the Spangler et al. (2004) work on leadership, for example, it is not only true that the nature of the organization moderates the personality–leadership relationship, but also that the relationship between organizational attributes and leadership effectiveness is moderated by the leader's personality. That we usually put the emphasis on the situation as the moderator is unfortunate because it makes it seem like personality depends on the situation for its relationship to behavior when it is just as true that the situation depends on personality for its relationship to behavior.

Stewart and Barrick (2004) reviewed the literature on the person–situation debate with regard to personality and reported four lessons had been learned from this research:

1. *Specific traits predict behavior only in relevant situations.* This is the issue we raised earlier with regard to behavior in a bar on Friday night compared with teaching Sunday school. We would not expect the same traits to be relevant across those situations. In summary, it is the traits that matter in a specific situation that are the ones that should be measured. Fortunately, we know that across most if not all work-for-pay situations, Extraversion and Neuroticism (Emotional Stability) matter. Unfortunately, we do not have a good way to classify situations so this makes job analysis more important than usual, especially with regard to the focus on personality attributes (see Table 9.2).

2. *All traits are more easily expressed in some situations than others.* This conclusion is based on Mischel's (1973) notion of the strength of a situation; some situations are strong and demand similar behavior from all, whereas other situations are weak and allow for the expression of individual differences. Mischel (1968) derived this idea after attacks on his conclusions that traits did not predict behavior, when it became clear that many of his conclusions were based on data collected in laboratory experiments—a strong situation. In the workplace, Barrick and Mount (1993) showed that traits relate to performance as a function of the autonomy offered workers in performing their jobs. Based on this and similar work, Stewart and Barrick (2004) made the appropriate prediction that, as jobs become increasingly less structured and decision making is pushed down the hierarchy in organizations, personality traits will be increasingly important as predictors of behavior at all levels of the organization.

3. *A person's traits can actually change a situation.* This appears to be easily understood when focusing on executive levels of management in a firm, but it is also true at lower levels of an organization. As an extreme example, consider the workers at Three Mile Island who failed to monitor the meters to which they were assigned or the rogue banker at Barings who brought down the venerable firm by making inappropriate investments (Goodman, 2000). When an organization does want change, it must carefully identify the kinds of changes it wants and then specify the kinds of traits the people hired must have to implement

change. Peterson, Smith, Martorana, and Owens (2003), for example, showed how CEO personality type affected top management team dynamics and how these dynamics in turn predicted top management team performance.

4. *People choose different settings to match their traits.* As we have noted several times, Schneider (1987) made this idea a key to his theory about how settings over time become relatively homogeneous with regard to the personalities of the people there. He is not the only one who has made such a prediction; Holland (1966) made it decades earlier with regard to the similarity of the interests of people in a given work setting. We review Holland's (1997) formulation on this issue in the section on interests and interest testing. Here, there is mounting evidence to support Stewart and Barrick's (2004) conclusion as summarized earlier in the section on personality and organizational culture. The important point is to focus on the word "match" in the title to this section because we turn to the issue of people matching or fitting in just a bit.

Prior to turning to that issue, it is useful to summarize that people do not behave consistently across situations, but they do behave in characteristic ways from setting to setting. Personality traits do not randomly correlate with behavior, but they do correlate with behavior when the appropriate traits(s) are assessed for a specific situation. Some situations are stronger than others, but the prediction is that work settings will expect and support more autonomy in the future so personality will be increasingly important as a predictor of behavior. In other words, taking into account both person and situation when thinking about the prediction of behavior via personality tests is necessary.

Person–Environment Fit

Fundamental to the study of personality has always been the idea that it is good fit that yields positive outcomes for people (Walsh, Craik, & Price, 2000). So, it is not just what a person's personality is that matters but how well the personality fits the setting (job, group, organization) that is also important—especially for understanding satisfaction and adjustment. Our reading of the research literature on fit suggests that

1. People choose situations they believe they fit. Other things being equal, people are attracted to and eventually choose settings where there is a better match or fit for them. For example, Turban and Keon (1993) found they could predict which of several organizations people

would be attracted to by knowing the pay plan (incentive versus salary) of the organization and knowing the achievement motivation of applicants. People with higher needs for achievement chose the organization with the incentive pay plan.

2. Fit is a useful predictor of satisfaction and adjustment, so it is an important issue when thinking about these outcomes (Judge & Kristof-Brown, 2004). This is especially true because we know that satisfaction is related to a host of other important outcomes such as absenteeism, turnover, and citizenship behavior (see chapter 5).

As can be seen, the notion of fit is consistent with a person–situation interactionist perspective (Judge & Kristof-Brown, 2004), but we wanted to make the important point that the issue of fit is salient when thinking especially about satisfaction.

The topic of fit is an appropriate one with which to close this discussion of personality and personality testing and move on to a discussion of interests and interest testing because fit has played such a central role in this area of research and practice.

INTERESTS AND INTEREST TESTING

It might be presumed that there is a strong relationship between people's vocational interests and their general personalities, but surprisingly there is not (Hogan & Blake, 1996). Indeed, Holland (1997) used the term "vocational personalities" in the title of his most recent book. If we reflect on what we learned in the prior section from the person–situation debate, this turns out to be completely understandable because we learned that general personality traits cannot be expected to correlate with all manner of criteria and that we may require specific personality traits if we are interested in making specific predictions. Vocational interests are a specific kind of outcome because the issue is one of choices and not of behavioral levels. So, the prediction issue in interest measurement is not how well someone will do but which among several alternative choices they will choose. In a real sense, vocational interest measurement is about binary choices, whereas the issue in personality testing is about predicting continua of behavior or performance.

However, we are getting ahead of ourselves. In this section of the chapter, we first present the theory of vocational interests and choices that has come to dominate that world and then discuss the measurement of those interests.

Holland's Theory of Vocational Choices

For more than four decades, John Holland has devoted his scholarly career to the conceptualization and measurement of vocational choice. On the surface, his theory is quite clear (as is his writing about it):

1. People have different kinds of vocational interests that are meaningfully summarized into six kinds of vocations: Realistic, Investigative, Artistic, Social, Enterprising, and Conventional. The vocations are arranged in a hexagon such that each of them occupies a node in the order just listed, and those closer to each other are more similar to each other than those farther away. Figure 9.1 shows this hexagonal feature of the theory, and there it is clear that Social and Artistic vocations are more like each other than are Social and Investigative.
2. People choose vocations based on their perception that they fit the career environments that characterize those vocations. Career environments, in turn, are characterized by the kinds of people in them. Thus, situations for Holland are a product of the vocational personalities in them.

Table 9.5 shows an adjectival summary of the behavioral attributes of Holland's six interest types that both define the kinds of interests of people in each category and define the career environments in which those types exist. We do not need to repeat the materials in Table 9.5,

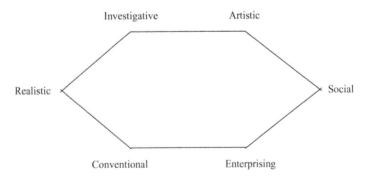

FIG. 9.1. Holland's RIASEC hexagon of vocational interests and career environments.

TABLE 9.5
Sample Adjectives Describing People and Environments for Holland's RIASEC Model

Realistic: conforming, genuine, inflexible, natural, persistent, realistic, robust, uninsightful

Investigative: analytic, complex, curious, intellectual, pessimistic, radical, reserved, unassuming

Artistic: complicated, emotional, idealistic, impractical, independent, intuitive, open, sensitive

Social: agreeable, empathic, generous, idealistic, patient, responsible, tactful, warm

Enterprising: acquisitive, ambitious, domineering, enthusiastic, exhibitionistic, forceful, resourceful, sociable

Conventional: careful, conscientious, efficient, inhibited, obedient, persistent, thorough, unimaginative

Source: Adapted from Holland (1997, pp. 44–48).

but it is important to note that these descriptors are of the same sort in many cases as those used to describe personality. One would suspect on the surface of it that it is logical to assume a factor analysis of these adjectives would also yield an FFM, but this is at best modestly true. Hogan and Blake (1996) showed the correlation of seven major personality inventories with a measure of Holland's scheme organized around the FFM and, although there are significant relationships shown and there is an observable pattern to those relationships, the correlations are modest (most in the 0.20–0.40 range even when statistically significant). The pattern that does emerge suggests the following:

- Realistic interests are mostly associated negatively with Agreeableness.
- Investigative and Artistic interests are mostly associated with Openness to Experience.
- Social and Enterprising interests are mostly associated with Extraversion.
- Conventional interests are mostly associated with Conscientiousness.

It is useful to note that none of Holland's categories relate to measures of Neuroticism (Emotional Stability), but this does not mean Neuroticism is irrelevant with regard to vocational interests. Tokar, Fischer, and Subich (1998) accomplished an extensive review of the FFM as it relates to various outcomes and showed that people high on

Neuroticism also engaged less frequently in job search activities and in general had less well-formed career decisions. This suggests that for people higher in Neuroticism, the prediction of vocational choice based on interest measurement should be less valid, but to our knowledge there is no research on this question. In this same vein, DeFruyt and Mervielde (1999) showed that although Holland's typology predicted career choice, the FFM was superior at predicting who was likely to be working at all with Neuroticism and Conscientiousness being a reliable predictor of this outcome.

Our conclusion with regard at least to Holland's (1997) theory is that if one is interested in vocational choices, one should likely use measures of vocational interests and not measures of the FFM or other personality tests. In contrast, if one is interested in predicting who will search for a job and who is likely to be employed, then Neuroticism and Conscientiousness from the FFM will be relevant.

We have written as if Holland's perspective on vocations is the only one that exists, and the fact is that it dominates thinking and research on vocational choice (see *Journal of Vocational Behavior*, 1999, for an entire issue devoted to the theory). Earlier theories of vocational choice (see Hall, 2002, for a review) have not been found to be a sound base for the development of predictors of choice, which is where the literature has put its major emphasis. We turn to that literature next.

Interest Measurement

Interest inventories are used to diagnose work and career preferences. In fact, the impetus for the construction of interest inventories came from vocational guidance, and this is still the locus of much of the research with such procedures (Landy & Conte, 2004). It is true that interest measurement has not played a prominent role in personnel selection, perhaps because the criterion for interest measurement has been individual job satisfaction. Thus, interest measurement has always implicitly or explicitly rested on the assumption that if people enter the career that measures of their interests show they should, then they will be satisfied for the long term and adjusted at work. There is good evidence to show that this is true, and the evidence has been gathering for more than a half-century (Hogan & Blake, 1996; Holland, 1997). Now that we understand the importance of job satisfaction and adjustment at work, and understand the role of personality in those, it may be useful for selection researchers to once again pay significant attention to interest measurement. That is, if interest measurement is only weakly or at best modestly related to personality measurement, perhaps interest measurement can add predictive capacity when attempting to understand job satisfaction.

Perhaps an equally interesting question is the degree to which personality measurements alone predict occupational interests and choice. We already have evidence that personality predicts organizational choice (Schneider et al., 1995, 1998). We also know that the logic Schneider (1987) used in developing his attraction-selection-attrition model was based on Holland's (1997) theory that people and environments tend to fit or match and that people choose to work in places and at occupations they believe they fit. In fact, Holland showed that over time as people switch jobs, the jobs they switch to become more and more like the jobs they should have taken in the first place. In any case, if personality predicts organizational choice, maybe it also predicts career choice. There is some evidence to support such an idea.

For example, Warr and Pearce (2004) simultaneously studied the predictive capabilities of a personality measure to predict both career and organizational choice. They studied 647 people in the United Kingdom, each of whom completed a personality questionnaire and supplied information about the preferences for the attributes of the careers in which they worked and the nature of the organizational culture they preferred. Warr and Pearce made explicit hypotheses about which of the personality attributes would likely relate to which features of work and organizational culture believing, as they did, that not all personality attributes predict career or organizational culture preferences. They report some modest correlations (in the high teens and low 0.20s) for personality as a correlate of career choices and organizational culture, with the correlations for career choices being a bit stronger.

One of the really interesting findings with regard to vocational interests is their durability over time. Many people begin taking such measures in their early teens, and the startling fact is that their scores remain relatively constant throughout life (Lubinski, 2000). Figure 9.2 shows some data from the Strong Vocational Interest Blank for Men (Vinitsky, 1973) in which people were tested first in 1927, then in 1948, and finally in 1968—40 years later. The reader can see the startling similarity in the results for the 37 psychologists who were tested. This is merely an example of the durability of vocational interests over time as Lubinski (2000) has shown. If interests and personality are durable over time, then using both forms of measurement to understand satisfaction at work would

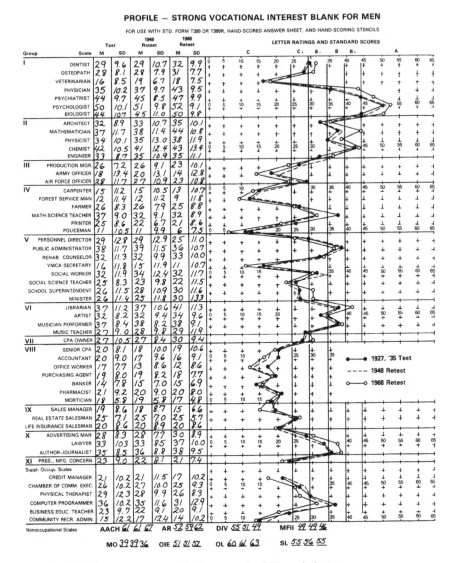

FIG. 9.2. Mean test-retest profiles for a sample of 37 psychologists. Source: From M. Vinitsky, "A forty-year follow-up on the vocational interests of psychologists and their relationship to career development," *American Psychologist*, 28 (1973), p. 1001.

potentially seem to be quite useful. However, as noted earlier, selection researchers have not been inclined to use such measurement.

In contrast to other forms of selection procedures, vocational interest measures are designed to differentiate in a binary sort of way (i.e., they are designed to separate those in one vocational area from those in all other vocational areas). So, Holland's Vocational Preference Inventory (VPI; Gottfredson, Holland, & Holland, 1978) is designed to separate the RIASEC types from each other, and Campbell's Interest and Skills Survey (CISS; Campbell, Hyne, & Nilsen, 1992) is designed to separate seven basic orientations from each other. Think of the problem this way: What set of questions about work activities could I ask 1,000 people in many different occupations so that I could, on the basis of their answers, identify who works in which? Note that the question is not who will be more satisfied or who will be a better performer, but how can I differentiate those in one occupation from those in other occupations. This measurement problem is a special case of the criterion-keying problem: You tell me the criterion (differentiation of people based on the occupations they are in), and I will key the questions so they maximize obtaining the criterion. It is a special case because the criterion is a categorical criterion (one area differentiated from another area) rather than a continuous criterion (referring to some scale of more or better). Interest inventories do this well, but the issue confronting selection researchers is that these measures are not predictive by themselves of even job satisfaction.

In addition, data from the VPI in particular has been extended to a classification scheme of the 12,860 occupations in the DOT (which preceded O*NET) resulting in the *Dictionary of Occupational Codes* (Gottfredson & Holland, 1996). This dictionary describes the career environments of all jobs that were listed in the DOT (i.e., it describes the kinds of RIASEC interests likely to be found in the people in those jobs).

Why Are Interest Inventories Not Used as Often as the Data Suggest They Should Be?

With all this good and accepted theory and good validity for the measurement of vocational interests, why have these not been used more in selection research? One explanation offered earlier was that the criterion for the development of such measures is job satisfaction and adjustment—and that this alone suggests the potential importance of such measures. However, with interest measures, one should not simply collect the data and run a correlation against job satisfaction or any other criterion because interest measures relate to job satisfaction only with regard to the fit or match between persons' interests and the career environment and/or occupation they enter. So, whether a person is high or low on any or all the RIASEC career areas is irrelevant; what is relevant is whether the person fits or matches the RIASEC areas corresponding to the career or occupation, entered. For example, we noted that Gottfredson and Holland (1996) developed RIASEC codes for the 12,680 occupations in the DOT. When people enter any of those occupations, their satisfaction is not their own RIASEC data but the degree to which their own RIASEC data match or fit the RIASEC data for that occupation. Fit is the predictor, not the RIASEC data. The bottom line is there is no g in interests, and raw scores on RIASEC dimensions are not relevant; fit is the issue.

We surmise that the use of interest measures would be more prominent if selection researchers and companies paid attention to the existing RIASEC codes for their own jobs and then used the VPI to test the fit of applicants for those jobs. Our one hesitancy here concerns the issue of fakability. Interest inventories ask people to express their interests in being in a particular occupation and/or doing things people do on occupation-specific jobs, making the items quite transparent. There has apparently been little recent research on this issue (Hogan & Blake, 1996), so it is needed prior to adopting our suggestion for additional use in personnel selection.

BIOGRAPHICAL DATA

Selection researchers deeply believe the best predictor of future behavior is past behavior. If you want to know how persons will perform on a specific job, find out how they performed that job somewhere else or, given that that is not possible, put them in a simulation to see them behave or, if that is also not possible, ask a series of questions about past behavior that might tell you something about their future behavior. One alternative is to just interview the person (see chapter 10); another alternative is to have some formal procedure(s) for gathering such information. Two such procedures ask about past experiences: the application blank and the Biographical Information Blank (BIB).

It is often difficult to tell an application blank from a BIB and a BIB from a personality inventory. However, as we have treated the latter already in this chapter, we also discuss the application blank and the BIB separately here.

Application Blanks

The usual first step in the employment process is to fill out an application blank. Application blanks typically request information about previous jobs held, educational level and type, and any special skills. The

BIOGRAPHICAL DATA 477

information about jobs held will include job title, a description of duties and responsibilities, period of employment, salary, name/telephone number of the organization and immediate supervisor, and reason for leaving or thinking of leaving. With respect to education, similar kinds of information are collected—schools attended and dates (including schools in the military and other training courses), major (for college) or program (e.g., college preparatory, commercial, or vocational if only a high school graduate) and specific training (e.g., engineering, accounting, plumbing, or auto repair), details about such courses.

Some application blanks also include requests for information about job-related hobbies and interests, as well as accomplishments the applicant thinks are relevant to the job being applied for. Examples of this kind of information include athletic team and/or club participation and offices held, volunteer work (sometimes especially important to have for homemakers returning to paid employment), awards and prizes, and so forth. Responses to these kinds of additional information on the application blank should be voluntary because, as in testing, job relevance is an important issue on the application blank. For example, a statement such as the following might be used:

> After listing your previous positions you may, if you wish, include in this space any pertinent civic, welfare, or organizational activity which you have performed, either with or without compensation.

Table 9.6 reproduces some information from the Michigan Department of Civil Rights regarding lawful and unlawful questions applicants can be asked. The legal issues noted in Table 9.6 apply to interview questions and the application blank. Making such issues illegal creates somewhat of a Catch-22 situation because organizations need to know whether they are meeting legal requirements regarding discrimination. If they do not collect such information on the application blank, how can they track their hiring practices? The issue here is one of when the information is collected and the purpose for collecting it. Data on age, race, and gender can be collected after an employment decision has been made (hire or not hire), and this information can be maintained in a separate file from actual application materials. Thus, the purpose for maintaining age, race, and gender data is to track decisions in the aggregate, that is, across all hiring decisions, so the information need not be maintained in a particular candidate's file.

On the surface, it is sensible to have an application blank because the information collected can be useful in screening candidates. However, we have found that application blanks are used quite subjectively and

TABLE 9.6
Examples of Lawful and Unlawful Preemployment Inquiries as Specified by the Michigan Department of Civil Rights

Subject	Lawful Inquiries	Unlawful Inquiries
Name	*Applicant's full name *Have you ever worked for this company under a different name? *Is any additional information relative to a different name necessary to check your work record?	*Original name of applicant whose name has been changed by court order or otherwise *Applicant's maiden name
Age	*Are you 18 years or older? *What is your date of birth	*How old are you?
Marital status		*No questions are legal about marital status, number of children, nature of spouse employment, spouse's name, and so forth
Gender		*No questions are legal
Memberships	*Questions about membership in organizations are legal as long as they in no way indicate anything about race, religion, color, national origin, or ancestry or its members.	
Arrests	*Have you ever been convicted of a crime? If so, when, where, and what was the nature of the offense? Are there any felony charges pending against you?	*No questions about being arrested are lawful

haphazardly by many organizations. By this we mean that the people who review application blanks as a way of screening for candidates are rarely provided with specific guidelines about what they should be looking for. Is length of time on previous jobs important? Is why they left jobs important? Are the job duties previously performed critical in the new job? How about majors or school programs?

Research shows, in fact, that people who screen application blanks think they know what they are looking for because they think they know what is important. What reviewers think is important will be the issues they assign the most weight to. The weight given to the various pieces of information are implicit weights if no guidelines are provided. Explicit weights are contained in a set of guidelines or rules for evaluating application blanks.

However, it is important to recognize that use of implicit weights can get you into trouble. Whenever background information, whether collected by resumes, application blanks, or biographical data, is used as a basis for making selection decisions, it falls under the various employment laws discussed in chapter 11. This means that we should only be weighting or focusing on those questions that are related to important KSAOs necessary for the job. For example, if we ask a question about highest level of education attained, it is vital that we can show the job relatedness of this question (either through content- or criterion-related validity). Here is another increasingly common example. Suppose you use an Internet application process and receive several thousand applications per month. You might decide to first consider all applicants who have at least a high school degree—this still results in several thousand applicants. So, with a simple click of the mouse you ask for all applicants with at least a college degree. This seemingly innocent act carries with it potentially serious professional and legal ramifications. In this instance, you would need to justify this decision, as well as the job relatedness of even asking about educational attainment.

Fortunately, selection researchers have devised procedures for weighting application blank information. When these procedures are used, we have a *weighted application blank* (WAB). These systems allow staffing people to literally score an application blank. Indeed, researchers have developed even more sophisticated ways of collecting and weighting background information on applicants. These procedures are called BIBs. Based on the premise that the best predictor of future performance is past performance, these WABs or BIBs have proven to be one of the most valid alternatives to standardized tests (see Stokes, Mumford, & Owens, 1994, for a review).

Biographical Information Blanks

A number of techniques are available for statistically weighting biographical items. Perhaps the most popular is the technique described by England (1971), a technique that mirrors the issues we raised under the title of criterion keying when discussing the development of interest measures. An illustration of the technique using sample data from a study by Schmitt and Pulakos (1983) is presented in Table 9.7.

TABLE 9.7
Example of Scoring Key Development for a Biographical Information Blank

Response Options	No. of Endorsing Options in Low Accident Group	No. of Endorsing Options in High Accident Group	Scoring Key
1	13	4	1
2	11	5	1
3	14	7	1
4	12	21	0
5	2	20	0
Total	52	57	

Note. For response options 1, 2, and 3, persons in the high accident group were less likely to endorse the option than those in the low accident group. The reverse was true for response options 4 and 5. Hence, those persons responding 1, 2, or 3 are scored 1, and those persons responding 4 or 5 are scored 0.

The development of scoring keys was begun by separating the BIB responses of 192 transport drivers (persons who deliver gas to local gas stations) into high and low accident groups on four different measures of accident behavior. The data in Table 9.7 represent transport drivers who had a high or low record of having made the error of blending two or more types of gas. Table 9.7 shows that persons in the low accident group are most likely to indicate option 1, 2, or 3 in answering this particular item. Hence, the scoring key for this item indicates that persons picking options 4 or 5 in response to this item get 0 points, those picking 1, 2, or 3 get 1 point. All items in the inventory are similarly keyed, and the persons' scores on all items are summed to produce a total score, which is used to predict a criterion. In the case of the example described in Table 9.7, the criterion is accidents involving gasoline blends. Obviously, this procedure will score items optimally and result in keys that produce significant relationships between total BIB scores and criteria in the sample on which the keys are developed. Hence, these keys must be evaluated, or cross-validated, in another sample.

Cascio (1982) summarized the necessary steps in choosing biodata items and developing their weights:

1. Choose an appropriate criterion (e.g., tenure, proficiency).
2. Identify criterion groups (low-versus high-performing individuals).
3. Select the items to be analyzed.
4. Specify the item response categories to be used in the analysis.

5. Determine item weights and/or scoring weights for each response option.
6. Apply the weights to a holdout group (cross-validation sample) and correlate total scores with criteria.
7. Set cutting scores for selection.

It is useful to focus on step 3, the selection of the items to be used in the BIB, because this issue has generated some controversy. The issue is that anyone could develop a series of questions about past activities and experiences people may have had and use those as items in a BIB—the statistics then tell you which have criterion-related validity and which do not, and you are done. There are a number of problems with this approach that rests completely on statistical determination of what "works" and what does not "work." First, if the BIB items do not have job relevance, it may be difficult to defend them. Thus, we have emphasized the use of job analysis findings as a basis for the design of other selection measures, and such data should also be used to frame the kinds of questions one would ask in a BIB. Second, and related to the first, the best tests are those that have conceptual clarity with regard to the criterion of interest and this applies to BIBs as well.

It is also useful to focus on step 5 concerning the development of weights for items. As Hogan (1994) noted, there are a host of critical decisions that must be made for the weights that are developed to reveal criterion-related validity, especially the choice of the criterion. To his way of thinking the key issue in item development, and thus eventual criterion-related validity, is to "better understand the psychological, experiential, and behavioral antecedents of the performance to be predicted" (p. 73). Again, we see the theme of not just proceeding blindly and having the statistics carry the day but doing the heavy lifting as background to writing relevant items to later submit to criterion keying. As in any test development, the criterion against which items are keyed, on the one hand, determines the items to be used, and on the other hand, the criterion itself must be as uncontaminated as possible, be relevant, and must be able to be reliably measured (see chapter 4). When these issues about item choice are sufficiently well addressed, BIBs demonstrate criterion-related validity.

One would think that there are good sources of information about how to write biodata items, but because biodata items can take many forms and refer to such different kinds of content, the rules are difficult to generalize. There are, however, sources to which one can turn for guidelines for the development of specific kinds of biodata items. An excellent beginning in this regard is to seek out Mumford and Owens (1987), who described six different sources of potential biodata items: (a) the human development literature, (b) life history interviews of incumbents, (c) typical factor loading of biodata items that exist in the literature, (d) known life history correlates of important criteria, (e) biodata items of known validity, and (f) items based on psychological theory.

Mael (1991) and Russell (1994) offer in-depth treatments of the kinds of items one might write to get at the different content that might comprise biodata items. For example, Russell showed how items may differ when one is testing a personality theory compared with vocational theory or human development theory. In other words, the variety of possible items is so large that specific guidelines are needed for writing the different kinds of items.

Criterion-Related Validity of Biographical Data

A number of rather comprehensive validity studies have been conducted over the years on the use of biographical data as a selection device. Although early on there was some inconsistent validity evidence in the literature, a more recent meta-analysis of BIBs reveals good validity generalization (corrected validity coefficient of 0.36) across a sample of 79 companies in which the same BIB was used (Schmidt & Rothstein, 1994). These results compare favorably to earlier non–meta-analytic reviews: 0.50 (Asher, 1972), 0.42 (England, 1971), 0.33 (Holberg & Pugh, 1978), and 0.35 (Reilly & Chao, 1982). The review by Reilly and Chao indicated that biodata is a valid predictor for military occupations (mean $r = 0.30$), clerical workers (mean $r = 0.52$), management (mean $r = 0.38$), sales (mean $r = 0.50$), and scientific/engineering occupations (mean $r = 0.41$). The authors located only two coefficients for nonmanagement jobs other than clerical, and these yielded the low average validity of 0.14. Reilly and Chao found that biodata successfully predicted a number of different measures of job success, including tenure (mean $r = 0.32$), performance in training (mean $r = 0.39$), performance ratings (mean $r = 0.36$), productivity (mean $r = 0.46$), and salary progress (mean $r = 0.34$).

Hopefully, the reason for the effectiveness of biodata is clear: The items that are scored are items determined to be correlated with the criterion or criteria of interest. As noted earlier, when criterion keying (empirical keying) is used, substantial validity is attainable. The use of

BIBs has proved to be both valid and to have less adverse impact than cognitive ability tests, making it useful in large-scale testing programs. Indeed, the federal government's Office of Personnel Management has a program of research on such measures (they call them the Individual Achievement Record or IAR), and significant validity has been demonstrated for the IAR in making employment decisions for the civil service (see Gandy, Dye, & MacLane, 1994, for a description of this impressive effort).

Faking Issues in Biographical Data

Lautenschlager (1994) provided a great service in reviewing the issues of accuracy and faking in these kinds of measures. He concluded as follows:

1. When instructed to "fake good," BIB items are fakable.
2. Factual items are responded to with more accuracy and less faking than more subjective items.
3. Warnings of potential response verification may reduce intentional distortion.

Schmitt and his colleagues (2004) addressed the issue of faking from a unique vantage point. They asked people completing a BIB to elaborate on their answers. For example, a sample question follows (Schmitt et al., 2004):

> In how many different languages besides English can you converse well enough to order a meal?
>
> a. None
> b. One
> c. Two
> d. Three
> e. Four or more
>
> If you answered either b, c, d, or e, please list the languages. Do not list more than four. (p. 988)

Schmitt et al. found that elaboration resulted in significantly lower mean scores for items in which elaboration was requested, but other items on the same form where elaboration was not requested did not have lower means. In addition, research indicates that neither the criterion-related validity nor mean subgroup differences were affected by elaboration.

Summary

Job-relevant BIBs have criterion-related validity when subjected to criterion keying prior to the development of scoring procedures. In other words, not all questions on a BIB reveal validity, so criterion keying is necessary. In addition, the adverse impact of BIBs is about half that of cognitive ability tests and verifiable items, and/or items on which elaboration is requested appear to be less susceptible to faking.

SUMMARY

We cover a lot of ground in this chapter with reference to the softer side of selection procedures. What we mean by the softer side is that the focus here has been on who people are and the experiences they have had rather than on what people can do. The criteria these measures seem to predict are also softer: job satisfaction, stress, and citizenship behavior, although not less important as we have learned. So, the nice thing is that the soft side reveals hard criterion-related validity. It is so nice that we have personalities again!

REFERENCES

Aldrich, H. (1999). *Organizations evolving*. Thousand Oaks, CA: Sage.
Allport, G. W., & Odbert, H. S. (1936). Trait-names: A psycho-lexical study. *Psychological Monographs, 47* (Serial No. 211).
Asendorpf, J. B. (1992). A Brunswickian approach to trait continuity: Application to shyness. *Journal of Personality, 60*, 53–77.
Asher, J. J. (1972). The biographical item: Can it be improved? *Personnel Psychology, 25*, 251–269.
Barrick, M. R., & Mount, M. K. (1991). The Big Five personality dimensions and job performance: A meta-analysis. *Personnel Psychology, 44*, 1–26.
Barrick, M. R., & Mount, M. K. (1993). Autonomy as a moderator of the relationship between the Big Five personality dimensions and job performance. *Journal of Applied Psychology, 78*, 111–118.
Bentz, V. J. (1990). Contextual issues in predicting high level leadership performance: Contextual richness as a criterion consideration in personality research with executives. In K. E. Clark & M. B. Clark (Eds.), *Measures of leadership* (pp. 131–143). West Orange, NJ: Leadership Library of America.
Brown, R. (1965). *Social psychology*. New York: The Free Press.
Buss, A. H. (1992, August). *A personal view of personality assessment*. Invited address to the meeting of the American Psychological Association, Washington, DC.
Campbell, D. P., Hyne, S. A., & Nilsen, D. L. (1992). *Manual for the Campbell Interest and Skill Survey*. Minneapolis: National Computer Systems.
Campbell, J. P., Dunnette, M. D., Lawler, E. E., III, & Weick, K. E., Jr. (1970). *Managerial behavior, performance, and effectiveness*. New York: McGraw-Hill.
Cascio, W. F. (1982). *Applied psychology in personnel management*. Reston VA: Reston.

REFERENCES

Cattell, R. B. (1945). The description of personality: Principles and findings in a factor analysis. *American Journal of Psychology, 58*, 69–90.

Clark, K. E., & Clark, M. B. (Eds.). (1990). *Measures of leadership.* West Orange, NJ: Leadership Library of America.

Colbert, A. E., Mount, M. K., Harter, J. K., Witt, L. A., & Barrick, M. R. (2004). Interactive effects of personality and perceptions of the work situation on workplace deviance. *Journal of Applied Psychology, 89*, 599–609.

Cornelius, E. T., III. (1983). The use of projective techniques in personnel selection. In K. M. Rowland & G. R. Ferris (Eds.), *Research in personnel and human resources management* (pp. 127–168). Greenwich, CT: JAI Press.

Costa, P. T., & McCrae, R. R. (1980). Influence of extraversion and neuroticism on subjective well-being: Happy and unhappy people. *Journal of Personality and Social Psychology, 38*, 668–678.

Costa, P. T., & McCrae, R. R. (1992). *Revised NEO Personality inventory (NEO-PI-R and NEO Five Factor Inventory (NEO-FFI) professional manual.* Odessa, FL: PAR.

DeFruyt, F., & Mervielde, I. (1999). RIASEC types of big five traits as predictors of employment status and nature of employment. *Personnel Psychology, 52*, 701–727.

Digman, J. M. (1990). Personality structure: Emergence of the five-factor model. *Annual Review of Psychology, 41*, 417–440.

England, G. W. (1971). *Development and use of weighted application blanks.* Dubuque, IA: Brown.

Fitzgerald, C. (1997). The MBTI and leadership development: Personality and leadership reconsidered in changing times. In C. Fitzgerald & L. K. Kirby (Eds.), *Developing leaders: Research and applications in psychological type and leadership development* (pp. 33–59). Palo Alto, CA: Davies-Black.

Frei, R. L., & McDaniel, M. A. (1998). Validity of customer service measures in personnel selection: A review of criterion and construct evidence. *Human Performance, 11*, 1–27.

Frisch, M. H. (1998). Designing the individual assessment process. In R. Jeanneret & R. Silzer (Eds.), *Individual psychological assessment: Predicting behavior in organizational settings* (pp. 135–177). San Francisco: Jossey-Bass.

Gandy, J. A., Dye, D. A., & MacLane, C. N. (1994). Federal government selection: The individual-achievement record. In G. S. Stokes, M. D. Mumford, & W. A. Owens (Eds.), *Biodata handbook: Theory, research, and use of biographical information in selection and performance appraisal* (pp. 275–309). Palo Alto, CA: CPP Books.

George, J. M., & Brief, A. P. (2004). Personality and work-related distress. In B. Schneider & D. B. Smith (Eds.), *Personality and organizations* (pp. 193–219). Mahwah, NJ: Erlbaum.

Ghiselli, E. E. (1966). *The validity of occupational aptitude tests.* New York: Wiley.

Ghiselli, E. E. (1973). The validity of aptitude tests in personnel selection. *Personnel Psychology, 26*, 461–478.

Goldberg, L. R. (1990). An alternative "description of personality": The Big Five factor structure. *Journal of Personality and Social Psychology, 59*, 1216–1229.

Goleman, D. (1998). *Working with emotional intelligence.* New York: Bantam.

Goleman, D., Boyatzis, R., & McKee, A. (2002). *Primal leadership: Realizing the power of emotional intelligence.* Boston: Harvard Business School Press.

Goodman, P. S. (2000). *Missing organizational linkages: Tools for cross-level research.* Thousand Oaks, CA: Sage.

Gottfredson, G. D., & Holland, J. L. (1996). *Dictionary of Holland occupational codes.* Odessa, FL: Psychological Assessment Resources.

Gottfredson, G. D., Holland, J. L., & Holland, J. E. (1978). The seventh revision of the Vocational Preference Inventory. *Psychological Documents, 8*, 98.
Guilford, J. P. (1959). *Personality.* New York: McGraw-Hill.
Guion R. M. (1965). *Personnel testing.* New York: McGraw-Hill.
Guion, R. M. (1998). *Assessment, measurement, and prediction for personnel decision.* Mahwah, NJ: Erlbaum.
Guion, R. M., & Gottier, R. F. (1965). Validity of personality measures in personnel selection. *Personnel Psychology, 18,* 49–65.
Hall, D. T. (2002). *Careers in and out of organizations.* Thousand Oaks, CA: Sage.
Hogan, J., & Hogan, R. (1998). Theoretical frameworks for assessment. In R. Jeanneret & R. Silzer (Eds.), *Individual psychological assessment: Predicting behavior in organizational settings* (pp. 27–53). San Francisco: Jossey-Bass.
Hogan, J. B. (1994). Empirical keying of background data measures. In G. S. Stokes, M. D. Mumford, & W. A. Owens (Eds.), *Biodata handbook: Theory, research, and use of biographical information in selection and performance appraisal* (pp. 69–107). Palo Alto, CA: CPP Books.
Hogan, R. (1982). Socioanalytic theory of personality. In M. M. Page (Ed.), *1982 Nebraska symposium on motivation: Personality—current theory and research* (pp. 55–89). Lincoln: University of Nebraska Press.
Hogan, R., & Blake, R. J. (1996). Vocational interests: Matching self-concept with the work environment. In K. R. Murphy (Ed.), *Individual differences and behavior in organizations* (pp. 89–144). San Francisco: Jossey-Bass.
Hogan, R., Raskin, R., & Fazzini, D. (1990). The dark side of charisma. In K. E. Clark & M. B. Clark (Eds.), *Measures of leadership* (pp. 343–354). West Orange, NJ: Leadership Library of America.
Holberg, A., & Pugh, W. M. (1978). Predicting Navy effectiveness: Expectations, motivation, personality, attitude, and background variables. *Personnel Psychology, 24,* 679–686.
Holland, J. L. (1966). *The psychology of vocational choice.* Waltham, MA: Blaisdell.
Holland, J. L. (1997). *Making vocational choices: A theory of vocational personalities and work environments* (3rd ed.). Odessa, FL: PAR.
Hough, L. M. (1992). The "Big Five" personality variables—construct confusion: Description versus prediction. *Human Performance, 5,* 139–155.
Hough, L. M. (1998). Effects of intentional distortion in personality measurement and evaluation of suggested palliatives. *Human Performance, 11,* 209–244.
Hough, L. M., Oswald, F. L., & Ployhart, R. E. (2001). Determinants, detection, and amelioration of adverse impact in personnel selection procedures: Issues, evidence, and lessons learned. *International Journal of Selection and Assessment, 9,* 152–194.
Hough, L. M., & Schneider, R. J. (1996). Personality traits, taxonomies, and applications in organizations. In K. R. Murphy (Ed.), *Individual differences and behavior in organizations* (pp. 31–88). San Francisco: Jossey-Bass.
Howard, A., & Bray, D. W. (1990). Predictions of managerial success over long periods of time: Lessons from the management progress study. In K. E. Clark & M. B. Clark (Eds.), *Measures of leadership* (pp. 113–130). West Orange, NJ: Leadership Library of America.
Hunthausen, J. M., Truxillo, D. M., Bauer, T. N., & Hammer, L. B. (2003). A field study of frame-of-reference effects on personality test validity. *Journal of Applied Psychology, 88,* 545–551.
James, L. R., & Mazerolle, M. D. (2002). *Personality in work organization.* Thousand Oaks, CA: Sage.
Journal of Vocational Behavior. (1999). Special issue on Holland's theory. 55.

REFERENCES 487

Judge, T. A., Bono, J. E., Ilies, R., & Gerhardt, M. W. (2002). Personality and leadership: A qualitative and quantitative review. *Journal of Applied Psychology, 87,* 765–780.
Judge, T. A., Heller, D., & Mount, M. K. (2002). Five factor model of personality and job satisfaction: A meta-analysis. *Journal of Applied Psychology, 87,* 530–541.
Judge, T. A., & Ilies, R. (2004). Affect and job satisfaction: A study of their relationship at work and at home. *Journal of Applied Psychology, 89,* 661–673.
Judge, T. A., & Kristof-Brown, A. (2004). Personality and person-organization fit. In B. Schneider & D. B. Smith (Eds.), *Personality and organizations* (pp. 87–110). Mahwah, NJ: Erlbaum.
Kinslinger, H. J. (1966). Application of projective techniques in personnel psychology since 1940. *Psychological Bulletin, 66,* 134–149.
Kluger, A. N., & Colella, A. (1993). Beyond the mean bias: The effect of warning against faking on biodata item variances. *Personnel Psychology, 46,* 763–780.
Korman, A. K. (1968). The prediction of managerial performance: A review. *Personnel Psychology, 21,* 295–322.
Korman, A. K. (1977). *Organization behavior.* Englewood Cliffs, NJ: Prentice Hall.
Landy, F. J., & Conte, J. M. (2004). *Work in the 21st century: An introduction to industrial and organizational psychology.* New York: McGraw-Hill.
Landy, F. J., & Trumbo, O. A. (1984). *Psychology of work behavior.* Homewood, IL: Irwin.
Lautenschlager, G. J. (1994). Accuracy and faking of background data. In G. S. Stokes, M. D. Mumford, & W. A. Owens (Eds.), *Biodata handbook* (pp. 391–420). Palo Alto, CA: CPP Books.
Law, K. S., Wong, C. S., & Song, L. J. (2004). The construct and criterion validity of emotional intelligence and its potential utility for management studies. *Journal of Applied Psychology, 89,* 483–496.
Lebreton, J. M., Barksdale, C. D., Burgess, J. R. D., & James, L. R. (2004). *Measurement issues associated with conditional reasoning tests of personality: Deception and faking.* Unpublished manuscript, Wayne State University, Department of Psychology, Detroit, MI.
Lievens, F., Decaester, C., Coetsier, P., & Geirnaert, J. (2001). Organizational attractiveness for prospective applicants: A person–organization fit perspective. *Applied Psychology: An International Review, 50,* 30–51.
Lubinski, D. (2000). Scientific and social significance of assessing individual differences: "Sinking shafts at a few critical points." *Annual Review of Psychology, 51,* 405–444.
Mael, F. A. (1991). A conceptual rationale for the domain and attributes of biodata items. *Personnel Psychology, 44,* 763–792.
Magnusson, D., & Endler, N. S. (1977). Interactional psychology: Present status and future prospects. In D. Magnusson & N. S. Endler (Eds.), *Personality at the crossroads: Current issues in interactional psychology* (pp. 3–35). Hillsdale, NJ: Erlbaum.
McClelland, D. C. (1961). *The achieving society.* Princeton, NJ: Van Nostrand.
McClelland, D. C., Atkinson, J. W., Clark, R. A., & Lowell, E. L. (1953). *The achievement motive.* New York: Appleton-Century-Crofts.
McClelland, D. C., & Boyatzis, R. E. (1982). Leadership motive pattern and long-term success in management. *Journal of Applied Psychology, 67,* 737–743.
McCrae, R. R., & Costa, P. T. (1989). Reinterpreting the Myers-Briggs Type Indicator from the perspective of the Five Factor model of personality. *Journal of Personality, 57,* 17–40.
McFarland, L. A. (2003). Warning against faking on a personality test: Effects on applicant reactions and personality test scores. *International Journal of Selection and Assessment, 11,* 265–276.

McFarland, L. A., & Ryan, A. M. (2000). Variance in faking across non-cognitive measures. *Journal of Applied Psychology, 85*, 812–821.
Miller, D. E., & Droge, C. (1986). Psychological and traditional determinants of structure. *Administrative Science Quarterly, 31*, 539–560.
Miner, J. B. (1960). The effect of a course in psychology on the attitudes of research and development supervisors. *Journal of Applied Psychology, 44*, 224–232.
Miner, J. B. (1978). Twenty years of research on role motivation theory of managerial effectiveness. *Personnel Psychology, 31*, 739–760.
Miner, J. B. (1993). *Role motivation theories*. London: Routledge.
Miner, J. B. (2002). The role motivation theories of organizational leadership. In F. J. Yammarino & B. J. Avolio (Eds.), *Transformational and charismatic leadership: The road ahead* (pp. 309–338). New York: Elsevier.
Mischel, W. (1968). *Personality and assessment*. New York: Wiley.
Mischel, W. (1973). Toward a cognitive social learning reconceptualization of personality. *Psychological Review, 80*, 252–283.
Moynihan, L. M., & Peterson, R. S. (2004). The role of personality in group processes. In B. Schneider & D. B. Smith (Eds.), *Personality and organizations* (pp. 317–346). Mahwah, NJ: Erlbaum.
Muchinsky, P. (1983). *Psychology applied to work*. Homewood, IL: Irwin.
Mumford, M. D., & Owens, W. A. (1987). Methodology review: Principles, procedures, and findings in the application of background data measures. *Applied Psychological Measurement, 11*, 1–31.
Organ, D. W., & McFall, J. B. (2004). Personality and citizenship behavior in organizations. In B. Schneider & D. B. Smith (Eds.), *Personality and organizations* (pp. 291–314). Mahwah, NJ: Erlbaum.
Peterson, R. S., Smith, D. B., Martorana, P. V., & Owens, P. D. (2003). The impact of chief executive office personality on top management team dynamics: One mechanism by which leadership affects organizational performance. *Journal of Applied Psychology, 88*, 795–808.
Ployhart, R. E., Weekley, J. A., & Baughman, K. (in press). The structure and function of human capital emergence: A multilevel investigation of the ASA model. *Academy of Management Journal*.
Reilly, R. R., & Chao, G. T. (1982). Validity and fairness of some alternative employee selection procedures. *Personnel Psychology, 35*, 1–62.
Riggio, R. E., & Fleming, T. (n.d.). *Validation of a measure of trait choosiness*. Unpublished manuscript, California State University, Fullerton.
Russell, C. J. (1994). Generation procedures for biodata items: A point of departure. In G. S. Stokes, M. D. Mumford, & W. A. Owens (Eds.), *Biodata handbook: Theory, research, and use of biographical information in selection and performance prediction* (pp. 17–38). Palo Alto, CA: CPP Books.
Sackett, P. R., Burris, L. R., & Callahan, C. (1986). Integrity testing for personnel selection: An update. *Personnel Psychology, 42*, 491–529.
Schein, E. H. (1992). *Leadership and organizational culture* (2nd ed.). San Francisco: Jossey-Bass.
Schmidt, F. L., & Rothstein, H. R. (1994). Application of validity generalization to biodata scales in employment selection. In G. S. Stokes, M. D. Mumford, & W. A. Owens (Eds.), *Biodata handbook: Theory, research, and use of biographical information in selection and performance appraisal* (pp. 237–260). Palo Alto, CA: CPP Books.

REFERENCES

Schmit, M. J., Ryan, A. M., Stierwalt, S. L., & Powell, A. B. (1995). Frame-of-reference effects on personality scale scores and criterion-related validity. *Journal of Applied Psychology, 80,* 607–620.
Schmitt, N. (2004). Beyond the Big Five: Increases in understanding and practical utility. *Human Performance, 17,* 347–357.
Schmitt, N., & Chan, D. (1998). *Personnel selection: A theoretical approach.* Thousand Oaks, CA: Sage.
Schmitt, N., Gooding, R., Noe, R. A., & Kirsch, M. (1984). Meta-analyses of validity studies published between 1964 and 1982, and the investigation of study characteristics. *Personnel Psychology, 37,* 407–422.
Schmitt, N., Oswald, F. L., Kim, B. H., Gillespie, M. A., Ramsay, L. J., & Yoo, T. Y. (2004). Impact of elaboration on socially desirable responding and the validity of biodata measures. *Journal of Applied Psychology, 88,* 979–988.
Schmitt, N., & Pulakos, E. D. (1983). *Evaluation of tests for the selection of transport drivers.* Report submitted to Marathon Oil Company, Findlay, OH.
Schneider, B. (1987). The people make the place. *Personnel Psychology, 40,* 437–453.
Schneider, B., Goldstein, H. W., & Smith, D. B. (1995). The attraction-selection-attrition framework: An update. *Personnel Psychology, 48,* 747–773.
Schneider, B., & Smith, D. B. (Eds.). (2004a). *Personality and organizations.* Mahwah, NJ: Erlbaum.
Schneider, B., & Smith, D. B. (2004b). Personality and organizational culture. In B. Schneider & D. B. Smith (Eds.), *Personality and organizations* (pp. 347–370). Mahwah, NJ: Erlbaum.
Schneider, B., Smith, D. B., Taylor, S., & Fleenor, J. (1998). Personality and organization: A test of the homogeneity of personality hypothesis. *Journal of Applied Psychology, 83,* 462–470.
Shaw, M. E. (1981). *Group dynamics: The psychology of small group behavior* (3rd ed.). New York: McGraw-Hill.
Smith, D. B., & Robie, C. (2004). The implications of impression management for personality research in organizations. In B. Schneider & D. B. Smith (Eds.), *Personality and organizations* (pp. 111–138). Mahwah, NJ: Erlbaum.
Smith, D. B., & Schneider, B. (2004). Where we've been and where were going: Some conclusions regarding personality and organizations. In B. Schneider & D. B. Smith (Eds.), *Personality & Organizations* (pp. 387–404). Mahwah, NJ: Erlbaum.
Spangler, W. D., House, R. J., & Palrecha, R. (2004). Personality and leadership. In B. Schneider & D. B. Smith (Eds.), *Personality and organizations* (pp. 251–290). Mahwah, NJ: Erlbaum.
Stagner, R. (1958). The gullibility of personnel managers. *Personnel Psychology, 11,* 347–352.
Stark, S., Chernyshenko, Chan, K., Lee, W., & Drasgow, F. (2001). Effects of the testing situation on item responding: Cause for concern. *Journal of Applied Psychology, 86,* 943–953.
Staw, B. M. (2004). The dispositional approach to job attitudes: An empirical and conceptual review. In B. Schneider & D. B. Smith (Eds.), *Personality and organizations* (pp. 163–192). Mahwah, NJ: Erlbaum.
Stewart, G. G., & Barrick, M. R. (2004). Four lessons learned from the person-situation debate: A review and research agenda. In B. Schneider & D. B. Smith (Eds.), *Personality and organizations* (pp. 61–85). Mahwah, NJ: Erlbaum.
Stogdill, R. M. (1948). Personal factors associated with leadership: A survey of the literature. *Journal of Psychology, 25,* 35–71.

Stokes, G. S., Mumford, M. D., & Owens, W. A. (Eds.). (1994). *Biodata handbook: Theory, research, and use of biographical information in selection and performance appraisal.* Palo Alto, CA: CPP Books.

Taylor, M. S., Locke, E. A., Lee, C., & Gist, M. E. (1984). Type A behavior and faculty research productivity: What are the mechanisms? *Organizational Behavior and Human Performance, 34,* 402–418.

Tett, R. P., Jackson, D. N., & Rothstein, M. (1991). Personality measures as predictors of job performance: A meta-analytic review. *Personnel Psychology, 44,* 703–742.

Thoreson, C. J., & Judge, T. A. (1997, August). *Trait affectivity and work-related attitudes and behaviors: A meta-analysis.* Paper presented at the annual convention of the American Psychological Association, Chicago, IL.

Tokar, D. M., Fischer, A. R., & Subich, L. M. (1998). Personality and vocational behavior: A selective review of the literature. *Journal of Vocational Behavior, 53,* 115–153.

Turban, D. B., & Keon, T. L. (1993). Organizational attractiveness: An interactionist perspective. *Journal of Applied Psychology, 78,* 184–193.

Vinitsky, M. (1973). A forty-year follow-up on the vocational interests of psychologists and their relationship to career development. *American Psychologist, 28,* 1000–1009.

Viteles, M. (1932). *Industrial psychology.* New York: Norton.

Walsh, W. B., Craik, K. H., & Price, R. H. (2000). Person–environment psychology: A summary and commentary. In W. B. Walsh, K. H. Craik, & R. H. Price (Eds.), *Person–environment psychology: New directions and perspectives* (2nd ed., pp. 297–326). Mahwah, NJ: Erlbaum.

Warr, P., & Pearce, A. (2004). Preferences for careers and organizational cultures as a function of logically related personality traits. *Applied Psychology: An International Review, 53,* 423–435.

Watson, D., Clark, L. A., & Tellegen, A. (1988). Development and validation of brief measures of positive and negative affect: The PANAS scales. *Journal of Personality and Social Psychology, 54,* 1063–1070.

Watson, D., & Tellegen, A. (1985). Toward a consensual structure of moods. *Psychological Bulletin, 98,* 219–225.

Winter, D. G., John, O. P., Stewart, A. J., Klohnen, E. C., & Duncan, L. E. (1998). Traits and motives: Toward an integration of two traditions in personality research. *Psychological Review, 105,* 230–250.

Zeidner, M., Matthews, G., & Roberts, R. D. (2004). Emotional intelligence in the workplace: A critical review. *Applied Psychology: An International Review, 53,* 371–399.

10

Hiring Procedures III: Interviews, Performance-Based Tests, Simulations, and Physical Ability

AIMS OF THE CHAPTER

In contrast to chapters 8 and 9, which focus primarily on predictor constructs, this chapter considers many kinds of predictor measures (including interviews, situational judgment, assessment centers, and work samples). These predictor methods are frequently called *performance-based tests* or *simulations*. We also consider the nature of physical ability tests, which are yet another variety of performance-based tests. As we saw in chapter 7, these various performance-based tests and simulations tend to have moderate to high criterion-related validity and favorable applicant reactions. This should come as no surprise because they often require applicants to perform the same (or highly similar) set of behaviors as those required on the job. For this reason, they have higher physical fidelity than traditional cognitive ability and personality tests. They also tend to produce smaller racial and gender subgroup differences than these other predictors (the exception is physical ability testing). However, they have two general limitations. The first is that many simulations are costly to develop and administer. The second is that, historically, they do not produce homogeneous scores for underlying psychological constructs. However, we see this is starting to change because research conducted over the last several years has begun to focus on enhancing the construct validity of these performance-based predictor measures.

This chapter is presented in three major sections. We first discuss research and issues surrounding interviews. The next section considers test methods that range from low (situational judgment) to high (assessment centers, work samples) physical fidelity. We then conclude with a discussion of physical ability testing.

INTERVIEWS

It is highly unlikely that anyone applying for a job will be able to escape the interviewing process. After all, who would not want to talk at least briefly with a person with whom one might be working for several years? The employment interview is probably more widely used than any other selection tool. In one survey of 852 organizations (Ulrich & Trumbo, 1965), 99% of the firms reported using interviews, and they remain frequently used to this day. As might be expected, it has also been the subject of an enormous amount of research. Summarizing across several major narrative reviews conducted since the mid-1970s (Arvey & Campion, 1982; Harris, 1989; Posthuma, Morgeson, & Campion, 2002; Schmitt, 1976), there have been an average of approximately 16 articles published on the interview every year (with the average increasing every decade). A book edited by Eder and Harris (1999) also summarizes a great deal of interview research.

As we might suspect from the widespread use of interviews, most managers and personnel workers—especially those who are actually interviewers—have a great deal of faith in the usefulness and validity of the selection interview. Hundreds of "how to" books have been written on the interview, purporting to tell the interviewer how best to conduct interviews. Also, there have been nearly as many books for applicants on how to "ace" or "beat" the interview (in fact, your college or university probably has a placement office with materials on how to behave in the interview).

However, what really is an interview? We saw in chapter 7 that it is a test method capable of measuring a variety of constructs. That is, we might use the interview to measure personality, social skills, or applicant–organization fit. At the same time, it is a test method, and there are consequently many different variations and formats to the interview. Furthermore, the interview serves many purposes for

INTERVIEWS

organizational decision makers, including selection and recruitment. Thus, to speak of the interview is to ignore a variety of subtle and not so subtle nuances across different types of interviews. To appreciate these differences and to structure the massive literature on interviews, we break our review into several subsections: purposes of interviews, interview structure, types of questions, interview validity, the constructs interviews measure, subgroup differences, applicant reactions, the interview process, contaminants of interview ratings, legal issues with the use of interviews, new technologies changing the way interviews are conducted, improving interview validity, and future directions.

Purposes of Interviews

We so often think of interviews as a basis for hiring that it may be surprising to learn they serve several different purposes. These purposes may be subsumed within recruitment, screening, and selection purposes (Schmitt, 1976). When the interview is used as a recruiting tool, it is primarily to "sell" the organization to applicants and provide information about the job and organizational culture. The focus is less on evaluating the applicant and more on conveying the qualities of the organization. Closely related to this purpose is the research on RJPs. We provided an overview of Realistic Job Previews (RJPs) in chapter 5 when we talked about recruitment. As they relate to the interview, an important point is that, although RJPs can help reduce turnover, the more qualified or experienced applicants most weight negative information (e.g., Meglino, Ravlin, & DeNisi, 1997).

The second interview purpose is the screening interview. Here, the goal is to conduct a brief interview, administered early in the selection process, and then make an early cut from which a smaller number of applicants will proceed to additional testing. The screening interview will typically ask only a few basic questions about the applicants, but questions that determine their status in the selection process. For example, suppose an office manager has 20 applicants for an administrative assistant position. The manager may use a brief screening interview to assess all 20 applicants and then invite the top three applicants back for the selection interview. Clearly, the screening interview is going to be used only in noncompensatory, multiple hurdle systems (see chapter 7). More recently, these screening interviews have been developed to be administered via telephone, where applicants for jobs dial in, the interview questions are presented aurally, applicants record their responses, those responses are scored later, and then feedback is provided.

The final interview purpose is the actual selection of applicants. This is the purpose we most think of when conducting an interview. The selection interview is a more intensive, data-gathering interview designed to produce a hiring decision. Although both the screening and selection interviews evaluate candidates and conclude with a hiring decision, the screening interview gathers specific information on only a few key points, whereas the appropriate selection interview gathers much more detail on job-relevant KSAOs of the candidate. Thus, the selection interview tends to be more comprehensive and inclusive.

Although there is not a great deal of research on how different interview purposes influence its administration and processes within the interview, it seems reasonable to draw from the performance appraisal literature (see chapter 4) and expect that the purpose of the interview will influence interview process and ratings. Preliminary research does point to such influences. Stevens (1998) demonstrated that interviewers in recruitment interviews provided more information and asked fewer questions than interviewers in selection interviews. Barber, Hollenbeck, Tower, and Phillips (1994) examined interviews designed for selection, recruitment, or both purposes. They found applicants in the mixed-purpose interview gathered and retained less information than in the recruitment interview, but conversely showed more persistence in the job search process. Research that squarely manipulates interview purpose to determine how this influences interview processes and outcomes would be a valuable addition to the literature. In the meantime, it seems safe to conclude the purpose of the interview should be carefully considered before development and administration occur.

Interview Structure

Early reviews lamented the sorry state of the interview, finding little criterion-related validity and hence little basis for using interviews in selection (Ulrich & Trumbo, 1965). This verdict has been largely overturned in more recent reviews, but with an important caveat—interviews can demonstrate useful levels of reliability and validity if they are appropriately structured. Indeed, structured interviews show among the strongest criterion-related validities of any predictor measure (Huffcutt & Arthur, 1994; Schmidt & Hunter, 1998).

Thus, the effectiveness of interviews is enhanced by incorporating various components of structure. Campion, Palmer, and Campion (1997) defined interview structure as "any enhancement of the interview that is intended to increase psychometric properties by increasing standardization or otherwise assisting the interviewer in determining what questions to ask or how to evaluate responses" (p. 656). In their review, Campion et al. (1997) identified 15 different components to interview structure:

1. Base the interview on a job analysis (most frequently, the critical incident method).
2. Ask the same questions of all applicants.
3. Limit the use of follow-up prompting and elaboration.
4. Use situational, past behavior, background, or job knowledge questions (discussed next section).
5. Ensure enough time, or number of questions, to fully assess the intended information.
6. Minimize the use of extraneous information (e.g., prior test scores).
7. Prohibit applicants from asking questions during the interview (save such questions until after the interview).
8. Use multiple ratings for each answer.
9. Use BARS.
10. Make detailed notes.
11. Use an interview panel (i.e., multiple interviewers).
12. Use the same interviewers across all candidates.
13. Avoid discussing candidates or ratings between interviewees.
14. Train the interviewers.
15. Statistically combine the interviewer ratings.

Notice that components 8 through 15 overlap substantially with good practice in conducting performance appraisal ratings (see chapter 4). Thus, many of the same recommendations we made for improving performance appraisal ratings apply to interview contexts (or for that matter, most any context that requires ratings, as we will see shortly). However, the first seven components are more novel. Hopefully after reviewing chapter 3, the importance of basing questions on a job analysis is obvious. This is simply the best and most legally defensible way to ensure the interviewer asks appropriate and relevant questions. Asking the same questions of all applicants is perhaps the best way to enhance structure. Although interviewers may not like the loss of control in directing the interview, it is clear that the best way to discriminate among applicants is to evaluate them on the same dimensions. Note there are four levels to using the same structured questions, such that the highest level is asking the same candidates identical questions, followed by asking candidates highly similar questions (but allowing small room for discretion), followed by having only an identical outline from which to base questions, and last, no consistency of questions across applicants (Campion et al., 1997; Huffcutt & Arthur, 1994). Finally, avoiding the use of extraneous information in the interview (including taking questions from applicants) helps the interviewer focus on the task at hand, minimizes the likelihood of confirmatory biases and expectancy effects (e.g., when test scores on applicants are made available to the interviewer before the interview), and ensures each applicant is evaluated on equal terms.

Although the 15 components of structure defined in Campion et al. (1997) help us understand the components that promote effective interviews, research has not clearly identified which of these components may be most important, or for that matter, critical versus desirable. For example, it may be that some of these components are necessary for the interview to show reasonable validity and reliability, whereas other components may merely be sufficient. This is important because not all elements of structure will be reacted to favorably by interviewers, management, or applicants (Latham & Finnegan, 1993). It will be important for future research to clearly identify the critical components of structure to maximize the application of structure and user reactions.

Types of Interview Questions

There is no shortage of books or Internet lists of "most asked" interview questions. The creativity we have seen from untrained interviewers in the questions they ask suggests no limit to the variety of questions that are used. One of the authors worked for a company where the favorite interview question was, "When can you start work?" Most applicants answered, "Right away" to which the hiring manager replied, "Okay, be here at 7 am tomorrow morning." It was usually at this point that applicants came up with reasons why they could not start

TABLE 10.1
Examples of Situational Interview Questions and Their Corresponding Critical Incidents Used for Selection of Emergency Telephone Operators

Interview Question	Critical Incident
1. Imagine that you tried to help a stranger, for example, with traffic directions or to get up after a fall, and that person blamed you for their misfortune or yelled at you. How would you respond?	1. Telephone operator tries to verify address information for an ambulance call. The caller yells at them for being stupid and slow. The operator quietly assures the caller an ambulance is on the way and that she is merely reaffirming the address.
2. Suppose a friend calls you and is extremely upset. Apparently, her child has been injured. She begins to tell you, in a hysterical manner, all about her difficulty in getting babysitters, what the child is wearing, what words the child can speak, etc. What would you do?	2. A caller is hysterical because her infant is dead. She yells incoherently about the incident. The operator talks in a clear calm voice and manages to secure the woman's address, dispatches the call, and then tries to secure more information about the child's status.
3. How would you react if you were a salesclerk, waitress, or gas station attendant and one of your customers talked back to you, indicated you should have known something you did not, or told you that you were not waiting on them fast enough?	3. A clearly angry caller calls for the third time in an hour complaining about the 911 service because no one has arrived to investigate a busted water pipe. The operator tells the caller to go to _____ and hangs up.

so quickly, leading the manager to end the interview and reject them for the job. It might seem impossible to categorize so many different interview questions, but fortunately, research has helped categorize the types of good structured interview questions. There are three general types.

The first is the use of future-oriented questions in what is called a *situational interview*. The nature of these questions presents applicants with hypothetical situations, and asks applicants how they believe they would handle these situations. The questions are most often based on a job analysis using the critical incident method (discussed in chapter 3). Table 10.1 gives an example of how one uses a critical incident to create an interview question. The questions are necessarily ambiguous, and because applicants would not have previously encountered the situation, they are based on the premise that intentions predict future performance (Latham, Saari, Pursell, & Campion, 1980). Therefore, situational interviews ask applicants what they intend to do in a hypothetical situation. Consider the following examples:

- Imagine your supervisor asks you to analyze some reports using a software system with which you are unfamiliar. In addition to your supervisor, there are a few coworkers around who are familiar with the software but who look very busy. How would you handle this situation?
- Your car breaks down on the way to visit an important client. When you call for roadside assistance, you realize you left your wallet/purse at home. It is unlikely that anyone will tow the car without a major credit card or cash. How would you handle this situation?

The second question type uses past or historical questions in what is called a *behavioral* or *experience-based interview*. Here, the questions specifically request the applicant to discuss how he or she has handled the situation in previous experience (Janz, 1982, 1989). Similar to the rationale for the predictive validity of biodata (see chapter 9), the principle is that past behavior predicts future behavior. Therefore, asking applicants how they have handled a given situation or problem provides some insight into how they will handle it in the future. For example:

- Tell me about a time you were asked to use a piece of software or equipment, but you had no experience and familiarity with the tools and you did not have anyone you could turn to for help. How did you handle this situation?
- Tell me about a time when you were left without the resources needed to perform your job. That is, when due to inadequate resources you were unable to accomplish what was expected. How did you handle this situation?

The third question type uses questions designed to identify the applicant's job knowledge and tend to be directly based on a job analysis. One might ask questions relating to declarative knowledge (knowledge of facts), such as, "What is the maximum speed limit on a residential road" or "What tax form is required to claim independent consultant wages?" Alternatively, one might ask questions relating to procedural knowledge (knowledge of how to do something), such as "Tell me how you would develop a homepage" and "Describe the steps necessary to correctly prepare a home mortgage application."

Interview Reliability and Validity

There have been numerous meta-analyses documenting the reliability and validity of different types of interviews (see Buckley & Russell, 1999). In terms of interrater reliability, a meta-analysis by Conway, Jako, and Goodman (1995) found the average reliability was 0.70, with higher reliability for panel interviews and more structured interviews. Schmidt and Hunter (1998) estimate the corrected validity of 0.51 for structured interviews and 0.38 for unstructured interviews. Huffcutt and Arthur (1994) found interview validity for entry-level jobs was comparable to overall cognitive ability ($r_{corrected} = 0.57$). McDaniel, Whetzel, Schmidt, and Maurer (1994) summarized findings based on 86,311 individuals to show a mean observed validity across all interview methods of 0.20 ($r_{corrected} = 0.37$). Situational and job-related interviews yielded uncorrected validities of 0.27 and 0.21 ($r_{corrected} = 0.50$ and 0.39, respectively). Interestingly, they also show interview validities are higher when no prior test information is available ($r_{uncorrected} = 0.25$, $r_{corrected} = 0.45$) than when it is available ($r_{uncorrected} = 0.14$, $r_{corrected} = 0.26$). They report median reliabilities of 0.74, 0.78, and 0.89 for psychological, job-related unstructured, and job-related structured interviews, respectively.

There have also been studies comparing situational to behavioral interviews. A meta-analysis by Taylor and Small (2002) showed the uncorrected validity for behavioral interviews was 0.31 ($r_{corrected} = 0.56$), and slightly greater than situational interviews ($r_{uncorrected} = 0.25$, $r_{corrected} = 0.45$). Interviews that used BARS showed validities from 0.01 to 0.07 higher than those that did not, and both interview questions showed similar interrater reliabilities (approximately 0.77–0.79). Pulakos and Schmitt (1995) speculated that situational interviews may show less validity for jobs that were more cognitively complex, higher-level positions (they found near zero validity for situational interviews, whereas behavioral interviews showed a validity of 0.32). This was replicated in two separate studies by Huffcutt, Weekley, Wiesner, Degroot, and Jones (2001). Thus, the behavioral interview appears to be more predictive of job performance.

Cortina, Goldstein, Payne, Davison, and Gilliland (2000) created a meta-analytic correlation matrix of interviews, cognitive ability, and conscientiousness. They demonstrated interviews exhibit incremental validity over these other predictor constructs. More important, they showed that interview structure relates positively to the degree of incremental validity, with more structured interviews exhibiting greater

> **Box 10.1 Sample Questions and Ratings From a Structured Interview Designed to Choose Retail Store Department Managers**
>
> Five sets of competencies were identified for this first-line supervisory job:
>
> 1. Leading and Influencing
> 2. Customer orientation
> 3. Self-management
> 4. Driving results through others
> 5. Following rules and procedures
>
> Each situation presented to applicants in the interview required them to answer the following questions:
>
> - What was the situation?
> - How did you handle it?
> - What were the outcomes?
>
> The five questions in the interview follow:
>
> 1. Briefly describe a situation or project that you completed where you were in charge and that resulted in a customer being extremely happy.
> 2. Tell me about a situation in which you won over a customer who was initially upset.
> 3. What was the most stressful situation you have ever been involved in when dealing with a customer?
> 4. Tell me about a situation that called for you to work with and through others to achieve your goals.
> 5. Briefly describe a situation where you were asked to follow a company rule or procedure that you did not agree with or think was necessary.
>
> Based on the replies to the five situations presented and the specific questions (what was the situation, how did you handle it, and what was the outcome), the interviewer makes a series of ratings for each dimension regarding the acceptability of the behavior reported. For this sample, we only show one of the ratings for each dimension:
>
> - Leading and influencing: Desires to take charge when things need to get done.

> - Customer orientation: Listens to complaints in a way that people feel they have been heard.
> - Self-management: Remains calm in the face of stressful circumstances.
> - Driving results through others: Enjoys working with others to achieve work goals.
> - Following rules and procedures: Accepts rules and policies at work.

incremental validity. Unstructured interviews exhibit near zero incremental validity, whereas highly structured interviews explained 17% incremental variance.

Note that in all these studies discussed so far, criteria were overall assessments of job performance. Using a telephone-administered interview, a meta-analysis by Schmidt and Rader (1999) extended findings to objective criteria such as production records ($r_{uncorrected} = 0.29$; $r_{corrected} = 0.40$), sales performance ($r_{uncorrected} = 0.15$; $r_{corrected} = 0.24$), and retention ($r_{uncorrected} = 0.10$; $r_{corrected} = 0.19$). In terms of interviewer factors affecting validity, a meta-analysis by Huffcutt and Woehr (1999) found the use of interviewer training and use of the same interviewers across applicants enhanced validity, but use of panel interviews did not.

Thus, after years of pessimism, it is clear that interviews, particularly structured interviews, can exhibit reliability and criterion-related validities on par with the best predictors currently available. Of course, keep in mind that many of the meta-analytic validities are based on multiple corrections, and use of different corrections may change these results. The next major task for improving the criterion-related validity of the interview will no doubt try to determine how various aspects of structure contribute to validity, or stated differently, to decompose validity into specific components of structure.

In Box 10.1, we provide some sample questions from a structured behavioral or experience-based interview that were developed after a job analysis of a retail store department manager. Also included are the ratings the interviewer is required to make at the conclusion of the information gathering. Standardizing both the question format and targets of the questions, as well as the ratings to be made at the conclusion of the interview, should enhance reliability and, hopefully, validity.

What Constructs Do Interviews Measure?

The impressive criterion-related validity evidence that has been accumulated for structured interviews must be tempered against the lack of interview construct validity and a disturbing lack of knowing what interviews measure. This is a problem that nags all performance measurement methods such as situational judgment tests and assessment centers, and it creates a major impediment toward developing a science of selection. Because interviews tend to be tailored to a particular job (often through job analysis procedures), it should come as no surprise that interviews do not measure a single construct but a large number of constructs. Reflecting back to the predictor response model in chapter 7, inferring what interviews measure is a difficult task because there will be multiple true scores that must be separated from multiple sources of systematic, but contaminating, variance. More recent research is starting to address the question of what interviews measure, whether different amounts of structure result in the assessment of different constructs and whether different interview questions (situational versus behavioral) assess the same constructs.

In one meta-analysis, Huffcutt, Roth, and McDaniel (1996) found a corrected validity of 0.40 between general cognitive ability and interview scores. A more comprehensive meta-analysis by Huffcutt, Conway, Roth, and Stone (2001) developed a taxonomy of constructs likely to be assessed in the interview. They identified the following latent dimensions:

1. General cognitive ability (including specific abilities and creativity)
2. Knowledge and skills (including experience)
3. Personality
4. Social skills (including leadership, communication, and negotiation)
5. Interests and preferences
6. Organizational fit
7. Physical abilities and attributes

Of these seven dimensions, personality and social skills were the most commonly assessed in interviews. Interestingly, high- and low-structure interviews assessed different constructs. High-structure interviews tended to focus more on job-relevant KSAOs, which the authors attribute to the greater reliance on job analysis in developing structured interviews. Theoretically, one might expect situational interviews to be measures of intention, whereas behavioral interviews are measures of experience and perhaps knowledge (Maurer, Sue-Chan, & Latham, 1999; Motowidlo, 1999). Moscoso (2000) cited several lines of empirical evidence, suggesting situational and behavioral interviews have some similar relationships with other constructs (e.g., cognitive ability), as well as important differences (e.g., personality, knowledge, experience). Indeed, a study by Huffcutt, Conway, et al. (2001) found the correlation between situational and behavioral interviews was only 0.09.

Schmitt (1999) made the important point that we still have a long way to go toward understanding what interviews measure. As he pointed out, a key task will be to design interviews to target particular constructs and train raters to identify these constructs in the interview. Although this seems rather challenging, it is not all that different from what many clinical and counseling psychologists are trained to do with the major difference being that these psychologists have much more training than the typical manager. We need to learn whether relatively brief training programs can actually lead to interviewers identifying constructs.

Of course, from a practical vantage point, the fact that structured interviews reveal considerable criterion-related validity is good news, and this good news should lead to more structure both in the questions asked and the ratings made. Our own impression is that interviews and other performance-based measures reveal significant validity because they tap into the complex real world of accomplishing work tasks that require the simultaneous activation of several or many KSAOs. So, rather than tapping into the specific psychological constructs that make researchers happy, performance-based measures tap into the way work gets done.

Subgroup Differences

Pulakos and Schmitt (1995) found experience-based interviews tend to show higher subgroup differences than situational interviews, although the subgroup differences are quite small. However, Roth, Van Iddekinge, Huffcutt, Eidson, and Bobko (2002) reminded us that when interpreting subgroup differences (d values), it is important to correct for range restriction. Failing to correct for range restriction (and unreliability) results in an underestimate of the true subgroup difference, and they showed subgroup differences for behavioral interviews were moderate in effect size (ranging from 0.36 to 0.56).

A large meta-analyses by Huffcutt, Conway, et al. (2001) examined White-Black and gender subgroup differences across the seven construct categories discussed earlier. In terms of race, most of the differences favored Whites, but none of the differences was greater than $d = 0.52$, suggesting small to moderate subgroup differences. Interestingly, racial differences were considerably greater using unstructured interviews. Gender differences were more variable (ranging from small to large) and tended to favor men on some constructs (e.g., work experience) and women on others (e.g., communication skills). The differences between structured and unstructured interviews were likewise ambiguous. The exception to these findings was for physical attributes, but the number of studies analyzed was so small that it is difficult to make any sense of the findings.

However, a different question is whether there is less of a subgroup difference when panels of interviewers (as opposed to a single interviewer) are used. The reasoning makes sense: If subgroup differences exist on interviews because of bias against minority groups (i.e., contamination in the ratings), then balancing out interview panels with both minority and majority raters should eliminate such bias. In practice, the results have not been so clear, with some studies finding complicated effects for race (e.g., McFarland, Ryan, Sacco, & Kriska, 2004) and others finding no such effects (e.g., Sacco, Scheu, Ryan, & Schmitt, 2003).

Applicant Reactions

Research has increasingly shown that applicants respond favorably to the interview in general. Steiner and Gilliland (1996) demonstrated students in the United States and France tend to consider interviews as fair methods for selection. The more the interview questions are tailored to the relevant job or business context, the more favorably applicants tend to respond (Rynes & Connerley, 1993; Smither, Reilly, Millsap, Pearlman, & Stoffey, 1993).

However, research also suggests the amount of interview structure is negatively related to applicant reactions (i.e., more structured interviews are perceived less favorably). Kohn and Dipboye (1998) found students favored unstructured interviews. Latham and Finnegan (1993) further demonstrated individuals respond more negatively to structured interview methods. They argued one reason may be an inability to control structured interviews (e.g., asking questions of the interviewer). Campion et al. (1997) made several specific predictions about how components of structure should relate to applicant reactions. Future research must identify how to enhance applicants' perceptions of structured interview methods. Kohn and Dipboye (1998) showed how providing information about the nature of the structured interview helped reduce negative reactions. Organizational justice theory predicts providing explanations for hiring procedures should promote more favorable reactions (Gilliland, 1993), and we suspect that if applicants are informed as to why the interview is structured, more positive reactions will result. For example, applicants might be told that the interview is based on a job analysis and that the questions are going to provide them with an opportunity to display their competencies with regard to the specific job for which they are applying. So, rather than asking questions about just anything—such as what they want to be doing 10 years from now—the questions will focus on the job. Gilliland and Steiner (1999) provided many interesting possibilities for research on reactions to interviews.

The Interviewer Judgment Process

Similar to research on the performance appraisal judgment process (see chapter 4), researchers have examined the interviewer judgment process. Although the models differ in terms of their comprehensiveness and types of variables examined, they are similar in trying to describe the cognitive processes through which interviewers make judgments about applicants. Several such models exist, with perhaps the most popular being Arvey and Campion (1982), Dipboye (1992), Dipboye and Gaugler (1993), Dreher and Sackett (1983), and Schmitt (1976). Because this process is so similar to the performance appraisal judgment process, we do not belabor the discussion here. Suffice it to say that the task of making ratings of interviews is no less difficult than the task of making performance ratings, and it is possibly more difficult because the interviewer must rely on much less behavioral observation when making the judgment.

Potential Contaminants of Interview Ratings

There can be many types of contaminating sources of variance on interview ratings: impression management based on the interviewee, individual differences among raters, confirmatory biases, the presence of extraneous information, and many of the same sources of contamination we discussed with performance ratings (e.g., "similar to me" effects).

As for individual differences of the applicant (e.g., anxiety, appearance) or the interviewer (e.g., cognitive ability), the data to date suggest such differences are found but tend to be small in magnitude and inconsistently favor different demographic groups, and we do not know the extent to which they actually influence selection outcomes (Graves & Karren, 1999; Posthuma et al., 2002). Likewise, although there is pretty clear evidence that interviewers will exhibit a confirmatory bias such that they act in ways consistent with their early impressions, it is not clear the extent this influences validity and other outcomes (Dougherty & Turban, 1999). It is quite likely that the small or inconsistent effects may be due to the use of structured interview methods (which, if appropriately implemented, should minimize these differences). Even interviewer experience, which one would expect to be positively related to interview outcomes (e.g., validity, accuracy), has not conclusively been shown to have these positive features (Dipboye & Jackson, 1999; Posthuma et al., 2002). Although more experienced interviewers show less leniency bias, they are still affected by many sources of contamination such as the attractiveness of the applicant (Marlowe, Schneider, & Nelson, 1996).

Applicant (and possibly interviewer) impression management tactics constitute another source of potential contamination in interviews. Impression management exists when applicants exhibit a variety of verbal (e.g., ingratiation) and nonverbal (e.g., nodding in agreement) attempts to promote a favorable impression. Research is starting to show how impression management can influence interviewer judgments and hiring decisions (McFarland, Ryan, & Kriska, 2003; Stevens & Kristof, 1995), particularly for such nonverbal behaviors as smiling, nodding, and making eye contact (Gilmore, Stevens, Harrell-Cook, & Ferris, 1999).

An interesting program of research by Motowidlo and colleagues (Burnett & Motowidlo, 1998; DeGroot & Motowidlo, 1999; Motowidlo & Burnett, 1995) has provided compelling evidence of the importance of nonverbal behavior on interviewer judgments. Across these studies, they have shown how (a) students role playing interviewers made accurate judgments when presented only with either visual or aural information, (b) both sources of information are important and necessary to make accurate judgments, and (c) these effects are in part due to variations in vocal (e.g., pitch) and visual cues (e.g., appearance). It is quite interesting to know that simply watching an applicant may result in similar evaluations as having full access to both visual and aural information. Although it is not clear whether these behaviors are truly impression management, the findings nonetheless speak to the importance of nonverbal information.

Future research must begin to examine the antecedents and consequences of impression management to accurately determine whether these tactics are deceptive in nature (contaminants) or simply represent different forms of true score variance. For example, impression management tactics might represent true score variance for sales positions, but represent contaminating sources of variance for a computer programmer. Gilmore et al. (1999; see also Ferris & Judge, 1991) provide a model of the impression management process in interviews that might help illuminate these concerns. They note how impression management is affected by characteristics of the applicant and interviewer (e.g., personality), situational characteristics (e.g., accountability), and also probably the type of job. They posit that impression management influences interviewer judgments and outcomes through three mediators: liking, perceived fit/similarity, and evaluation of the applicant. Measuring all such variables in this kind of a process model will be necessary to determine whether impression management is a source of contamination or true score variance in interviewer ratings. It will also be necessary to manipulate or at least identify the components of structure in the interview because McFarland et al. (2003) demonstrated that structure influences the manifestation of impression management.

Legal Issues With the Use of Interviews

Like any predictor, the interview must hold up to legal scrutiny. We talk about legal and political issues in chapter 11, but for now let us summarize the recommendations of others (e.g., Roehling, Campion, & Arvey, 1999; Williamson, Campion, Malos, Roehling, & Campion, 1997). Roehling et al. (1999) suggested three dimensions of interviews that are most closely scrutinized with respect to cases of discrimination: consistency of applying the interview across applicants, the job relatedness of the interview questions, and the extent to which the interview process was designed to be objective.

What enhances these three characteristics? First, it is clear that ensuring interview questions are job related and based on a sound job analysis helps defend the legality of the interview. Second, anything that can be done to increase the objectivity of the interview ratings is evaluated favorably by the courts. This means using structured interviews (particularly asking the same questions of all applicants), well-defined (and documented) scoring systems, interviewer training, and related features will likely enhance the defensibility of the interview. Third, using demographically mixed panel interviews will if nothing else contribute to positive reactions from the courts and most likely minority applicants themselves. Finally, demonstrating criterion-related validity for the interview should make it more defensible. Williamson et al. (1997) demonstrated empirically how many of these interview features are related to favorable legal outcomes.

Technology and Interviews

Technology is influencing all assessment methods, and the interview is no exception. Organizations are increasingly conducting telephone interviews, using IVR interviews and videoconferencing technology. Research is just starting to accumulate on how these methods compare with traditional face-to-face interviews (e.g., validity, reliability) and are perceived by applicants. For example, Schmidt and Rader (1999) described the validity of an interview that is conducted entirely over the telephone. The interview is tape recorded, and a separate group of individuals scores the transcript of the interview. This was done to save administration costs—applicants can be from anywhere in the world, and one central office can conduct the interview. A study by Chapman and Rowe (2001) compared interviewer ratings when applicants participated in a face-to-face interview versus over videoconference software. They found applicants were rated higher in the videoconference condition.

To date, there is insufficient evidence to provide clear conclusions about the effects of such technologies. Given the previously mentioned findings by Motowidlo and colleagues, we may very well find the validity of such methods is not dramatically affected. In contrast, user reactions may be either more or less favorable (e.g., convenience versus impersonal). It is likely that practitioners will increasingly use these technologies because of their high cost savings, but this proceeds without a clear understanding of the benefits and trade-offs. Here is an area where there is a great need for solid research.

Improving Interview Validity

To this point, we have discussed a number of variations and issues with interviews. It may be helpful to summarize this information into a prescriptive set of guidelines to be followed when developing and using an interview. Table 10.2 lists the sequence of steps we recommend when developing interviews. Although not exhaustive, it captures the essence of the interview development process.

The suggestions shown in Table 10.2 should do more than just promote high reliability and validity from the interview. They should also help ensure the legal appropriateness of the interview. However, as we have noted previously, these suggestions may run counter to promoting favorable applicant reactions. Although we have seen only preliminary

TABLE 10.2
Suggestions for Developing Effective Selection Interviews

1. Conduct a thorough job analysis.
2. Base the interview questions on this job analysis. Keep the questions focused on job-relevant behaviors and processes, perhaps using the critical incident technique. It would be additionally helpful to target the questions to particular constructs.
3. Depending on the nature of the job and purpose of the interview, ensure a high degree of structure (most important, ask identical questions of all applicants).
4. Depending on the nature of the job and purpose of the interview, use situational, experience-based, or job-related questions.
5. Develop (at a minimum) two rating dimensions for each question (or more if not using interview panels).
6. Develop behaviorally anchored rating scales for each dimension and for each question. Leave space for interviewers to take notes or document behaviors that justify each rating.
7. Train interviewers on how to conduct the interview and use the rating forms.
8. Use interview panels of demographically diverse interviewers.
9. Require the interview panel to make an independent set of ratings, and then have the panel discuss and reach consensus to provide a final panel rating (this latter rating is ultimately to be used for the hiring decision).
10. Hold individual interviewers and interview panels accountable for their ratings. Monitor the process to ensure certain interviewers and panels are administering the interview correctly. An effective way to accomplish this is to audio- or videotape the interviews.
11. Unless consistent with the purpose of the interview, eliminate the use of extraneous information such as test scores and status in the selection process.
12. Explain to applicants the nature of the interview, why it has to be structured, and the roles and responsibilities taken by the applicant and interviewer.

research on this issue (Kohn & Dipboye, 1998), we propose that providing applicants with detailed explanations about the interview will promote more favorable reactions (Gilliland, 1993; Ryan & Ployhart, 2000).

Future Directions

The interview has and will continue to be a central part of staffing. We have learned a great deal about how to best design and use the interview, but as times change so does our need to adapt what we have learned to new contexts. We see two major thrusts for future interview research. The first is to resolve the dilemma between using structure for enhancing validity and reliability versus the negative effect this produces on applicant reactions. We strongly believe the resolution lies in the presentation of well-developed explanations for why the interview must be structured. However, a more recent article by Schmidt and Zimmerman (2004) may challenge some of the thinking about the role of structure in interview validity. After conducting several meta-analyses of both unstructured and structured interviews, they found it takes about four independent administrations of unstructured interviews to produce a validity similar to a single structured interview. This idea is interesting because when organizations or applicants resist structured interviews, a potential solution will be to conduct multiple—at least four—unstructured interviews. It will be fun to see how research addresses this issue. The second is to understand how to best merge the interview with technology (telephone, videoconferencing). There are a host of unanswered questions that are in dire need of research. For example, how do we know that persons on the phone are who we think they are? Are these technologies more or less useful for situational, behavioral, or job-based interviews? How do applicants respond to these technologies, and does their inclusion help or hurt organizational attractiveness? So, despite decades of research and hundreds of published papers on the interview, we are still far from having a total understanding of them.

PERFORMANCE-BASED AND SIMULATION TESTING

In this section, we consider performance-based and simulation testing. To facilitate our review, we compare these simulations on a continuum of physical fidelity, which refers to how well the tasks and cognitive

PERFORMANCE-BASED AND SIMULATION TESTING

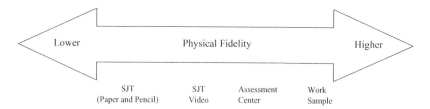

FIG. 10.1. The physical fidelity of various simulation predictor methods SJT, situational judgment test.

processes performed in the assessment correspond to the tasks and cognitive processes performed on the job. The situational judgment test (SJT) is often referred to as a low physical fidelity simulation. Assessment centers show higher degrees of physical fidelity, but the tasks are usually somewhat removed from the specific job tasks required for the job. The highest fidelity assessments are work samples because they require the completion of tasks nearly identical (if not identical) to those performed on the job. Figure 10.1 places SJTs, assessment centers, and work samples on a continuum of physical fidelity. In practice, these lines are often blurred, but we attempt to keep them distinct as has been most common in the extant literature.

Situational Judgment Tests

SJTs are becoming an increasingly popular predictor measure. Indeed, they have become so popular that an edited book has been devoted entirely to this topic (Weekly & Ployhart, in press). Although there are several variations of SJTs, the most common type will present applicants with a variety of work situations and ask them to choose from several behavioral responses how they would respond to each situation. These tests are called tests of *situational judgment* because they require applicants to use judgment in determining the appropriate (and inappropriate) responses in each situation. They are also called "low-fidelity" simulations (Motowidlo, Dunnette, & Carter, 1990) because although a work context is presented, the response mode is most frequently in a paper-and-pencil, multiple-choice format. However, the fidelity of SJTs is increasingly getting higher due in part to the use of video formats, where the situations are presented visually and aurally (although responses are still frequently recorded in a written format).

To illustrate an SJT, consider the following student situation described in Ployhart and Ehrhart (2003) to predict college GPA. The brackets indicate the correct [+1], incorrect [−1], and neutral [0] scores discussed shortly:

1. You are working on a group project with three other people. One of the members of the group has rarely been showing up to group meetings. She took on a lot of responsibility early in the project but is not producing any results. The paper is to be graded for group effort, not individual performance. What would you (a) most likely do and (b) least likely do?

   A. Go to the professor and report the problem group member. [0]
   B. Get together with the other two people and divide the work to make sure it all gets done. [0]
   C. Get together with the other two people, do only the work you were originally supposed to do, and include a note when you turn the project in that explains why it is incomplete. [−1]
   D. Confront the problem group member and ask him/her to start contributing to the project. [1]

SJTs have been available for decades, but their use in research and practice was restimulated in the 1990s by a series of studies by Motowidlo and colleagues (Motowidlo et al., 1990; Motowidlo & Tippins, 1993). Their research showed SJTs exhibit high criterion-related validity, but with much smaller subgroup differences than cognitive ability tests. Since that time, several studies have shown SJTs can have considerable validity. A meta-analysis by McDaniel, Morgeson, Finnegan, Campion, and Braverman (2001) found a corrected criterion-related validity of 0.34 (uncorrected 0.26), and Clevenger, Pereira, Wiechmann, Schmitt, and Harvey (2001) and Weekley and Ployhart (2005) showed that SJTs have incremental validity over other cognitive, experience, and noncognitive predictors. Lievens, Buyse, and Sackett (in press) further showed that the validity of SJTs is higher when the test is matched to the criterion. A meta-analysis by Hough, Oswald, and Ployhart (2001) reported standardized White–Black differences (d) favoring Whites 0.61 for paper and 0.43 for video-based formats; gender differences tend to favor women (0.26 paper, 0.19 video-based formats).

Developing SJTs typically requires a three-step process (Motowidlo, Hanson, & Crafts, 1997). Table 10.3 gives examples of each step for the job of a customer service retail position. The first step is to collect critical incidents of work behaviors from a sample of job incumbents

PERFORMANCE-BASED AND SIMULATION TESTING 513

(we discuss the critical incident method in chapter 3). This step identifies the types of situations that will comprise the test inventory. The critical incident method is ideal because it helps identify the context or situation in which the behavior occurred. The second step is to ask a different group of SMEs to review and edit the situations for their appropriateness. At the same time, they will write as many behavioral response options as appropriate. When writing the response options, SMEs should attempt to write behaviors that are behavioral, specific, and unambiguous. Further, they should write responses that job incumbents would reasonably perform, but that are not so obvious that all test takers would pick that option. A good way to get SMEs to generate these kinds of responses is to have them think about how they would respond in that situation, and then three different employees: an exceptional employee, an average employee, and a below average employee. This helps ensure a range of effective and ineffective responses, which is important for eliciting variance in applicants' responses. The final step involves developing the scoring key for the SJT, where SMEs (usually expert job incumbents or supervisors) will determine which responses are the appropriate and inappropriate ones. They will also review the situations and response options to ensure they adequately tap the nature of the job. The SMEs will read each situation and then indicate (in their own opinion) which response options are effective and ineffective. Situations are retained when there is high agreement on the correct and incorrect options; situations are eliminated when there is disagreement or no clear correct responses. Table 10.4 gives some guidelines that in our experience have proven useful for generating effective SJT items.

Because SJTs are measures of judgment and "correct" responses are usually a matter of degree rather than absolute, there are a variety of additional scoring issues to consider. The first is whether test instructions ask what applicants "would do," "should do," or "is most effective" in each situation (see Weekley, Ployhart, & Holtz, in press, for many more variations). Ployhart and Ehrhart (2003) showed this choice is not arbitrary, and asking different questions produced differences in SJT reliability, construct validity, and criterion-related validity. They speculated "would do" instructions tap behavioral intentions, personality, and/or past behavior, whereas "should do" and "rate the effectiveness" instructions tap job knowledge and cognitive ability. For example, consider the situations in Table 10.3. There might be a big difference between what you would do, what you should do, and what

TABLE 10.3

Steps for Developing a Situational Judgment Test for a Retail Service Position

Step	Subject-Matter Experts	Examples
Step 1. Generate situations using the critical incident technique. Create situations that are fairly common, but where there is no clear course of action.	Job incumbents who currently perform the job.	It has been a busy day in the store and you are tired. You finally get a few minutes to relax because there are only a few customers in the store. However, you are not scheduled for a break and so you cannot leave the customer area. In your store, there are many small tasks (e.g., putting away inventory) that are not part of anyone's specific job but are something that everybody is expected to do. Many of your coworkers do not do these tasks, but if they are to get done somebody will have to do them.
Step 2. Edit situations and create response options. Attempt to generate four to six options for each situation. No option should be obviously correct or incorrect, but would realistically be performed by at least some job incumbents.	Job incumbents who currently perform the job.	It has been a busy day in the store and you are tired. You finally get a few minutes to relax because there are only a few customers in the store. However, you are not scheduled for a break and so you cannot leave the customer area. a. Leave the customer area and take a break. b. Continue working but pace yourself. c. Take a break in the customer area. d. Ask a coworker to cover for you. e. Request a break from your supervisor. f. Keep working no matter how tired you are. In your store, there are many small tasks (e.g., putting away inventory) that are not part of anyone's specific job but are something that everybody is expected to do. Many of your coworkers do not do these tasks, but if they are to get done somebody will have to do them. a. Perform all the tasks, even though they are not part of your job.

TABLE 10.3
(continued)

Step	Subject-Matter Experts	Examples
		b. Do not perform the tasks because they are not part of your job.
		c. Give the impression you are performing these tasks, but in reality leave them for someone else.
		d. Inform your boss about this problem.
		e. Try to convince your coworkers to share the tasks equally.
		f. Get the newer employees to do these tasks.
Step 3. Determine the scoring key. Refine and revise items. Delete items that show low agreement. Think carefully about what type of test instructions will be used, how subject-matter experts will be asked to evaluate the options, and how situational judgment test scores will be created and combined.	Job incumbents who currently perform the job; supervisors familiar with the job.	It has been a busy day in the store and you are tired. You finally get a few minutes to relax because there are only a few customers in the store. However, you are not scheduled for a break and so you can not leave the customer area.
		a. Leave the customer area and take a break. [0]
		b. Continue working but pace yourself. [+1]
		c. Ask a coworker to cover for you. [0]
		d. Request a break from your supervisor. [−1]
		In your store there are many small tasks (for example, putting away inventory) that are not part of anyone's specific job but are something that everybody is expected to do. Many of your coworkers do not do these tasks, but if they are to get done somebody will have to do them.
		a. Perform all the tasks, even though they are not part of your job. [0]
		b. Make a small attempt to perform these tasks. [0]
		c. Inform your boss about this problem. [−1]
		d. Try to convince your coworkers to share the tasks equally. [+1]

Note: The numbers following the options in Step 3 indicate a scoring system based on Motowidlo et al. (1990). That is, +1 is most appropriate and 1 is least appropriate.

TABLE 10.4
Guidelines for Writing Effective Situational Judgment Test Items

---

1. Write the item with the minimum level of reading difficulty required for the job. Sacco et. al. (2000) demonstrated how SJT reading level was related to Black–White subgroup differences, such that larger differences were found with higher-level reading requirements.
2. Minimize the use of job- or context-specific jargon.
3. Ensure the response options reflect specific behaviors.
4. Minimize the use of "leading phrases" in the response options. "You ask the supervisor for assistance" is a good example; "You ask the supervisor for assistance because you are tired" is a bad example.
5. Ensure each response option is likely to be exhibited by at least a few employees.
6. When generating situations, ask SMEs to generate situations that are specific, difficult, and challenging. Ensure not everyone performs the same in the situation.
7. When generating response options, ask SMEs to first think of how they would handle the situation. Then ask them to think of how a highly effective, average, and ineffective employee would handle the situation.
8. When generating situations using the critical incident technique, first ask SMEs to freely recall the situation, and then give them examples of performance dimensions to stimulate the generation of new situations.
9. Work with SMEs in small groups, but conduct multiple sessions to ensure a wide range of responses (sometimes a group will get fixated on a particular type of situation or response).
10. Choose the appropriate form of instruction for the purpose of the test. "Would do" instructions tend to produce more variability in responses, but that may only be true when applicants respond honestly. "Should do" instructions may be more faking resistant (McDaniel & Nguyen, 2001), but may not exhibit as much variability in responses.

Note: SJT, situational judgment test; SME, subject-matter expert.

might be most effective. In practice, the "would do" instructions are most frequently used (McDaniel & Nguyen, 2001) and appear to show the most desirable psychometric properties (Ployhart & Ehrhart, 2003), but effects on validity are unclear.

A second issue to consider is how SMEs create the scoring key. The most common method asks SMEs to indicate which of the options is most appropriate and least appropriate (McDaniel & Nguyen, 2001). SMEs may simply be asked to indicate which is most and least effective or may be asked to rate each option on a 5-point effectiveness scale, and then the highest and lowest responses are by default the best and worst options. Alternatively, one may simply score SJTs based on a single correct response as would be done on a knowledge test, but this approach appears to be less frequently used in practice (Motowidlo et al., 1997).

A final scoring issue involves how the scores will be combined. This decision is clearly affected by the type of SME scoring method used (e.g., identify best and worst options, rate the effectiveness of each option). When using a "choose what you would most and least likely do" instruction set, Motowidlo et al. (1990) scored the SJT as follows:

- Correctly identifying the most appropriate response is worth +1 point.
- Correctly identifying the least appropriate response is worth +1 point.
- Incorrectly picking the least appropriate response as the most appropriate response is worth –1 point.
- Incorrectly picking the most appropriate response as the least appropriate response is worth –1 point.
- Choosing all other options is worth 0 points.

Thus, in this system, an SJT item can range from −2 to +2. This scoring system has been among the most widely used (McDaniel & Nguyen, 2001; Motowidlo et al., 1997), and is illustrated in Step 3 in Table 10.3. Yet another approach asks applicants to rate all options on a 5-point effectiveness scale. This latter approach can be quite burdensome for applicants (e.g., a 20-item SJT with four options each will require 80 ratings). In our experience, rating the effectiveness of each option produces scores highly related to the Motowidlo et al. (1990) scoring method and does not produce meaningful differences in criterion-related validity. However, the Motowidlo approach is considerably easier for applicants and less cognitively taxing.

As you might guess, an important question with using SJTs in practice is whether SJT scores are affected by experience and/or job knowledge. That is, one cannot very well ask a person what he or she would do in a situation if that person had never experienced it (and if all applicants had not experienced it, the SJT scores would be essentially random variance). For this reason, there has been some concern over whether SJTs are simply surrogate measures of knowledge, experience, cognitive ability, or even personality. Although experience and/or job knowledge appear to be tapped by SJTs, the data do not support claims that SJTs are redundant with knowledge, experience, cognitive ability, or personality (Clevenger et al., 2001; Weekley & Ployhart, 2005).

Thus, research to date suggests that SJTs are clearly not redundant with job knowledge, experience, cognitive ability, or personality, but this does not necessarily tell us what SJTs measure. The current state of the literature argues SJTs are measurement methods that capture a broad array of KSAOs (Chan & Schmitt, 1997). Meta-analyses have supported this assertion, finding that SJTs correlate with cognitive ability, personality, experience, and knowledge (e.g., McDaniel et al., 2001; McDaniel & Nguyen, 2001). Although the notion that SJTs are multidimensional measures may be a sufficient explanation in the short term, it is less than desirable from a scientific perspective. A major task for future research is to better understand the nature of what SJTs measure and whether they can be designed to target specific unidimensional constructs (Schmitt & Chan, in press). Future research must also examine whether different scoring methods and systems produce differences in validity, subgroup differences, and construct validity (Weekley et al., in press). Likewise, we need to understand how the validity and subgroup differences of SJTs vary across different administration formats (e.g., paper, video). For example, Chan and Schmitt (1997) compared the subgroup differences between a paper SJT and a video SJT and found dramatic reductions in Black–White differences (see also chapter 8) when using the video SJT.

To disentangle what SJTs really measure, it may be necessary to adopt methods from cognitive psychology and conduct a "cognitive task analysis" on the task of completing an SJT (see chapter 3). Consider the predictor response process model discussed in chapter 7. What KSAOs represent the "true score" for an SJT? It is quite possible that different KSAOs would influence the interpretation and response phases. For example, what KSAOs are used when one completes an SJT containing situations one has never experienced? Research needs to move from examining correlates of SJTs to experimentation testing predictor response theories.

This leads to our last question with SJTs—the extent to which they may be faked. Considering the response process model in chapter 7, SJT responses may be affected by motivation and impression management. To a certain extent, faking could be mitigated by the type of instruction used. Asking "would do" instructions leaves a lot of opportunity for applicants to distort their responses. Currently, the verdict is out on whether and how much SJTs may be affected by socially desirable responding because so few studies have been conducted (see Hooper, Cullen, & Sackett, in press). In terms of actual applicant contexts,

PERFORMANCE-BASED AND SIMULATION TESTING

Weekley, Ployhart, and Harold (2004) found that SJTs demonstrate similar levels of criterion-related and construct validity across applicant and incumbent contexts. However, this may be affected by administration format. Ployhart, Weekley, Holtz, and Kemp (2003) examined the equivalence of SJT scores administered in paper and Internet formats across applicants and incumbents. They found the applicant Web-based version showed lower mean differences from incumbents than the applicant paper version, perhaps indicating less response distortion in the Internet version. Considerably more research is necessary to answer these questions, and Hooper et al. (in press) provide many excellent suggestions. Box 10.2 describes a project by one of the authors to develop an SJT for college admissions.

Assessment Centers

The assessment center is widely used for selecting such personnel as diverse as managers, firefighters, police officers, and soldiers. Like other simulation methods, assessment centers are based on the assumption that the best predictor of future behavior is past behavior. Because they are a method of measurement rather than an attempt to assess a single construct, assessment centers take a variety of forms and may be used to measure a variety of constructs. However, a typical assessment center works something like this: Potential candidates participate in the assessment center either individually or in small groups. They participate in a variety of exercises that represent different aspects of the job; these exercises often demonstrate high physical and psychological fidelity. The assessment center may last anywhere from a few hours to a few days. As they participate in the exercises, several trained raters will evaluate the performance of the applicant by using ratings. Table 10.5 gives three examples of assessment centers.

Earlier when we discussed interviews, we divided the literature into subsections to facilitate our review. The literature on assessment centers is likewise large, and we follow a similar set of topics to help structure our review: guidelines, purpose, exercises, constructs assessed, validity, subgroup differences, utility, applicant reactions, the rating process, legal issues, new technologies, and future directions. However, unlike the interview literature, there is a strong network of professional organizations that advocate good assessment center practice. For example, a group of practitioners developed a document of "best practices" for assessment centers called the *Guidelines and Ethical Considerations*

> **Box 10.2 Development of a Situational Judgment Measure for College Students**
>
> One of the authors has worked on the development of a situational judgment measure designed to measure college student potential. The project began with a content analysis of university goal statements. This analysis identified 12 dimensions that were common across various universities. They included knowledge, ethics/integrity, perseverance, leadership, multicultural appreciation, social responsibility, and continuous learning activity, to name a few. Development of situational judgment items relevant to these dimensions included the following steps:
>
> 1. Students were provided with definitions of each dimension and asked to generate critical incidents (effective and ineffective behavior) relevant to each dimension.
> 2. The researchers used these incidents to develop item stems (questions) concerning each incident. That is, what would you do if you saw students steal an exam?
> 3. Another group of students was asked to provide answers to these questions.
> 4. These answers were edited by the research team so there were four to eight plausible and nonredundant answers to each question.
> 5. A group of junior and senior student experts were then asked to judge the effectiveness of each alternative action and to indicate which alternative they would be most and least likely to take if confronted with each situation. Their answers were used to develop a scoring key for this situational judgment measure using the Motowidlo et al. (1990) approach.
>
> The end result was an instrument that posed problems of relevance to a wide domain of college student performance in language generated by the students themselves. In three separate studies, the instrument has been correlated with first-year college GPA (0.15–0.20) and self-ratings of the 12 dimensions (0.4–0.6).

TABLE 10.5
Brief Descriptions of Three Assessment Centers

	POSITION FOR WHICH CANDIDATES ARE EVALUATED		
	School Administrator	First-line Supervisor	Middle Manager
Length of Time for Candidate Evaluation	2 days	1 day	3 days
Length of Time for Assessor Integration Meetings	2 days	1 day	2 days
Assessor/Candidate Ratio	1 : 2	1 : 3	1 : 4
Exercises	• In-basket • Personal interview • Leaderless group discussion of case • Assigned role group discussion • Fact-finding exercise with oral presentation	• In-basket • Interview • Group exercise involving human relations problem • Problem-solving group exercise • Biographical questionnaire • Mechanical test • Situational questionnaire	• Intelligence test • Projective test • In-basket • Group discussion of case study • Assigned role exercise • Self-report • Personality tests • Oral report
Sample of Rated Dimensions	• Problem analysis • Judgment • Organizational ability • Sensitivity • Personal motivation • Oral and written communication	• Organizing and planning • Analyzing • Decision-making • Controlling • Influencing • Oral communications • Interpersonal relations • Flexibility	• Judgment • Initiative • Adaptability • Planning and organizing • Originality • Abstract reasoning • Interpersonal sensitivity • Impact • Technical expertise • Oral and written communication

for *Assessment Center Operations* (Task Force on Assessment Center Guidelines, 2000). This document and earlier versions have served as the source for effective assessment center development for decades; hence, it makes sense to start from this perspective.

Guidelines

There are 10 guidelines established as a minimum for adequate assessment center practice (Task Force on Assessment Center Guidelines, 2000). The key recommendations are as follows:

1. The assessment center exercises and dimensions should be based on a job analysis.
2. Candidates' behavior during the assessment center must be able to be classified into appropriate categories (i.e., dimensions).
3. The assessment center exercises must be designed to appropriately measure the dimensions identified in the job analysis.
4. Multiple assessments and exercises must be used to elicit relevant, objective, and reliable information.
5. Multiple simulations must be used to allow adequate opportunity to observe behavior. These should parallel actual work behaviors and tasks as much as possible.
6. Use several assessors for evaluating each assessee, unless a computer will be doing the scoring. The candidate's immediate supervisor should be excluded, and consideration of demographic factors should influence assessor composition.
7. The assessors should be fully trained.
8. Use a measurement system that accurately assesses candidate behavior.
9. Reports of the candidate's observations must be made during each exercise.
10. Use a valid and reliable system for integrating candidate information (e.g., decide the consensus process raters will use to make a final decision).

These guidelines have had a fundamental influence on assessment center design, implementation, and administration. Yet, they are much broader than even these 10 suggestions; they also provide guidance linking assessment centers into HR practices, how to convey the purpose of the assessment center to candidates, and how to convey results to various constituencies.

Spychalski, Quiñones, Gaugler, and Pohley (1997) asked whether HR personnel actually use assessment centers in a manner similar to these guidelines. Although they focused only on organizations within the United States, they obtained responses from 215 organizations. Interestingly, they found organizations followed the guidelines pretty carefully with respect to assessment center design and training (relating to the 10 guidelines summarized previously). Importantly, most assessment centers were based on a job analysis. However, organizations less frequently informed participants about the purpose and nature of the assessment center (although they did provide fairly detailed information about how they performed after completing the assessment center). They also insufficiently demonstrated the validity evidence of the assessment center and did not regularly evaluate the performance of the assessors. Thus, we find assessment center users tend to pay careful consideration to development and administration issues, but less to evaluation issues and concerns about the candidate's perspective.

Purpose

Because they are a comprehensive evaluation of an individual's behavior, assessment centers may serve multiple HR purposes. By far, the most common is for selection and promotion (Spychalski et al., 1997). Here, the goal is to evaluate candidates' potential to successfully perform the job in question. Assessment centers can be expensive to develop and administer, so they are frequently used during later stages of selection (often only after applicants have passed multiple hurdles). Furthermore, they are primarily used for selection and promotion into jobs that have important consequences, such as higher-level managerial positions, and for police and firefighter positions.

However, unlike many other selection methods, assessment centers also serve developmental or training purposes. In fact, during World War II, the British and Germans were the first to use something like our current assessment centers for the selection of spies. Potential spies were placed in realistic situations requiring them to cope, reason, solve, and work under the stress of fear and time deadlines. Today, military divisions throughout the world use assessment centers for training and development, often in a team environment. For example, U.S. Special Forces will participate in exercises that last several days in the field.

They are presented with nearly impossible tasks, and their performance is evaluated continuously by trained assessors who observe—but do not participate—in the field exercises. Similarly, Lievens, Harris, Van Keer, and Bisqueret (2003) demonstrated how the assessment center dimensions of adaptability, teamwork, and communication proved useful for selecting managers into cross-cultural training.

It is safe to say that for jobs where the consequences are great, assessment centers have been one of the most favored selection, promotion, and development methods.

Exercises

We mentioned earlier that assessment centers are really a collection of various exercises that assess multiple constructs (discussed next). Woehr and Arthur (2003) reported the average number of exercises used in assessment centers is 4.73 ($SD = 1.47$). Developers of assessment centers have shown real creativity in the types of exercises that are used. However, a few exercises are so common that they exemplify the typical assessment center (Spychalski et al., 1997):

- *In-Basket.* Here, the task requires candidates to review a set of tasks sitting in their "in-baskets," and prioritize and complete these tasks in the most efficient and effective manner. The in-basket will typically comprise letters, memos, reports, and related documentation. Examinees are asked to act on these items as though they were on the job. These actions are scored as a test. Figure 10.2 gives an example of an in-basket exercise for school administrators.
- *Leaderless Group Discussion.* A group of participants is presented with a problem that needs to be solved (e.g., production scheduling, product development). The participants do not know each other, and nobody is instructed to be the leader of the group (although sometimes a person is randomly chosen to be the leader). Assessors observe the group interactions to rate the leadership behaviors of the participants.
- *Role Play.* A role play asks candidates to put themselves into a particular role and situation, and either act or describe what they would do. For example, McFarland et al., (2003) described role plays used by firefighters for promotional purposes. Candidates are asked to take the role of a lieutenant who must address a

PERFORMANCE-BASED AND SIMULATION TESTING 525

EAGLETOWN HIGH SCHOOL

TO: Joe
FROM: Bill Smith
RE: Sex Education Classes
DATE: January 25

    As you know, a few of the counselors and Mary Brown, our Home Economics teacher, have been tossing around the idea of completely revamping our sex ed. classes. We have researched the curriculum content with several Class A communities and feel confident it will be a plus to all involved. We're ready to move on it but realize we've got some convincing ahead—it's still controversial in some parts of the community. What do you suggest be the first step?

Dear Mr. Heckle:

    As you know, I take pride in my students' ability to achieve. Having to deal with behavioral problems decreases my effectiveness in assisting the rest of the class successfully.

    Perhaps you haven't heard of what a troublemaker Tom Miller is. He continually disrupts the class during student participation in delivering oral book reports (an area by the way, which I emphasize for building reading comprehension and communication skills). He has never given a report and when I asked him to read aloud a passage from our text, he barely could get through one sentence.

    I want Tom to be reassigned to a different class—anything but mine. I'm bringing this to your attention first, because the sooner the better.

                                   Sincerely,

                                   Viola Twig

FIG. 10.2. In-Basket Examples.

behavioral dispute among subordinate firefighters; their task in the role play is to explain what they would tell these firefighters.

- *Simulations/Business Game.* This generic set of exercises will require candidates to evaluate, analyze, and synthesize various sources of information. For example, they may be asked to look at a series of financial reports to determine what steps the organization should take. This kind of situation is often referred to as a business game. Alternatively, the simulation may involve solving a challenging task. An example of such a simulation is shown in Fig. 10.3. This simulation was developed as a promotional exam for team leaders in a manufacturing setting. The aspiring team leaders had to

10. HIRING PROCEDURES III

Provide a performance evaluation for a below-average associate

Overview

An associate who has been on your team for some time is starting to have performance problems. Although this associate has typically been a strong performer and hard worker in the past, the last few months have shown a noticeable decline in performance. In fact, this associate is producing only about half as well as s/he has in the past. Unfortunately, you believe the associate does not seem too concerned about the change in performance, if s/he even notices it at all.

Your task is to inform this associate about the decline in performance and to try to determine why it is occurring. You should also try to solve the problem, if possible, or take disciplinary action if required.

Simulation Constraints

Preparation Time: 15 minutes
Simulation Time: 10 minutes
Questions: 5 minutes

Situation

An associate on your team has been performing poorly for the last several months. This associate has worked for the company for a little more than 2 years, and although you are not close friends with the associate, you both seem to get along quite well with each other. The associate has always been a hard worker and has not had any problems before. However, over the past few months you have noticed the associate performing worse than usual. In fact, the associate's performance is about half of what it used to be. This can be seen in the graph below:

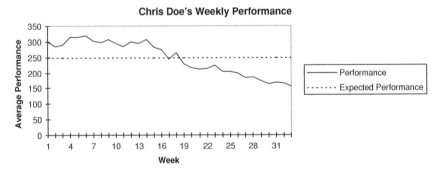

You have been watching the associate for some time and have documented in your own records the following major incidents and behaviors:

January 9. Associate fails to detect a defective piece of equipment. The equipment breaks down, causing a 5-hour delay for several teams. At that time, it was the associate's task to monitor the equipment and s/he had performed this task very effectively in the past.

January 11. Associate trips over a cord. Falls onto another associate, causing her to fall to the ground and sprain her ankle.

FIG. 10.3. Sample of simulation exercise.

PERFORMANCE-BASED AND SIMULATION TESTING 527

January 28. Associate's performance falls below the expected performance level. Speaking casually with the associate, s/he suggested it was a temporary situation.

February 2. Associate is talking to a coworker and breaks a torquethrust spanner wrench. Follow-up examination of the wrench found that the power controls were set at too high a level for the task. Associate is somewhat argumentative during conversation.

February 9. Associate has not improved performance since casual conversation held on January 28.

February 12. Associate is clearly performing at about half his/her previous rate.

February 13. Several complaints received over last week from other teams about poor workmanship. All complaints relate to projects with which the associate has been involved.

February 20. Associate identified by other associates as making several mistakes on assigned tasks. Most mistakes were a result of carelessness.

February 25. Associate's performance was about 70% of expected rate.

March 1. Associate reprimanded about joking too much with other team members during operating hours. The joking was causing the team's performance to decrease.

March 3. Associate had a conversation with a member from another team while the team leader and subleader were away. Upon the subleader's return, the other team member left but the conversation had apparently taken place for some time (at least 20 minutes), as the associate was very far behind in his/her tasks.

You are sure the decline in performance is not due to faulty equipment or a difficult task because the associate has been performing the same task for most of the year and is very familiar with it. Even more troubling is that the associate does not seem too concerned about the decline in her performance. In fact, s/he appears to be in high spirits because this associate seems to be joking around quite a bit lately, although not to the extent that it is the main cause for the decline in performance.

You speak to your team leader about this associate, and s/he suggests you speak to the associate directly because you know him/her best. Your team leader also tells you that unless this associate's performance improves, disciplinary action will be required.

You have heard second-hand information from other coworkers that this associate may be having personal problems, but you do not know whether this information is accurate.

Your task is to formally address this associate's performance problem and attempt to solve it. You have 15 minutes to review these materials and to prepare what you would say to this associate. Scrap paper is available if you would like to write down notes or ideas. Then, you will have 10 minutes to formally present this information to the personal interviewers. Note that you are provided with a transparency and overhead projector to use in your presentation, if you want.

Your presentation should include what you would say to the associate and the types of questions you might ask in this situation. However, you will not directly ask questions in this simulation. For example, instead of asking, "Why are you doing this?", you would say, "I would ask the associate why he or she was doing this." Thus, state what types of questions you would ask and how you would ask them. Remember that you want to be sure the associate is aware of the severity of his/her performance decline. After your presentation, one rater, acting as this associate, will ask you some questions to which you are to respond. These questions will last no longer than 5 minutes.

FIG. 10.3. (*continued*)

demonstrate how they would conduct a performance evaluation for an underperforming employee.
- *Mock Presentations.* Presentation skills are a critical component of many managerial jobs, and assessment centers provide a useful means for assessing those skills. For example, one of the authors developed a mock presentation exercise as part of a promotional exam for state troopers. A key part of the trooper's job was to interact with the press and media. The exercise was developed such that the trooper had to convey the details of a suspect connected with a bank robbery. Trained assessors took the role of media personnel by asking questions and details about the event.
- *Interviews, paper-and-pencil tests, and related measures.* Many of the other predictor constructs and measures we have discussed are frequently used in conjunction with the aforementioned exercises.

This list gives a sampling of the types of exercises frequently used in assessment centers. Clearly, there are many similarities across these exercises, and it sometimes becomes difficult to differentiate between a role play and a simulation. Such differences are not that important; what does matter is whether the exercises are based on a solid job analysis and reflect important aspects of the job.

Constructs Assessed (Dimensions)

If there is variability in assessment center exercises, there is even more variability in the types of constructs assessed. In the language of assessment centers, constructs are usually referred to as *dimensions.* A study by Arthur, Day, McNelly, and Edens (2003) reviewed research on assessment center dimensions. Their comprehensive review finds approximately 129 different labels given to dimensions, such as aggressiveness, career motivation, primacy of work, self-confidence, and personal impact. A major contribution of their work was to collapse these dimensions into six core dimensions:

- Consideration and awareness of others (largely interpersonal skills and social sensitivity)
- Communication (oral and written)
- Drive (reflecting high motivation and energy)
- Influencing others (persuasion and influence)

PERFORMANCE-BASED AND SIMULATION TESTING

- Organizing and planning (for both self and others)
- Problem solving (gathers, interprets, analyzes, and synthesizes information)

In their review, Woehr and Arthur (2003) found the average number of dimensions evaluated in a given assessment center was 10.60 ($SD = 5.11$). Surprisingly, the number of dimensions ranged from 3 to 25. Given these findings, it should be clear why assessment centers are considered a test method rather than a construct—the average assessment center is expected to assess almost 11 constructs. It is also informative to examine how little overlap exists between the dimensions assessed in interviews with the dimensions assessed in assessment centers. Despite many similarities, these two test methods have not frequently been used to assess the same constructs.

Validity

The issue of validity has been both the strength and the weakness of the assessment center. Early research on the validity of the assessment center conducted at AT&T (Bray & Grant, 1966) was impressive. In one study, 123 newly hired college graduates and 144 noncollege, first-level supervisors were assessed, but the assessment center reports were filed away and not used. Eight years later it was found that 82% of the college graduate candidates and 75% of the noncollege candidates who had been promoted to middle management had been correctly identified by the assessment center as persons likely to be promoted to middle management. The ability of the assessment center to identify those who would not be promoted was even greater. Eighty-eight percent of the "unsuccessful" college graduates and 95% of the "unsuccessful" noncollege candidates were correctly identified.

These optimistic findings helped stimulate considerable research on assessment center validity. Meta-analyses of these primary studies have indeed found high criterion-related validity for assessment centers, with more recent estimates of overall ratings correlating 0.37 (corrected) with overall job performance (Gaugler, Rosenthal, Thornton, & Bentson, 1987; Schmidt & Hunter, 1998).

We have also learned about the limits of assessment center validity. For example, Schmitt, Schneider, and Cohen (1990) documented that when local assessors alter the procedures to be followed in an assessment center, the validity of the method declines. Lievens and Goemaere

(1999) interviewed assessment center administrators and found considerable variability in administrative procedures, due in part to problems with dimension definitions, exercise design, and the content of training. Not all findings are so discouraging. Jansen and Stoop (2001) found assessment centers predict performance up to 7 years later, although there is considerable variability in these relationships over time and across dimensions. It has also been found that assessment centers exhibit incremental validity over other predictors (e.g., cognitive ability, personality), or at least tap different aspects of the performance domain (e.g., Goffin, Rothstein, & Johnston, 1996).

Unfortunately, why assessment centers relate to performance has been unclear because they have been said to lack construct validity (Klimoski & Brickner, 1987); that is, assessment centers tend to show poor convergent and discriminant validity. Consider an assessment center with multiple exercises and dimensions. Convergent validity would occur when the same dimensions correlate highly across exercises (latent dimension factors), and discriminant validity would occur when the within exercise ratings are relatively uncorrelated with each other. However, this is not the typical finding. Rather, exercise factors tend to emerge such that ratings are more correlated within exercises than across exercises (thus contributing to latent exercise factors; Sackett & Dreher, 1982). This finding has historically been robust (Schneider & Schmitt, 1992). For example, ratings of assessees on a given dimension of behavior (e.g., Flexibility) in one part of the assessment center (e.g., the In-Basket) do not tend to correlate well with ratings made of Flexibility in the Leaderless Group Discussion. The finding of latent exercise factors has led to many pessimistic conclusions regarding the construct validity of assessment centers. The review by Klimoski and Brickner (1987) stimulated considerable research to better understand this problem, often resulting in the manipulation of various assessment center design factors (e.g., training, rating formats, integration method) to enhance construct validity. However, within the last few years, there have been several important advancements to our understanding of assessment center construct validity.

One major source of advancement has come from recent reviews and meta-analyses examining the impact of assessment center design features on validity. These studies have looked at such features as number of dimensions, number of exercises, type of consensus procedure, and type of assessor. A review by Lievens (1998) indicated construct validity would be enhanced when:

- A smaller number of dimensions are used that are highly dissimilar to each other
- Raters are trained
- Psychologists play a major role in the evaluation of candidates
- Exercises maximize dimension-related variance
- Rating aids such as note-taking or behavioral checklists are used

Large-scale reviews and meta-analyses help refine these suggestions. For example, Arthur, Woehr, and Maldegen (2000) used generalizability theory to estimate the amount of variance due to dimensions, exercises, applicants, and raters. They found most of the variance in ratings was due to applicants, dimensions, and the interaction between dimensions and applicants. Woehr and Arthur (2003) meta-analyzed a number of methodological factors and found that rating approach, assessor occupation, and assessor training helped enhance convergent validity. In contrast, Lance, Lambert, Gewin, Lievens, and Conway (2004) found evidence across 24 studies that exercise variance is still considerably greater than dimension variance.

Although the dominance of exercise factors may still occur, primary research is showing more optimistic signs of assessment center construct validity. In a thoughtful study, Lievens (2002) manipulated assessor type (student, manager, I/O psychologist) and assessee performance. Assessee performance consisted of four types, varying on how differentiated performance was across dimensions and consistent across exercises (thus a 2 × 2 design). Assessors watched videotapes and made their ratings. The results indicated construct validity (convergent and discriminant) can only occur when candidates exhibit high cross-exercise consistency and high differentiation across dimensions. This is a critical finding because it argues that assessment center validity may be most determined by assessee behavior (a finding consistent with Arthur et al., 2000). In practice, it is unlikely for applicants to display such high consistency across situations or differentiation across exercises (e.g., McFarland et al., 2003). The implication is not only that assessment centers may have had more construct validity than we previously believed, but also we should not expect to find such high convergent and low discriminant validities due simply to the nature of candidate behavior.

A program of research by Lance and colleagues (Lance, Foster, Gentry, & Thoresen, 2004; Lance et al., 2000) offers related support. They argued that the presence of exercise effects does not represent a lack construct validity (i.e., method effects), but rather situational specificity. That is, candidates do not perform identically across exercises, but they do tend to perform better on some exercises than others. In a compelling study, Lance, Lambert, et al. (2004) demonstrated that exercise factors showed significant correlations with measures of knowledge; something that should not happen if they merely represented method bias. Thus, it appears that exercise factors may capture substantive variance related to job performance.

Given these more encouraging findings, researchers are starting to reexamine the role of dimensions relative to an overall assessment center rating. Earlier we mentioned how Arthur et al. (2003) identified the six primary assessment center dimensions. In their review, they also demonstrated the criterion-related validity of these dimensions, finding they ranged from 0.20 to 0.30 uncorrected (0.25 to 0.39 corrected). In a multiple regression, four of these dimensions (problem solving, influencing others, organizing and planning, communication) accounted for the majority of variance and predicted performance more strongly than an overall composite rating. Thus, there appears to be some unique information about performance that is gathered from dimensions that is lost when collapsing across dimensions.

So, after years of trying to understand the construct validity of assessment centers, it appears we finally have some reasons to be optimistic. It is clear that despite attempts to structure, develop, and administer the assessment center following best practices, the nature of assessee behavior may be the one feature that most impacts construct validity—but it is also beyond the control of the assessment center developer. This is not a bad thing, and in fact, it helps explain why assessment centers relate to performance even in the apparent "absence" of construct validity.

Subgroup Differences

Hough et al. (2001) reported relatively small subgroup differences in assessment centers, with the sole exception being that older individuals tend to be rated less favorably. Part of the reason may be due to the lower reliability of such measures (which attenuates subgroup differences). It may also be due to the multidimensional nature of the KSAOs required to perform in the simulation. However, focusing on overall ratings may mask important differences. Goldstein and colleagues (Goldstein, Yusko, Braverman, Smith, & Chung, 1998; Goldstein, Yusko, & Nicolopoulos, 2001) found that assessment center dimensions vary in their relationships with cognitive ability. A consequence is that exercises with more of a cognitive "loading" should exhibit more of a subgroup difference; conversely, if one statistically removes cognitive ability from the exercise, the subgroup difference should decrease. This is what they found. The more cognitively loaded exercises (e.g., problem solving) exhibited larger Black–White subgroup differences, and these differences were reduced when removing the variance shared with cognitive ability. However, criterion-related validity was not dramatically reduced when controlling for cognitive ability, and the assessment center dimensions still predicted performance incrementally beyond cognitive ability. The implication is that exercises should be developed so their cognitive loading does not exceed that necessary for the job.

The Utility of Assessment Centers

An obvious concern with the use of assessment centers is their development and administrative costs. For example, Hoffman and Thornton (1997) provided an illustration of assessment centers costing between $500 to $1,000 per candidate, and that is in 1997 dollars. One may ask, "Is this cost worth it?" Previous research would tend to suggest that given the high criterion-related validity of the assessment center, the method demonstrates reasonable utility (e.g., Cascio & Silbey, 1979). This is particularly true when adverse impact is an important consideration. Hoffman and Thornton (1997) demonstrated that assessment centers can exhibit higher utility than cognitive ability tests because of their smaller subgroup differences and hence ability to set a higher cut score without producing adverse impact.

Applicant Reactions

Assessment centers have a number of advantages in addition to the considerable evidence supporting their validity. One of the most important is high "face validity." Because the nature of these exercises overlaps so much with the nature of the job, applicants can easily see how the exercises relate to performance (Cascio & Phillips, 1979; Macan, Avedon, Paese, & Smith, 1994; Schmidt, Greenthal, Hunter, Berner, & Seaton, 1977). This high face validity may do more than make the test more acceptable, it may contribute to fewer legal challenges and perhaps reduce subgroup differences. This high face validity may be why organizations do not spend much time explaining to test takers what the assessment center is designed to accomplish (Spychalski et al., 1997). However, we believe organizations should convey the purpose and procedures of the assessment center.

Assessment centers have also been favorably accepted by large numbers of managers. The fact that some of these managers have served as assessors has aided this process, of course, but this factor alone probably does not explain the high level of acceptance. Responses of assessees have been equally positive—even those who do not perform well feel the assessment process has value. From the viewpoint of assessees, the assessment center experience often stimulates self-development efforts by focusing attention on their training and development needs. Assessees might learn, for example, that their major weakness is a tendency to jump to decisions before gathering adequate information. As a result, they might develop a well thought out plan for changing their behavior.

The Rating Process

Because assessment centers ultimately rely on trained assessor ratings, issues with contamination and bias discussed in connection with performance ratings in chapter 4 and earlier with interviews are relevant here. For example, Schleicher, Day, Mayes, and Riggio (2002) reported that rater frame of reference training had positive effects on construct validity, criterion-related validity, and accuracy. Likewise, Reilly, Henry, and Smither (1990) demonstrated that assessors who used behavioral checklists showed more construct valid ratings. Early studies concerning the internal validity of the assessment center (Huck, 1976; Mitchel, 1975; Schmitt, 1977; Schmitt & Hill, 1977; Thornton & Byham, 1982) indicated that interrater reliability is high and that overall assessment judgments are indeed more highly correlated with performance on individual assessment dimensions than are test scores or other available information.

However, there are some important differences with assessment centers that deserve mention. First, the type of assessor can have a substantial influence on the quality of the ratings. Research suggested it is not only important to have trained assessors, but that assessors such as psychologists provide the best ratings (Lievens, 2002; Lievens & Conway, 2001; Woehr & Arthur, 2003). The reason is presumed to be due the fact that psychologists have more training in objectively observing and evaluating behavior. Unfortunately, the review of common assessment center practice by Spychalski et al. (1997) suggested most assessors are line or staff managers; psychologists served as assessors in less than 10% of the cases. However, these managers tend not to be the candidate's immediate supervisor. Second, across-exercise rating systems, where evaluation of candidates does not occur until all exercises have been completed, seems to show higher levels of construct validity (Robie et. al., 2000; Woehr & Arthur, 2003). This is likely due to the multiple chances assessors have to observe the various behaviors that constitute the full range of the constructs. The logic here is the same logic for having many items on a personality inventory for assessing each construct in the FFM of personality. More opportunities to observe behavior are like more items for assessing a personality construct, and the result is the increased reliability necessary to define a construct. To date, there is simply not enough data to determine how composition of assessment center panels (e.g., demographic) or consensus methods relate to ratings such as those we have seen in the interview and performance appraisal literature. However, our own experience suggests that the following rules yield both interrater reliability and construct validity:

- There must be three or more trained assessors for the entire assessment center, with two assessors per exercise.
- Assessors each take independent notes for the behavior of the participants in each exercise.
- At the conclusion of the assessment center, each assessor independently rates the candidates they observed on the various dimensions of interest.
- Each dimension being rated should have at least three (five is better) specific behaviors associated with the dimension being rated and then an overall rating for the dimension is made.

- Discussions based on notes taken are used to resolve rating differences of more than 2 points (assuming a 5-point scale) for the overall ratings for each dimension.

Legal Issues

Probably due in part to their high face validity and development based on a job analysis yielding formal content validity, assessment centers have usually been favorably reviewed by courts concerned with violations of EEO. In fact, the courts have, in at least one case, required that assessment centers be used to eliminate possible bias against women and to increase their movement into managerial positions. A consent decree (described in chapter 11) involving AT&T, the Equal Opportunity Employment Commission (EEOC), and the U.S. Department of Labor resulted in the evaluation of 1,634 women during a 15-month period and the judgment that 42% possessed the requisite management skills. In other consent decrees, the development of assessment centers for promotion purposes have been viewed favorably by the court overseeing the case based on the demonstrated job relevance of the center. Indeed, validity evidence generalized from other studies and the contention that assessment center exercises are content valid have been enough to save from challenge even the centers for which no criterion-related validity evidence exists. In those court cases where assessment centers have been challenged, they have usually been upheld.

New Technologies

We have seen how computers and the Internet have dramatically altered the nature of psychological measurement and the assessment center is no exception. Assessment exercises are becoming increasingly administered via computer, and scoring may take place automatically by the computer or later by trained raters. For example, we have seen administrative assistant positions be screened using a computerized assessment center. The exercise involves having the person sit at the computer with a headset. Phone calls and e-mails start to pour in, and the person must sift through these and prioritize the most important ones. Interestingly, the exercise even involves the use of audio soundbites played over the headset. A benefit of this technology is that the person is performing the identical behaviors on the identical equipment used on the job. Thus, physical and psychological fidelity are maximized.

Summary and Future Directions

For jobs that have serious consequences, such as police officers, firefighters, managers, and military personnel, the assessment center remains an important and valuable predictor method. We have seen how they have long been known to be predictive of job performance and generally well perceived by applicants. However, we have also seen years of pessimism regarding their construct validity, a conclusion that is slowly starting to change. In fact, the wealth of more recent research paints an optimistic picture of assessment center construct validity. We anticipate even more favorable research findings to emerge as this new thinking about assessment centers takes hold.

In Table 10.6, we provide some recommendations for creating effective assessment centers. These suggestions come from more recent

TABLE 10.6
Suggestions for Developing Effective Assessment Centers

1. Conduct a thorough job analysis.
2. Use the minimum number of exercises necessary to fully capture the critical aspects of the job, and use exercises that maximize variance in these aspects.
3. Use the minimum number of dimensions necessary to fully capture the critical aspects of the job and try to ensure these dimensions are maximally dissimilar.
4. Develop behaviorally anchored rating scales for each exercise and for each dimension. Leave space for assessors to take notes or document behaviors that justify each rating, and use observational aids such as behavioral checklists.
5. Use psychologists as assessors whenever possible or at least have a psychologist train and oversee the assessors.
6. Train assessors on how to conduct the assessment center and use the rating forms.
7. Use an across-exercises rating approach, where assessors only provide ratings after candidates have completed all exercises.
8. Use panels of demographically diverse assessors.
9. Require the assessor panel to first make an independent set of ratings, then have the panel discuss and reach consensus to provide a final panel rating (this latter rating is ultimately to be used for the hiring decision).
10. Hold individual assessors and assessor panels accountable for their ratings. Monitor the process to ensure certain assessors and panels are administering the assessment center correctly.
11. Unless consistent with the purpose of the assessment center, eliminate the use of extraneous information such as test scores and status in the selection process.
12. Explain to applicants the nature of the assessment center, why it is being used and how it was developed, and the roles and responsibilities taken by the applicant and assessor.

research, particularly the work of Woehr and Arthur (2003) and Lievens (1998). Note that not all the suggestions are based on empirical research. For example, we advocate the use of demographically mixed assessor panels based on research from interviews, even though no research has addressed this question with assessment centers. We suspect the findings from interview panels will generalize to this context, although it admittedly remains an empirical question and deserves further research.

There are other important areas for future research. One is based on the exciting research that suggests assessment centers appear to lack construct validity not due to problems with the method, but with "true" variability among candidates. The studies by Lievens (2001), Lance et al. (2000), and Lance, Lambert, et al. (2004) provide optimism, but clearly more research will be necessary. One possibility examines the role of technology in assessment centers. Can such exercises as the In-Basket be entirely administered over computer? Can a leaderless group discussion take place over the Internet using videoconferencing software? Technology may help reduce the cost of assessment centers, but by doing so also change the nature of what is being assessed. Yet, a third area of research is to examine assessment centers within the context of international selection and executive development (e.g., Howard, 1997; Lievens et al., 2003). These are important positions that have high strategic value to organizations, and the assessment center seems well suited for purposes of international selection and development. They may also be used in more educational environments such as the assessment of student performance (Riggio, Aguirre, Mayes, Belloli, & Kubiak, 1997). However, we have seen less research on these issues, even though such practices are fairly common.

Work Samples

Perhaps the most objective method of gathering job-relevant information on applicants is the work sample (also known as the job sample or job simulation). A work sample is a test in which the applicant performs a selected set of actual tasks that are physically and psychologically similar to those performed on the job. For example, airline pilots who participate in a flight simulator with an identical cockpit layout are participating in a work sample. Job analysis and other techniques are used to ensure the tasks selected are representative of important tasks or problems actually encountered on the job. Procedures are standardized

PERFORMANCE-BASED AND SIMULATION TESTING

and scoring systems are worked out with the aid of experts in the occupation in question. A job sample test for electricians, for example, might involve, among other things, the wiring of a switch box for a standard house.

Work sample testing has been used for years, and studies going back to the first quarter of the twentieth century can be found (see Asher & Sciarrino, 1974). One of the more interesting tests was that used by the U.S. Public Health Service in selecting dentists (Newman, Howell, & Cliff, 1959). One of the tasks in this test was preparing and filling a cavity in a synthetic tooth, and another involved constructing a gold inlay. Work sample tests have been common for many years in clerical areas. Among the early tests are the Seashore-Bennett Stenographic Proficiency tests (Bennett & Seashore, 1946), the Blackstone Stenographic Proficiency tests (1932), and the Thurstone Examination in Typing (Thurstone, 1922), They are also widely used as part of licensing and certification exams (Shimberg & Schmitt, 1996).

Work samples demonstrate the highest physical fidelity of any predictor measure because they involve the same tasks and processes necessary for performance on the job (see Fig. 10.1). A work sample is different from an assessment center because assessment centers are comprised of multiple exercises that vary in terms of their physical fidelity and tap a variety of dimensions. A work sample may comprise one or more exercises in an assessment center, but a work sample may also be a stand-alone test. Thus, assessment centers are typically broader and encompass more diverse exercises than work samples.

The interest in work sample tests is partly due to an article by two industrial psychologists who criticized the assumptions of conventional paper-and-pencil aptitude testing. Wernimont and Campbell (1968) argued that "samples" of the kinds of behaviors or performances actually required on the job will be better predictors of future job performance than scores on aptitude or ability tests. Scores on aptitude and ability tests, they argued, are merely "signs" of potential that are statistically but imperfectly related to future job performance. They recommended that we strive to make the performances sampled by our tests as similar as possible to the performances required on the job. Assessment centers, written knowledge tests, and achievement tests come closer to meeting this stipulation than aptitude and ability tests. Work sample tests come even closer.

This recommendation has received empirical support. Meta-analytic research suggests work samples are among the best predictors of job performance we currently have available, demonstrating corrected criterion-related validities ranging from .33 (Roth, Bobko, & McFarland, 2005) to .54 (Schmidt & Hunter, 1998). Work samples generally show higher validity than personality and biodata tests. They have played a major role in the development of theoretical models of supervisory ratings because Hunter (1983) demonstrated how cognitive ability drives job knowledge and work sample performance, which subsequently influence supervisory ratings. Thus, the validity and importance of work samples in staffing is well documented.

There is another important reason for using work sample tests—the increasing level of public criticism of the paper-and-pencil aptitude and ability tests. Much of this criticism is based on the fact that the content of paper-and-pencil aptitude and ability tests often appears to have little to do with the content of the jobs on which performance is being predicted. That is, they lack face validity and perceived job relatedness. Work sample tests, in contrast, are not subject to this criticism because the tasks and behaviors are nearly (if not entirely) identical to those on the job. Because staffing experts must frequently justify their techniques to the general public and courts who have had little appreciation for statistical arguments related to validity, such perceptions are important. In other words, "signs" require more of an inference about psychological fidelity than "samples"; samples are more content relevant and perceived as such.

Yet another reason for interest in job sample tests is that they may increase employment opportunities for minorities and the disadvantaged. That is, they may show smaller subgroup differences and thus reduced adverse impact. Research in the area of selection has repeatedly shown that the differences between majority and minority group members is typically smaller on appraisals of actual job performance than on tests (e.g., Sackett & Wilk, 1994). The content of work sample tests is psychologically more like the content of the job itself. Therefore, we would expect majority–minority differences on the work sample to be closer in size to those observed on the job and smaller than those observed on paper-and-pencil tests. Obviously, this would lead to more minority group members being hired.

Several empirical studies support these predictions. An early study documenting this finding was shown by Schmidt et al. (1977). A job sample test of apprentice skills in the machine trades was carefully constructed and administered to a group of White and Black apprentices.

Paper-and-pencil tests were also administered. All the paper-and-pencil tests, including a well-constructed achievement test in the metal trades, showed larger and significant subgroup differences between minority and majority groups. The work samples either showed no difference or much smaller differences. More recently, Pulakos, Schmitt, and Chan (1996) compared racial subgroup differences on three work samples and a paper-and-pencil cognitive ability test. Blacks scored lower than Whites on all tests, but whereas the d value for the cognitive test was 1.25, the d values for the work samples ranged from 0.35 to 0.83. Hispanics also scored lower than Whites, but again work samples showed smaller differences (ds = 0.05 − 0.59) than the cognitive ability test ($d = 0.92$). Two reviews of subgroup differences on work samples, Hough et al. (2001) and Schmitt, Clause, and Pulakos (1996), report Black–White subgroup differences of only around one-third of a standard deviation favoring Whites, and Hispanics performing similarly to Whites. Women frequently perform better than men on work samples, $d = 0.38$ (Schmitt et al., 1996).

The likely theoretical reason work samples show lower subgroup differences may be understood within the context of the predictor-response model described in chapter 7. Because the work sample is a performance-based method that captures a variety of KSAOs, performance depends less on cognitive abilities (although these abilities may still be required). That is, there are likely to be less reading and verbal requirements on work samples because individuals are asked to physically complete the task. For example, a work sample test for an electrician may require some cognitive ability (primarily memory and declarative knowledge), but procedural and tacit knowledge will also influence performance. It is likely that work samples are more "compensatory" types of assessments, such that experience may make up for lower cognitive ability, for example. However, it is important to remember that reliability may be lower for work samples, so the reported subgroup differences may be underestimated. Further, Bobko, Roth, and Buster (2005) have found that work samples can show subgroup differences much larger than previously suggested, when the tests are not range-restricted (i.e., based on applicants). It is likely that when the nature of the tasks performed in the work sample are cognitive in nature, subgroup differences will be larger.

Finally, it is interesting to note that work samples are frequently used as criteria. Jackson, Harris, Ashton, McCarthy, and Tremblay (2000) suggested that work samples make useful criteria in instances where standardization is important, face validity is necessary, and range restriction is a concern (because all participants can be tested using the work sample). For example, Hattrup and Schmitt (1990) used work samples as criteria for trade apprentices to help establish the criterion-related validity of various aptitude tests. These criteria tended to show reasonable internal consistency reliability, although test–retest reliability was lower. Hedge and Teachout (1992) described a work sample–interview hybrid known as the "walk-through performance test." The idea was to reduce the cost of hands-on work sample tests by having individuals complete a smaller set of tasks, and then use interview questions to ask how they would complete the remaining tasks. Although the work sample and interview components produced some differences, the authors concluded the results were sufficiently similar to use the new criterion for selection, training, and certification.

As with the assessment center, the major changes to work sample testing come from technological innovations. For many jobs and tasks (particularly those already dependent on computer technology), it makes sense to administer and score the work sample directly on the computer. For example, many software designers are given a simple software design task as part of the application process (e.g., create a Web site conforming to a given set of specifications). The applicants must then use the programming language to create the software, which is scored by SMEs. The big question is whether work samples administered in such computerized formats show similar levels of validity and subgroup differences as has been found in previous research.

PHYSICAL ABILITY TESTING

Physical ability testing is common in many occupations, including construction and trades occupations, public safety (firefighter and police), and obviously the military. Psychomotor abilities typically include the more refined senses (vision, hearing), reaction time, dexterity (eye–hand, finger), control, and precision. Physical abilities are usually thought of more in terms of strength and endurance and are subsumed within three meta-categories: cardiovascular endurance, muscular strength, and movement quality (Hogan, 1991a).

As with any other human characteristic, there are individual differences in psychomotor and physical abilities. This fact was demonstrated nearly a century ago when scientists administered psychomotor

# PHYSICAL ABILITY TESTING 543

and physical tests at fairs and carnivals. The tests were a novelty, and people were so excited about being measured they would stand in long lines for the opportunity to be tested (what a different world it is today!). Hunter and Hunter (1984) showed using meta-analysis that these differences are reflected in job performance. Hogan (1991a) conducted an even more extensive review of physical testing and found substantial criterion-related validity for objective, subjective, and work sample criteria. Across the three criteria, muscular strength, anthropometric, muscular endurance, and muscular power were the most consistent predictors, with corrected validities ranging from 0.23 to 0.82. Given these findings, which in fact summarize many years of research, it is useful to describe some procedures that have been used for assessment.

Psychomotor Abilities

Psychomotor abilities tests are often of two major types—sensory and dexterity.

Sensory Tests (Based on Aiken, 1982)

Reliable and useful sensory tests of visual acuity, hearing, and color vision are available, and as a grammar school child you probably encountered them all. The chart you had to read from when you took your "eye test" is called a Snellen chart. In this test, examinees stand 20 feet from the chart and read off that series of smallest letters they can see. When the letters they can read are those a normal person can read at 20 feet, their vision is said to be 20/20. When they can read only those letters a normal person could read at 40 feet, their vision is said to be 20/40. Usually, this is done for each eye and for both eyes combined.

Audiometers are used to measure auditory acuity. These instruments present pure tones of varying intensities and frequencies to the listener. The typical examining sequence is to do one ear at a time, starting above the examinee's auditory level, moving to below acuity and then coming back up to the level of audition. By doing this for various frequencies, we obtain a person's auditory acuity for different frequencies; jobs in which information is presented through the ears at particular frequencies are obviously good candidates for this procedure.

Color vision is assessed in various ways, most often with the Ishihara Test for Color Blindness. This test presents a person with a series of cards, and each card is comprised of colored dots. Depending on the level and kind of color blindness, different numbers embedded in the dots are reported. Some people are said to be color weak, not color blind because, like all kinds of acuity, color "blindness" is a matter of the degree of deficiency, rather than an all-or-none phenomenon. Also, as in tests of audition where people can vary based on the frequencies they can hear, color blindness exists for different color combinations. Red-green confusion is most frequent, females are rarely color blind, and some people see no colors at all. These people, called monochromates (one color), see only different shades of gray. Color vision can be critical for some occupations, and staffing specialists must ensure that when this is true, such color vision is tested. Color blindness was not actually identified until the 1600s, and tests for it were not implemented widely until much later. Numerous accidents at sea and on the road were later discovered to be due to color blindness or color weakness.

There are some multipurpose sensory tests, especially for visual abilities, for use when sophisticated assessment of complex visual abilities are required. One such test, the Frostig Developmental Test of Visual Perception, permits assessment of eye–motor coordination, perception of figure–ground relationships, and spatial relationships, among others. In addition, Bausch & Lomb and the American Optical Company have quick tests (about 5 minutes) that permit assessment of visual acuity (near and far) and color discrimination, as well as eye muscle control for pursuing objects both horizontally and vertically. The Bausch & Lomb procedure has a booklet that comes with the test that includes norms for *visual job families*.

It is too easy to forget about these basic sensory issues in selection. Perhaps one reason for this is the tendency to forget about them in job analysis. The point is that whenever vision, audition, and color perception are important abilities for performance, they should be so identified because testing for them is quick and reliable and failure to test for them can be dangerous.

Dexterity Tests (based on Aiken, 1982)

Basically, dexterity tests are tests of *manipulation*—using tweezers, fingers, and hands in a coordinated way to move, place, or put together objects or things. For example, the Crawford Small Parts Dexterity Test has two parts. In Part 1, examinees use tweezers to insert pins in close-fitting holes and then they place a small collar over each pin. In Part 2, small screws are screwed into threaded holes with a screwdriver. There are similar small parts dexterity tests in which only fingers are used and others in which larger tools are required.

One popular test requiring tools is the Bennett Hand-Tool Dexterity Test. The Bennett test uses a three-sided open wood frame. Two sides have holes in them into which large screws and bolts can be inserted. Then, using a screw driver, pliers, and wrench, washers and nuts are placed on the screws. The test requires disassembling the screws, bolts, washers, and nuts and then reassembling them.

The fine and moderate dexterity tests have a reaction time and precision component, as well as a dexterity component, because time to completion is critical and actually doing the task as prescribed is required. Because many jobs, especially in high-technology electronics, require fine dexterity/precision, these tests will continue to be usefull, although most were originally developed in the late 1920s and early 1930s.

All the assessment procedures described are assessed with great reliability and, because they are performance tests, they reveal considerable physical and psychological fidelity—and they have been shown to be consistently predictive of job performance, where these abilities are identified through job analysis to be important.

Physical Ability Performance Testing

The interest in physical ability testing since the mid-1960s has been spurred primarily by EEO considerations, especially as they relate to women entering nontraditional jobs, and even more recently, because of concern for individuals with disabilities. When physical ability was assessed previously, it was usually done by a physician who made a judgment concerning the possibility that persons would experience risks to their health as a result of employment. The research by Fleishman (1964, 1975) on physical ability measurement began decades ago, but it took many years before his work was used as a basis for developing selection procedures (e.g., Arnold, Rauschenberger, Soubel, & Guion, 1982; Cooper & Schemmer, 1983; Reilly, Zedeck, & Tenopyr, 1979). Fleishman's great contribution was the development of a taxonomic structure of physical abilities.

Traditionally, organizations such as police have used height and weight requirements as indices of strength. However, these standards were typically set arbitrarily, and their job relatedness was not demonstrated; hence, in most cases, they have been ruled illegal (e.g., see *Blake v. City of Los Angeles*, 1979). Given their expected adverse impact against females and some ethnic groups, the need to validate physical abilities selection procedures against job performance is especially important. In this section, we discuss methods of assessing physical abilities and the physical requirements of jobs and then describe an example of the use of physical abilities testing for selection purposes.

Types of Physical Abilities

Hogan (1991a, 1991b) refined the taxonomic structure of physical abilities and collapsed them into three meta-categories of cardiovascular endurance, muscular strength, and movement quality. However, nested within these three general categories are seven specific types of physical abilities that serve as useful predictors in selection contexts:

- Muscular strength:
  - Muscular tension (pulling, pushing, lifting, carrying)
  - Muscular power (overcoming static resistance, such as loosening a bolt)
  - Muscular endurance (physical tasks performed repeatedly and continuously, but localized to a particular muscle group)
- Cardiovascular endurance (physical tasks that are not localized and require prolonged emphasis on respiration and blood circulation)
- Movement quality:
  - Flexibility (stretching, twisting, bending)
  - Balance (able to maintain body in a particular position)
  - Neuromuscular coordination (skillful synthesis of neurological and muscle abilities; coordination required to use tools in the appropriate sequence)

A wide variety of tests have been used to assess these various constructs, including measures of strength, speed, flexibility, isometric tension, ergometers, and aerobic capacity, to name just a few examples. Many of these tests come from the medical and fitness disciplines.

Justifying Physical Abilities in Selections

Like all good predictor development, appropriate use of physical abilities tests depends on a sound job analysis. We discuss several types of job analysis in chapter 3; here, we focus on issues specific to physical ability testing.

Fleishman's approach to the measurement of human physical ability resulted in a method to measure the physical requirements of jobs called the Physical Abilities Analysis (Fleishman, 1978). BARS are constructed to assess the job requirements along each physical ability dimension identified in his earlier research, plus measures for upper and lower body strength. One such BARS used to assess flexibility requirements is illustrated in Fig. 10.4. Note the definition of flexibility and the behavioral anchors describing various points on the scale of flexibility. Also note the behavioral anchors describing various points on the scale of flexibility requirements. The scales are easy to use in a field setting; they cover a broad range of physical ability requirements, the requirements are linked to job tasks, and they relate to abilities for which specific tests are available. However, this approach is most useful for either a criterion-related or construct validity approach; it is less defensible for a content strategy (Hogan, 1991a).

Fleishman's (1979) approach is among the few specific to the assessment of physical abilities. Hogan (1991a) noted two other job analytic methods that may be useful. These include FJA and the CIT. Because both approaches are discussed at length in chapter 3, we do not spend more time with them here. However, once the job analysis information is collected, one must be careful to ensure the physical ability really is required for the job. If it is not really required and is used as a selection hurdle, it will be discriminatory, usually against women. For example, in the installation and repair of telephone lines, it is frequently necessary to remove a ladder from the truck to make a repair. Tall strong males were easily able to perform this crucial task and women were not. However, when installer repair people were provided with a hook to use to remove the ladder from the truck, height and upper body strength were no longer a requirement for the job and women began entering this occupation.

Once the job analysis is completed, the next step is to determine criterion dimensions and measures. Hogan (1991a) noted that although subjective performance ratings are often used, archival data can often be a better source of physical performance information. For example,

548　　　10. HIRING PROCEDURES III

FLEXIBILITY

This is the ability to bend, stretch, twist, or reach out with the body, arms, or legs.

HOW FLEXIBILITY IS DIFFERENT FROM OTHER ABILITIES:

FLEXIBILITY:	vs.	STRETCH FACTORS:
Involves the ability of the arms, legs, and back to move in all directions without feeling "tight" or being able to move to a desired position (e.g., toe-touching, reaching high above one's head, crawling through a very small space).		The ability of the muscles to exert a force.

Requires a high degree of bending, stretching, twisting, or reaching out into unusual positions.

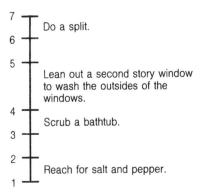

Requires a low degree of bending, stretching, twisting, or reaching out.

FIG. 10.4.　PAA Flexibility Scale.
Source: From Fleishman, (1977).

accidents, injuries, days sick, and related criteria may be highly dependent on physical fitness. She also noted the widespread use of work samples as criteria, and in fact, work samples may be most frequently used for physical types of assessment. The same concerns and procedures noted in the earlier section on work samples are relevant here.

Throughout all this development, concerns about demographic subgroup differences and adverse effects on those with disabilities, women,

PHYSICAL ABILITY TESTING

and the elderly must be paramount. Not surprisingly, physical ability tests demonstrate very large subgroup differences against women, although racial differences are much smaller. Hough et al. (2001) reported gender differences to range from 0.20 to 2.10 *SDs*—among the highest shown for any predictor construct and measure. Because these differences are so large, it is critical that only those physical abilities truly required for effective job performance be present in the predictor battery. A more recent study by Terpstra, Mohamed, and Kethley (1999) found physical ability tests were among the most frequently challenged in U.S. federal court cases. It is extremely difficult to defend physical abilities tests when they are challenged, but ultimately it will be the quality of the job analysis that determines one's success. The Americans with Disabilities Act of 1990 (discussed in chapter 11) has several implications for when and how physical ability testing must be conducted. Arvey, Nutting, and Landon (1992) and Hogan (1991a) listed several other concerns and means to enhance the defensibility of physical ability testing. In the next section, we discus an excellent example for how one can justify using a physical ability test.

Example of Construct Validation for Physical Abilities Tests

An excellent example of a construct-oriented approach to physical ability test validation was conducted by Arvey, Landon, Nutting, and Maxwell (1992). They examined the physical abilities required for police officers. Step 1 involved specifying the physical components of job performance. This required the usual job analysis and review of archival records, but with an emphasis on physical requirements truly necessary for the job. Step 2 specified a theory of the physical abilities necessary for effective performance. Here, it was determined that strength and endurance were the two key latent constructs of most relevance. Step 3 required the identification of manifest tests to assess the two latent constructs. The tests included a 100-yard dash, dummy drag, obstacle course, grip strength, dummy wrestle, sit-ups, bench dips, and a 1-mile run. A concurrent study was first conducted, and supervisory ratings of physical performance on the job and physiological data supplemented the physical tests. Step 4 sampled the incumbents and tested them. Step 5 subjected the test scores and ratings to CFA (discussed in chapter 2), a powerful method for determining the adequacy of the latent factor model. They found latent strength and endurance constructs adequately explained variance in both the tests and performance ratings. Step 6 replicated the confirmatory factor model with the test scores obtained from a real applicant sample. Step 7 explored the nomological network of the physical manifest measures and latent factors. Step 8 examined gender subgroup differences on the physical tests. Although there were sizeable differences on many measures, the supporting construct and criterion-related data were compelling in demonstrating the abilities were necessary for the job.

SUMMARY

This chapter discusses various alternatives to traditional paper-and-pencil tests that demonstrate higher physical and often psychological fidelity. The rationale for the use of these alternatives is usually that they represent actual job samples (as opposed to signs) or that they measure past behavior of the type required in a new situation on the hypothesis that past behavior is the best predictor of future behavior. In this context, we describe and review interviews, SJTs, assessment centers, work samples, physical ability tests, and licensing exams. It is clear that many types of simulation methods exist, and technology is blurring many of the historical distinctions between these methods—and creating new ones.

REFERENCES

Aiken, L. R. (1982). *Psychological testing and assessment* (4th ed.). Boston: Allyn & Bacon.

American Educational Research Association (AERA), American Psychological Association, & National Council on Measurement in Education. (1999). *Standards for educational and psychological testing*. Washington, DC: AERA.

Arnold, J. D., Rauschenberger, J. M., Soubel, W. G., & Guion, R. M. (1982). Validation and utility of a strength test for selecting steelworkers. *Journal of Applied Psychology, 67*, 588–604.

Arthur, W., Jr., Day, E. A., McNelly, T. L., & Edens, P. S. (2003). A meta-analysis of the criterion-related validity of assessment center dimensions. *Personnel Psychology, 56*, 125–153.

Arthur, W., Jr., Woehr, D., & Maldegen, R. (2000). Convergent and discriminant validity of assessment center dimensions: A conceptual and empirical reexamination of the assessment center construct-related validity paradox. *Journal of Management, 26*, 813–835.

Arvey, R. D., & Campion, J. E. (1982). The employment interview: A summary and review of recent research. *Personnel Psychology, 35*, 281–322.

Arvey, R. D., Landon, T. E., Nutting, S. M., & Maxwell, S. E. (1992). Development of physical ability tests for police officers: A construct validation approach. *Journal of Applied Psychology, 77*, 996–1009.

REFERENCES

Arvey, R. D., Nutting, S. M., & Landon, T. E. (1992). Validation strategies for physical ability testing in police and fire settings. *Public Personnel Management, 21*, 301–312.
Asher, J. J., & Sciarrino, J. A. (1974). Realistic work sample tests: A review. *Personnel Psychology, 27*, 519–538.
Barber, A. E., Hollenbeck, J. R., Tower, S. L., & Phillips, J. M. (1994). The effects of interview focus on recruitment effectiveness: A field experiment. *Journal of Applied Psychology, 79*, 886–896.
Bennett, G. K., & Seashore, H. G. (1946). *The Seashore-Bennett Stenographic Proficiency.* New York: Psychological Corporation.
Blackstone Stenographic Proficiency tests. (1932). New York: Harcourt, Brace, & World.
Blake v. City of Los Angeles. (1979). 19 FEP 1441 (9th Cir.).
Bobko, P., Roth, P. L., & Buster, M. A. (2005). Work Sample selection tests and expected reduction in adverse impact: A cautionary note. *International Journal of Selection and Assessment, 13*, 1–10.
Bray, D. W., & Grant, D. L. (1966). The assessment center in the measurement of potential for business management. *Psychological Monographs, 80*, Whole No. 625.
Buckley, M. R., & Russell, C. J. (1999). Validity evidence. In R. W. Eder & M. M. Harris (Eds.), *The employment interview handbook* (pp. 35–48). Thousand Oaks, CA: Sage.
Burnett, J. R., & Motowidlo, S. J. (1998). Relations between different sources of information in structured selection interview. *Personnel Psychology, 51*, 963–983.
Campion, M. A., Palmer, D. K., & Campion, J. E. (1997). A review of structure in the selection interview. *Personnel Psychology, 50*, 655–702.
Cascio, W. F., & Phillips, N. F. (1979). Performance testing: A rose among thorns? *Personnel Psychology, 32*, 751–766.
Cascio, W. F., & Silbey, V. (1979). Utility of the assessment center as a selection device. *Journal of Applied Psychology, 64*, 107–118.
Chan, D., & Schmitt, N. (1997). Video-based versus paper-and-pencil method of assessment in situational judgment tests: Subgroup differences in test performance and face validity perceptions. *Journal of Applied Psychology, 82*, 143–159.
Chapman, D. S., & Rowe, P. M. (2001). The impact of video conference technology, interview structure, and interviewer gender on interviewer evaluations in the employment interview: A field experiment. *Journal of Occupational and Organizational Psychology, 74*, 279–298.
Clevenger, J., Pereira, G. M., Wiechmann, D., Schmitt, N., & Harvey, V. S. (2001). Incremental validity of situational judgment tests. *Journal of Applied Psychology, 86*, 410–417.
Conway, J. M., Jako, R. A., & Goodman, D. F. (1995). A meta-analysis of interrater and internal consistency reliability of selection interviews. *Journal of Applied Psychology, 80*, 565–579.
Cooper, M., & Schemmer, F. M. (1983, August). *The development of physical ability tests for industry-wide use.* Paper presented at the national convention of the American Psychological Association, Anaheim, CA.
Cortina, J. M., Goldstein, N. B., Payne, S. C., Davison, H. K., & Gilliland, S. W. (2000). The incremental validity of interview scores over and above cognitive ability and conscientiousness scores. *Personnel Psychology, 53*, 325–351.
DeGroot, T., & Motowidlo, S. J. (1999). Why visual and vocal interview cues can effect interviewers' judgments and predict job performance. *Journal of Applied Psychology, 84*, 986–993.
Dipboye, R. L. (1992). *Selection interviews: Process perspectives.* Cincinnati, OH: South-Western.

Dipboye, R. L., & Gaugler, B. (1993). Cognitive and behavioral processes in selection interviews. In N. Schmitt & W. C. Borman (Eds.), *Personnel selection in organizations* (pp. 135–171). San Francisco: Jossey-Bass.

Dipboye, R. L., & Jackson, S. J. (1999). Interviewer experience and expertise effects. In R. W. Eder & M. M. Harris (Eds.), *The employment interview handbook* (pp. 259–278). Thousand Oaks, CA: Sage.

Dougherty, T. W., & Turban, D. B. (1999). Behavioral confirmation of interviewer expectancies. In R. W. Eder & M. M. Harris (Eds.), *The employment interview handbook* (pp. 217–228). Thousand Oaks, CA; Sage.

Dreher, G. F., & Sackett, P. R. (1983). *Perspectives on employee staffing and selection: Readings and commentary*. Homewood, IL: Irwin.

Eder, R.W., & M. M. Harris (Eds.). (1999). *The employment interview handbook*. Thousand Oaks, CA: Sage.

Ferris, G. R., & Judge, T. A. (1991). Personnel/human resources management: A political influence perspective. *Journal of Management, 17*, 447–488.

Fleishman, E. A. (1964). *The structure and measurement of physical fitness*. Englewood Cliffs, NJ: Prentice Hall.

Fleishman, E. A. (1975). Toward a taxonomy of human performance. *American Psychologist, 30*, 1127–1149.

Fleishman, E. A. (1978). *Physical abilities analysis manual*. Washington, DC: Advanced Resources Research Organization.

Fleishman, E. A. (1979). Evaluating physical abilities required by jobs. *Personnel Administrator, 24*, 82–91.

Gaugler, B. B., Rosenthal, D. B., Thornton, G. C., & Bentson, C. (1987). Meta-analysis of assessment center validity. *Journal of Applied Psychology, 72*, 493–511.

Gilliland, S. W. (1993). The perceived fairness of selection systems: An organizational justice perspective. *Academy of Management Review, 18*, 694–734.

Gilliland, S. W., & Steiner, D. W. (1999). Applicant reactions. In R. W. Eder & M. M. Harris (Eds.), *The employment interview handbook* (pp. 69–82). Thousand Oaks, CA: Sage.

Gilmore, D. C., Stevens, C. K., Harrell-Cook, G., & Ferris, G. R. (1999). Impression management tactics. In R. W. Eder & M. M. Harris (Eds.), *The employment interview handbook* (pp. 321–336). Thousand Oaks, CA: Sage.

Goffin, R. D., Rothstein, M. C., & Johnston, N. G. (1996). Personality testing and the assessment center: Incremental validity for managerial selection. *Journal of Applied Psychology, 81*, 746–756.

Goldstein, H. W., Yusko, K. P., Braverman, E. P., Smith, D. B., & Chung, B. (1998). The role of cognitive ability in the subgroup differences and incremental validity of assessment center exercises. *Personnel Psychology, 51*, 357–374.

Goldstein, H. W., Yusko, K. P., & Nicolopoulos, V. (2001). Exploring Black–White subgroup differences of managerial competencies. *Personnel Psychology, 54*, 783–807.

Graves, L. M., & Karren, R. J. (1999). Are some interviewers better than others? In R. W. Eder & M. M. Harris (Eds.), *The employment interview handbook* (pp. 243–258). Thousand Oaks, CA: Sage.

Harris, M. M. (1989). Reconsidering the employment interview: A review of recent literature and suggestions for future research. *Personnel Psychology, 42*, 691–726.

Hattrup, K., & Schmitt, N. (1990). Prediction of trades apprentices' performance on job sample criteria. *Personnel Psychology, 43*, 453–466.

Hedge, J. W., & Teachout, M. S. (1992). An interview approach to work sample criterion measurement. *Journal of Applied Psychology, 77*, 453–461.

REFERENCES

Hoffman, C. C., & Thornton, G. C., III. (1997). Examining selection utility where competing predictors differ in adverse impact. *Personnel Psychology, 50*, 455–470.
Hogan, J. (1991a). Physical abilities. In M. D. Dunnette & L. M. Hough (Eds.), *Handbook of industrial and organizational psychology* (2nd ed., Vol. 2, pp. 753–831). Palo Alto, CA: Consulting Psychologists Press.
Hogan, J. (1991b). Structure of physical performance in occupational tasks. *Journal of Applied Psychology, 76*, 495–507.
Hooper, A. C., Cullen, M. J., & Sackett, P. R. (in press). Operational threats to the use of SJTs: Faking, coaching, and retesting issues. In J. A. Weekley & R. E. Ployhart (Eds.), *Situational judgment tests*. Mahwah, NJ: Erlbaum.
Hough, L. M., Oswald, F. L., & Ployhart, R. E. (2001). Determinants, detection, and amelioration of adverse impact in personnel selection procedures: Issues, evidence, and lessons learned. *International Journal of Selection and Assessment, 9*, 152–194.
Howard, A. (1997). A reassessment of assessment centers: Challenges for the 21st century. *Journal of Social Behavior and Personality, 12*, 13–52.
Huck, J. R. (1976). The research base. In J. L. Moses & W. C. Byham (Eds.), *Applying the assessment center method* (pp. 261–292). New York: Pergamon.
Huffcutt, A. I., & Arthur, W. (1994). Hunter and Hunter (1984) revisited: Interview validity for entry-level jobs. *Journal of Applied Psychology, 79*, 184–190.
Huffcutt, A. I., Conway, J. M., Roth, P. L., & Stone, N. J. (2001). Identification and meta-analytic assessment of psychological constructs measured in employment interviews. *Journal of Applied Psychology, 86*, 897–913.
Huffcutt, A. I., Roth, P. L., & McDaniel, M. A. (1996). A meta-analytic investigation of cognitive ability in employment interview evaluations: Moderating characteristics and implications for incremental validity. *Journal of Applied Psychology, 81*, 459–473.
Huffcutt, A. I., Weekley, J. A., Wiesner, W. H., Degroot, T. G., & Jones, C. (2001). Comparison of situational and behavior description interview questions for higher-level positions. *Personnel Psychology, 54*, 619–644.
Huffcutt, A. I., & Woehr, D. J. (1999). Further analysis of employment interview validity: A quantitative evaluation of interviewer-related structuring methods. *Journal of Organizational Behavior, 20*, 549–560.
Hunter, J. E. (1983). A causal analysis of cognitive ability, job knowledge, job performance, and supervisory ratings. In F. Landy, S. Zedeck, & J. Cleveland (Eds.), *Performance measurement and theory* (pp. 257–266). Hillsdale, NJ: Erlbaum.
Hunter, J. E., & Hunter, R. F. (1984). Validity and utility of alternative predictors of job performance. *Psychological Bulletin, 96*, 72–95.
Jackson, D. N., Harris W. G., Ashton, M. C., McCarthy, J. M., & Tremblay, P. F. (2000). How useful are work samples in validation studies? *International Journal of Selection and Assessment, 8*, 29–33.
Jansen, P. G. W., & Stoop, B. A. M. (2001). The dynamics of assessment center validity: Results of 7-year study. *Journal of Applied Psychology, 86*, 741–753.
Janz, J. T. (1982). Initial comparisons of patterned behavior description interviews versus unstructured interviews. *Journal of Applied Psychology, 67*, 577–580.
Janz, J. T. (1989). The patterned behavior description interview: The best prophet of the future is the past. In R. W. Eder & G. R. Ferris (Eds.), *The employment interview: Theory, research, and practice* (pp. 158–168). Newbury Park, CA: Sage.
Klimoski, R., & Brickner, M. (1987). Why do assessment centers work? The puzzle of assessment center validity. *Personnel Psychology, 40*, 243–260.
Kohn, L. S., & Dipboye, R. L. (1998). The effects of interview structure on recruiting outcomes. *Journal of Applied Psychology, 28*, 821–843.

Lance, C. E., Foster, M. R., Gentry, W. A., & Thoresen, J. D. (2004). Assessor cognitive processes in an operational assessment center. *Journal of Applied Psychology, 89*, 22–35.

Lance, C. E., Lambert, T. A., Gewin, A. G., Lievens, F., & Conway, J. M. (2004). Revised estimates of dimension and exercise variance components in assessment center post-exercise dimension ratings. *Journal of Applied Psychology, 89*, 377–385.

Lance, C. E., Newbolt, W. H., Gatewood, R. D., Foster, M. S., French, N. R., & Smith, D. E. (2000). Assessment center exercises represent cross-situational specificity, not method bias. *Human Performance, 13*, 323–353.

Latham, G. P., & Finnegan, B. J. (1993). Perceived practicality of unstructured, patterned, and situational interviews. In H. Schuler, J. L., Farr, & M. Smith (Eds.), *Personnel selection and assessment: Individual and organizational perspectives* (pp. 41–55). Hillsdale, NJ: Erlbaum.

Latham, G. P., Saari, L. M., Pursell, E. D., & Campion, M. A. (1980). The situational interview. *Journal of Applied Psychology, 65*, 422–427.

Lievens, F. (1998). Factors which improve the construct validity of assessment centers: A review. *International Journal of Selection and Assessment, 6*, 141–152.

Lievens, F. (2001). Assessor training strategies and their effects on accuracy, interrater reliability, and discriminant validity. *Journal of Applied Psychology, 86*, 255–264.

Lievens, F. (2002). Trying to understand the different pieces of the construct validity puzzle of assessment centers: An examination of assessor and assessee effects. *Journal of Applied Psychology, 87*, 675–686.

Lievens, F., Buyse, T., & Sackett, P. R. (in press). The operational validity of a video-based situational judgment test for medical college admissions: Illustrating the importance of matching predictor and criterion construct domains. *Journal of Applied Psychology*.

Lievens F., & Conway, J. M. (2001). Dimension and exercise variance in assessment center scores: A large-scale evaluation of multitrait-multimethod studies. *Journal of Applied Psychology, 86*, 1202–1222.

Lievens, F., & Goemaere, H. (1999). A different look at assessment centers: Views of assessment center users. *International Journal of Selection and Assessment, 7*, 215–219.

Lievens, F., Harris, M. M., Van Keer, E., & Bisqueret, C. (2003). Predicting cross-cultural training performance: The validity of personality, cognitive ability, and dimensions measured by an assessment center and a behavior description interview. *Journal of Applied Psychology, 88*, 476–489.

Macan, T. H., Avedon, M. J., Paese, M., & Smith, D. E. (1994).The effects of applicants' reactions to cognitive ability tests and assessment center. *Personnel Psychology, 47*, 715–738.

Marlowe, C. M., Schneider, S. L., & Nelson, C. E. (1996). Gender and attractiveness biases in hiring decisions: Are more experienced managers less biased? *Journal of Applied Psychology, 79*, 599–616.

Maurer, S. D., Sue-Chan, C., Latham, G. P.(1999). The situational interview. In R. W. Eder & M. M. Harris (Eds.), *The employment interview handbook* (pp. 159–177). Thousand Oaks, CA: Sage.

McDaniel, M. A., Morgeson, F. P., Finnegan, E. B., Campion, M. A., & Braverman, E. P. (2001). Use of situational judgment tests to predict job performance: A clarification of the literature. *Journal of Applied Psychology, 86*, 730–740.

McDaniel, M. A., & Nguyen, H. T. (2001). Situational judgment tests: A review of practice and constructs assessed. *International Journal of Selection and Assessment, 9*, 103–113.

McDaniel, M. A., Whetzel, D. L., Schmidt, F. L., & Maurer, S. D. (1994). The validity of employment interviews: A comprehensive review and meta-analysis. *Journal of Applied Psychology, 79*, 599–616.

REFERENCES

McFarland, L. A., Ryan, A. M., & Kriska, S. D. (2003). Impression management use and effectiveness across assessment methods. *Journal of Management, 29*, 641–661.
McFarland, L. A., Ryan, A. M., Sacco, J. M., & Kriska, S. D. (2004). Examination of structured interview ratings across time: The effects of applicant race, rater race, and panel composition. *Journal of Management, 30*, 435–452.
Meglino, B. M., Ravlin, E. C., & DeNisi, A. S. (1997). When does it hurt to tell the truth? The effect of realistic job reviews on employee recruiting. *Public Personnel Management, 26*, 413–422.
Mitchel, J. O. (1975). Assessment center validity: A longitudinal study. *Journal of Applied Psychology, 60*, 573–579.
Moscoso, S. (2000). Selection interview: A review of validity evidence, adverse impact and applicant reactions. *International Journal of Selection and Assessment, 8*, 237–247.
Motowidlo, S. J. (1999). Asking about past behavior versus hypothetic behavior. In R. W. Eder & M. M. Harris (Eds.), *The employment interview handbook* (pp. 179–190). Thousand Oaks, CA: Sage.
Motowidlo, S. J., & Burnett, J. R. (1995). Aural and visual sources of validity in structured employment interviews. *Organizational Behavior and Human Decision Processes, 61*, 239–249.
Motowidlo, S. J., Dunnette, M. D., Carter, G. W. (1990). An alternative selection procedure: The low-fidelity simulation. *Journal of Applied Psychology, 75*, 649–647.
Motowidlo, S. J., Hanson, M., & Crafts, J. L. (1997). Low-fidelity simulations. In D. L. Whetzel & G. R. Wheaton (Eds.), *Applied measurement methods in industrial psychology* (pp. 241–260). Palo Alto, CA: Davies-Black.
Motowildo, S. J., & Tippins, N. (1993). Further studies of the low-fidelity simulation in the form of a situational inventory. *Journal of Occupational and Organizational Psychology, 66*, 337–344.
Newman, S. H., Howell, M. A., & Cliff, N. (1959). The analysis and prediction of a practical examination in dentistry. *Educational and Psychological Measurement, 19*, 557–568.
Ployhart, R. E., & Ehrhart, M. G. (2003). Be careful what you ask for: Effects of response instructions on the construct validity and reliability of situational judgment tests. *International Journal of Selection and Assessment, 11*, 1–16.
Ployhart, R. E., Weekley, J. A., Holtz, B. C., & Kemp, C. F. (2003). Web-based and paper-and-pencil testing of applicants in a proctored setting: Are personality, biodata, and situational judgment tests comparable? *Personnel Psychology, 56*, 733–752.
Posthuma, R. A., Morgeson, F. P., & Campion, M. A. (2002). Beyond employment interview validity: A comprehensive narrative review of recent research and trends over time. *Personnel Psychology, 55*, 1–81.
Pulakos, E. D., & Schmitt, N. (1995). Experience-based and situational interview questions: Studies of validity. *Personnel Psychology, 48*, 289–308.
Pulakos, E. D., Schmitt, N., & Chan, D. (1996). Models of job performance ratings: An examination of ratee race, ratee gender, and rater level effects. *Human Performance, 9*, 103–119.
Reilly, R. R., Henry, S., & Smither, J. W. (1990). An examination of the effects of using behavior checklist on the construct validity of assessment center dimensions. *Personnel Psychology, 43*, 71–84.
Reilly, R. R., Zedeck, S., & Tenopyr, M. L. (1979). Validity and fairness of physical ability tests for predicting performance in craft jobs. *Journal of Applied Psychology, 64*, 262–274.
Riggio, R. E., Aguirre, M., Mayes, B. T., Belloli, C., & Kubiak, C. (1997). The use of assessment center methods for student outcome assessment. *Journal of Social Behavior and Personality, 12*, 1–15.

Robie, C., Adams, K. A., Osburn, H. G., Morris, M. A., & Etchegaray, J. M. (2000). Effects of the rating process on the construct validity of assessment center dimension evaluations. *Human Performance, 13*, 355–370.
Roehling, M. V., Campion, J. E., & Arvey, R. D. (1999). Unfair discrimination issues. In R. W. Eder & M. M. Harris (Eds.), *The employment interview handbook* (pp. 49–68). Thousand Oaks, CA: Sage.
Roth, P. L., Bobko, P., & McFarland, L. A. (2005). A meta-analysis of work sample test validity: Updating and integrating some classic literature. Paper presented at the annual conference of the Academy of Management, August, Honolulu, HI.
Roth, P. L., Huffcutt, A. I., & Bobko, P. (2003). Ethnic group differences in measures of job performance: A new meta-analysis. *Journal of Applied Psychology, 88*, 694–706.
Roth, P. L., Van Iddekinge, C. H., Huffcutt, A. I., Edison, C. E., & Bobko, P. (2002). Corrections for range restriction in structured interview ethnic group differences: The values may be larger than researchers thought. *Journal of Applied Psychology, 87*, 369–376.
Ryan, A. M., & Ployhart, R. E. (2000). Applicants' perceptions of selection procedures and decisions: A critical review and agenda for the future. *Journal of Management, 26*, 565–606.
Rynes, S. L., & Connerley, M. L. (1993). Applicant reactions to alternative selection procedures. *Journal of Business and Psychology, 7*, 261–277.
Sacco, J. M., Scheu, C. R., Ryan, A. M., & Schmitt, N. (2003). An investigation of race and sex similarity effects in interviews: A multilevel approach to relational demography. *Journal of Applied Psychology, 88*, 852–865.
Sacco, J. M., Scheu, C. R., Ryan, A. M., Schmitt, N., Schmidt, D., & Rogg, K. (2000, April). *Reading level and verbal test scores as predictors of subgroup differences and validities of situational judgment tests.* Poster presented at the annual conference of Society for Industrial and Organizational Psychologists, New Orleans, LA.
Sackett, P. R., & Dreher, G. F. (1982). Constructs and assessment center dimensions: Some troubling empirical findings. *Journal of Applied Psychology, 67*, 401–410.
Sackett, P. R., & Wilk, S. L. (1994). Within-group norming and other forms of score adjustment in preemployment testing. *American Psychologist, 49*, 929–954.
Schleicher, D. J., Day, D. V., Mayes, B. T., & Riggio, R. E. (2002). A new frame for frame-of-reference training: Enhancing the construct validity of assessment centers. *Journal of Applied Psychology, 87*, 735–746.
Schmidt, F. L., Greenthal, A. L., Hunter, J. E., Berner, J. G., & Seaton, F. W. (1977). Job samples vs. paper-and-pencil trades and technical tests: Adverse impact and examinee attitudes. *Personnel Psychology, 30*, 187–197.
Schmidt, F. L., & Hunter, J. E. (1998). The validity and utility of selection methods in personnel psychology: Practical and theoretical implications of 85 years of research findings. *Psychological Bulletin, 124*, 262–274.
Schmidt, F. L., & Rader, M. (1999). Exploring the boundary conditions for interview validity: Meta-analytic validity findings for a new interview type. *Personnel Psychology, 52*, 445–464.
Schmidt, F. L., & Zimmerman, R. D. (2004). A counterintuitive hypothesis about employment interview validity and some supporting evidence. *Journal of Applied Psychology, 89*, 553–561.
Schmitt, N. (1976). Social and situational determinants of interview decisions: Implications for the employment interview. *Personnel Psychology, 29*, 79–101.
Schmitt, N. (1977). Interrater agreement in dimensionality and combination of assessment center judgments. *Journal of Applied Psychology, 62*, 171–176.

REFERENCES

Schmitt, N. (1999).The current and the future status of research on the employment interview. In R. W. Eder & M. M. Harris (Eds.), *The employment interview handbook* (pp. 355–367). Thousand Oaks, CA: Sage.

Schmitt, N., & Chan, D. (in press). Situational judgment tests: Method or construct? In J. A. Weekley & R. E. Ployhart (Eds.), *Situational judgment tests*. Mahwah, NJ: Erlbaum.

Schmitt, N., Clause, C. S., & Pulakos, E. D. (1996). Subgroup differences associated with different measures of some common job-relevant constructs. *International Review of Industrial and Organizational Psychology, 11*, 115–139.

Schmitt, N., & Coyle, B. W. (1976). Applicant decisions in the employment interview. *Journal of Applied Psychology, 61*, 184–192.

Schmitt, N., & Hill, T. E. (1977). Sex and race composition of assessment center groups as a determinant of peer and assessor ratings. *Journal of Applied Psychology, 62*, 261–264.

Schmitt, N., Schneider, J. R., & Cohen, S. A. (1990). Factors affecting validity of a regionally administered assessment center. *Personnel Psychology, 43*, 1–12.

Schneider, J. R., & Schmitt, N. (1992). An exercise design approach to understanding assessment center dimension and exercise constructs. *Journal of Applied Psychology, 77*, 32–41.

Shimberg, B., & Schmitt, K. (1996). *Demystifying licensing and professional regulation: Answers to questions you may have been afraid to ask*. Lexington, KY: Council on Licensing, Enforcement and Regulation.

Smither, J. W., Reilly, R. R., Millsap, R. E., Pearlman, K., & Stoffey, R. W. (1993). Applicant reactions to selection procedures. *Personnel Psychology, 46*, 49–76.

Spychalski, A. C., Quiñones M. A., Gaugler, B. B., & Pohley, K. (1997). A survey of assessment center practices in organizations in the United States. *Personnel Psychology, 50*, 71–90.

Steiner, D. D., & Gilliland, S. W. (1996). Fairness reaction to personnel reaction techniques in France and United States. *Journal of Applied Psychology, 81*, 134–141.

Stevens C. K. (1998). Antecedents of interview interactions, interviewers' rating, and applicants' reactions. *Personnel Psychology, 51*, 55–85.

Stevens, C. K., & Kristof, A. L. (1995). Making the right impression: A field study of applicant impression management during job interviews. *Journal of Applied Psychology, 80*, 587–606.

Task Force on Assessment Center Guidelines. (2000). Guidelines and ethical considerations for assessment center operations. *Public Personnel Management, 18*, 457–470.

Taylor P., & Small B. (2002). Asking applicants what they would do versus what they did do: A meta-analytic comparison of situational and past behaviour employment interview questions. *Journal of Occupational and Organizational Psychology, 75*, 277–294.

Terpstra, D. E., Mohamed, A. A., & Kethley, R. B. (1999). An analysis of federal court cases involving nine selection devices. *International Journal of Selection and Assessment, 7*, 26–34.

Thornton, G. C., III, & Byham, W. C. (1982). *Assessment centers and managerial performance*. New York: Academic Press.

Thurstone, L. L. (1922). *Thurstone employment tests: Examination in typing*. New York: Harcourt, Brace, and World.

Ulrich, L., & Trumbo, D. (1965). The selection interview since 1949. *Psychological Bulletin, 63*, 100–116.

Weekley, J. A., & Ployhart, R. E. (2005). Situational judgment: Antecedents and relationships with performance. *Human Performance, 18*, 81–104.

Weekley, J. A., & Ployhart, R. E. (in press). *Situational judgment tests*. Mahwah, NJ: Erlbaum.

Weekley, J. A., Ployhart, R. E., & Harold, C. (2004). Personality and situational judgment tests across applicant and incumbent contexts: An examination of validity, measurement, and subgroup differences. *Human Performance, 17*, 435–464.

Weekley, J. A., Ployhart, R. E., & Holtz, B. C. (in press). Scaling, scoring, and developing situational judgment tests. In J. A. Weekley & R. E. Ployhart (Eds.), *Situational judgment tests*. Mahwah, NJ: Erlbaum.

Wernimont, P. R., & Campbell, J. P. (1968). Signs, samples, and criteria. *Journal of Applied Psychology, 52*, 372–376.

Williamson, L. G., Campion, J. E., Malos, S. B., Roehling, M. A., & Campion, M. A. (1997). Employment interview on trial: Linking interview structure with litigation outcomes. *Journal of Applied Psychology, 82*, 900–912.

Woehr, D. J., & Arthur, W., Jr. (2003). The construct-related validity of assessment center ratings: A review and meta-analysis of the role of methodological factors. *Journal of Management, 29*, 231–258.

11

The Practice of Staffing: Legal, Professional, and Ethical Concerns

AIMS OF THE CHAPTER

Throughout this book, we discuss staffing principles and provide examples of how these staffing principles are applied to real work situations. For example, we present ideas about how staffing fits within the larger job and organizational design principles of an organization and then describe ways in which job analyses might be accomplished to identify the KSAOs required to do the work and perform it in an acceptable fashion. Or, consider the issue of personality at work, where we identify the ways in which personality has been conceptualized and the measures that have been developed—and then explicate the ways these personality measures might be employed to predict OCB, job satisfaction, and turnover in a real organization. So, in each previous chapter, the focus is primarily on how a particular aspect of staffing (e.g., job analysis, criterion development, predictor development) was applied in solving real world problems.

In contrast to the prior chapters where we focus on specific facets of the staffing specialist role, here we take a broader perspective and consider the challenges that staffing specialists face every time they carry out a staffing project. Thus, the present chapter examines the various constraints and opportunities that are present in the practice of staffing and the role of being a staffing practitioner.

Too often, the treatment of staffing practice emphasizes the constraints staffing practitioners face in attempting to be in compliance with various laws and regulations. The discussion of these laws is frequently one of grudging acceptance. We believe this alone is not a useful perspective from which to proceed. As we will see, many of the laws and regulations have been put into place for good reasons. They are well intentioned and are designed to create a society in which all have equal opportunity to use their capabilities to contribute in meaningful ways to themselves, their families, organizations, and society as a whole. We will also see that these laws frequently contribute to better staffing procedures because they require such procedures to be job related. However, conforming and complying with these laws is sometimes difficult, particularly when the laws are at odds with employers' desires. This can obviously be frustrating. However, for the organization that goes beyond merely complying with such regulation to adopting the laws as a source of business strategy, the laws can be sources of competitive advantage. That is, whenever a new law is enacted, it produces a change to the competitive landscape, and consequently, offers a new strategic opportunity for organizations. We therefore view the legal context as one that offers both opportunities and constraints.

We will also see that a focus on laws and regulations neglects other important aspects of staffing practice, namely, professional guidelines and ethical issues. There may be practices that are legal but are not necessarily ethical or consistent with good practice. Thus, our treatment of staffing practice considers the staffing specialist within this multidimensional and ever-changing world. We explain this perspective in the following section.

A TRIPARTITE VIEW OF STAFFING PRACTICE

Our view of staffing practice, and the role of the staffing practitioner, can be summarized in Fig. 11.1. We call this the *tripartite view* because it illustrates that staffing practitioners must achieve their staffing/organizational goals within three important boundaries. First, there is the legal boundary. The legal boundary includes all laws, regulations, and federal orders that impact the legal acceptability of staffing practices. Second are professional guidelines. These are not laws but more like codes of conduct. Several organizations, such as the Society for Industrial and Organizational Psychology (SIOP) and the American

A TRIPARTITE VIEW OF STAFFING PRACTICE

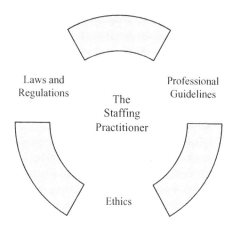

FIG. 11.1. Tripartite view of staffing practice.

Psychological Association (APA), have developed sets of principles and standards for using tests in employment contexts, and staffing experts must be sure to adhere to these professional standards. Third are ethical considerations. Ethics transcend both legal and professional guidelines and provide a frame of reference for evaluating the staffing experts' behavior, judgment, and practices.

Staffing practitioners operate at the intersection of these three boundaries, and as such, must be knowledgeable and/or proficient in each. This means they must not only be expert in the details of staffing practices, but also familiar with legal, professional, and ethical issues. It is a challenging role that requires continuous learning because staffing practices, laws, and professional guidelines are constantly evolving. Sometimes the changes are small and incremental, whereas sometimes they are large and redefine the nature of staffing practice.

The tripartite model is necessary because neither legal, professional, nor ethical considerations fully capture the complexity of staffing practice. For example, it may be the case that a particular staffing practice is illegal in the United States but legal in countries that are just starting to modernize. If the organization develops a venture in this modernizing country (perhaps because of the lax legal system), opportunities for exploiting employees can exist. The staffing specialist must be aware that even though the practice is legal in the particular country, it may not be consistent with professional or ethical guidelines. Likewise, a particular practice may be legally and professionally appropriate but unethical. The staffing specialist is called on to make numerous judgment calls, and the tripartite model helps ensure these judgment calls are evaluated in the proper perspective.

In the next three sections, we discuss each of the legal, professional, and ethical boundaries in turn.

I. LEGAL ISSUES

The legal boundary reflects all laws, regulations, and federal orders that relate to staffing practice. This includes not only laws about employment and discrimination, but also federal guidelines such as those concerning affirmative action programs. Every major area of staffing is affected by these laws and regulations, and many of these laws are constantly being reinterpreted and revised by the courts.

The top and left side of Fig. 11.2 shows how these laws are enacted. As noted in Ledvinka and Scarpello (1991), the process starts when there is a pressing societal need, such as when there are high levels of unemployment for certain groups of people, unfair or unequal access to employment opportunities, or discrimination. When such problems become sufficiently important to catch the attention of federal authorities (whether through lobbying or public outcry), legislation is passed that regulates or prohibits a particular practice with the goal of addressing the societal problem. This then frequently leads to the creation of regulatory agencies to enforce and ensure compliance with the laws. Two of the main federal regulatory agencies are the Equal Employment Opportunity Commission (EEOC) and the Office of Federal Contract Compliance Programs (OFCCP). These regulatory agencies monitor compliance, and when there is a problem, they take action to address the problem. However, it is important to recognize that these laws and agencies are designed to enforce general principles, but it is state and federal courts that actually interpret these laws in specific situations. This is why the legal context is constantly changing; a court's interpretation of a given law sets a precedent for other courts to interpret the law. The precedent may be overturned in other courts, or the outcomes of court cases may be appealed to higher levels (culminating with the Supreme Court).

It is important to recognize that this process unfolds over time, and while it unfolds organizations may or may not perceive and respond to such impending changes. Therefore, the right side of Fig. 11.2 shows the consequences for organizations that do not respond and those that respond in a proactive manner. As laws are being passed, many organizations choose to do nothing. For these organizations, possible

I. LEGAL ISSUES

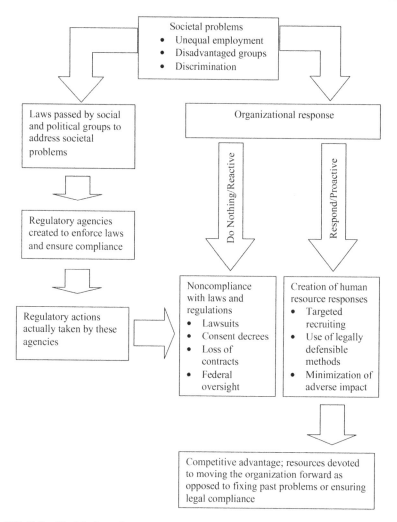

FIG. 11.2. Model of regulatory action and consequences of organizational responses.

consequences are that their noncompliance is identified by applicants who pursue litigation due to discriminatory practices or a regulatory agency enacts a *consent decree* (i.e., a court mandate for how staffing will change to address past discrimination) or removes federal contracts. In contrast, the organization that is proactive sees the signs of change and enacts procedures to capitalize on the potentially changing legal environment to achieve competitive advantage. This organization may enact HR practices that maximize adherence to the laws, perhaps before such laws have even been passed. Such organizations frequently benefit from "good press" and recognition by regulatory agencies for exemplary practices. As such, these organizations use the legal system as a source of competitive advantage, and while other organizations struggle to comply, the proactive organization is devoting its resources to being more competitive.

This simple regulatory model underscores the importance of viewing legal regulation as an opportunity rather than a threat. An organization, or even an entire industry, may be opposed to a particular law, but the fact is that laws and regulations are not going away. When it is apparent such laws are going to be written and enforced, organizations—and staffing specialists—would be well advised to proactively ensure compliance with such regulation. Doing so leaves the organization to pursue competitive opportunities rather than being "stuck" with the process of ensuring compliance.

Since the late 1800s, but starting most seriously in the 1960s, various employment laws have been interpreted in the courts. Although the goal of these laws and court decisions is to balance the concerns of employers, labor, and society, the actual cases tend to swing the pendulum of "burden" more toward the applicants or the organization. That is, over time we have seen how the burden was greater for employers or applicants, and that later court interpretations move the pendulum back toward the other group. It is also true that lower courts and higher courts may disagree, and that through the appeals process, court decisions can be overturned. All this leads to a rather shifting and dynamic process that represents the country trying to decide what it values—and that may evolve over time.

Table 11.1 provides a summary of key U.S. laws, regulations, and orders, and Table 11.2 provides a summary of key U.S. court cases and decisions. Reading through this table and the brief descriptions of the various laws and cases that follow show this evolution.

Civil Rights Acts of 1964 and 1991

Although employment legislation had existed prior to the 1960s, the modern start to this process began with the 1964 Civil Rights Act. Prior to this act, there was little recourse for applicants who believed they were discriminated against as part of a hiring process. Title VII of the 1964 Civil Rights Act basically made discrimination based on race, color, religion, sex, or national origin illegal for employers with more than 15 employees. Note that these various groups are known

I. LEGAL ISSUES

TABLE 11.1
Key Federal Acts and Orders

Law/Regulation/Order	Major Features
Civil Rights Act, Title VII (1964)	For employers with more than 15 employees, Title VII makes it illegal to refuse to hire or discharge any person on the basis of race, color, religion, sex, or national origin. This act created the EEOC to oversee and enforce this act.
Executive Order 11246 (1965)	For employers with more than 15 employees who have contracts with the federal government, it is illegal to refuse to hire or discharge any person on the basis of race, color, religion, sex, or national origin. This order enacted affirmative action programs and led to formation of the OFCCP to enforce such programs.
Age Discrimination Act (1967)	Prohibits discriminating against employees on the basis of age, specifically persons age 40 and older.
Americans with Disabilities Act (1990)	For employers with 15 or more employees, prohibits discriminating against individuals who are qualified but are judged to have a physical or mental disability.
Civil Rights Act (1991)	Negates the excessive burden of proof placed on the plaintiff in the *Ward's Cove* decision and allows plaintiffs to apply for punitive damages and have a jury trial. Outlawed score adjustments based on protected subgroup status (e.g., race, sex).
Proposition 209, California (1996)	A California state law that prohibits subgroup preferences in hiring and promotion. Judged to be "harmful" to affirmative action programs.

EEOC, Equal Employment Opportunity Commission; OFCCP, Office of Federal Contract Compliance Programs.

as protected groups or subgroups, and it is only discrimination based on membership in subgroups defined by these variables to which the law applied. The list of legally recognized subgroups was expanded with the 1967 Age Discrimination Act, the 1973 Rehabilitation Act, and the 1990 Americans with Disabilities Act (described shortly). Perhaps an equally important aspect of the 1964 Civil Rights Act was the establishment of the EEOC, a federal agency charged with ensuring and monitoring fair employment practices. The EEOC Web site contains a wealth of free information about employment law and can be viewed at http://www.eeoc.gov/.

In terms of defining employment discrimination, the EEOC considers two major types. First, *disparate treatment* occurs when individuals from a protected subgroup are treated differently than those of another subgroup. An example from an interview would be if the interviewer

TABLE 11.2
Key Court Cases and Decisions

Case	Major Features
Griggs v. Duke Power Company (1971)	Required an employer to demonstrate the job relatedness of a procedure if it resulted in adverse impact against a protected subgroup.
United States v. Georgia Power (1973)	Required an employer to conduct a validation study according to professional guidelines.
Regents of the University of California v. Bakke (1978)	A reverse discrimination case filed by a White applicant. The applicant won the case, but a consequence was the court endorsing a view that diversity could at times be a compelling interest in instances of past discrimination.
Connecticut v. Teal (1982)	Required employers to demonstrate that each part of a selection process was not discriminatory, even if the overall process was not.
Watson v. Fort Worth Bank & Trust (1988)	Required both subjective (judgmental) and objective (standardized) selection procedures to meet similar validity standards.
Ward's Cove Packing Company, Inc. v. Antonio (1989)	Required comparisons of adverse impact to be those qualified in the labor pool. Plaintiff also had to identify the specific selection practice causing the discrimination.
Gratz v. Bollinger (2003)	Supreme Court rejected the University of Michigan's use of diversity as a basis for undergraduate admissions, arguing it was not narrowly tailored.
Grutter v. Bollinger (2003)	Supreme Court upheld the University of Michigan's use of diversity as a basis for law school admissions, arguing it was narrowly tailored.
United States v. Garland, Texas (2004)	Upheld the city's use of standardized cognitive ability testing and denied the potential requirement of supplementing cognitive tests with noncognitive measures.

asked women if they are able to work weekends, but did not ask this question of men (perhaps because the interviewer inappropriately assumes women have more child care responsibilities). In this instance, there is a disparate impact on women because they may be hired less frequently. Second, *disparate impact* occurs when subgroups are treated equally during the hiring process, but the outcome of the selection procedures favors one group over another. For example, we may administer cognitive ability tests equally to all subgroups, but because of the large subgroup test score difference (see chapters 7 and 8), we know Whites are more likely to be hired than Blacks and Hispanics.

I. LEGAL ISSUES

In both instances, there is a specific three-step sequence that is followed when presenting evidence of discrimination:

1. The process starts with the applicant (called the *plaintiff*) forwarding the discrimination claim. In a disparate treatment case, plaintiffs must show they belong to a protected group, were qualified for the job, were rejected for the job, and the position remained unfilled. In a disparate impact case, plaintiffs must identify the specific staffing practice that resulted in unequal selection outcomes.

2. If the plaintiff is successful, the organization (called the *defendant*) must demonstrate a business rationale for the staffing practice. This frequently involves the presentation of validity evidence.

3. If the defendant is successful, the plaintiff has one final chance to refute this evidence. In the case of disparate treatment, plaintiffs must demonstrate that the real basis for hiring/rejection was discrimination. In the case of disparate impact, plaintiffs must show that a different staffing practice exists that is equally valid and appropriate, but results in less adverse impact. If the plaintiff cannot make these claims, the defendant will win the case.

In the determination of discrimination cases, the *Uniform Guidelines on Employee Selection Procedures* (UGESP) is given great weight. The UGESP is not a law, but a document published in 1978 that was prepared jointly by the EEOC, U.S. Department of Justice, U.S. Department of Labor, and Office of Personnel Management to aid in the interpretation of the 1964 Civil Rights Act. It defines good conduct for validity studies, defines discrimination, suggests ways for identifying adverse impact, and provides suggestions for ensuring the appropriateness of a staffing process. The entire UGESP can be viewed for free on the Department of Labor's Web site (http://www.dol.gov/dol/allcfr/ESA/Title_41/Part_60-3/toc.htm). The UGESP has dominated the staffing landscape for more than 25 years and continues to do so. However, some parts of the UGESP are outdated and do not address testing problems that occur today. In particular, the explosion in Internet testing has made it difficult to interpret some of the guidelines from the UGESP. Of most importance is the recognition that the UGESP applies to Internet testing just as it has to more traditional forms of testing. The full description of how the UGESP applies to Internet testing can be seen at http://www.dol.gov/dol/allcfr/ESA/Title_41/Part_60-3/toc.htm.

One final consideration with respect to employment discrimination is how one documents disparate impact, particularly adverse impact. There are a variety of methods available (see Bobko & Roth, 2004), but two are most common and are given the greatest deference in the courts. One is to use statistical significance to estimate whether the differences in hiring rates between two subgroups is sufficiently large that it could not have occurred by chance. Because it is a statistical significance test, many of the same issues noted in chapter 2 are relevant (e.g., statistical power, effect size, sampling variability). The second major index is known as an *adverse impact ratio*. This ratio is frequently used to assess violations of something called the *four-fifths rule* (or 80% rule). The four-fifths rule states that a prima facie (on the face of it) instance of adverse impact exists when the hiring rate for minority group members is less than 80% of the hiring rate for majority group members. That is, if the hiring rate of minority group members is less than 80% of the hiring rate for the majority group, it is taken as an indication that disparate impact may be present. Specifically,

$$\frac{\left(\frac{\#Minority\ Applicants\ Hired}{Total\#\ of\ Minority\ Applicants}\right)}{\left(\frac{\#\ Majority\ Applicants\ Hired}{Total\#\ of\ Majority\ Applicants}\right)}$$

For example, suppose we have 1,000 applicants, 800 are White and 200 are Black. If 500 Whites are hired and 75 Blacks are hired, then we have an indication that discrimination may be present because we violated the four-fifths rule:

$$\frac{\left(\frac{75}{200}\right)}{\left(\frac{500}{800}\right)} = \frac{(0.375)}{(0.625)} = 0.60$$

That is, the ratio of hiring rates is less than 0.80.

Notice that we said discrimination may be present. Violation of the four-fifths rule does not automatically mean the organization is using discriminatory practices. If the organization can show the job relatedness and business necessity of the selection procedure (usually through validity information) and demonstrate that there are no equivalent selection procedures that produce less adverse impact, the organization may continue to use this selection practice. Indeed, the *United States v. City of Garland, Texas* (2004) case described in Table 11.2 is one such example. In that case, the city was sued because it used a cognitive ability test that produced adverse impact against Blacks and Hispanics.

I. LEGAL ISSUES

The city was able to show the job relatedness of this procedure and also successfully refuted claims that other selection procedures would produce equivalent prediction but less adverse impact.

In 1991, the Civil Rights Act was modified and several important changes were made. First, it reduced the burden of proof placed on plaintiffs. Second, it allowed plaintiffs to sue for punitive damages and have a trial by jury. Finally, it explicitly forbade using protected subgroup status as a means for adjusting test scores to equalize the groups, a practice known as *within-group norming*. Prior to 1991, it was not unusual for an employer who wanted to meet specific diversity goals to use subgroup membership as a means for hiring. For example, suppose an organization tests 500 applicants, of whom 80% are White and 20% Black. If 50 applicants can be hired, the organization might choose to allot 40 positions to Whites and 10 to Blacks because 40/50 positions is 80% and 10/50 positions is 20% (thus ensuring the subgroup hiring rates are equal to their representation in the applicant pool). Now, if this organization uses test scores in a top-down manner within each group as a basis for making decisions, it would not have been an illegal practice because the test was used the same way for both groups. Further, within-group norming would result in the optimal balance of statistical prediction (because test scores were used in a top-down manner) and diversity (because specific percentages of subgroups could be hired). However, as we know from previous chapters, for many valid tests (e.g., cognitive ability) there are large subgroup differences. With such tests, use of within-group norming would result in Blacks being hired before Whites who have higher test scores. This was often seen as reverse discrimination. The 1991 Civil Rights Act made within group norming illegal, and by doing so, made solving the "prediction versus diversity" problem foremost on the minds of many staffing researchers (not to mention much of this book).

Executive Order 11246 and Affirmative Action

In 1965, President Lyndon B. Johnson signed Executive order 11246. Executive orders are regulations directed to federal government contractors and are put forth by the executive branch of the U.S. government. Executive Order 11246 essentially applied the same rationale as the 1964 Civil Rights Act to federal contractors and subcontractors. However, it also set the foundation for affirmative action programs (AAPs). This means employers (currently, those with at least

50 employees and who earn at least $50,000 per year from federal contracts) must develop action plans for the recruitment, selection, promotion, and retention of minorities and protected subgroups. As it turns out, this affects more than 20% of the U.S. workforce. The idea behind affirmative action (AA) is that, through proactive attempts to attract and retain individuals from disadvantaged groups, an organization's workforce will reflect the demographic composition of its labor pool. The AAP is a written document that describes how the organization would attempt to achieve this goal. The typical AAP will involve the following (specific information about AAPs can be found at http://www.dol.gov/dol/allcfr/ESA/Title_41/Part_60-2/toc.htm):

- *Organizational profile*: information about the staffing patterns within the organization
- *Job group analysis*: identifying job groups for reporting purposes
- *Placement of incumbents in job categories*: reporting the percentages of subgroups in each job category
- *Availability of positions*: estimates of the number of subgroup members qualified for each job category
- *Comparing incumbency to availability*: comparing the percentages of subgroup members in each job category to the availability of positions and qualified members
- *Placement goals*: goals for ensuring equal representation across job categories
- *Establishment of AA compliance officer role*: a person with a designated responsibility to oversee the AAP
- *Regular government survey to audit AA practices and goals*: an annual survey that evaluates the aforementioned elements

There are several misconceptions about AA. First, it is important to recognize that at no time does an AAP require a quota hiring system. In fact, quota systems were made illegal in the 1991 Civil Rights Act. AAPs emphasize the setting of hiring and promotion goals for different subgroups, but they do not require hiring members of such groups who are unqualified. Unless court-ordered under a consent decree to address past discrimination, quota systems are simply illegal. Second, AA is more than establishing hiring and promotion goals or race/gender-based initiatives. AA programs may include onsite day care facilities

I. LEGAL ISSUES 571

(to help with single parents and/or women), offering flexible working schedules, or even offering elderly care for employees taking care of their parents.

Whether for business necessity or because of severe labor shortages, organizations are increasingly adopting AA-like policies to reach out to neglected subgroups. Many organizations have realized that promoting diversity can provide a strategic advantage because it allows them to reach customers previously unobtainable or to develop new products in more creative ways. IBM, for example, implemented a major diversity initiative in the mid-1990s. They saw diversity as a strategic initiative to better identify and reach diverse customers. To accomplish this goal, they developed several task forces to represent the concerns and challenges faced by various subgroups (Asians, Blacks, persons with disabilities). These task forces introduced changes to the processes, structure, and climate of IBM. As a result, the number of female executives increased 370% and the number of ethnic minority executives increased 233%, and IBM expects this entire effort to generate more than $1 billion in revenue within the next 10 years (Thomas, 2004).

Organizations such as these put diversity as a business objective not only because of social reasons, but also because of economic reasons. A major review of AA provided by Kravitz et al. (1997) found no negative monetary consequences for organizations implementing AA programs. Yet interestingly, at the same time there is a backlash against AAP. Much of this may be attributable to the self-serving nature of people's perceptions of AA. A review of this research (Kravitz, 2004) suggests that in general, Whites most oppose race-based AAPs relative to other forms of AA. The strength of the AAP also matters, such that stronger programs are met with more resistance from Whites. The attitudes toward AA largely drive people's responses to such programs, but these attitudes are frequently based on incorrect or unfounded information. The entire report of Kravitz et al. (1997) can be seen free of charge at http://www.siop.org/AfirmAct/siopsaartoc.html. It is clear that until attitudes toward AA change, perhaps through people becoming more familiar with the fact that AA programs do not mean quotas, it is likely to remain under close scrutiny and skepticism from the general public.

The OFCCP, a division of the Department of Labor, is in charge of reviewing contractors and evaluating their compliance with the order. They will review the AAP survey provided by organizations contracting with the federal government Their Web site provides a lot of useful information and can be viewed at http://www.dol.gov/esa/ofccp/.

Americans With Disabilities Act of 1990

One final act deserves some attention, the Americans with Disabilities Act of 1990 (ADA). The ADA was proposed because a large percentage of the U.S. population (by some reports, up to 25%) can claim to have a disability, and yet employment of these individuals has been below what would be expected given their level of qualifications. That is, the issue was that otherwise qualified individuals were being denied employment because they had a disability. The ADA caused quite a stir when it was introduced because it was written in very ambiguous terms (resulting from numerous compromises that were necessary to get the act passed). Some feared the ADA would make employment testing legally impossible because of needs for special accommodation to those being tested who had disabilities. However, it is clear that the act only applies to qualified individuals with disabilities; it does not require the hiring of a disabled person who cannot perform the job.

As defined by the ADA, a person with a disability is "a person who has a physical or mental impairment that substantially limits one or more major life activities, a person who has a history or record of such an impairment, or a person who is perceived by others as having such an impairment." Furthermore, this impairment must limit a major life activity—for our purposes, job-related tasks. Blindness is an example of a disability, as are recovering drug users and alcoholics (but not current drug users/alcoholics). Fortunately for many staffing specialists, test anxiety is not a legitimate disability. We note that the ADA does not require the hiring of an unqualified person, but the act made it difficult to understand what this means because it stated, a "qualified individual with a disability means an individual with a disability who, with or without reasonable accommodation, can perform the essential functions of the employment position that such individual holds or desires." However, what are the essential functions of the job, and what is a reasonable accommodation? An "essential function of the job" is not necessarily what is identified in a job analysis because the ADA requires the possibility of a reasonable accommodation, meaning the "job" may need to be changed. Job analysis can help in identifying these essential job functions. A reasonable accommodation means that the employer may be asked to change the nature of the tasks/job to accommodate the person with the disability. For example, a blind applicant may need to be administered a written test in Braille, or an individual with a wheelchair may need a desk under which the chair can fit. Note

I. LEGAL ISSUES

that reasonable accommodation does not mean the job must change to fit the person with the disability; no accommodations are necessary if they create an *undue hardship* on the organization. Judgments of undue hardship are based on the size of the organization, expense of the accommodation, and impact of the accommodation on other employees and work processes.

It should now be obvious why, based on the wording and provisions in the ADA, many staffing specialists were nervous. However, the courts have been very reasonable in their interpretation of the act. According to the federal ADA Web site (http://www.usdoj.gov/crt/ada/pubs/mythfct.txt), relatively few cases have been brought to court and the costs of reasonable accommodation have been modest. Employers are not required to hire persons with disabilities or to create a job for such persons. Yet, the ADA has changed staffing practice. For example, it is now illegal to ask employees if they have a disability—such information must be self-disclosed. Likewise, medical exams must be job related and can only be given after the person would otherwise be hired (hence, it must occur at the end of the staffing process). Applicants can ask for an accommodation in the testing process; for example, a person with one arm requesting not to take an Internet test that requires two hands for typing. This is not necessarily a problem because as we note in chapter 7, we should strive to reduce confounding sources of variance in our tests. If the test method introduces variance not relevant to the KSAO we want to assess, we should use a format that minimizes such confounds. Understanding the ADA provisions is an important job for the staffing specialist. Fortunately, the federal government provides a comprehensive Web site with detailed information (http://www.ada.gov/). An excellent review of the ADA and its implications for staffing is provided by Campbell and Reilly (2000).

Individual Court Cases

The 1964 Civil Rights Act and Executive Order 11246 laid a foundation for employment law, but such law must be interpreted in specific situations. Therefore, it has been the charge of the courts to interpret these laws. Such court decisions are important because they set a precedent for future interpretations, unless the interpretation is overturned by a later decision or a higher court. There have been literally thousands of employment cases heard since the mid-1960s. As noted by Sharf and Jones (2000), the total number of EEOC filings grew from 62,135 in 1990 to 80,680 in 1997. However, they noted that most court cases are not brought forward by the EEOC but by private attorneys, roughly at a rate of 1 EEOC charge to 400 private lawyer charges. Furthermore, for plaintiffs who win common law claims, the average "win" is $408,719.

Table 11.2 shows some key decisions, and here we see the pendulum shifting back and forth, favoring either employers or applicants. Early case law focused on defining the organization's responsibilities, such as needing to demonstrate the job relatedness of a procedure (*Griggs v. Duke Power Company*, 1971), and requiring that validity information be of reasonable quality (*United States v. Georgia Power*, 1973). Later cases emphasized the role of the plaintiff in identifying what specific selection practice was discriminatory (*Ward's Cove Packing Company, Inc. v. Antonio*, 1989).

Earlier we noted how organizations are increasingly adopting practices that are consistent with AA, yet at the same time the general public appears to be resistant to AA. California Proposition 209 explicitly forbade the consideration of subgroup status in promotion and hiring in state employment and education. This is obviously at odds with Executive Order 11246, although it is limited to the state of California. However, after the passing of this proposition in California, Texas and Florida passed similar legislation. AA cases frequently occur in educational selection, as opposed to employment selection. *Regents of the University of California v. Bakke* (1978) made it allowable for universities to use diversity as a component in making admissions decisions, if there had been clear prior discrimination. Two more recent cases have made this a very visible topic and are likely to impact the practice of AA and staffing for years.

Both cases come from lawsuits filed against admissions processes at the University of Michigan. The two cases have many similarities. Both cases were appealed several times and were ultimately settled by the Supreme Court. Both admissions processes were modeled after the vision set forth in the *Bakke* case, so the Supreme Court in deciding the University of Michigan cases basically determined whether the earlier decision was appropriate (see Guttman, 2003a). Both cases involved the admission of minority students over majority students who scored better on standardized tests and had higher GPAs (i.e., the *Gratz* and *Grutter* cases that follow). Yet, they also had some important differences. In *Gratz v. Bollinger* (2003), the issue involved the undergraduate admissions process. The university used a points system, with some of the points being devoted to standardized test scores and GPA, and other

I. LEGAL ISSUES 575

points (about 20% of the total) being used for such factors as disadvantaged status, race, and ethnicity. In *Grutter v. Bollinger* (2003), the issue involved the graduate law admissions process. The university emphasized the creation of a *critical mass* of diversity through the admission of underrepresented groups. The argument was that such a critical mass contributes to better educational and learning opportunities and better prepares students for the reality of a diverse workforce. Thus, a critical mass of diversity creates a compelling state interest (the educational equivalent of a business necessity). The focus of the law school admissions process was not only on individuals, but also the composition of the entering class, though test scores and GPA was factored.

These were highly visible cases, and proponents on both sides offered briefs to the court supporting their position. The National Association of Scholars, President George W. Bush, and John Ellis Bush (Governor of Florida) sided with Gratz and Grutter. Siding with the University of Michigan were 65 Fortune 500 organizations, U.S. Senators Kennedy and Daschle, and the APA (see Guttman, 2003b, for descriptions of these briefs). This was a highly publicized case that showed the polarization between government, academics, and business on this issue. It is important to recognize that neither side questioned the value of diversity, but rather how to achieve it—through race-neutral or race-conscious selection procedures.

The Supreme Court sided with Grutter and upheld the law school admissions process, but it rejected the undergraduate admissions process and hence negated Gratz. The key distinction between these two cases was that the law school admissions process was deemed to be *narrowly tailored*. That is, the process was focused on the individual qualifications of each applicant, was flexible, did not harm nonminority applicants, presented the rationale for the process being used, and was on a limited timeframe. In contrast, the undergraduate admissions process was not narrowly tailored. There was no compelling rationale offered for a need for diversity, the points were applied to entire members of minority groups (and hence not on an individual-by-individual basis), and there was no clear time limit.

Thus, the key global legal issue defining the appropriateness of subgroup preferences in university admissions, and quite possibly AAPs in business, too, is whether the preference is narrowly tailored. The keys for ensuring narrow tailoring involve the provision of race-neutral alternatives to race-based programs, time limits on the race-based program, flexibility in the program, a compelling interest for using the program, and minimizing the burden on other subgroups (Guttman, 2002). However, perhaps the more important lesson from these highly visible cases is how divided the United States is on issues of the management of diversity. Politicians and academics were divided, yet industry seemed to argue for the need to incorporate diversity into staffing. Perhaps this is because industry most feels the challenges of employing a diverse workforce.

Summary

We provide a lot of information about the legal environment in which staffing practice operates. The reader can see that this environment has changed over time and, we predict, will continue to evolve. The key legal issue concerns whether the outcomes of the hiring process for an organization match the applicant pools for jobs in that organization. When there is a disparity, a judgment is made about whether there is adverse impact (e.g., the four fifths rule), and then judgments are made about whether any adverse impact shown was due to illegal development and use of hiring procedures. It is important to note that just the demonstration of adverse impact does not ipso facto prove the organization's practices were illegal, but the burden of proof falls on the organization to defend its practices in terms of job relevance for the procedures used and other forms of validity evidence. The fact of the matter is that the various acts and court cases have emphasized good practice (the next topic we attend to), and it is our clear impression that the practice of staffing has become greatly improved at a national level due to the legal issues we discuss here.

It is also important to note that adverse impact can occur in hiring even when the best practices are used. So, it is not the practice itself that is to blame for adverse impact. A goal of staffing however is to promote individual, organizational, and societal welfare so we are bound by the laws to practice in ways that ensure each applicant is given a fair chance to succeed. In the next section, we review some of the practitioner guidelines that have been developed to enhance the probability that people will have fair access to jobs—and these practices are what we have presented earlier in the book.

II. PROFESSIONAL STANDARDS

The second major boundary in the tripartite model involves professional standards for conduct. These are not laws in the traditional

II. PROFESSIONAL STANDARDS

sense, but guidelines for conducting the practice of staffing. We previously note the UGESP is one such set of professional guidelines. The UGESP is unique in that it was developed jointly by several government agencies. In contrast, most professional guidelines follow prescriptions developed by more narrowly focused professional organizations and societies. These guidelines exist because there are many gray areas of the law, and even though a particular practice may not be illegal, staffing specialists are expected to adhere to a defined set of professional standards.

This is most true for those who develop and administer psychological tests. There have been instances where those untrained or unqualified for administering such tests have done so in a manner that caused harm to the test taker or organization. There are also many examples of consultants or management "consultants" or management "coaches" who administer tests without knowing what they measure, their validity, or how to interpret them. For example, the growth of Internet testing has led to "test brokers" who provide Web sites with large numbers of tests available for purchase. An individual need only pay the fee to download and administer the test. The problem is that unless the person who does so is qualified, there is no way to evaluate the adequacy and appropriateness of the test, whether it is used correctly, or whether it is interpreted appropriately. Further, there are instances where people have developed their own tests without an understanding of psychometrics. This is similar to asking unstructured interview questions, and we know that such methods are unlikely to predict job performance and could be harmful. In all cases, the test user, organization, and test takers run serious legal, physical, and psychological risks that can easily be avoided by relying on experts knowledgeable about test development and use for making hiring decisions. If all this was easy, we would not have written this book.

As a staffing practitioner, one needs to be familiar with these issues. Many staffing experts are trained in HR, and as such may not have the appropriate background in test theory and psychology to develop and administer tests. Frequently, individuals with an HR background will be trained as managers from business schools. They may have a bachelor's degree, a Master's degree in HR, or an MBA. Their training will be broad toward an understanding of business, how various business units fit together, and how to administer HR programs and processes (e.g., determining staffing needs in the aggregate, running compensation programs) and broadly managing HR to be aligned with the strategic direction of the firm. Even though these individuals may be in HR and specialize in managing recruitment and staffing, none of these degrees provides sufficient training to actually develop interpret, and evaluate selection procedures such as those required by professional standards and the law. Rather, for this role one must rely on psychologists, usually with a Master's degree or PhD in I/O psychology, psychometrics, individual differences, or education. Such psychologists work in organizations in HR departments, as consultants, or as independent contractors. They frequently work with HR managers who are trained more broadly in organizational issues. All the authors of this book are I/O psychologists. Although we have specialized in staffing, I/O psychologists also study a range of workplace issues, including training, motivation, job attitudes, and groups and teams. The key distinction between I/O psychologists and HR managers is that I/O psychologists are trained in psychology, science, and research methodology, whereas HR managers will be trained in business-related disciplines, such as managing the HR function, marketing, and finance (of course, the distinctions get blurred in the real world).

Regardless of one's degree, there are several professional organizations that have produced guidelines and professional standards for conduct in staffing. The largest of these is the APA. The APA represents all types of psychologists, providing both member services and acting as an advocate on political issues relevant to psychology and testing. The APA has produced a number of guidelines for the appropriate use of tests in employment and nonemployment settings. Perhaps the most important is a document known as the *Standards for Educational and Psychological Tests* (American Educational Research Association, American Psychological Association, & National Council on Measurement in Education 1999; known more commonly as the *Standards*). This nearly 200-page book is one of the most important sets of testing guidelines. It describes the responsibilities of test users, test construction, evaluation, and documentation, ensuring fairness in testing, validity, and reliability. It even has an entire chapter devoted specifically to employment testing.

The APA also produces reports more specific to a particular topic as practice creates a need for such specialized reports. Most of these reports can be obtained for free from their Web site (http://www.apa.org/science/testing.html). For example, the unique issues with Internet testing led the APA to produce a document that describes the association's stance on good and ethical practice with Internet

II. PROFESSIONAL STANDARDS 579

testing (Naglieri et al. 2004). Although not prepared by the APA, members of APA and Division 14 of the APA (I/O psychology) have also recently produced a report on record keeping and defining job applicants in Internet testing (see Reynolds, 2004).

The APA has several other important documents and reports relevant to staffing. They have produced a report on test user qualifications (Turner, DeMers, Fox, & Reed, 2001). This document is helpful because it lists the qualifications and concerns that must be met when using psychological tests, including knowledge of how to select tests, administer tests, and test diverse groups. For example, to administer tests the staffing specialist must have knowledge of

- Legal rights of test takers
- Standardization and scoring procedures
- Ensuring the confidentiality of testing materials and respondents' test scores
- How to provide feedback to test takers

The APA has also prepared a document that describes the rights and responsibilities of test takers. The entire document can be found at http://www.apa.org/science/ttrr.html. Some of the key rights involve being informed of your test results in a timely manner, receiving a brief explanation about the purpose of the test, and having your test scores kept confidential. However, test takers also have a responsibility to ask questions prior to testing, represent themselves honestly during testing, and accept the consequences of not taking the test.

More specific to staffing specialists is a set of guidelines called the *Principles for the Validation and Use of Personnel Selection Procedures* (Society for Industrial and Organizational Psychology 2003; commonly called the *Principles*), prepared by SIOP. SIOP represents I/O psychologists and is the main professional organization representing staffing practitioners. It is affiliated with the APA and represents Division 14 (the APA is composed of multiple divisions, including social psychology, cognitive psychology, and experimental psychology). As taken from their Web site (http://www.siop.org), "The Society's mission is to enhance human well-being and performance in organizational and work settings by promoting the science, practice, and teaching of industrial-organizational psychology." The *Principles* are important because they are devoted specifically to the staffing contexts we have been concerned with in this book. They were written by practicing I/O psychologists experienced with real world staffing in business settings. They provide such information as the types of evidence necessary to demonstrate various forms of validity, the proper plan and conduct of a validation study, fairness and bias in testing, when meta-analysis might be a preferred alternative to an empirical validation study, and the development of criteria. The *Principles* can be freely obtained from SIOP's Web site (http://www.siop.org/_Principles/principlesdefault.htm).

Although the APA and SIOP represent testing specialists, there are other professional organizations relevant to staffing practitioners. The Society for Human Resource Management (SHRM) represents nearly 200,000 HR practitioners. It provides a host of member services, offers resources and publications, and advocates the HR function to politicians. Their Web site (http://www.shrm.org/) is also informative.

Summary

At this point you might ask, why do we need so many professional organizations' guidelines? There are two main reasons. First, each organization is devoted primarily to a particular group of professionals. For example, SIOP primarily advocates for psychologists, whereas SHRM primarily supports HR managers. These two groups advocate for their profession to different constituencies in government and business. Likewise, the *Standards* address testing across many different areas (e.g., employment, educational, clinical), whereas the *Principles* are devoted entirely to employment testing. These two documents are largely consistent with each other, but the *Principles* will cover some topics unique to business contexts. These various professional organizations also help determine the appropriate content and type of education necessary to be a member of the profession. Second, it sometimes happens that the guidelines are in conflict with each other or members of a particular profession do not agree with the guidelines set forth by another professional organization. This frequently happens because members of the two organizations may operate in different contexts with different issues, such as testing issues in diagnosing clinical depression versus testing issues in employment testing.

There is a final reason these professional organizations and guidelines are necessary. Quite simply, experts frequently disagree, and the guidelines at least help provide a level playing field and a forum for comparing the opinions of experts. At a minimum, professional

III. ETHICAL CONSIDERATIONS

guidelines help ensure technical adequacy even though the values and opinions of experts may differ. For example, among I/O psychologists, there are debates over the relative importance of societal issues concerning diversity in the workplace versus ensuring valid prediction. Some argue diversity is an important goal and that it is appropriate to use tests that are slightly less predictive if it means diversity goals can be achieved. Others argue that this mixes values with science because any reduction in validity will result in a reduction in utility and lower predicted performance (see chapter 6). These differences in opinions often manifest themselves when staffing experts serve as expert witnesses in court cases (many I/O psychologists have served as experts in the cases described earlier). However these differences are resolved, at least both parties agree to the set of professional guidelines that bind members to a common set of technical standards.

The bottom line is that staffing practitioners, whether they are consultants or part of the internal HR department of a company, must be familiar with the various guidelines available so when they make decisions they at least are aware of the different positions available to guide their practice. For the most part, there is consistency on these guidelines and that is the good news. In addition, we have tried throughout the book to adhere to our interpretation of the highest professional standards.

III. ETHICAL CONSIDERATIONS

The final boundary in the tripartite model shown in Fig. 11.1 represents ethical considerations. The professional guidelines discussed previously are guidelines relevant to the practice of staffing. However, we distinguish ethics from these professional guidelines because there may be instances where professional guidelines are silent on a particular issue. This most frequently happens when there is a change in the way staffing is practiced; therefore, ethical decisions must be made before the professional societies have had time to respond.

This occurred more recently with Internet testing. Because such testing is still relatively new, but grew rapidly in popularity, there were several years where the *Standards* and *Principles* did not directly address Internet testing (and because it was so new, there were no court cases to help interpret the relevant employment laws). Although most of the *Standards* and *Principles* were still relevant, there were some unique aspects of Internet testing that posed challenges. For example, Internet testing allows applicants to take tests in the privacy of their homes. But how can we ensure the person completing the test is the applicant? How can we ensure consistency and standardization? How can we be sure the applicant has read the instructions? Is the person an applicant if the test is begun but never completed? There are many such judgment calls that must be made, and these are frequently ethical judgments.

Fortunately, both the APA and SHRM have provided ethical guidelines to help staffing experts in such situations. The APA more recently revised their code of ethics, which is now available on their Web site (http://www.apa.org/ethics/code2002.html). They cover such issues as respect for persons, integrity, treating people with dignity, and beneficence. The APA ethics are primarily for psychologists and those involved with testing and assessment. For I/O psychologists and testing specialists, there are books describing the ethical practice of applying psychology in organizations (Lefkowitz, 2003; Lowman, 1998). These books are excellent and provide examples of ethical dilemmas, relevant guidelines, and resolutions for such dilemmas. For managers, SHRM has a code of ethics relevant to HR practitioners (http://www.shrm.org/ethics/code-of-ethics.asp). In this code are guidelines for acting in a socially responsible manner, complying with the law, and promoting organizational and individual well-being. Thus, staffing specialists have a general set of ethical guidelines presented by the APA and SHRM to help them navigate the ever-changing world of staffing. The staffing specialist will still need to use judgment in applying these ethical guidelines in specific novel situations that occur in organizations.

CURRENT ISSUES IN STAFFING PRACTICE

Staffing practitioners face many challenges. As if developing and administering effective staffing systems were not challenging enough, experts must also balance legal, professional, and ethical considerations. Yet, changes in the law, politics, industry, and professional practice are always occurring, resulting in a fairly continuous balancing act. There have been several good discussions of staffing practitioner challenges (e.g., Anderson, in press; Kehoe, 2000). In this section, we discuss some of the key issues and challenges faced by staffing practitioners. To illustrate these issues, we focus on describing the multiple roles of the staffing expert and the often competing demands when trying to

CURRENT ISSUES IN STAFFING PRACTICE 583

balance these roles. Furthermore, we lay out some of the judgment calls that must be made in these instances.

The Staffing Specialist as Practitioner

One challenge faced by many staffing practitioners is trying to balance diversity with optimal prediction. Again, the challenge is that many of our most valid tests (e.g., measures of cognitive ability) produce significant subgroup differences. Using such tests in practice will produce adverse impact against minority applicants. From a scientific perspective, this is not an issue because we are using the tests in a statistically optimal manner. We may be legally justified because the cognitive tests have demonstrable criterion-related validity and hence meet business necessity standards [e.g., *United States v. City of Garland, Texas* (2004)]. However, what if the organization wants a more diverse group of hires, whether for business necessity, to improve public relations, or simply to avoid future legal scrutiny? Strong pressure from organizational decision makers is often placed on the staffing specialist to enhance diversity.

In this case, the staffing specialist must balance prediction with diversity. The expert must determine whether other equally valid selection procedures can be used that produce less adverse impact. This is often difficult to do and may be so resource intensive that the organization is unwilling to participate. So, the organization may "encourage" the use of a less valid test or a battery of tests that also shows less adverse impact. Should the staffing expert oblige? If so, how much of a reduction in criterion-related validity is reasonable? There is no clear legal or clear professional guidance on this issue; arguments can be made on both sides. Irrespective of such external guidelines, what does this person believe is correct? In this common situation, staffing experts must balance the organization's goals, professional guidelines and ethics, and their own beliefs about what is appropriate. Box 11.1 gives an example from one of the author's experiences in this regard.

Staffing specialists who are consultants often face a related kind of pressure. Frequently, the employing organization may want to save time and money by cutting corners in a staffing procedure. We frequently hear such things as, "Is a job analysis really necessary?", "We already have a performance appraisal system," and "Our interview works fine the way it is." Staffing practitioners must be able to know what can and cannot be eliminated. Here, professional and legal

Box 11.1 Balancing Subgroup Differences With Prediction

One of the authors was asked to help an organization in which he had developed and validated several staffing procedures previously. In this instance, organizational personnel had developed a clerical test battery that had six parts. They had tested more than 300 applicants, but had not yet used any of the test scores to make employment decisions. Preliminary analyses by the company's affirmative action officer indicated that there were mean differences between minority and White candidates' scores on the test, especially on the part of the test identified as a measure of organizational skills. This skill was defined as the applicant's ability to "analyze, summarize, and/or organize information to solve a problem or follow an instruction." Items used to assess this skill involved answering questions based on reading a paragraph of information and some word problems of an arithmetic nature. The company now wanted help in determining how to use the information on this test in a way that would not result in adverse impact on protected groups.

The author asked for a validation study and the applicant data the company had collected. Based on the organization's description of its job analysis and the care with which the test was developed, the author believed defending the measures on content valid grounds was defensible. An analysis of the applicant scores, along with an estimate of the cutoff scores the company traditionally used to select employees, revealed that adverse impact was likely only if the organizational skills test were used. These data were discussed with company personnel, and they agreed that they would not object to removing the organizational skills component of the test.

However, removal of the organizational skills component would eliminate consideration of a skill critical to the performance of a significant part of the job. Instead, the author proposed the development of a series of interview questions designed to assess the degree to which the individuals had previously engaged in tasks that required organizational skills of the type typically required in the target jobs and required evidence of the adequacy of the performance of these tasks.

The organization agreed and a new interview measure of the type suggested was developed. Subsequent analysis of this measure indicated minimal subgroup differences. In addition, the author worked with company experts to develop cutoff scores that were defensible based on the experts' judgments of what was required of minimally competent newly hired individuals.

guidelines are helpful, but the problem is that many managers do not care about such guidelines. The staffing expert must be able to convince management about the need for doing things right. For example, Wal-Mart is currently facing the largest class action lawsuit brought against an employer. Nearly 1.6 million current and former female employees may be affected. The suit is essentially that Wal-Mart uses discriminatory practices against women in promotion and internal recruitment (see chapter 5). Many of the alleged discriminatory practices involved not giving equal access to promotional and training opportunities (a "test" in the legal sense of the word), not making job postings public (potential adverse impact on recruiting applicants), and not having clear standards for promotion (permitting biases in appraisals that are used as a basis for promotion). Developing these practices in a standardized manner and based on a job analysis with clear links between KSAO requirements for promotion could well keep organizations such as Wal-Mart out of court. It usually takes the threat of litigation for some organizational decision makers to recognize the value in what I/O psychologists do, giving highly visible examples such as Wal-Mart help to make the point.

The Staffing Specialist as Scientist

Many staffing practitioners, particularly those trained as I/O psychologists, are also scientists. That is, we are frequently asked to develop an entirely new selection procedure or solve some pressing organizational problem for which ready-made solutions do not exist. Perhaps there is no existing test available, or the job is so unique that there is no prior method one can adapt to the current situation. For example, the U.S. Army is currently involved in a project called Select21, where they are developing an entirely new selection and classification battery that incorporates cognitive and noncognitive predictors. Many of the predictors and criteria need to be developed from scratch, yet the particular context is fairly unique and using existing methods is frequently not possible. Similarly, many college and graduate testing organizations such as the College Board are trying to incorporate noncognitive assessments to help determine college admissions, and the development of such measures is an entirely new undertaking.

In these instances, staffing practitioners must also be scientists. They must be able to correctly analyze the nature of the problem; draw from existing theory, research, and literature to identify potential solutions; conduct studies to evaluate the adequacy of the solutions; and interpret the results to make recommendations. This is neither easy nor inconsequential. The results may affect thousands or even hundreds of thousands of individuals. The expert may work with a large team of other experts with diverse sets of expertise. Yet, all the major testing programs you have probably experienced, such as achievement tests in high school, the SAT or ACT, and the certification tests you may need to take after graduation, were developed this way. As such, staffing specialists have done much to improve national productivity and equality by developing reliable and valid procedures.

However, most staffing specialists do not work at such macro levels of the society; rather, they work in organizations facing specific problems that can benefit from a scientific perspective. Box 11.2 gives an example of a such a project conducted by one of the authors.

Box 11.2 Using Staffing to Improve Customer Satisfaction

One of the authors did some work for a supermarket chain and studied the relationship between the ways employees said they were treated in their departments (meat, bakery, deli, and so forth) and the reports customers provided of the satisfaction they experienced when dealing with the people in those departments. This kind of work relating employee experiences to customer satisfaction has come to be called "linkage" research, and the findings are quite robust (i.e., they have been repeated in branches of banks, insurance agencies, hotels, and so forth; Schneider & White, 2004).

However, the management of this supermarket was not only interested in this linkage, but also wanted to do something to improve customer satisfaction. So they asked the author for his ideas on what could be leading to positive employee experiences that were subsequently reflected in customer satisfaction. Based on prior research by Schneider and his colleagues (Schneider, Salvaggio, & Subirats, 2003; Schneider, White, & Paul, 1998), we had some conceptual ideas of what those "causes" might be. Indeed, a close examination of the data collected in the supermarket departments revealed that in those departments where the department manager was reported by employees to have a positive service quality management style and emphasis, the customers reported higher levels of satisfaction. We proposed to the company that a project to hire and promote department managers who would behave in these service quality ways could be a key to improving customer satisfaction.

> The next step was to do a thorough job analysis of the supermarket department manager position. In the job analysis, we had department managers rate the extent to which a long list of KSAOs was important for carrying out their job. The results of the job analysis would be used to form the basis for the development of selection and promotion procedures. However, prior to doing this we wanted to test the validity of the job analysis procedure itself and, because we already had customer satisfaction data for the department managers who completed the job analysis questionnaire, we simply correlated their KSAO importance ratings with the customer satisfaction data. By doing so, we were able to identify 17 KSAOs that significantly distinguished departments on the basis of customer satisfaction. What this means is that we now had a measure that might be used as a selection tool—ratings of how important KSAOs were to the job, as well as a measure to be used as a basis for the design of other selection tools. In fact, we designed for the organization a role-play exercise in which applicants would have an opportunity to behaviorally reveal the presence of the KSAOs. We also designed a structured interview to get at the KSAOs that were more difficult to assess in the period of time we were given by the organization. Criterion-related validity studies are under way, with the content validity of the procedures having already been demonstrated.
>
> The point is that a conceptual approach to research can stimulate ideas extremely relevant for staffing applications. Such research on organizational processes should be encouraged, and staffing specialists would do well to follow and explore the results for potential staffing implications.

The Staffing Specialist as Expert Witness

Staffing specialists live in a highly litigious world, and they are frequently asked to serve as experts in court cases. As if this were not interesting enough, they also frequently serve on opposing sides of court cases. For example, one expert may represent the plaintiff and another the defense. These two individuals may very well be good colleagues, but their task in the courtroom is to provide expert testimony and evaluate the adequacy of the selection procedure. So here is yet another instance where one finds considerable variance in judgment, even though the laws and professional guidelines apply equally to both sides. The *United States v. City of Garland, Texas* case described earlier was this kind of situation, with members of SIOP representing both the Department of Justice and the city.

In these situations, there is often considerable difference in the opinions of two sets of experts. Their judgment is of course informed by training, experience, and knowledge of staffing practices, but the application of general staffing principles to specific situations requires judgment nonetheless. For example, what if we administer a personality battery that is based on solid development procedures and assesses KSAOs identified in a job analysis, but the validity of the test is only 0.05. This is not unusual in practice. Suppose the battery is used operationally because even though the validity is low, it is still greater than zero and hence produces some positive utility. Furthermore, the personality test does not produce subgroup differences in scores; hence, its use may have some positive impact on hiring members of minority groups. Yet, an opponent might argue that validity is too low, suggesting a different personality test should have been used. There are lots of reasons for and against such a recommendation. It is open to debate and examination by experts on both sides, but frequently the side with the lawyers who best understand employment testing will win.

The Staffing Specialist as Political Advocate

Staffing specialists who serve as expert witnesses are also considered to be political advocates because their role in the case helps determine who wins, setting a precedent for future court cases. However, I/O psychologists can also serve as political advocates in the types of research they undertake and how they might convey these research findings to political decision makers. This is a point where the balance between science and practice becomes delicate. Some views of science say it should be devoid of values, that it should not be political or serve political means. In this orientation, science exists for the sake of generating knowledge. Others, particularly those who conduct science in the real world on real issues, recognize that whether the people who conduct the research want it to or not, their results may be used for political purposes. It is probably impossible to truly separate the values of scientists from the types of questions and interpretations they provide.

A clear example of this involves research on subgroup differences. We know such differences exist on many predictor constructs, but we also know that differences do not appear to be caused by the tests

CURRENT ISSUES IN STAFFING PRACTICE

themselves. Thus, for example, claims about tests being culturally biased or unfair simply do not have a great deal of empirical support. However, those in more fundamental areas of science often do not believe these results. They frequently discourage discussion of such differences in courses and even in research. Yet, they present no new knowledge about why such differences exist or how they can be reduced—instead, they choose to ignore the problem. Alternatively, staffing research has studied how to minimize these differences, and as such has led to advancements in understanding and reducing subgroup differences. Here is a situation where there is a pressing societal issue that has stimulated research that ultimately improves opportunities for minorities. However, it is also a situation where the research was politically motivated.

Zedeck and Goldstein (2000) wrote a thoughtful essay on the relationship between advocacy in public policy and being an I/O psychologist. They argue the domain of our research—the world of work—cannot be distinguished from public policy because work is simply too vital for societal well-being. A key point in their argument is that staffing specialists can and should inform public policy, as long as they act in accordance with professional guidelines and rely on honest evaluations of scientific evidence. Here, we see how the technical aspects of staffing, which most of this book is about, provide the common ground from which to advocate a particular position. Our opinions are grounded in data and stringent methods, so they are at least informed opinions.

Can (or Should) Staffing Experts Reach Consensus?

It might seem discouraging that staffing experts, trained in similar ways and using agreed-on methodologies, cannot come to a stronger consensus about some key legal, practitioner, and ethical issues. It may seem counterproductive to have staffing experts opposing each other in courts of law or advocating different opinions based on the same data. This is actually not a problem, but rather the way in which complex issues are resolved. When we reviewed legal cases in an earlier section, we noted how the pendulum of burden tends to shift back and forth between employers and applicants. For every solution that is reached, new problems appear that were not previously identified. As laws change to reflect the changing makeup of our society, new challenges appear that staffing practice must now solve. This process of debate and exchange is for the most part a useful means to ensure balance and acceptability in staffing procedures for many constituencies. It ensures all perspectives are equally considered and that individuals are treated as fairly as possible.

SUMMARY

In this chapter, we discuss the challenging role of the staffing practitioner. We discuss the tripartite view of staffing practice, noting how staffing practitioners must balance legal, professional, and ethical guidelines. We describe the variety of laws and regulations that apply to staffing and present cases just recently decided that are likely to have a profound impact on staffing practice. We note the important role of professional guidelines for ensuring good practice and recognize the role of ethics in those situations where regulations and guidelines do not yet exist. We then conclude by illustrating some of the roles played by staffing experts, including practitioner, scientist, expert witness, and advocate. It is hoped that this chapter leaves the reader with an appreciation of the dynamic and exciting work of the staffing practitioner. Our job is often challenging, yet rewarding, because it ultimately contributes to better-functioning organizations and societies.

REFERENCES

Age Discrimination in Employment Act of 1967, 29 U.S.C. Sec. 621 et Seq. (1967).
American's with Disabilities Act of 1990, 42 U.S.C. Sec. 933.
American Educational Research Association (AERA), American Psychological Association, & National Council on Measurement in Education. (1999). *Standards for educational and psychological testing*. Washington, DC: AERA.
Anderson, N. R. (in press). Relationships between practice and research in personnel selection: Does the left hand know what the right is doing? In A. Evers, O. Smit-Voskuyl, & N. R. Anderson (Eds.), *Handbook of personnel selection*. Chichester/London: Wiley.
Bobko, P., & Roth, P. L. (2004). The four-fifths rule for assessing adverse impact: An arithmetic, intuitive, and logical analysis of the rule and implications for future research and practice. In J. Martocchio (Ed.), *Research in personnel and human resources management* (Vol. 23, pp. 177–198). Oxford, UK: Elsevier.
Campbell, W. J., & Reilly, M. E. (2000). Accommodations for persons with disabilities. In J. F. Kehoe (Ed.), *Managing selection in changing organizations* (pp. 319–367). San Francisco: Jossey-Bass.
Civil Rights Act of 1964, 42 U.S. Code, Stat 253 (1964).
Civil Rights Act of 1991, Pub. L. No. 102–166, 105 Stat. 1075 (1991).
Connecticut v. Teal, 457 U.S. 440 (1982).
Equal Employment Opportunity Commission, Civil Service Commission, Department of Labor, & Department of Justice. (1978). Uniform guidelines on employee selection procedures. Federal Register, 43 (166), 38290–38315.

REFERENCES

Gratz v. Bollinger 539 U.S. (2003).
Griggs v. Duke Power Company, 401 U.S. 424 (1971).
Grutter v. Bollinger 539 U.S. (2003).
Guttman, A. (2002). Affirmative action: What's going on? *The Industrial-Organizational Psychologist, 40*, 59–68.
Guttman, A. (2003a). The administration's position on *Gratz* and *Grutter*: Too many inconsistencies. *The Industrial-Organizational Psychologist, 40*, 60–69.
Guttman, A. (2003b). More excerpts from the *Gratz* and *Grutter* briefs. *The Industrial-Organizational Psychologist, 41*, 144–153.
Kehoe, J. F. (2000). Research and practice in selection. In J. F. Kehoe (Ed.), *Managing selection in changing organizations* (pp. 397–437). San Francisco: Jossey-Bass.
Kravitz, D. A. (2004). *Enhancing diversity through staffing: The role of affirmative action*. Manuscript submitted for publication.
Kravitz, D. A., Harrison, D. A., Turner, M. E., Levine, E. L., Chaves, W., Brannick, M. T., Denning, D. L., Russell, C. J., & Conrad, M. A. (1997). *Affirmative action: A review of psychological and behavioral research*. Bowling Green, OH: Society for Industrial and Organizational Psychology.
Ledvinka, J., & Scarpello, V. (1991). *Federal regulation of personnel and human resource management*. Boston: PWS-Kent.
Lefkowitz, J. (2003). *Ethics and values in industrial-organizational psychology*. Mahwah, NJ: Erlbaum.
Lowman, R. (1998). *Ethical practice of psychology in organizations*. Washington, DC: American Psychological Association.
Naglieri, J. A., Drasgow, F., Schmit, M., Handler, L., Prifitera, A., Margolis, A., & Velasquez, R. (2004). Psychological testing on the Internet: New problems, old issues. *American Psychologist, 59*, 150–162.
Regents of the University of California v. Bakke, 438 U.S. 265 (1978).
Reynolds, D. (2004). EEOC and OFCCP guidance on defining a job applicant in the Internet age: SIOP's response. *The Industrial-Organizational Psychologist, 42*, 127–138.
Schneider, B., Salvaggio, A. N., & Subirats, M. (2003). Climate strength: A new direction for climate research. *Journal of Applied Psychology, 87*, 220–229.
Schneider, B., & White, S. S. (2004). *Service quality: Research perspectives*. Thousand Oaks, CA: Sage.
Schneider, B., White, S., & Paul, M. (1998). Linking service climate and customer perceptions of service quality: Test of a causal model. *Journal of Applied Psychology, 83*, 150–163.
Sharf, J. C., & Jones, D. P. (2000). Employment risk management. In J. F. Kehoe (Ed.), *Managing selection in changing organizations* (pp. 271–318). San Francisco: Jossey-Bass.
Society for Industrial and Organizational Psychology. (2003). *Principles for the validation and use of personnel selection procedures*. Bowling Green, OH: Author.
Thomas, D. A. (2004). Diversity as strategy. *Harvard Business Review, September*, 99–108.
Turner, S. M., DeMers, S. T., Fox, H. R., & Reed, G. M. (2001). APA's guidelines for test user qualifications: An executive summary. *American Psychologist, 56*, 1099–1113.
United States v. Garland, Texas (2004).
United States v. Georgia Power 474 U.S. 906 (1973).
Ward's Cove Packing Company, Inc. v. Antonio, 490 U.S. 642 (1989).
Watson v. Fort Worth Bank & Trust, 487 U.S. 977 (1988).
Zedeck, S., & Goldstein, I. L. (2000). The relationship between I/O psychology and public policy: A commentary. In J. Kehoe (Ed.), *Managing selection in changing organizations* (pp. 371–396). San Francisco: Jossey-Bass.

12

Staffing Organizations: Review and Implications

AIMS OF THE CHAPTER

At this point, we must face the problem of review and integration. In this chapter, we

1. Summarize some of the major points made throughout the book.
2. Note how some of the procedures and processes we outline can be applied in specific circumstances.
3. Speculate regarding the next big discoveries in staffing research.
4. Present ways by which job seekers can take advantage of the material to find a job in which they can both be satisfied and productive.

Prior to discussing each of these elements, however, it is important to place the staffing of organizations in context, that is, to repeat some of the material presented in chapter 1. Specifically, it is important to remember that any staffing system cannot resolve all organizational problems. Although we note throughout the book that people make the organization what it is, most situations do not permit starting from scratch. The reality is that organizations can be helped to be more effective through a variety of strategies, including hiring new people

AIMS OF THE CHAPTER

and/or promoting incumbents. Training, management development, and education; changes in organizational design and structure; adopting new accounting systems; creating innovations; and acquiring or merging with other companies are all ways by which organizations can be made more productive and more satisfying—and the selection and placement of new and current employees can also help.

How do we know which way to go? This is a key issue for organizations, especially when some unforeseen turbulence either within or outside the organization emerges. The turbulence might be conflict between the heads of marketing and HR, or it could be the announcement of a major innovation by a competitor. When such turbulence exists, management obviously attempts to grapple with it and, if possible, prevent recurrences. When fighting fires, for example, we put them out and try to prevent them in the future. In organizations, this means the emergency is handled and attempts are made to ensure at least this particular form of emergency does not occur again.

Most frameworks for organizational effectiveness emphasize the necessity to focus on the many systems and subsystems of organizations as a remedy for preventing unexpected turbulence. For example, Katz and Kahn (1978) said there are five major subsystems in all organizations, and for the total organization to be effective, all the subsystems must be functioning effectively. The five subsystems they specify are presented in Table 12.1.

What is important in Table 12.1 is the variety of issues that require attention for an organization to function effectively. We know from HR strategy research that organizations who adopt high-performance work systems, or bundles of complementary HR practices such as staffing, training, and compensation, perform better than those who do not (Huselid, 1995; MacDuffie, 1995). This means that staffing must be in alignment with the organization's strategy, HR strategy, and other HR practices.

Obviously, if the organization has the right people in place in the various subsystems, then the probabilities for success are improved. That is why good staffing is so important. Organizational diagnosis might reveal, however, that people who could do the job are in place, but they have not been provided with the kinds of skills necessary to do so—a case for training. Or, suppose we confront a competitor's innovation—a solution might be to acquire or merge with the competitor. The point here is that there is a tendency in organizations to put all the blame for

TABLE 12.1

Five Subsystems of Organizational Functioning

1. **Production Subsystems:** The primary cycles of activity that yield the eventual output of organizations; the production process; technology, division of labor, design of jobs.

2. **Supportive Subsystems:** Providing the materials to be produced and the delivery of products and services; supporting actual productivity; interacting with the larger environment for procurement of materials and disposition of goods produced.

3. **Maintenance Subsystems:** Maintaining both personnel and equipment; recruitment, socialization, training, compensation and reward (prestige, job satisfaction) systems; provisions for keeping equipment up and running.

4. **Adaptive Subsystems:** Monitoring of external changes and translating that information so the meaning for the organization is clear; product research, market research, long-range planning, research and development.

5. **Managerial Subsystems:** The activities for controlling, coordinating, and directing the other organizational subsystems; decision making; conflict resolution; coordinating internal structures to meet external requirements; the authority subsystem.

Source: Adapted from Katz & Kahn (1978).

ineffectiveness on the people in the organization and to fail to realize that those people may simply not have created the kinds of structures, processes, and procedures under which they could have been more successful. Just because our opening framework emphasizes the reciprocal nature of individuals and organizations does not mean that the most appropriate or best structures and processes will emerge in an organization. Indeed, we have written elsewhere (Schneider, Smith, & Sipe, 2000; Ployhart & Schneider, 2005) that a key criterion of effectiveness in the validation of selection procedures for executives and managers should be the degree to which they in fact create conditions in which the effectiveness of workers is maximized.

The reason for this background is to caution against having a simplistic view of what is required to make an organization effective. As should be clear from Table 12.1, the number and variety of issues requiring attention in any organization is astounding. We carry around an image of the managers in organizations as fantastic jugglers, keeping all the organization's subsystems in the air at the same time, treating each as it comes up (or down) for attention, and trying to have the whole thing look like a polished act. It is an enormous task and a good staffing process can help, but it certainly is not the whole answer to organizational problems.

EFFECTIVE STAFFING: EMPHASIZING SOME KEY POINTS

This entire book is about effective staffing—how to conduct staffing and how it relates to individual and organizational effectiveness. Sometimes one can get bogged down in the details and forget the big picture. Therefore, let us take a step back and reiterate a few key points:

- Effective staffing begins with a job analysis. Whether this is done using a homegrown approach, an existing methodology, or O*NET, we cannot know what to predict and how to predict it without understanding the basics of the job—the tasks and the KSAOs required to do those tasks. The job analysis unambiguously defines the content of the criterion and predictor domains, and hence, is the foundation of staffing.
- Predictors derive their importance from criteria. Once we have defined the nature of effective performance, there should be little mystery about how to best predict it. The nature of performance dictates the relevant KSAO characteristics that contribute to effective performance. Thus, we must have a conceptual understanding of the important aspects of performance and good measures of each dimension of performance.
- Choose the measurement method best suited for measuring the KSAOs. Although paper-and-pencil testing remains a dominant administration method, we should not adopt this method in an unthinking manner. It may be better to use a computerized or Web-based test to measure some KSAOs, and an interview or interactive voice recording (IVR) for others. Because each method carries with it certain strengths and weaknesses, it is a decision that must be carefully considered. Our own predilection is to use performance testing whenever the investment can be justified. Such tests reveal considerable criterion-related validity, can usually reduce adverse impact, are well received by applicants, and are easily defended in court.
- Recruitment is the means through which we retain and attract talent to the organization. It is always difficult to remember to think about recruitment as both internal and external recruitment; however, it is important to remember that most jobs are filled with internal not external candidates, and it is worthwhile to keep the

best talent in an organization to fill openings. In times of a strong economy, when there are many job openings and few applicants, recruitment may be even more important than selection. The organization's potential largely lies in the quality of the people it can attract to the organization, convince to accept a position, and then retain.
- The legal system has a fundamental impact on staffing. We cannot separate staffing from the legal system, and as laws change so will staffing. The staffing expert must be familiar with these laws, but in doing so, the organization will also be more likely to be competitive with other organizations because staffing will be done correctly and validly.
- Effective staffing contributes to better individual and organizational performance. Although we have much more data on the former than the latter, it makes sense that effectiveness across multiple levels should improve if we hire people who can do their jobs and work with others as good citizens of the organization.

RECOMMENDATIONS FOR PRACTICAL STAFFING PROBLEMS

Every organization must make hiring decisions. In fact, all of you will apply for jobs and be hired. Staffing affects millions of people every year—the better the economy, the more people it affects—and has a profound influence on individuals, organizations, and societies. It is important we do it right, and in this book we give you the steps for making this happen. Yet, surprisingly, many organizations do not see staffing as either important or strategic. They sometimes believe formal staffing does not apply to their situation. In this section, we consider the most frequent of such arguments and show how staffing—as we describe it in this book—does in fact apply. Let us start in some detail with a most frequent situation—staffing the small business.

"We're Too Small for This Stuff"

Frequently, small employers, those ranging from a few employers to maybe around 100, do not believe formal staffing procedures "work" for them. On the surface, it would obviously be ridiculous to conduct a job analysis on five employees. However, our points raised in chapters 3 and 6 bear consideration—staffing will help these small organizations, even though the procedures are not applied in the same way. For example, Box 3.1 (chapter 3) described how a small business owner could use O*NET to identify the KSAO requirements for a job, and Box 6.1 (chapter 6) described various validity options (e.g., synthetic validity, job analysis and content validity, validity generalization, use of an existing test). The key here is to be systematic and have a reasonable basis for choosing particular staffing procedures.

Let us give you an example of how a content validity approach might work. One of your authors (Schmitt & Ostroff, 1986) helped develop selection procedures for a relatively small group of people, 39 emergency telephone operators in a medium-size urban area. These operators were required to answer phone calls from people in emergency situations requiring police, fire, or medical attention. The project consisted of three major components: a job analysis to determine the tasks involved in performing the job of emergency telephone operator and the KSAOs necessary to perform those tasks, the development of exercises that sample the KSAOs needed for adequate job performance, and the evaluation of the psychometric properties of the exercises and their content validity. These components are discussed in more detail as follows:

1. A homegrown job analysis procedure was used similar to that outlined in chapter 3. Task statements and KSAOs were generated in group meetings. Both tasks and KSAOs were then rated by SMEs, and the interrater consistency of those ratings was assessed. Six major KSAO dimensions were identified: communication skills, cooperativeness, emotional control, judgment, memory, and clerical/technical skills.

Both KSAOs and task statements played a role in the construction of a variety of selection tests. Each test devised centered around two or more of the KSAO dimensions. An attempt was made to devise a test of a given KSAO dimension that resembled the actual telephone operator tasks for which that KSAO was required. The tests constructed were based on tasks rated as most important and resembled components of the job as closely as possible. Expert job incumbents who helped in test administration were asked to review each test item and indicate whether it was essential to the performance of each major task dimension. These data generally confirmed the content validity of the various test components, although some minor additions and deletions were indicated by this review.

2. The examination plan consisted of three consecutive phases: (a) an oral direction/typing test, (b) a situational interview,

TABLE 12.2
Examination Plan of Skill Dimension Matrix

Dimension	Oral Directions	Interview	Simulation
Communication skills		X	X
Emotional control		X	X
Judgment		X	X
Cooperativeness		X	X
Memory	X		
Clerical/technical skills	X		

and (c) a phone call simulation test. The examination plan by skills matrix is given in Table 12.2.

a. The oral directions/typing test was designed to measure the applicant's memory ability and technical/clerical skills. This test consisted of four components: spelling, telephone call recordings, monitoring, and typing. The first three components of the exam were administered via a tape recorder; applicants listened to the information and questions presented on tape and responded in writing on answer forms.

The spelling test required applicants to properly spell common street names and places presented on the tape, an obviously critical skill for an emergency operator. The telephone call portion of the test was actually a series of scripted conversations between complainants (callers) and operators. Here the candidate had to listen to the conversations and accurately record the pertinent information, as shown in Fig. 12.1.

The third portion of the oral directions test was a monitoring exercise resembling tasks required for monitoring police units. Applicants

Address of incident	City	Location/building name
Caller's name	Phone no.	Caller's address
Nature of incident		Description of subjects/vehicles
Complaint against		Other information

FIG. 12.1. Form candidates used to record information from tape-recorded calls.

listened to a series of statements that gave information about police units, their location, and activities. Following the presentation of these statements, questions were asked about the location of a unit, what type of call the officers were responding to, and so on. Applicants recorded the information on their answer forms. Scores were based on the accuracy of the information recorded regarding the police units' locations and activities.

The final portion of the exam was a typing test. The typing test matched the kind of typing performed on the job. For this test, applicants typed information printed on standardized forms into blank forms. The forms were exactly like the ones used in the phone call portion of the exam (see Fig. 12.1). Scoring was based on speed (the number of blanks typed in each form) and accuracy (the number of errors made while typing).

b. The situational interview, described in chapter 10, is a structured interview in which applicants respond to a series of job-related incidents. In developing our modification of situation interview questions for the emergency operator's job, we began with the collection of critical incidents (see chapter 3). Job incumbents were asked to identify incidents in which particularly good or poor job behavior had been exhibited by some operator. The list of critical incidents was analyzed and edited by the investigators and several SMEs. Incidents were then translated into relevant interview questions in which job applicants were asked to indicate how they would behave in each situation. Because applicants were unlikely to have the job experience necessary to deal with the critical incidents, it was essential to translate these critical incidents into questions with which job candidates would have some knowledge and/or experience.

As an illustration of the transformation of critical incidents into interview questions, consider the following example of a question developed for the situational interview. To demonstrate communication skill, the following critical incident was suggested by experienced workers:

> A caller becomes abusive when talking to an operator. The operator gets mad and verbally abuses the caller using derogatory language.

This incident was transformed into the following interview question:

> How would you react if you were a sales clerk, waitress, or service station attendant and one of your customers talked back to you, indicated you should have known something you did not, or told you that you were not waiting on them fast enough?

12. STAFFING ORGANIZATIONS

Seventeen interview questions were developed to evaluate the applicants on KSAO dimensions of communication skills, emotional control, judgment, and cooperativeness. Because the interview questions were based on incidents of critical job behavior, tasks important to the job were the items in the interview questions. All these interview questions posed situations that were familiar to inexperienced job applicants.

A set of standardized BARS-type rating scales (see chapter 4) was devised to aid the interviewers in making more objective judgments of applicants' performance in the situational interview. The interview rating guide consisted of a set of 12 scales. On each scale, the dimension was listed, along with a set of interview questions that were likely to elicit candidate responses relevant to that particular dimension. Also listed were potential good and poor answers to each interview question within each dimension. Interviewers were instructed to review the interview questions relevant for that particular scale, consider the examples of good and poor answers to those questions, and then make a rating on a 5-point scale ranging from excellent to poor based on the applicant's responses. A total interview score for the applicant was computed by adding the scores received on each scale. One example of a rating scale is reproduced as Fig. 12.2.

c. The phone call simulation placed job candidates in a role-playing exercise (see chapter 10). The applicants played the role of an operator taking calls from complainants. An experienced operator played the role of a caller. This exercise assessed applicants' communication skills, emotional control, and judgment by focusing on the important tasks related to obtaining and recording critical information accurately.

Applicants talked with the caller and obtained the information necessary to send help. Some callers were hysterical or emotional. The candidates had a variety of questions they needed to ask each caller to elicit the appropriate information. The information obtained was recorded on a standardized form, the same form used during the oral directions/typing test; hence, all candidates were familiar with this aspect of the simulation.

The technician, playing the caller, was in the next room using special phone equipment to speak with the applicant. A series of six phone call scripts (see Fig. 12.3) were written to guide the technician in making calls. One practice call was given to allow applicants to acquaint themselves with the procedure. The phone call conversations were tape

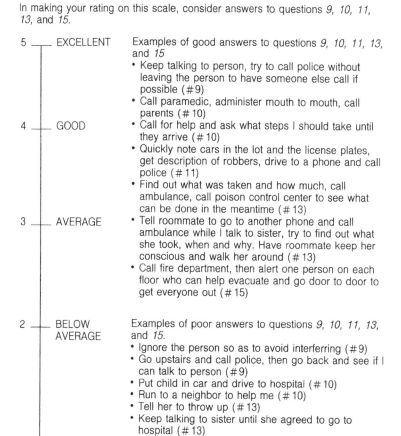

FIG. 12.2. Judgment/common sense: ability to think and act clearly during emergencies.

recorded so raters could later listen to the tape and judge applicants' abilities. The role-playing operator was given extensive practice and instruction regarding the necessity to give the same information to all candidate callers. Because all conversations were recorded and evaluations were based on those recordings, nonstandard behavior on the part of the role-playing technician could be taken into account.

Scoring for the phone call simulation consisted of two parts: (a) scoring of the standardized forms applicants completed; and (b) ratings of the applicants' communication skill, emotional control, and judgment.

Caller:	(Angry) I'm really mad and I've had enough of this. I have to work the morning shift and I get up really early. The morning shift starts at 6 o'clock. I work at Olds and I need my sleep. This is just ridiculous and it goes on every night. I can't believe it…
CBO:	Should interrupt at some point and ask what the problem is, get address, caller's name, etc.
Caller:	(Gives this information to caller *only when asked*): • Name: Henry Abbott (or Harriete Abbot) • Incident: neighbor's dog barking furiously • Caller's address: 2728 W. Landsdown • Phone no.: 555-5493 • City: Okemos • Neighbor's address: 2792 W. Landsdown (when caller asks for this address, also say): I wish you'd hurry up and take care of this. I really need my sleep. You know, I have to work very early in the morning. This goes on every night and I'm tired of it…
CBO:	Should interrupt and continue getting information by asking questions—complaint against: don't know neighbor's name…

FIG. 12.3. Sample phone call script.

3. Psychometric evaluation of the tests included a review of their content validity and reliability. The development of these exercises clearly relied on a content-oriented validation technique. The procedure used was in accordance with the standards established in the *Uniform Guidelines on Employee Selection Procedures* (1978; see chapter 11). As stated in the *Guidelines*, a selection procedure can be validated by a content-oriented strategy if it is representative of the important aspects of performance on the job. The tests developed matched the important tasks performed on the job based on information provided by SMEs in

the job analysis. Further, the *Guidelines* specify that a job analysis that focuses on work behaviors and associated tasks is required. The job analysis procedure outlined previously provided specific job behaviors and their importance and necessity to the job.

Content-oriented validity is usually not an appropriate strategy when the selection procedure KSAOs will be expected to be learned on the job. The initial job analysis ratings of the KSAO statements by job incumbents provided a basis for delineating KSAOs necessary for newly hired workers and those KSAOs learned on the job or learned through training. The tests devised concentrated only on those dimensions that were necessary for new workers to possess. Furthermore, skills and abilities that were necessary, but required some training, were modified to accommodate inexperienced workers. For example, the standardized forms used to record information in the oral directions test and the phone call simulation were modified and used in written form rather than on a computer system; the situational interview questions were translated from critical job incidents to experiences a person could encounter in everyday life.

The selection strategy involved multiple hurdles and both compensatory and noncompensatory scoring (see chapters 2 and 7). Applicants who met or exceeded the specified cutoff scores in the oral directions/typing test phase of the examination process proceeded with the situational interview and phone call simulation in which a compensatory scoring scheme was employed. The use of minimum cutoff scores was justified by information gathered in job analysis interviews that indicated that people needed minimal memory and technical/clerical skills (primarily assessed in the oral directions test) and that other skills were more important in separating the marginal employee from the truly superior worker. Hence, the interview and phone call simulation were used to rank order candidates who had passed the earlier oral directions test.

Another advantage of this type of selection procedure is that it provides a more realistic preview of the job (or RJP; see discussion of RJPs in chapter 5). Applicants gain a clearer understanding of the job in question as they perform behaviors that are required on the actual job.

A final attraction of this project was the extensive involvement of supervisory personnel and job incumbents in all phases. Although this is a useful strategy in any project, this involvement ensured all proposed selection instruments were job related. The involvement was also necessary to meet the administrative requirements of the emergency center in which the jobs existed. That is, police personnel had previously conducted the selection interview of candidates, and they wanted to have continued input into the hiring decision. To accommodate these desires, civil service personnel screened most of the low qualified applicants, making the interview procedure less burdensome. Of course, the usual affirmative action demands were also present and needed consideration. Because of these requirements, applicants proceeded through two phases of testing. Civil service personnel administered an oral directions/typing test designed for ease in administration to small groups of people. Police and supervisory personnel administered the situational interview and the phone call simulation.

In this example, we see a comprehensive staffing procedure being developed for a job employing relatively few people. Note the central role of the job analysis, the heavy use of simulations, the use of rating scales for judging candidate performance even in an interview, the involvement of supervisors and incumbents, and the care taken to adhere to legal issues in selection. Thus, even in a small organization, the staffing practices discussed in this book are quite applicable and relevant.

"We Need to Hire Someone Now"

No matter what the problem, always start with at least a cursory job analysis. Pick the procedure from chapter 3 with which you feel most comfortable and that best suits your purpose. Specify the variety of tasks to be done and the standards of performance to which you will hold the employee. Think of activities you could ask a person to do that correspond to the job activities required. If time is truly short, consider using O*NET as the basis for the job analysis—it is comprehensive, available online 24/7, and free.

Make the content of your assessment like the job content. For example, say you need a short-order cook. Would you trust an interview to hire a cook? Before reading this book you may have, but now you would have applicants cook the kinds of things you want them to do on the job. The logic here is that you do not want to fly on an airplane on which the pilot was selected based only on an interview, so why should you serve food cooked by a person hired only by an interview?

The same principle applies to hiring salespersons, bricklayers, dishwashers, software designers, secretaries, and office managers. Let content validity be your guide in designing the exercise, simulation, or

PRACTICAL STAFFING PROBLEMS 605

activity. However, be sure to test the applicant on not only the core task (e.g., laying bricks), but also on other tasks required for effective performance (e.g., reading blueprints and being conscientious about safety on the job).

Another possibility is to call the local state employment service office after doing the job analysis. Describe the kind of person desired—remember to concentrate on only the important KSAOs. Because state employment services typically have KSAO data on people, they may have applicants available who fit the descriptions you provide.

While waiting for an applicant, devise some weights for the importance of the different components of the job you have open. You will probably do some implicit weighting in the presence of an applicant, so why not be as explicit as possible? Such weighting will also serve as a guideline for the kinds of information you will want an interview to generate if you decide to do one (and everyone seems to).

After developing these weights, think carefully about the kinds of rewards the job offers. Again, chapters 4 and 5 can be used here. Providing applicants with such information may help make the decision more two way and also places some of the burden of responsibility for the decision on the applicant.

The clue to making such immediate kinds of decisions is to be systematic. Although decisions made in this fashion may be of unknown validity, they are more apt to be valid because they are based on principles discussed throughout this book that have proven important in relatively rigorous research efforts. Of course, organizations faced with a consistent need for making fast hiring decisions may consider adopting online screening and assessment systems. When used appropriately, they can greatly reduce administration and scoring time and result in near instant hiring recommendations.

"We Can't Attract Enough Qualified Applicants"

This is perhaps the most common complaint we hear from employers; the problem is that for many jobs there are not enough qualified applicants. Many small businesses face this situation and report they cannot grow or handle as much work as they want to because they are continually short staffed. The issue here is not so much selection—we must have enough applicants to make selection viable (see chapter 6)—but rather attracting enough people to apply to the organization. Thus, this is really a recruitment problem.

So, how can we increase applicant attraction? Most organizations erroneously think the only way they can increase attraction is to increase pay—something that is frequently not an option. However, pay is just one of several reasons why people choose jobs and organizations, and we note several additional reasons in chapters 4 and 5. We have seen how several less "tangible" features, such as fit (person–job, person–workgroup, and person–organization), organizational climate, and working conditions can also contribute to applicant attraction.

In terms of emphasizing fit, the key is to ensure the recruitment message emphasizes trait or value images. Because it appears that applicants tend to infer "personalities" of organizations (Lievens & Highhouse, 2003) and these personalities may in fact be the modal personality of employees within the organization (Schneider, Smith, Taylor, & Fleenor, 1998), organizations may want to make this information salient because applicants find organizations with similar personalities more attractive. In particular, we would want to emphasize fit information in a manner similar to that discussed in Table 5.4 (chapter 5) so the information conveyed maximally distinguishes the organization from competitors in the most favorable light. The fact is that, perhaps especially in smaller organizations, the environment created in an organization for employees not only yields increased retention of incumbents, but also, through word of mouth, makes the organization more attractive to potential applicants. In this way, the rich actually do get richer.

In terms of improving the working conditions, the task is a little more involved. Chapter 3 introduces the idea of worker reward attributes analyses as part of job and organization analyses. The idea was that as part of the selection procedure, we should analyze what the company offers its employees by way of reward for their effort and commitment. Frequently, such analysis may lead to the conclusion that some significant rewards are lacking and company resources are simply not adequate to supply these rewards. As shown in chapters 4 and 5, dissatisfaction and low commitment can yield turnover. Although it is beyond the scope of this text to discuss in detail all potential solutions to this problem, we suggest exploring some of the alternatives that may be relatively low in cost:

1. Redesigning jobs or rotating jobs may be rewarding to some individuals, particularly if employees want to learn the business. A job in a small company, then, could clearly have some long-term experiential and educational value to the employee and, of course, the organization as well.

2. Flexibility in working hours, or the option for telework, may be another significant low-cost reward to employees who have nonwork obligations they must fulfill. One of the authors worked with a small (seven-person) company that served as a clearinghouse, matching companies that needed deliveries with deliverers who wanted the work. The seven employees were the dispatchers and were constantly harassed by both the customers who wanted their delivery and the delivery people who wanted more work (they were paid on commission). Turnover was out of sight. After diagnosing the situation, it was recommended that the company employ eight people for the seven positions instead of the seven currently in place. This would permit each worker to have a 2-hour break per day instead of 1 hour to decompress and take care of personal issues (all were women with children). Of course, the company believed the suggestion was crazy but was shown that if turnover was cut by 50% the salary of the new person would be a wash; if turnover was cut by more, the savings would be gravy. Turnover was cut by two-thirds, the company had more applicants for the job than it knew what to do with (via word of mouth of present employees), and everyone lived happily ever after.

3. Decision making regarding the work environment may help increase morale and a spirit of teamwork. In the emergency telephone operator example described earlier, the supervisor frequently organized efforts to increase the "livability" of the work area. A break area with refrigerator and microwave was contributed by employees; the group organized itself to paint and redecorate their workspace and to set up an aquarium. These must be activities the workers themselves want to do, or they can be seen as manipulations by management to get them to contribute their time and money.

4. Of course, input to decisions about how work is done and organized may also be solicited from employees in firms of any size. Appropriately administered and followed up, such "participative systems" can be a powerful force in building a more productive company as well as a more satisfied and committed workforce (Lawler, Mohrman, & Ledford, 1998).

"The Job Doesn't Exist Yet"

Occasionally, in both large and small organizations, we encounter the need to hire people for a job that is just being created because of technological change and/or company expansion. How do we analyze a job that does not exist and establish the requisite KSAOs to do it? We discussed these issues previously in chapter 3, but let us give an example of how this might work.

One of the authors worked on such an interesting project (Schneider & Schechter, 1986). What made the effort particularly interesting was that it concerned supervisors, rather than operative employees. The logic underlying the project was that the job of the operating employees (telephone salespersons) was changing in a number of significant ways, so the question we asked was: If the job changes and the kinds of people doing the new job are different from the people who did the old job, how does the supervisor's job change? To answer this question we went through the following steps:

1. A traditional job analysis was accomplished for the present supervisory job, including specification of the tasks to be accomplished (e.g., incumbent ratings of their importance and the amount of time spent doing them) and the KSAOs necessary to perform the important tasks. For each KSAO identified, SME judgments were, in turn, made about the importance of the KSAO for doing the tasks (see Goldstein, 1986, for details about this form of job analysis).

2. A workshop of SMEs was planned to accomplish three goals:

   a. Define the nature of the job being supervised 3 to 5 years from the time of the workshop.

   b. Rate the importance of and time spent on the existing task statements for the supervisory job.

   c. Rate the importance of the KSAOs to accomplish the important tasks of the supervisory job.

3. Attending the workshop was the team of job analysis consultants and five persons in the organization knowledgeable about the current supervisory job and the kinds of changes likely to occur in the job being supervised. These two groups of SMEs generated a list of the changes in the job and the organization that had to be considered in doing the job analysis of the future:

   a. Increased computerization, permitting accurate monitoring of performance and improved efficiency

   b. Increased complexity, requiring more highly skilled people being hired and more intensive training of those hired; a change

PRACTICAL STAFFING PROBLEMS 609

        in the job from order taking to active selling, requiring more assertive people being hired

    c. Increased competition on price in the marketplace, making service a major goal of the salesperson

Given these kinds of predictions about the nature of the technology of the job, the people who will be doing the job, and the market in which the organization functions, the chore of the workshop was to literally predict what the supervisor of this job would be doing and the KSAOs needed to do it.

    4. As part of the job analysis, not only were the KSAOs rated for their importance, but also the SMEs made judgments about how the KSAOs were to be acquired. That is, were supervisors to be selected already having the KSAOs, or were those selected going to be trained in the KSAOs? If the latter case, then those selected would have to be competent enough to be trainable. In this particular project, the SMEs agreed that candidates should be hired who already possessed many of the skills and abilities of supervisors (e.g., define goals, plan, coordinate others) and that the task-specific knowledge required would be primarily provided in training.

Analyses of the data generated in the workshop against the data coming from the "job of the present" analysis revealed that those KSAOs currently the most important were going to increase in importance for the "job of the future."

This project suggests an important issue in job analysis: Job analysis techniques uniformly address the status quo in that they focus on the world of today. In organizations that operate in an ever-changing world, jobs will also be ever changing. It seems that any help we can get in defining what jobs will look like and, thus, what kinds of people they will require, could be of considerable strategic advantage.

In summary, it is true that in this book we present what we know are the best and most appropriate ways of carrying out the staffing of organizations. By doing so, we present what is maximally possible and doable, but we understand it may not be possible to do everything "right" in any specific circumstance. What you have here are standards to try to reach, approaches that have been demonstrated to work effectively, ways to think about predictors and criteria, and how to go about doing things—even when you may have believed you cannot and/or the circumstances just are not right. In this brief section, we are telling our readers that you can do it and do it well even when the circumstances are not optimal. As they say in the Nike ad, "Just do it."

SPECULATION ON THE FUTURE INNOVATIONS IN STAFFING

Throughout this book, we indicate areas where there is a need for more research. Despite a wealth of knowledge about staffing, the kinds of workplace changes discussed in chapter 1 and described in Table 1.1 require that staffing researchers continue to refine and improve staffing practices and procedures. We have seen how technological, global, cultural, legal, demographic, and work-based changes have influenced the science and practice of staffing. In this last section, we conclude with a few final thoughts—some bets—on where we believe the next big innovations will occur and where staffing research should be directed.

Multilevel Staffing Models

From chapter 1 through the present chapter, we note how multilevel processes pose several challenges for staffing. Whether it is staffing in a team environment to showing how individual-level staffing contributes to organizational effectiveness, the traditional individual-level staffing model has a difficult time incorporating these perspectives. This is not to say the individual-level model is outdated—far from it. However, it is to say the individual-level model may not be fully, or even partially, adequate for addressing questions where the criterion of interest is above the individual level.

If so, then one might think the staffing research conducted at the organizational level, typically done by "macro" HR researchers, might better address this question. However, a review by Ployhart (2004) suggested that this approach does not link individual differences to organizational differences. Rather, it focuses on how HR practices in the aggregate relate to organizational effectiveness and performance (e.g., Huselid, 1995).

What is missing is an explicit consideration of how hiring individual people contributes to better functioning organizations. Utility analysis, discussed in chapter 6, provides one way to estimate such consequences. However, we believe it is time for scholars to examine how individual differences in the aggregate—what we called human capital in chapter 1—contribute to organizational effectiveness. This requires a consideration of multilevel theories and models so we can build a theory of staffing that links individual, organizational, and intermediate levels. We have written a fair amount on this topic and tried to emphasize how staffing must take a multilevel perspective (Ployhart, 2004; Ployhart & Schneider, 2002, 2005; Schneider, 1987; Schneider, Smith, & Sipe, 2000). Here, we summarize the main features of these arguments:

1. Organizations look and feel the way they do because of the people in them. The founder, top management team, and supervisors/leaders have a fundamental influence on the structure, processes, and culture of the organization. As Schneider (1987) noted, "The people make the place." This means we understand organizations by understanding the people in them.

2. Multilevel theory suggests single-level research may not generalize to other levels of analysis. This means we must question whether organizations composed of more cognitively able and conscientious individuals do in fact perform better, as we have seen at the individual level. Research on groups and teams suggests there are both similarities and differences from individual-level findings, and it appears social mediating processes may be more important at higher levels. Hence, our individual-level validity studies and meta-analyses may not necessarily tell us much about how human capital creates organizational value. Or maybe they do; we simply do not know.

3. There is no point in arguing whether micro or macro perspectives are more important. Both are necessary to fully understand the staffing system, and research that neglects the other's perspective provides an incomplete picture of staffing organizations.

4. To study multilevel staffing, our research must be longitudinal and adopt multilevel methods. It takes time for aggregation in KSAOs to occur, and modeling time must be a formal element of the research design.

In our approach, we suggest researchers look for composition (homogeneity) and compilation (heterogeneity) of KSAOs within jobs, departments, and organizations. Some research has already supported such composition, such as Schneider et al. (1998), who found personality was nested within organizations—and within industries, too. What we then do is aggregate the KSAOs to the job, department, or organizational level and correlate the aggregate KSAOs (i.e., the human capital) with the organizational criteria of interest. We can take this a step further and link this KSAO composition to HR practices, such that human capital mediates the relationship between HR practices and effectiveness.

The Role of Meta-Analysis and Primary Validity Studies

It is obvious that meta-analysis has had a fundamental impact on staffing research and practice. It led to the death of the situational specificity hypothesis and the rebirth of personality employment testing. Yet, today we are becoming concerned about how meta-analysis is being used relative to other research methodologies. We do not suggest there are problems with meta-analysis. Rather, for some questions, we believe researchers would be better served by conducting new primary studies rather than relying on another meta-analysis of old data. What we need is some creative thinking operationalized via experimental and field studies. Consider two examples.

First, in chapter 9 we note how personality testing has enjoyed a rebirth due largely to meta-analyses finding stronger validities with performance than were those conducted a decade before. What has actually changed is our willingness to use statistical and measurement corrections. However, the uncorrected validities remain consistently modest and to some rather discouraging. There is only one way to improve the validity and reduce potential threats of faking, and that is to design better measures and implement better measurement strategies. We need people researching how to improve the predictiveness of personality. The conditional reasoning work of James (1998) discussed in chapter 9 is noteworthy in this regard, but we have seen few other such attempts. This situation must change, and we need researchers improving personality measurement.

Second, SJTs continue to capture a lot of current attention (see chapter 10), but what these tests measure remains unclear. To date, we have relied on meta-analyses of correlations to answer this question, but it seems to us that some well-designed experimental studies that test theories of SJT response processes will give us a much more clear sense of SJT construct validity.

Our point is that we seem to have less focus today on measurement. As a profession, we have not kept pace with the measurement technology used in cognitive, educational, and social psychology. They

SPECULATION ON FUTURE INNOVATIONS 613

have developed sophisticated measurement systems that are more objective and rely less on self-reports. These technologies have not seen widespread use in practice, and that is a situation that must surely change.

Subgroup Differences Must Be Understood Through Cognition

A reoccurring issue has been the balance between optimal prediction and enhancing diversity, an issue most salient for cognitive ability tests but relevant for all predictors (see chapters 6–10). We have seen many attempts to reduce these differences (see chapters 7 and 8), and none are entirely successful. It has been over a decade since the prohibition of within-group norming (in the 1991 Civil Rights Act, see chapter 11), and yet we still live with large subgroup differences in several of our tests. We believe there is only one solution to this problem—we must adopt a cognitive orientation and identify what forms of cognition are contributing to these differences. For example, what cognitive process(es) contribute to the difference? Are they working memory, processing speed, or something similar? Knowing the sources of cognition that contribute to these differences would go a long way toward improving the measurement of cognitive and noncognitive constructs. Our field must learn the theories and methods of cognitive psychology if we are to ultimately understand and eliminate these differences, when we know that the magnitude of the differences on cognitively loaded tests is not reflected equivalently in on-the-job performance.

Increased Use of Technology

At several places in this book, we describe the manner in which predictors and criteria have been measured using computerized or Web-based versions of what have been usually administered as paper-and-pencil measures or job samples. We expect that this use of technology will accelerate and that there will be increasingly sophisticated techniques (e.g., matching retinas of respondents) that will allow us to solve some of the difficult problems associated with documenting who takes these measures and how to maintain their security. In addition to changing the format with which selection instruments are delivered, technology can also be used to measure perceptual or psychomotor abilities that have proven difficult to assess using paper-and-pencil tests (McHenry, Hough, Toquam, Hanson, & Ashworth, 1990; McHenry & Schmitt, 1994). We also describe CATs; we believe this technology will become more widely accessible and will be adapted to measure a wider variety of constructs (e.g., Drasgow, Olson, Keenan, Moberg, & Mead, 1993).

Global Concerns

Most large corporations are now global in nature and will become increasingly so. This has at least two important implications for staffing. If we are to use the instruments that have proven useful here or develop new procedures, they must be sensitive to the cultures in which they are used (Levy-Leboyer, 1994). Many research studies have documented that the usual back translation techniques used in adapting a test to a new language are not sufficient to eliminate cultural nuances (e.g., Budgell, Raju, & Quartetti, 1995; Schmit, Kihm, & Robie, 2000). However, on the bright side, there is considerable evidence to show that the results obtained in the United States with selection tests are replicated quite consistently in other countries, especially in Europe and the United Kingdom (Salgado, Anderson, Moscoso, Bertua, & Fruyt, 2003).

The second staffing problem that global organizations face is the recruitment, selection, and support of individuals who must work in foreign countries. These organizations will want persons who can adapt and appreciate cultural differences, tolerate the security problems that work in some countries represents, and deal competently and sensitively with individuals of a different culture. There are relatively few research studies that address these concerns with expatriates (see Black, Mendenhall, & Oddou, 1991; Shaffer & Harrison, 1998).

IMPLICATIONS FOR THE JOB SEEKER

Throughout this book, we discuss selection from the organization's viewpoint. That is, we discuss how organizations organize and operate to try to ensure the recruitment and selection of a competent work force. Those same procedures, however, also have implications for the way in which an individual chooses a job or organization. The stakes for the individual are just as significant, and in this last section of the book, we provide suggestions for selecting a job and organization.

IMPLICATIONS FOR THE JOB SEEKER

Job and Organization Analysis

Before or during the application process, job seekers should do their homework regarding the job for which they apply. What are the job responsibilities? Whom would I report to? What are the possibilities for promotions? What are the pay rates and fringe benefits and of what value are they to me? What are expectations in the company regarding job performance, and how will I be evaluated? What are the organization's expectations regarding dress, community involvement, attendance at company functions, and so on? What is the status of the organization—financially and technologically—and what are its HR policies? Certainly, we should have answers to as many of these questions as possible before accepting a position, but research done early in the application process (before a job interview) can do much to direct our information search during the interview and other components of the job application process.

What kinds of sources of information exist? Perhaps the best source of information about a particular job and organization is incumbents. People who work at a job in a company not only have general opinions about them, but can also be a source of explicit, critical information. Use of the word "critical" here should serve as a reminder of the CIT described so many times, especially in chapter 3. For example, ask people to describe actual examples that occur in the organization about issues that concern you. How do supervisors treat you? Is hard work rewarded? What are the attendance/punctuality rules and procedures? Is promotion from within? Remember, have them cite specific incidents: what were the circumstances, who was involved, what happened, and so on.

Information about specific jobs/organizations will also be available from one or more of the following sources:

1. Web sites and the Internet
2. Friends/family
3. Clergy
4. State employment services
5. Employment agencies
6. School counselors (both high school and college)
7. News articles in newspapers (e.g., *Wall Street Journal*) and magazines (e.g., *Fortune, Business Week*) for larger companies (and *Inc.* for smaller companies).

The message is that you should have knowledge when you apply for a job. Indeed, if the RJP literature is correct (see chapter 5), the more realistic your information is for the job you take, the more satisfied you will be at it. Be prepared with knowledge.

Recruitment

In searching for a job, it is sometimes a problem just finding appropriate job opportunities. This process of finding job opportunities is similar to an organization looking for job applicants. In recruitment of job opportunities, the best way to start is with a realistic assessment of your interests and capabilities. In this regard, you should examine your educational and experience background; even standardized interest or aptitude batteries can be used to make these assessments. Most high schools and colleges have testing facilities and experienced test administrators and interpreters. As we show in chapter 9, interest inventories, in particular, can be useful data in making broad occupational choices; people, not only organizations, can use them. These standardized tests and an interview with a trained career counselor may suggest occupations that were not previously considered.

The next step in recruitment is to consider the jobs, organizations, and careers available that may fit your interests and capabilities. We emphasize the word "fit" because how well you fit the job, work group, and environment will be an important influence on your job choice. A standard source of information about various jobs that we discuss in chapter 3 was O*NET. O*NET was constructed so it could serve a career counseling function, and it has a wealth of information, including job descriptions for nearly every job in the U.S. economy. In addition to the narrative job descriptions, it also presents data on the abilities, interests, and temperaments of individuals who occupy those jobs. Most people looking for jobs are tied to a specific geographic area. The next step would be to locate companies that locally employ individuals in these jobs. In this context, the Internet is an obvious source, but so are local chambers of commerce, state employment agencies, or even the telephone book.

Once you have identified companies that may have jobs for which you believe you may qualify, you may apply online or in person by contacting the personnel officer in the company. Online applications are fast and easy, but frequently end up in a digital black hole. Many of our students report better success the old-fashioned way, by showing up in person or using the telephone. Describe the job or jobs you are looking for and for which you believe you are qualified, and express why you are particularly interested in/qualified to work in their organization. The more information you have about the job or organization, the more likely it is that they will be impressed by and interested in you.

Remember, recruitment goes on internally and externally. This means that it is important for you to continually process information about opportunities inside your company. This is accomplished by making yourself known to people who (a) have information and (b) have access to jobs. This may sound manipulative, but, as the expression goes, whose career is this, anyhow? An important means to ensure you are qualified for such positions is to engage in a process of continual learning and development. Given the rate of change, it is important for you to stay current in your knowledge, skills, and expertise.

The Selection Process

Assuming one or more companies decide to pursue your application and request that you go through their formal selection process, there are a number of precautions suggested by the research on hiring procedures outlined in chapters 7 to 10. If asked to respond to biographical forms, be honest but also be sure to mention all possible job-relevant experiences and why you perceive them to be relevant. If you are asked for names of individuals who can write a letter of reference, be sure to ask those individuals if they are willing to serve in that role. If standardized tests are required, find out, if possible, what those tests are and do some research regarding the test. If you are familiar with the test, its general content and format, you should be more relaxed when taking it.

An interview is part of the selection process in every company. If you have done your background research on the company, you will be able to ask informed questions if the opportunity arises and you may be able to anticipate and prepare for the kinds of questions that may be asked. Recall the similar-to-me effect on rater judgments; this effect suggests you be concerned about your dress and physical appearance. You should view the interview as an opportunity to find out about the company and whether you want to work in the organization. The degree to which the interviewer is informative, professional, and forthright in answering your questions can be an important source of information about how the organization views its human resources (Schmitt & Coyle, 1976).

618　　　　　　　　　　　12. STAFFING ORGANIZATIONS

Finally, and perhaps most important, most interviewers are untrained and most interviews are open ended. Both facts allow you, as the interviewee, to almost guarantee that the interviewer will focus on what you say. What you say is a function of how well prepared you are regarding the points you want to get across. For example:

1. Your strengths, with examples of situations in which you have displayed them
2. Your hopes, with examples of what you know you will have to do to make them reality
3. How you work with others and the kinds of people with whom you particularly enjoy working
4. Your short- and long-range plans

The motto is: Treat the interview as an opportunity to say the things you would have said if the interviewer had been a well-trained one who knew what she or he was looking for. Try not to be shocked or offended by some of the inappropriate interview questions that are frequently used (e.g., "If you could be a dog, what kind would you be?"). Simply do your best to answer the question, feeling smug that if you are hired you will probably be running the company in a month!

Systematic Evaluation and Decision Making

Just as organizations should not be making decisions based on hunches, job applicants should not be haphazard or rely on their "gut" impressions of organizations in deciding where to work. As organizations are evaluated, you should keep notes on those things we suggested in the job/organization analysis. When job offers are made, you can then review each opportunity on the dimensions critical to you as a person and decide on the relative merits of each organization in an informed manner. If you have done a careful analysis of what is important to you (e.g., flexible hours, pay, promotion opportunities, coworkers) and if you have gathered information on these dimensions for each opportunity that presents itself, then organization and job choices can be optimal.

Here is one simple but effective way you can do this that is based on subjective expected utility theory. First, list as many attributes of jobs and organizations that you value (see also next section on utility). Second, rate the importance of each attribute on a 1 (low value) to

7 (high value) point scale. Third, create a matrix of the job offers in the columns and the list of attributes you value in the rows. Fourth, rate each job offer on each attribute you listed using a 1 (low on this attribute) to 7 (high on this attribute) point scale. Fifth, multiply each importance rating with each attribute rating, and then add the total scores for each job offer. The offer with the highest overall score is your best choice—provided you have been honest with yourself.

Utility

For an organization, consideration of utility most often involves an economic analysis of the contributions of employee skill levels relative to the costs incurred in making decisions. For an individual, utility considerations involve a consideration of the input (e.g., work, commitment) relative to the return (e.g., challenging and interesting work, pay). In choosing an organization wisely, it is important to consider what you value most and how employment in various organizations will have an impact on securing those things. In this context, both work and nonwork values will be considered. Certainly, issues relating to commuting time, quality of community life, possibility of spouse's employment, affordability of housing, quality of schools, and other issues are of significant concern to individuals and, by necessity, to organizations as well. Do not minimize the importance of these "nonwork" factors because they have an important impact on your overall life satisfaction, which will influence your work satisfaction.

SUMMARY

In this last chapter, we summarize the major issues dealt with in previous chapters. We also outline and provide examples of ways in which organizations with limited resources and numbers of people can apply the techniques discussed in this book. We provide some bets for where the next big movements will occur in staffing. Finally, we discuss ways in which individuals can benefit from the same procedures and research when they pursue the selection of a job/organization/career.

By concluding with the role of individual responsibility in accomplishing a match of individual to organization, we come full circle to the idea that individual and organization effectiveness are inextricably intertwined. Both the individual and the organization have responsibilities to and for each other, and together they can pursue effectiveness through goal accomplishment. Just as organizations must carefully plan for the kinds of staff they need, so must individuals plan for the environment in which they can achieve. The list is long for both parties, but dedication to the specification of goals and objectives on both parts, followed by pursuit of likely avenues to achievement, can surely yield both individual and organizational success.

REFERENCES

Black, J. S., Mendenhall, M., & Oddou, G. (1991). Toward a comprehensive model of international adjustment: An integration of multiple theoretical perspectives. *Academy of Management Review, 16,* 291–317.

Budgell, G. R., Raju, N. S., & Quartetti, D. A. (1995). Analysis of differential item functioning in translated assessment instruments. *Applied Psychological Measurement, 19,* 309–321.

Drasgow, F., Olson, J. B., Keenan, P. A., Moberg, P., & Mead, A. D. (1993). Computational assessment. *Personnel and Human Resources Management, 11,* 163–206.

Goldstein, I. L. (1986). *Training in work organizations: Needs assessment, development, and evaluation.* Pacific Grove, CA: Brooks-Cole.

Huselid, M. A. (1995). The impact of human resource management practices on turnover, productivity, and corporate financial performance. *Academy of Management Journal, 38,* 635–672.

James, L. R. (1998). Measurement of personality via conditional reasoning. *Organizational Research Methods, 1,* 131–163.

Katz, D., & Kahn, R. L. (1978). *The social psychology of organizations* (2nd ed.). New York: Wiley.

Lawler, E. E., III, Mohrman, S. A., & Ledford, G. E., Jr. (1998). *Strategies for high performance organizations: Employee involvement, TQM, and reengineering programs in Fortune 500 corporations.* San Francisco: Jossey-Bass.

Levy-Leboyer, C. (1994). Selection and assessment in Europe. In H. C. Triandis, M. D. Dunnette, & L. M. Hough (Eds.), *Handbook of industrial and organizational psychology* (pp. 173–190). Palo Alto, CA: Consulting Psychologists Press.

Lievens, F., & Highhouse, S. (2003). The relation of instrumental and symbolic attributes to a company's attractiveness as an employer. *Personnel Psychology, 56,* 75–102.

MacDuffie, J. P. (1995). Human resource bundles and manufacturing performance: Organizational logic and flexible production systems in the world auto industry. *Industrial and Labor Relations Review, 48,* 197–221.

McHenry, J. J., Hough, L. M., Toquam, J. L., Hanson, M. A., & Ashworth, S. D. (1990). Project A validity results: Relationships between predictor and criterion domains. *Personnel Psychology, 43,* 335–354.

McHenry, J. J., & Schmitt, N. (1994). Multimedia testing. In M. G. Rumsey, C. B. Walker, & J. H. Harris (Eds.), *Personnel selection and classification* (pp. 193–232). Hillsdale, NJ: Erlbaum.

Ployhart, R. E. (2004). Organizational staffing: A multilevel review, synthesis, and model. In J. Martocchio (Ed.), *Research in personnel and human resource management* (Vol. 23, pp. 121–176). Oxford, UK: Elsevier.

Ployhart, R. E., & Schneider, B. (2002). A multilevel perspective on personnel selection: Implications for selection system design, assessment, and construct validation. In

REFERENCES

F. J. Dansereau & F. Yammarino (Eds.), *Research in multi-level issues. Volume 1: The many faces of multi-level issues* (pp. 95–140). Oxford, UK: Elsevier Science.

Ployhart, R. E., & Schneider, B. (2005). Multilevel selection and prediction: Theories, methods, and models. In A. Evers, O. Smit-Voskuyl, & N. R. Anderson (Eds.), *Handbook of personnel selection* (pp. 495–516). Chichester/London: Wiley.

Salgado, J. F., Anderson, N., Moscoso, S., Bertua, C., & Fruyt, F. D. (2003). International validity generalization of GMA and cognitive abilities: A European community meta-analysis. *Personnel Psychology, 56,* 573–605.

Schmit, M. J., Kihm, J. A., & Robie, C. (2000). Development of a global measure of personality. *Personnel Psychology, 53,* 153–193.

Schmitt, N., & Coyle, B. W. (1976). Applicant decisions in the employment interview. *Journal of Applied Psychology, 61,* 184–192.

Schmitt, N., & Ostroff, C. (1986). Operationalizing the "behavioral consistency" approach: Selection test development based on a content-oriented strategy. *Personnel Psychology, 39,* 91–108.

Schneider, B. (1987). The people make the place. *Personnel Psychology, 40,* 437–453.

Schneider, B., & Schechter, D. (1986). *Job analysis of the future job: Procedure and implications.* Unpublished manuscript, University of Maryland, Department of Psychology, College Park.

Schneider, B., Smith, D. B., & Sipe, W. P. (2000). Personnel selection psychology: Multilevel considerations. In K. J. Klein & S. W. J. Kozlowski (Eds.), *Multilevel theory, research, and methods in organizations* (pp. 91–120). San Francisco: Jossey-Bass.

Schneider, B., Smith, D. B., Taylor, S., & Fleenor, J. (1998). Personality and organizations: A test of the homogeneity of personality hypothesis. *Journal of Applied Psychology, 83,* 462–470.

Shaffer, M. A., & Harrison, D. A. (1998). Expatriates' psychological withdrawal from international assignments: Work, nonwork, and family influences. *Personnel Psychology, 51,* 87–118.

Uniform guidelines on employee selection procedures. (1978) *Federal Register, 43,* 38290–38309.

Author Index

Page numbers followed by *b* indicate boxed material.
Page numbers followed by *t* indicate tables.
Page numbers in *italics* indicate pages with complete bibliographic information.

A

Abraham, L. M., 221, *225*
Academy of Management Journal, 1996, 25, *30*
Ackerman, L., 136, *147*
Ackerman, P. L., 199, *225*, 366, 371, *399*
Ackerman, T. A., 374, *401*
Adams, K. A., 535, *556*
Age Discrimination Act of 1967, 565, *590*
Aguirre, M., 538, *555*
Aguinis, 425
Aiken, L. R., 544, *550*
Aiken, L. S., 79, *98*
Aldage, R. J., 277, *298*
Aldrich, H., 447, *484*
Alexander, R. A., 198, 199, *225*, 304, *360*, 427, *430*
Allen, N. J., 240, 249, 254, *297*
Allen, T. D., 163, 188, *226*
Allport, G. W., 437, *484*
Allscheid, S. P., 218, *234*
Alvarez, K. M., *225*
American Educational Research Association (AERA), *398*
American Educational Research Association (AERA), American Psychological Association, & National Council on Measurement in Education, 79, *98*, 301, 313, 319, 340, 341, *360*, 416, 429, *430*, 578, *590*
American Psychological Association, 300, *360*
Americans with Disabilities Act of 1990, 565, 572, *590*
Anastasi, A., 342, *360*
Anderson, N., 393, *398*, *400*, 414, 415, 432, 614, *621*
Anderson, N. R., 582, *591*
Anderson, S. E., 372, *401*
Arad, S., 213, 214, 215, *233*

623

Argyris, C., 21, *30*
Arnold, J. D., 327, *360*, 545, *550*
Aronson, J., 422, *433*
Arthur, W., 494, 499, *553*
Arthur, W. Jr., 528, 529, 531, 532, 535, 538, *550*, *558*
Arvey, R. D., 221, *225*, 287, *294*, 384, *398*, 492, 505, 507, 549, *550*, *551*, *556*
Asendorpf, J. B., 451, *484*
Ash, R. A., 129, *149*
Ash, S. R., 283, *295*
Asher, J. J., 482, *484*, 539, *551*
Ashton, M. C., 541, *553*
Ashworth, S., 410, *431*
Ashworth, S. D., 614, *620*
Atkinson, J. W., 443, *487*
Austin, J. T., 157, *225*, 379, *398*
Avedon, M. J., 534, *554*
Avolio, B. J., *207*, *235*

B

Bachrach, D. G., 249, *298*
Baker, B. O., 41, *98*
Baker, R. C. IV, 278, *295*
Banas, J. T., 214, *235*
Baratta, J. E., 199, 201, *229*
Barber, A. E., 254, 269, 270, 271, 273, 275, 289, *294*, 494, *551*
Barksdale, C. D., 464, *487*
Barling, J., 164, *230*
Baron, H., 156, 216, *233*, 393, *400*, 415, *432*
Barrett, G. V., 198, 199, 222, *225*, *230*, 304, *360*, 427, *430*
Barrick, M. R., 203, *231*, 390, 392, *398*, 435, 438, 439, 448, 450, 460, 467, 468, *484*, *485*, *489*
Bartlett, C. J., 79, *98*, 185, *236*, 342, *360*
Bartol, K. M., 179, *225*
Bateman, T. S., 249, *294*
Battista, M., 129, *149*
Bauer, T. N., 287, *294*, 461, *486*
Baughman, K., 312, *363*, 447, *488*
Baysinger, B. D., 239, *294*
Beatty, R. W., 178, 181, 203, *225*
Beaubien, J. M., 390, *398*
Becker, B. E., 358, *360*
Beehr, T. A., 222, *225*
Belloli, C., 538, *555*
Bennett, G. K., 539, *551*
Bennett, M., 287, *297*
Bennett, N., 201, *227*
Bennett, R. J., 164, 177, *225*, *233*
Bentson, C., 529, *552*
Bentz, V. J., 450, *484*
Bernardin, H. J., 178, 181, 205, *225*
Berner, J. G., 534, 540, *556*

Berry, L. L., 211, *232*
Bertua, C., 393, *398*, *400*, 414, 415, *432*, 614, *621*
Bevier, C. A., *382*, *400*, 416, *432*
Binet, A., 13, 14, 28, *30*
Bisqueret, C., 524, 538, *554*
Black, J. S., 614, *620*
Blackstone Stenographic Proficiency tests, 539, *551*
Blake, R. J., 469, 471, 472, 476, *486*
Blake, R. R., 421, *430*
Blake v. City of Los Angeles, 546, *551*
Blanz, F., 182, *225*
Blau, G., 173, *225*
Blau, P. M., 267, *294*
Bliese, P. D., 88, 89, *98*, *147*, 390, *398*
Bobko, P., 79, *98*, 207, *233*, 309, 340, 342, 355, *360*, *361*, 374, 382, 383, 386, *398*, *400*, 416, 419, *430*, *432*, 503, 540, 541, *556*, 568, *590*
Bock, R. D., 94, *99*
Bokemeier, J. L., 276, *297*
Bollen, K. A., 84, 87, *98*
Bommer, W. H., 195, *225*, *233*
Bono, J. E., 222, 230, 243, *296*, 442, 444, 445, 460, *487*
Boodoo, G., 340, *363*, 416, *431*
Boring, E. G., 10, *30*
Borman, W. C., 120, 122, 136, 137, 140, *147*, *149*, 162, 163, 172, 178, 181, 186, 188, 193, 207, 213, 214, 217, *226*, *232*, 233, 243, 248, *294*, *297*, 309, *361*, 415, *430*
Boswell, W. R., 177, *226*
Bouchard, T. J., 221, *225*
Bouchard, T. J. Jr., 340, *363*, 416, *431*
Boudreau, J. W., 177, *226*, 253, *294*, 345, 355, *358*, *360*, *361*
Bowen, D. E., 211, 212, 218, *226*, *234*
Boyatzis, R., 458, *485*
Boyatzis, R. E., 436, 444, 452, *487*
Boykin, A. W., 340, *363*, 416, *431*
Brandt, C. J., 137, *149*
Brannick, M. T., 128, 142, *147*, 314, *361*, 571, *591*
Braverman, E. P., 512, 518, 533, *552*, *554*
Bray, D. W., 450, *486*, 529, *551*
Brayfield, A. H., 243, *294*
Breaugh, J. A., 270, 289, *294*
Brett, J. M., 241, 283, *295*
Bretz, R. D., 179, 186, *226*, 272, 273, *295*, *298*
Brewster, C., 156, 216, *226*, 230, *231*
Brickner, M., 530, *553*
Bridges, W., 131, *147*
Brief, A. P., 221, 222, *226*, 247, *295*, 441, 450, *485*

AUTHOR INDEX

Brody, N., 340, 416, *431*
Brogden, H. E., 345, 348, 354, *361*, 353, 350
Brown, C. W., 351, *361*
Brown, D. F., 203, *230*
Brown, D. J., 284, 285*b*, *295*
Brown, K. G., 214, *230*
Brown, R., 443, *484*
Bryan, J., 217, *226*
Bryk, A. S., 88, 89, *98*
Buck, D. E., 188, *226*
Buckley, M. R., 499, *551*
Budgell, G. R., 614, *620*
Burger, G. K., *147*
Burgess, J. R. D., 464, *487*
Burke, R. J., 278, *295*
Burke, W. W., 21, *30*
Burket, G. R., 70, *99*
Burnett, J. R., 506, *551*, *555*
Burns, T., 108, *147*
Burris, L. R., 448, *488*
Busch, C. M., 212, *229*
Buss, A. H., 451, *484*
Buster, M. A., 383, *398*, 541, *551*
Buxton, V., 218, *234*
Buyse, T., 512, *554*
Byham, W. C., 534, *557*

C

Cable, D. M., 268, 277, 289, *295*
Cabrera, E. F., 355, *361*
Cafferty, T. P., 185, *227*
Caldwell, M. S., 198, *225*
Callahan, C., 448, *488*
Campbell, D. P., 475, *484*
Campbell, D. T., 329, 330, 332, *361*, 372, *398*, 452, *484*
Campbell, J. P., 157, 160, 161, 164, 186, 194, 202, 207, *226*, *227*, *231*, *232*, 242, *298*, 436, 453, *484*, 539, *558*
Campbell, W. J., 573, *590*
Campion, J. E., 492, 495, 496, 505, 507, 508, *550*, *551*, *556*, *558*
Campion, M. A., 120, 122, 128, 132, 140, *147*, *149*, *150*, 240, 287, *294*, *297*, 425, 492, 495, 496, 497, 505, 506, 507, 508, 512, 518, *551*, *554*, *555*, *558*
Caretta, T. R., 403, 408, 410, *432*
Carlson, J. E., 374, *401*
Carr, L., 129, *149*
Carroll, J. B., 404, 405, *430*
Carter, G. W., 407, *431*, 511, 512, 517, 520*b*, *555*
Cascio, W. F., 213, 227, 345, 358, 360, *361*, 427, *430*, 480, *484*, 533, 534, *551*
Cattell, R. B., 437, *484*

Cattell, J. M., 13, 14, *30*
Ceci, S. J., 340, *363*, 416, *431*
Chan, D., 186, 203, 213, *227*, *233*, 287, *295*, 309, 314, *363*, 375, 386, *398*, *401*, 419, 421, 422, *430*, *433*, 460, *489*, 518, 541, *555*, *557*
Chan, K., 463, *489*
Chan, K. Y., 203, *232*
Chao, G. T., 209, *230*, 482, *488*
Chapman, D. S., 508, *551*
Chaves, W., 571, *591*
Chen, G., 390, *398*
Chernyschenko, Chan, 463, *489*
Chesney, M. A., 278, *295*
Cheung, G. W., 209, *227*
Christiansen, N. D., 329, *362*
Chung, B., 533, *552*
Civil Rights Acts of 1964, 564, 565, 569, *590*
Civil Rights Acts of 1991, 564, 569, 570, *590*
Clark, K. E., 436, *485*
Clark, L. A., 440, 450, *490*
Clark, M. B., 436, *485*
Clark, R. A., 443, *487*
Clause, C. S., 137, *149*, 287, *295*, 340, *363*, 383, *401*, 416, 420, 421, 422, *430*, *433*, 541, *557*
Cleary, T. A., 341, *361*
Cleveland, J. N., 153, 185, 193, 209, 210, 217, 219, *232*
Clevenger, J., 512, 517, *551*
Cliff, N., 539, *555*
Clothier, R. C., 15, *31*
Cober, A. B., 284, 285*b*, *295*
Cober, R. T., 284, 285*b*, *295*
Coetsier, P., 446, *487*
Cohen, J., 381, 382, *398*
Cohen, S. A., 136, *150*, 529, *557*
Colbert, A. E., 289, *297*, 448, *485*
Colella, A., 463, *487*
Colihan, J., *147*
Colligan, M. J., 221, *234*
Colquitt, J. A., 214, *231*
Conley, P. R., 137, *147*
Conn, A. B., 133, *148*
Connecticut v. Teal, 566t, *591*
Connerley, M. L., 384, *400*, 504, *556*
Conrad, M. A., 571, *591*
Conte, J. M., 462, 472, *487*
Conway, J. M., 208, *227*, 499, 502, 503, 504, 531, 532, 535, 538, *551*, *553*, *554*
Cooper, L. A., 327, *364*
Cooper, M., 545, *551*
Coovert, M. D., 202, *235*
Cornelius, E. T. III, 181, *228*, 436, 453, *485*
Cortina, J. M., 366, 372, 377, *401*, 499, *551*

Costa, P. T., 437, 450, 451, *485*, *487*
Costa, P. T. Jr., 393, *399*
Coyle, B. W., 70, *99*, 617, *621*
Crafts, J. L., 512, 516, 517, *555*
Craik, K. H., 468, *490*
Cranny, C. J., *147*
Crites, J. O., 267, *295*
Crockett, W. H., 243, *294*
Cronbach, L. J., 75, *98*, 202, 227, 345, 348, 350, 353, 354, *361*, 372, *399*
Cronshaw, S. F., 358, *364*
Cudeck, R., 202, *231*
Cullen, M. J., 423, *430*, 518, 519, *553*

D

Dalessio, A. T., 219, *227*
Das, J. P., 407, *431*
Davies, P. G., 422, 423, *433*
Davison, H. K., 499, *551*
Dawes, R. M., 64b, *99*
Dawis, R. V., 4, *30*
Day, D. V., 534, *556*
Day, E. A., 528, 532, *550*
Deadrick, D. L., 201, *227*
Dean, J. W. Jr., 110, *150*
Decaester, C., 446, *487*
DeCorte, W., 420, *430*
Decotiis, T. A., 181, *235*
DeFruyt, F., 472, *485*
de Fruyt, F., 414, 415, *432*
DeGroot, T., 506, *551*
DeGroot, T. G., 329, *362*, 499, *553*
Delaney, J. T., 358, *361*
Delaney-Klinger, K., 140, *149*
Delbridge, K., 287, *295*, 422, *430*
Delery, J. E., *31*
DeMers, S. T., 579, *591*
DeNisi, A., 220, *230*
DeNisi, A. S., 185, 217, *227*, 271, 297, 493, *555*
Denning, D. L., 571, *591*
DeShon, R. P., 208, 209, *227*, 232, 287, *295*, 422, *430*
Deszca, E., 278, *295*
DeVore, C. J., 163, *234*
Dew, A. F., 278, *295*
Dickson, W. J., 17, 18, *31*
Digman, J. M., 437, *485*
Dineen, B. R., 283, *295*
Dipboye, R. L., 504, 505, 506, 510, *551*, *552*, *553*
Doherty, M. E., *147*
Dolen, M. R., 287, *294*
Donnelly, J. M., 278, *296*
Donovan, M. A., 213, 214, 215, *233*
Dorsey, D., 136, 137, *147*

Dorsey, D. W., 186, 207, 213, 214, *226*, *233*
Dougherty, T. W., 506, *552*
Drasgow, F., 374, 394, 397, *398*, *399*, *400*, 407, *430*, 463, *489*, 579, *591*, 614, *620*
Drauden, G., 287, *294*
Dreher, G. F., 505, 530, *552*, *556*
Droge, C., 446, *488*
DuBois, C. L. Z., 202, *227*
DuBois, D., 141, *147*
Duncan, L. E., 443, 452, *486*, *490*
Dunnette, M. D., *148*, 181, 196, 219, *226*, 227, 407, *431*, 436, 453, *484*, 511, 512, 517, 520*b*, *555*
Durham, C. C., 179, *225*
Dvorak, P., 292, *295*
Dye, D. A., 483, *485*
Dye, D. M., 120, 122, 140, *149*

E

Earles, J. A., 411, *432*
Earley, P. C., 216, *227*
Eaton, N. K., 217, *226*
Ebel, R. L., 427, *431*
Eden, D., 22, 23, *30*, 159, 160, *227*, *235*
Edens, P. S., 528, 532, *550*
Eder, R. W., 492, *552*
Edison, C. E., 503, *556*
Ehrhart, M. G., 512, 513, 516, *555*
Ellingson, J. E., 386, *400*, 417, 421, *432*
Ellis, R. A., 278, *295*
Elman, B. A., 11, *30*
Embertson, S. E., 92, *99*
Endler, N. S., 465, 466, *487*
England, G. W., 479, 482, *485*
Epstein, D., 201, *231*
Equal Employment Opportunity Commission, 429, *430*, 565, *590*
Erez, A., 214, *231*
Erez, M., 241, 249, 276, *297*
Etchegaray, J. M., 535, *556*
Eyde, L. D., 129, *149*

F

Fames, H. R., 267, *294*
Farh, J. L., 216, *227*
Farr, J. L., 178, 179, 185, 186, *230*, 424, *431*, *433*
Farrell, D., 222, *228*
Fazzini, D., 444, *486*
Feldman, D. C., 265, *295*
Feldman, J. M., 185, *228*
Ferrara, P., 140, *149*
Ferris, G. R., 185, *230*, 506, 507, *552*
Field, H. S., 255, *295*, 387, *398*

AUTHOR INDEX

Fine, S. A., 115, *148*
Finnegan, B. J., 496, 504, *553, 554*
Finnegan, E. B., 512, 518, *554*
Fischer, A. R., 471, *490*
Fiske, D. W., 329, 330, 332, *361, 372, 398*
Fitzgerald, C., 451, *485*
Fitzgerald, M., 221, *234*
Flanagan, J. C., 118, *148*
Fleenor, J., 112, *150,* 268, *298,* 446, 473, *489,* 606, *621*
Fleischman, E. A., 210, *228*
Fleishman, E. A., 122, 132, *148, 149,* 545, 547, *552*
Fleming, T., 451, *488*
Flesch, R., 317, *361*
Fogli, L., 202, *227, 234*
Ford, J. K., 207, *228,* 318, *362*
Foster, M. R., 532, *554*
Foster, M. S., 532, 538, *554*
Fox, H. R., 579, *591*
Frederiksen, N., 321, *361*
Frei, R. L., 388, *398,* 448, 461, *485*
French, N. R., 532, 538, *554*
Frese, M., 222, *235*
Friedman, B. A., 181, *228*
Friedman, M., 278, 279, *295, 296*
Frisch, M. H., *485*
Fruyt, F. D., 393, *400, 432,* 614, *621*
Fulkerson, J. R., 216, *228*

G

Gael, S., 342, *361*
Gaertner, S., 222, *228,* 239, *295*
Gandy, J. A., 483, *485*
Ganster, D. C., 6, *31*
Garcia, M., 182, *232*
Gardner, J. N., 220, *228*
Gasser, M. B., 160, 161, *227*
Gatewood, R. D., 255, *295,* 387, *398,* 532, 538, *554*
Gaugler, B., 505, *552*
Gaugler, B. B., 523, 524, 529, *552, 557*
Geirnaert, J., 446, *487*
Gentry, W. A., 532, *554*
George, J. M., 441, 450, *485*
Gerhardt, M. W., 442, 445, 460, *487*
Gerhart, B., *30,* 146, *148,* 273, *298*
Gerras, S. J., 176, 199, 201, *229*
Gewin, A. G., 531, 532, 538, *554*
Ghiselli, E. E., 182, 198, *225, 228,* 332, 351, *361, 362,* 403, *430, 431,* 436, *485*
Gibson, W. M., 186, *228*
Gillespie, M. A., 459, 483, *489*
Gilliland, S. W., 278, 286, 289, *295,* 393, *401,* 499, 504, 505, 510, *551, 552, 557*
Gilmore, D. C., 506, 507, *552*

Gist, M. E., 436, *490*
Gleser, G. C., 75, *98,* 345, 348, 350, 353, 354, *361*
Goemaere, H., 529, *554*
Goff, M., 208, *235*
Goffin, R. D., 530, *552*
Goldberg, L. R., 323, *362,* 437, *485*
Goldstein, H. W., 446, 473, *489,* 533, *552*
Goldstein, I. L., 138, 140, *148,* 318, *362,* 377, *398,* 608, *620*
Goldstein, N. B., 499, *551*
Goleman, D., 458, *485*
Gooding, R., 304, *363,* 436, 460, *489*
Goodman, D. F., 499, *551*
Goodman, P. S., 8, *30,* 467, *485*
Gottfredson, L. S., 177, *226,* 403, 415, *431*
Gottier, R. F., 436, 460, *486*
Gottredson, G. D., 475, 476, *485, 486*
Gowing, M. K., 120, 122, 140, *149*
Grant, D. L., 342, *361,* 529, *551*
Gratz v. Bollinger, 566t, 574, *591*
Graves, L. M., 506, *552*
Green, M., 171, *236*
Greenberg, J., 145, *148*
Greenthal, A. L., 534, 540, *556*
Greguras, G. J., 208, 219, *228,* 423, *432*
Griffeth, R. W., 222, *228*
Griffith, R. W., 239, *295*
Griggs v. Duke Power Company, 566t, 574, *591*
Groth, M., 278, *295*
Grutter v. Bollinger, 417, *431,* 566t, 575, *591*
Gruys, M., 163, *228*
Guilford, J. P., 187, *228,* 405, *431,* 451, *486*
Guion, R. M., 111, 134, *148, 149,* 186, 196, 197, 198, *228,* 303, 327, *360, 362,* 425, 436, 447, 451, 453, *486,* 545, *550*
Gully, S. M., 214, *230,* 390, *398*
Gunnarson, S. K., 239, *298*
Gupta, N., *31*
Gustad, J. W., 267, *294*
Guttman, A., 574, 575, 576, *591*

H

Haaland, S., 329, *362*
Hachiya, D., 172, 229, 240, 254, *296*
Haire, M., 198, *228*
Hakel. M. D., 199, 201, *232*
Hall, D. T., 218, 221, *228,* 234, 252, 255, 256, 265, *295, 296,* 472, *486*
Halpern, D. F., 340, *363,* 416, *431*
Hambleton, R. K., 92, 97, *99*
Hammer, L. B., 461, *486*
Handler, L., 394, *399,* 579, *591*

Hanges, P. J., 22, *30*, 199, 222, *228*, *234*, 244, *298*
Hannan, R., 342, *360*
Hannon, R., 79, *98*
Hanson, M., 512, 516, 517, *555*
Hanson, M. A., 410, *431*, 614, *620*
Hardison, C. M., 423, *430*
Hardyck, C. D., 41, *98*
Harold, C., 519, *558*
Harrell-Cook, G., 506, 507, *552*
Harris, D. H., 143, *148*, 217, *228*
Harris, M. M., 374, *399*, 492, 524, 538, *552*, *554*
Harris, W. G., 541, *553*
Harrison, D. A., 175, *228*, 571, *591*, 614, *621*
Harter, J. K., 222, *228*, 244, 246, *296*, 448, *485*
Hartigan, J. A., 340, *362*, 408, *431*
Harvey, R. J., 100, 138, 139, 140, *148*
Harvey, V. S., 512, 517, *551*
Hastie, R., 64*b*, *99*
Hattrup, K., 196, *229*, 420, *431*, 542, *552*
Hauenstein, N. M. A., 192, 193, *229*
Hausknecht, J. P., 424, *431*, *433*
Hayes, T. L., 222, *228*, 244, 246, *296*
Hedge, J. W., 202, 213, *233*, *235*, 542, *552*
Hedlund, J., 392, *399*
Heller, D., 221, *230*, 245, *296*, 440, 448, 460, *487*
Heneman, H. G. III, 181, *235*
Heneman, R. L., 255, *257*, *296*
Henneman, R. C., 195, *229*
Henry, R. A., 199, *229*
Henry, S., 534, *554*
Hernstein, R., 414, *431*
Herzberg, F., 21, *30*
Hesketh, B., 129, *149*
Hezlett, S. A., 208, *231*
Highhouse, S., 290, 292, *296*, 297, 606, *620*
Hill, T. E., 534, *557*
Hoffman, D. A., 195, 199, 201, *229*
Hofman, C. C., 533, *553*
Hofmann, D. A., 176, *229*
Hofstede, G., 394, *399*
Hogan, J., 212, *229*, 389, *399*, 452, *486*, 542, 543, 546, 547, *553*
Hogan, J. B., 481, *486*
Hogan, J. C., 340, *363*
Hogan, R., 212, *229*, 389, *399*, 438, 444, 469, 471, 472, 476, *486*
Holberg, A., 482, *486*
Holden, L. M., 195, *229*
Holland, J. E., 475, *486*

Holland, J. L., 262, 265, 267, *296*, 344, *362*, 447, 468, 469 471*t*, 472, 473, 475, 476, *485*, *486*
Holland, P. W., 96, *99*
Hollenbeck, J. R., 133, 139, *148*, 163, *229*, 392, *399*, 494, *551*
Hollingsworth, A. T., 222, *231*
Holtom, B. C., 241, 249, 276, *297*
Holtz, B. C., 374, *400*, 513, 519, *555*, *558*
Hom, P. W., 222, *228*, 239, *295*
Hooper, A. C, 518, 519, *553*
Horner, S. O., 222, *231*
Hough, L. M., 383, 388, 389, *399*, 410, 421, *431*, 436, 437, 438, 451, 459, 461, 462, 463, *486*, 512, 532, 540, 541, 549, *553*, 560, 614, *620*
House, R., 21, *30*
House, R. J., 442, 443, 444, 453, 465, 466, *489*
Howard, A., 450, *486*, 538, *553*
Howell, M. A., 539, *555*
Hoyt, W. T., 207, 208, 209, *229*
Huck, J. R., 534, *553*
Huffcutt, A. I., 207, 208, 227, *233*, 329, *362*, 494, 496, 499, 501, 502, 503, 504, *551*, *553*, *556*
Hui, C., 216, *230*
Hulin, C. L., 172, 173, 174, 175, 176, 199, 222, *229*, 239, 240, 242, 254, *296*, *298*
Hull, C. L., 15, *30*, 347, *362*
Human Performance, 2003, 423, *431*
Humphreys, L. G., 371, *398*
Humphries, L. G., 414, *431*
Hunt, S. T., 379, *398*
Hunter, J. E., 18, *30*, *31*, 72, *99*, 185, 186, 208, *229*, *234*, 302, 305, 307, 309, 332, 335, 337, 345, 346, 354, 355, 360, *363*, 386, *400*, 403, 408, 411, 412, 413, 414, 415, 417, *431*, *432*, *433*, 494, 499, 529, 534, 540, 543, *553*, *556*
Hunter, R. F., 408, 411, 412, 413, 414, 415, *431*, *432*, 543, *553*
Hunthausen, J. M., 461, *486*
Huse, E. F., 174, *229*
Huselid, M., 358, *362*, 610, *620*
Huselid, M. A., 103, 145, *148*, 593, *620*
Hyland, A. M., 428, *431*
Hyne, S. A., 475, *484*

I

Iaffaldano, M. T., 243, *296*
Ilgen, D. R., 133, 139, *148*, 163, 213, *229*, 392, *399*
Ilies, R., 441, 442, 444, 445, 460, *487*

AUTHOR INDEX

Incalcaterra, K. A., 390, *398*
Ingerick, M. J., 366, 372, 377, *401*
Iris, B., 222, *230*
Ivancevich, J. M., 278, *296*

J

Jackson, D. N., 442, 451, *490*, 541, *553*
Jackson, S. E., 107, *148*, 259, 260, *297*
Jackson, S. J., 506, *552*
Jacobs, R., 199, 201, *229*
Jacoby, J., 179, *231*
Jako, R. A., 499, *551*
James, L. R., 455, 456, 457t, 464, *486*, *487*, 612, *620*
Jansen, K. J., 289, *297*
Jansen, P. G. W., 530, *553*
Janz, J. T., 498, *553*
Jeanneret, P. R., 120, 122, 140, *149*
Jenkins, C. D., 279, *296*
Jenkins, G. D., *31*
Jenkins, J. J., 11, *30*
Jennings, D., 309, *363*, 386, *401*, 419, *433*
Jensen, A. R., 382, 393, *399*, 414, 415, 416, *431*
Jesson, R., 267, *294*
Jex, S. M., 222, 225, *230*
John, O. P., 443, 452, *490*
Johnson, J. L., 195, 225, *233*
Johnson, N. B., 359, *363*
Johnson, R., 218, 221, *233*
Johnson, R. H., 329, *363*
Johnston, N. G., 530, *552*
Jones, C., 212, *236*, 329, *362*, 499, *553*
Jones, D. P., 573, *591*
Jones, J. R., 6, *31*
Jones, R. G., 132, *148*, 390, *399*
Joshi, A., 390, *398*
Journal of Vocational Behavior, 471, *486*
Judge, T. A., 177, 185, 208, 214, 221, 222, 226, *230*, *231*, 243, 245, 255, 257, 268, 272, 277, *295*, *296*, 440, 441, 442, 444, 445, 448, 460, 469, *487*, *490*, 507, *552*
Jungeblut, A., 424, *431*

K

Kabin, M. B., 417, 421, *432*
Kahn, R. L., 154, *230*, 248, *296*, 593, 594, *620*
Kahn, W. A., 256, *296*
Kaplan, L. B., 197, *234*
Karr, A. R., 283, *296*
Karren, R., 355, *360*
Karren, R. J., 506, *552*
Katz, D., 154, *230*, 248, *296*, 593, 594, *620*

Keaveny, T. J., 181, *230*
Keenan, P. A., 614, *620*
Keeping, L. M., 284, 285b, *295*
Kehoe, J., 129, *149*
Kehoe, J. F., 425, 582, *591*
Kelloway, E. K., 164, *230*
Kemp, C. F., 374, *400*, 519, *555*
Kendall, L. M., 119, *150*, 180, *235*, 239, 242, *298*
Keon, T. L., 268, *299*, 446, 468, *490*
Kephart, N. C., 188, *231*
Kerns, M. D., 207, 208, *229*
Kethley, R. B., 549, *557*
Kiesler, S., 374, *400*
Kihm, J. A., 393, *401*, 614, *621*
Kim, B. H., 459, 483, *489*
King, J., 284, *299*
Kinney, T. B., 209, *232*
Kinslinger, H. J., 453, *487*
Kirk, A. K., 203, *230*
Kirsch, M., 304, *363*, 436, 460, *489*
Klein, K. J., 8, *31*, 133, *148*, 311, 312, *362*
Klimoski, R., 132, *148*, 178, 186, *236*, 390, *399*, 530, *553*
Klimoski, R. J., 366, 379, *398*, *399*
Klohnen, E. C., 443, 452, *490*
Klube, R. B., 87, *98*
Kluger, A. N., 220, *230*, 384, *399*, 463, *487*
Koch, M. J., 145, *148*
Kohn, L. S., 504, 505, 510, *553*
Konz, A. M., 141, *150*
Korman, A. K., 238, 252, *296*, 436, 453, *487*
Kornhauser, A. W., 222, *230*
Kozlowski, S. W., 209, 214, *230*
Kozlowski, S. W. J., 311, 312, *362*
Kozlowski, S.W. J., 8, *31*, 214, *230*
Kraiger, K., 207, *228*, *230*
Kraut, A. I., 240, 246, *296*, *297*
Kravitz, D. A., 571, *591*
Kriska, S. D., 291, *298*, 504, 506, 507, 524, *555*
Kristof, A. L., 277, *297*, 506, *557*
Kristof-Brown, A. L., 289, *297*, 469, *487*
Kubiak, C., 538, *555*

L

Lacy, W. B., 276, *297*
Laczo, R. M., 104, 112, 123, *149*
Lam, S. S. K., 216, *230*
Lambert, T. A., 531, 532, 538, *554*
Lancaster, A. R., 120, 122, 140, *149*
Lance, C. E., 332, *362*, 531, 532, 538, *554*
Landis, R. S., 137, *149*
Landon, T. E., 549, *550*, *551*
Landy, F. J., 136, *148*, 178, 179, 185, 186, *230*, 436, 462, 472, *487*

Langdon, J. C., 278, *295*
Lappin, M., 193, 221, *234*
Larsen, H. H., 156, 216, *226*, *230*, *231*
Larson, R. J., 221, *230*
Latham, G. P., 103, 119, *148*, *149*, 182, 192, *231*, 358, *362*, 496, 497, 503, 504, *553*, *554*
Lautenschlager, G. J., 483, *487*
Law, K. S., 216, *230*, 459, *487*
Lawler, E. E., 145, *148*
Lawler, E. E. III, 218, *228*, 329, *362*, 436, 453, *484*, 607, *620*
Lawshe, C. H., 188, *231*, 318, *362*
Lebreton, J. M., 464, *487*
Ledford, G. E. Jr., 145, *148*, 607, *620*
Ledvinka, J., 562, *591*
Lee, C., 436, *490*
Lee, T. L., 175, 222, *231*
Lee, T. W., 241, 249, 276, *297*
Lee, W., 463, *489*
Lefkowitz, J., 582, *591*
Lepak, D. P., 109, 110, *149*, 284, *299*
LePine, J. A., 214, *231*, 392, *399*
Levin, K. Y., 120, 122, 140, *149*
Levine, E. L., 128, 139, 140, 142, *147*, *149*, 314, *361*, 571, *591*
Levy, P. E., 284, 285b, *295*
Levy-Leboyer, C., 614, *620*
Lewin, K., 19, *30*
Lievens, F., 290, 292, *297*, 374, *399*, 446, *487*, 512, 524, 529, 531, 532, 535, 538, *554*, 606, *620*
Lim, B. C., 203, *231*, 232, 392, *399*
Lin, S. C., 216, 227
Lindell, M. K., 137, *149*
Linn, R. L., 341 362
Livingston, S. A., 316, *362*, 427, 428, *431*
Locke, E. A., 103, *149*, 243, 244, *297*, 436, *490*
Loehlin, J. C., 340, *363*, 416, *431*
Lofquist, L. H., 4, *30*
Lombardo, K., 208, *227*
London, M., 218, 219, 220, *231*
Long, J. S., 84, *98*
Lord, R. G., 22, *30*
Loughlin, C., 164, *230*
Louis, M., 253, *297*
Lowell, E. L., 443, *487*
Lowman, R., 582, *591*
Lubinski, D., 372, *399*, 414, *431*, 472, 473, *487*
Lyall, J. T., 292, *299*

M

Macan, T. H., 534, *554*
MacCallum R. C., 202, *235*

MacDuffie, J. P., 144, *149*, 593, *620*
Mack, M. J., 417, *433*
MacKenzie, S. B., 195, *225*, *233*, 244, 249, *298*
MacLane, C. N., 483, *485*
Mael, F. A., 482, *487*
Maertz, C. P., 240, 287, *294*, *297*
Magnusson, D., 40, *99*, 465, 466, *487*
Maldegen, R., 531, *550*
Malos, S. B., 205, 206, *231*, 507, 508, *558*
March, J. G., 240, 267, *297*
Margolis, A., 394, *399*, 579, *591*
Marlowe, C. M., 506
Marrow, A. J., 19, *30*
Martin, C., 182, *232*, 287, *294*
Martin, J., 145, *148*, 154, *231*
Martorana, P. V., 468, *488*
Maslow, A. H., 21, *30*
Matell, M. S., 179, *231*
Matthews, G., 397, *399*, 459, *490*
Maurer, S. D., 499, 503, *554*
Mausner, B., 21, *30*
Maxwell, S. E., 549, *550*
Mayes, B. T., 534, 538, *555*, *556*
Mayfield, M. S., 120, 122, 140, *149*
Mayrhofer, W., 156, *226*
Mazerolle, M. D., 455, 456, 457t, 464, *486*
McArdle, J. J., 200–201, *231*
McCarthy, J. M., 541, *553*
McClelland, D. C., 436, 443, 444, 452, *487*
McCloy, R. A., 157, 160, 194, 202, *227*, *231*
McCormick, E. J., 188, *231*
McCrae, R. R., 393, *399*, 437, 450, 451, *485*, *487*
McDaniel, M. A., 321, *362*, 388, *398*, 448, 461, *485*, 499, 502, 512, 516, 517, 518, *553*, *554*
McFall, J. B., 5, *31*, 440, *488*
McFarland, L., 393, *400*, 415, *432*
McFarland, L. A., 156, 216, *233*, 273, 290, 291, *297*, *298*, 463, 464, *487*, *488*, 504, 506, 507, 524, 531, 540, *555*, *556*
McGann, A. F., 181, *230*
McGourty, J., 210, *233*
McGrath, R. G., 145, *148*
McGregor, D. M., 21, *31*
McHenry, J. J., 410, *431*, 614, *620*
McIntyre, R. M., 210, *231*
McKee, A., 458, *485*
McKenzie, R., 345, 354, 355, 360, *363*
McLaughlin, G. H., 317, *362*
McNelly, T. L., 528, 532, *550*
Mead, A. D., 374, *399*, 614, *620*
Meehl, Paul, 64b, *99*
Meglino, B. M., 185, *227*, 271, *297*, 493, *555*

AUTHOR INDEX

Mendenhall, M., 614, *620*
Mervielde, I., 472, *485*
Messick, S., 302, 339, *362*
Messick, S. M., 424, *431*
Meyer, J. P., 240, 249, 254, *297*
Miles, G. H., 15, *32*
Milkovich, G. T., 179, 186, *226*
Miller, D. E., 446, *488*
Millsap, R. E., 273, *298*, 384, *401*, 504, *557*
Miner, J. B., 436, 443, 448, 453, 454, 455, *488*
Mischel, W., 465, 467, *488*
Mislevy, R. J., 94, *99*
Mitchel, J. O., 534, *555*
Mitchell, T. R., 21, *30*, 175, 222, *231*, 241, 249, 276, *297*
Moberg, P., 614, *620*
Mobley, W. H., 222, *231*, 239, 267, *294*, *297*
Mohamed, A. A., 549, *557*
Mohrman, S. A., 145, *148*, 607, *620*
Mone, E. M., 218, 219, *231*
Morgeson, F. P., 120, 122, 128, 139, 140, *149*, 176, *229*, 492, 506, 512, 518, *554*, *555*
Morris, M. A., 535, *556*
Morrison, R. F., 209, *230*
Morrow, P. C., 249, *297*
Moscoso, S., 374, 393, *400*, 414, 415, *432*, 503, *555*, 614, *621*
Mosier, C. I., 69, *99*
Mosier, S. B., 79, *98*, 342, *360*
Motowidlo, S. J., 161, 162, 163, *226*, *231*, 243, 248, *294*, *297*, 407, *431*, 503, 506, 511, 512, 516, 517, 520b, *551*, *555*
Moulton, J. S., *430*
Mount, J. B., 448, *487*, *488*
Mount, M. K., 203, 208, 221, *230*, *231*, 235, 245, *296*, 390, 392, *398*, 435, 438, 439, 440, 448, 450, 460, 467, *484*, *485*, *487*
Mouton, 421
Mowday, R. T., 174, *231*
Moynihan, L. M., 445, 448, *488*
Muchinsky, P., 436, *488*
Muchinsky, P. M., 243, *296*
Muldrow, T., 345, 354, 355, 360, *363*
Mumford, M. D., 120, 122, 140, *149*, 479, 482, *488*, *490*
Munchinsky, P. M., 165, *232*, 428, *431*
Murphy, K. A., 358, *363*
Murphy, K. R., 153, 182, 185, 193, 196, 199, 208, 209, 210, 217, 219, *232*, 291, 292, *297*, *299*, 385, *399*, 417, 424, 425, *431*, *433*
Murray, C., 414, *431*
Murray, M. A., *222*, 225

N

Nagel, B. F., 196, *232*
Naglieri, J. A., 394, *399*, 407, *431*, 579, *591*
Nanda, H., 75, *98*
Nason, E., 372, *401*
Nason, E. R., 214, *230*
Nathan, B. R., 195, *229*
National Academy of Sciences, 1982, 26, *31*
National Alliance of Business, 270, *297*
Nault, A., 164, *230*
Neisser, U., 340, *363*
Nelson, C. E., 506, *554*
Neubert, M. J., 390, 392, *398*, *399*
Neuman, G. A., 390, *399*
Newbolt, W. H., 532, 538, *554*
Newman, S. H., 539, *555*
Nguyen, H. T., 516, 516t, 517, 518, *554*
Nicolopoulos, V., 533, *552*
Niesser, U., 416, *431*
Niles, K., 199, *228*
Nilsen, D. L., 475, *484*
Noble, C. L., 332, *362*
Noe, R. A., 283, *295*, 304, *363*, 436, 460, *489*
Noon, S. L., 199, *229*
Nutting, S. M., 549, *550*, *551*
Nygren, H. T., 221, *234*

O

O'Connor, E. J., 146, *149*
Odbert, H. S., 437, *484*
Oddou, G., 614, *620*
Olea, M. M., 411, *432*
Olson, J. B., 614, *620*
Ones, D., 168, 197, 208, *235*
Ones, D. S., 209, *234*, 388, 389, *399*
Oppler, S. H., 157, 160, 186, 194, 207, *227*, *232*, 309, *361*, 415, *430*
Organ, D. W., 5, *31*, 162, 163, 222, *232*, 243, 248, 249, *294*, *297*, 440, *488*
Osburn, H. G., 535, *556*
Ostroff, C., 221, *232*, 597, *621*
Oswald, F. L., 160, 161, *227*, 383, *399*, 459, *483*, *486*, *489*, 512, 532, 541, 549, *553*
Outerbridge, A. N., 186, 208, *234*
Outtz, J. L., 417, 425, *432*
Owens, P. D., 468, *488*
Owens, W. A., 479, 482, *488*, *490*

P

Packard, P. M. A., 151, *233*
Paese, M., 534, *554*
Page, R., 156, 216, *233*, 393, *400*, 415, *432*

Paine, J. B., 249, *298*
Palmer, D. K., 495, 496, 505, *551*
Palrecha, R., 442, 443, 444, 453, 465, 466, *489*
Parasuraman, A., 211, *232*
Parkington, J. J., 218, *234*, 355, *360*
Parson, C. K., 268, *295*
Parsons, C. K., 289, *295*
Paterson, D. G., 11, *30*
Patton, G. K., 222, *230*, 243, *296*
Paul, M., 586*b*, *591*
Paul, M. C., 218, *235*
Paulhus, D. L., 327, *363*
Payne, S., 499, *551*
Pearce, A., 473, *490*
Pearlman, K., 120, 122, 129, 140, *149*, 273, *298*, 384, *401*, 403, 411, *432*, *433*, 504, *557*
Penner, L. A., 163, 188, *226*
Pereira, G. M., 512, 517, *551*
Perloff, R., 340, *363*, 416, *431*
Peters, L. H., 146, *149*
Peterson, N. G., 120, 122, 140, *149*
Peterson, R. S., 445, 448, 468, *488*
Petrinovich, L. F., 41, *98*
Phillips, J. M., 271, 297, 494, *551*
Phillips, J. S., 304, *360*
Phillips, N. F., 534, *551*
Plamondon, K. E., 213, 214, 215, *233*
Plato, 10, *31*
Ployhart, R. E., 24, 29, *31*, 90, *99*, 199, 201, 203, 211, 212, 216, 217, *231*, *232*, *233*, 287, 290, 297, *298*, 312, *363*, 367, 368, *374*, 383, 384, 392, *399*, *400*, 423, *432*, 447, 459, *486*, *488*, 510, 511, 512, 513, 516, 517, 519, 532, 541, 549, *553*, *555*, *556*, *557*, *558*, 594, 610, 611, *620*, *621*
Pohley, K., 523, 524, *557*
Poksakoff, P. M., 195, *225*, *233*, 249, *298*
Polly, L. M., 278, *295*
Poon, J. M. L., 179, *226*
Porter, L. W., 174, 222, *231*, *232*
Posakoff, P. M., 244, *298*
Posthuma, R. A., 492, 506, *555*
Potosky, D., 309, 340, *361*, 386, 416, 419, *430*
Potosky, E., 374, *400*
Powell, A. B., 461, *489*
Power, D. J., 277, *298*
Powers, D. E., 403, *432*
Premack, S. L., 271, *298*
Price, J., 254, *298*
Price, R. H., 468, *490*
Prien, E. P., 129, *149*
Prifitera, A., 579, *591*

Prifiteral, A., 394, *399*
Psychological Services, Inc., 409, *432*
Pucik, V., 214, *230*
Pugh, W. M., 482, *486*
Pulakos, E. D., 186, 192, 193, 206, 207, 213, 214, 215, 221, *226*, *229*, *232*, *233*, 309, 340, *361*, *363*, *372*, *401*, 415, 416, 419, 420, *430*, *432*, *433*, 479, *489*, 499, 503, 541, *555*, *557*
Pulkos, E. D., 383, *401*
Pursell, E. D., 496, *554*

Q

Quartetti, D. A., 614, *620*
Quiñones, M. A., 523, 524, *557*

R

Radar, M., 501, 508, *556*
Rajaratnam, N., 75, *98*
Raju, N., 355, *361*
Raju, N. S., 614, *620*
Ramsay, L. J., 483, *489*
Ramstad, P. M., 355, 358, 360, *361*
Randell, G. A., 151, *233*
Raskin, R., 444, *486*
Raudenbush, S. W., 88, 89, *98*
Rauschenberger, J. M., 70, *99*, 327, 346, 358, *360*, *362*, *363*, 545, *550*
Ravlin, E. C., 493, *555*
Raymark, P. H., 134, *149*
Read, W., 179, *226*
Reckase, M. D., 91, *99*, 374, *401*
Ree, M. J., 403, 408, 410, 411, *432*
Reed, G. M., 579, *591*
Regents of the University of California v. Bakke, 566*t*, 574, *591*
Reilly, A. H., 241, *295*
Reilly, M. E., 573, *590*
Reilly, R. E., 545, *555*
Reilly, R. R., 210, *233*, 273, *298*, 384, *401*, 482, *488*, 504, 534, *554*, *557*
Reise, S. P., 92, *99*
Reynolds, D., 579, *591*
Rhodes, S. R., 174, *231*
Rich, G. A., 195, *225*, *233*
Richard, O. C., 359, *363*
Richman, W. L., 374, *400*
Riggio, R. E., 451, *488*, 534, 538, *555*, *556*
Ritchie, R. L., 342, *361*
Roberts, R. D., 459, *490*
Robertson, D. U., 304, *364*
Robie, C., 208, 219, *228*, 393, *401*, 462, 463, *489*, 535, *556*, 614, *621*
Robinson, S. L., 164, 177, *225*, *233*

AUTHOR INDEX

Rock, J., 196, 229, 420, *431*
Roe, A., 265, *298*
Roehling, M. A., 507, 508, *558*
Roehling, M. V., 507, *556*
Roethlisberger, F. J., 17, 18, *31*
Rogan, R., 451, *486*
Rogers, H. J., 92, 97, *99*
Rogers, W., 309, *363*, 386, *401*, 419, *433*
Rogg, K., 216, *232*
Rogosa, D. R., 200, *233*
Rolland, J., 415, *432*
Rosenman, R. H., 278, 279, *295*, *296*
Rosenthal, D. B., 529, *552*
Ross, J., 221, *235*
Rosse, R. L., 217, *226*
Roth, L., 386, *400*
Roth, P. L., *207*, *233*, 309, 340, *361*, 382, 383, 386, *399*, *400*, 416, 419, 425, *430*, 502, 503, 504, 540, 541, *553*, *556*, 568, *590*
Rothstein, H. R., 168, 185, *233*, 384, *399*, 482, *488*
Rothstein, M., 442, 451, *490*
Rothstein, M. C., 530, *552*
Rotundo, M., 197, 207, *233*
Rousseau, D. M., 249, 251, *298*
Rousseau, D.M., 7, *31*
Rowe, P. M., 508, *551*
Rozell, E. J., 23, *31*, 145, *148*, 358, *364*
Roznowski, M., 172, *229*, 240, 254, *296*
Russell, C. J., 201, 227, 482, *488*, 499, *551*, 571, *591*
Russell, J. T., 345, 348, *364*
Ryan, A. M., 156, 211, 212, 216, 218, 221, *233*, 287, 288, 291, *298*, 384, 393, *400*, 415, 423, *432*, 461, 464, *488*, *489*, 504, 506, 507, 510, 516, 524, *555*, *556*
Ryan, K., 162, 222, *232*
Rynes, S. L., 273, *298*, 384, *400*, 504, *556*

S

Saari, L. M., 497, *554*
Sablynski, C. J., 241, 249, 276, *297*
Sacco, J., 291, *298*
Sacco, J. M., 216, *232*, 422, *430*, 504, 516t, *555*, *556*
Sackett, P. R., 104, 112, 123, 137, *147*, *149*, 163, 164, 172, 176, 177, 197, 202, 207, 227, *233*, 234, 341 *363*, 383, 384, 386, *398*, *399*, *400*, 416, 417, 421, 423, 425, *430*, *432*, 448, *488*, 505, 512, 518, 519, 530, 540, *552*, *553*, *554*, *556*
Sager, C. E., 157, 160, 194, *227*
Salas, E., 210, 214, *230*, *231*

Salgado, J. F., 374, 393, *398*, *400*, 414, 415, *432*, 614, *621*
Salvaggio, A. N., 222, *234*, 244, *298*, 586*b*, *591*
Sanchez, J., 129, *149*
Sanchez, J. I., 139, 140, *149*
Sanders, K. C., 208, 227
Sayer, A. G., 199, *236*
Scalia, C., 196, *229*, 420, *431*
Scarpello, V., 242, *298*, 562, *591*
Schaubroeck, J., 6, *31*
Schechter, D., 608, *621*
Schechtman, S. L., 207, *228*
Schein, E. A., 8, *31*, 145, *148*
Schein, E. H., 152, *234*, 446, *488*
Schemmer, F. M., 545, *551*
Scheu, C. R., 504, 516t, *556*
Schippman, J. S., 129, 142, *149*
Schleicher, D. J., 534, *556*
Schmidt, F. L., 30, 72, *99*, 168, 186, 197, 208, 209, 222, *228*, *234*, *235*, 244, 246, *296*, 302, 305, 307, 309, 329, 332, 335, 337, 345, 346, 354, 355, 358, 360, *363*, 386, *400*, 403, 411, 414, 415, 416, 417, 425, *431*, *432*, *433*, 482, *488*, 494, 499, 501, 508, 510, 534, 540, *554*, *556*
Schmit, M., 393, 394, *399*, *401*, 579, *591*
Schmit, M. J., 134, *149*, 162, 218, 221, *231*, *233*, 243, 291, *297*, *298*, 423, *432*, 461, *489*, 614, *621*
Schmitt, K., 426, 429, 433, 539, *557*
Schmitt, N., 70, *99*, 132, 136, *148*, *150*, 186, 193, 213, 214, 216, 221, *232*, *233*, *234*, 287, *295*, 304, 309, 314, 340, *363*, 366, 372, *375*, 377, 383, 386, *398*, *401*, 416, 417, 419, 420, 421, 422, *430*, *432*, 436, 460, 479, 483, *489*, 492, 493, 499, 503, 504, 505, 512, 516, 517, 518, 529, 530, 534, 540, 541, 542, *551*, *552*, *555*, *556*, *557*, 597, 614, 617, *620*, *621*
Schneider, B., 6, 8, 21, 23, 24, 29, *31*, 81, 90, *99*, 112, 138, 140, 142, *148*, *150*, 154, 156, 169, 199, 211, 212, 217, 218, 221, 222, *226*, *228*, *234*, *235*, 239, 244, 269, *298*, 312, *363*, 377, 392, *398*, *400*, 439, 446, 447, 451, 468, 473, *489*, 586*b*, *591*, 594, 606, 608, 611, *620*, *621*
Schneider, J. R., 529, 530, *557*
Schneider, R. J., 388, 389, *399*, 436, 437, 438, 461, *486*
Schneider, S. L., 506, *554*
Schuler, R. S., 107, *149*, 216, *228*, 258, 259, 260, *296*, *298*
Schwab, D. P., 181, *235*

Schwartz, S. H., 394, *401*
Sciarrino, J. A., 539, *551*
Scott, W. D., 15, *31*
Scullen, S. E., 208, *231*, 235, 332, *362*
Seashore, H. G., 539, *551*
Seaton, F. W., 534, 540, *556*
Segal, N. L., 221, *225*
Shaffer, M. A., 614, *621*
Shalin, U. L., 141, *147*
Shane, G. S., 403, *433*
Sharf, J. C., 573, *591*
Shaw, J. D., *31*
Shaw, M. E., 445, *489*
Shaw, R. L., 151, *233*
Shein, E. H., 190, *234*
Shepard, J. M., *276*, 297
Sheppard, L., 309, *363*, 386, *401*, 419, *433*
Shiarella, A. H., 196, *232*
Shimberg, B., 426, 428, 429, *433*, 539, *557*
Silbey, V., 533, *551*
Silver, M. B., 120, 122, 140, *149*
Silverman, S. B., 137, *150*
Simon, H. A., 8, *31*, 240, 267, *297*
Simon, T., 13, 14, 28, *30*
Sin, H. P., 424, *433*
SIOP. *See* Society for Industrial Organizational Psychology (SIOP)
Sipe, W. P., 24, 29, *31*, 217, *234*, 594, 611, *621*
Skattbo, A. L., 209, *232*
Slater, A. M., 151, *233*
Slaughter, J., 292, *296*
Small, B., 499, *554*
Smith, D. B., 24, 29, *31*, 112, 133, *148*, *150*, 217, 222, *234*, 244, 269, *298*, 439, 446, 451, 462, 463, 464, 468, 473, *488*, *489*, 533, *552*, 594, 606, 611, *621*
Smith, D. E., 192, *235*, 532, 534, 538, *554*
Smith, E. M., 214, *230*
Smith, P., 119, *150*
Smith, P. C., 157, 180, *235*, 239, 242, *298*
Smither, J. W., 273, *298*, 384, *401*, 504, 534, *554*, *557*
Snell, S. A., 109, 110, *150*
Sniezek, J. A., *276*, *299*
Snow, C. C., 109, *150*
Snyderman, B., 21, *30*
Society for Industrial Organization, *401*
Society for Industrial Organizational Psychology (SIOP), 313, *363*, 579, *591*
Soelberg, P. O., 277, *298*
Song, L. J., 459, *487*
Sonnentag, S., 222, *235*
Sorra, J. S., 133, *148*
Soubel, W. G., 327, *360*, 545, *550*

Spangler, W. D., 442, 443, 444, 453, 465, 466, *489*
Spearman, C., 403, 404, *433*
Spector, P. E., 242, 244, *298*
Spencer, S. J., 422, *433*
Spray, J. A., 374, *401*
Spychalski, A. C., 523, 524, 534, 535, *557*
Stacy, B. A., 222, *225*
Stagner, R., 483*b*, *489*
Stalker, G. M., 108, *147*
Stamm, C. L., 222, *228*
Stark, S., 463, *489*
Starke, M., 289, *294*
Staw, B., 245, 247, *299*
Staw, B. M., 221, *235*, 440, 441, 450, 465, *489*
Steele, C. M., 422, 423, *433*
Steele, R. P., 241, *299*
Steers, R. M., 174, 222, *231*, *232*
Steinberg, R. J., 340, *363*
Steiner, D., 393, *401*
Steiner, D. D., 504, *557*
Steiner, D. W., 505, *552*
Sternberg, R. J., 407, 416, *431*, *433*
Stetzer, A., 176, *229*
Stevens, C. K., 494, 506, 507, *552*, *557*
Stevens, M. J., 132, *150*
Stevens, S. S., 37, *99*
Stewart, A. J., 443, 452, *490*
Stewart, G. G., 467, 468, *489*
Stewart, G. L., 201, *235*, 390, 392, *398*
Stierwalt, S., 292, *296*
Stierwalt, S. L., 461, *489*
Stine, W. W., 41, *99*
Stoffey, R. W., 273, *298*, 384, *401*, 504, *557*
Stogdill, R. M., 436, 442, *489*
Stokes, G. S., 327, *364*, 479, *490*
Stone, N. J., 502, 503, 504, *553*
Stoop, B. A. M., 530, *553*
Strickland, W., 287, *294*
Stroh, L. K., 241, *295*
Sturman, M. C., 200, *235*, 358, *364*
Subich, L. M., 471, *490*
Subriats, M., 586*b*, *591*
Sue-Chan, C., 503, *554*
Super, D. E., 264, 265, *299*
Sussmann, M., 304, *364*
Swaminathan, H., 92, 97, *99*
Switzer, F. S. III, 382, *400*, 416, *432*
Sytsma, M. R., 208, *231*

T

Tam, A. P., 292, *299*
Tanaka, J. S., 84, *99*
Task Force on Assessment Center Guidelines, 522, *557*

AUTHOR INDEX

Taylor, E. K., 174, 229
Taylor, H. C., 345, 348, 364
Taylor, M. S., 276, 278, 298, 299, 436, 490
Taylor, P., 499, 554
Taylor, S., 112, 150, 269, 298, 446, 473, 489, 606, 621
Teachout, M. S., 411, 432, 542, 552
Tellegen, A., 440, 450, 490
Tenopyr, M. L., 545, 555
Terpstra, D. E., 23, 31, 145, 148, 358, 364, 549, 557
Tett, R. P., 442, 451, 490
Thierry, H., 21, 32
Thomas, D., 571, 591
Thompson, J. D., 8, 31
Thoresen, C. J., 214, 222, 230, 243, 296, 440, 490
Thoresen, J. D., 532, 554
Thorndike, R. L., 16, 31, 306, 364, 411, 433
Thornton, G. C., 529, 552
Thornton, G. C. III, 533, 534, 553, 557
Thorsteinson, T., 292, 296
Thurstone, L. L., 404, 405, 433, 539, 551
Timms, H., 321, 362
Tippins, N., 512, 555
Tokar, D. M., 471, 490
Toquam, J. L., 410, 431, 614, 620
Tornow, W. W., 219, 235
Tower, S. L., 494, 551
Tremblay, P. F., 541, 553
Trevor, C. O., 200, 235, 424, 431
Trumbo, D., 492, 494, 557
Trumbo, O. A., 436, 487
Truxillo, D. M., 461, 486
Turban, D. B., 268, 299, 446, 468, 490, 506, 552
Turner, M. E., 571, 591
Turner, S. M., 579, 591
Tyler, P., 382, 400, 401, 416, 432
Tziner, A., 159, 235

U

U. S. Department of Labor, 113, 119, 150
Ulrich, L., 492, 494, 557
Uniform guidelines on employee selection procedures, 27, 31, 602, 621
United States v. Garland, Texas, 566t, 568, 583, 588, 591
United States v. Georgia Power, 566t, 574, 591
Urbina, S., 340, 342, 360, 363, 416, 431
Urry, V. W., 305, 307, 309, 363

V

Vance, R. J., 202, 235
Van Eerde, W., 21, 32
Van Iddekinge, C. H., 503, 556
Van Keer, E., 524, 538, 554
Vasey, J., 136, 148
Velasquez, R., 394, 399, 579, 591
Villanova, P., 157, 225
Vinitsky, M., 473, 490
Viswesvaran, C., 168, 176, 197, 208, 209, 234, 235, 388, 389, 394, 399, 401
Viteles, M., 441, 490
Viteles, M. S., 15, 32, 245, 299
Vroom, V. H., 20, 21, 32, 243, 277, 299

W

Wagner, R. K., 407, 433
Wainer, H., 96, 99
Waldman, D. A., 207, 211, 226, 235
Wallace, S. R., 152, 202, 225, 235
Walsh, W. B., 468, 490
Wanberg, C. R., 214, 235
Wanous, J. P., 271, 298
Ward's Cove Packing Company, Inc. v. Antonio, 566t, 574, 591
Wareham, J., 282, 299
Warr, P., 473, 490
Watson, D., 440, 450, 490
Watson v. Fort Worth Bank & Trust, 566t, 591
Wechsler, D., 403, 433
Weekley, J. A., 212, 236, 312, 329, 362, 363, 374, 400, 447, 488, 499, 511, 512, 513, 517, 519, 553, 555, 557, 558
Weick, K. E. Jr., 436, 453, 484
Weisband, S., 374, 400
Welbourne, T. M., 214, 230
Welch, H. J., 15, 16, 32
Wernimont, P. R., 539, 558
West, S. G., 79, 98
Wexley, K. N., 119, 137, 148, 150, 171, 178, 182, 186, 192, 231, 236
Wheeler, J. K., 239, 298
Wherry, R. J., 185, 236
Whetzel, D. L., 499, 554
White, L. A., 186, 207, 226, 309, 361, 415, 430
White, S. S., 112, 150, 169, 218, 235, 586b, 591
Whitney, D. J., 372, 401
Whyte, G., 358, 362
Wigdor, A. K., 408, 431

Wiechmann, D., 216, 232, 371, 372, 377, 401, 512, 517, 551
Wiesner, W. H., 329, 362, 499, 553
Wigdor, A. K., 171, 236, 340, 362, 408
Wilcox, R. C., 267, 294
Wiley, W. W., 115, 148
Wilk, S. L., 341, 363, 383, 399, 416, 417, 432, 540, 556
Willett, J. B., 199, 236
Williams, K. J., 271, 297
Williams, L. J., 372, 401
Williamson, I. O., 284, 299
Williamson, L. G., 507, 508, 558
Wilson, M. A., 100, 139, 140, 148
Wingersky, M. S., 428, 431
Winter, D. G., 443, 452, 490
Witt, L. A., 448, 485
Woehr, D., 529, 531, 535, 538, 550, 558
Woehr, D. J., 501, 531, 553, 558

Wong, C. S., 459, 487
Wright, J., 390, 399

Y

Yoo, T. Y., 459, 483, 489
Youngblood, S. A., 271, 297
Yusko, K. P., 533, 552

Z

Zaccaro, S. J., 132, 148, 210, 228
Zedeck, S., 138, 140, 148, 202, 227, 234, 377, 398, 425, 545, 555, 589, 591
Zeidner, M., 459, 490
Zeithaml, V. A., 211, 232
Zickar, M., 292, 296
Zieky, M. J., 316, 362
Zimmerman, R. D., 510, 556
Zippo, M., 282, 299

Subject Index

Page numbers followed by *b* indicate boxed material
Page numbers followed by *f* indicate a figure
Page numbers followed by *t* indicate a table

A

AA (Affirmative action), 569–571
A (Achievement), 330*f*, 331
AAPs (Affirmative action programs), 569–571
Ability
  paper-and-pencil tests of, 402–416
  perceptual, 412*t*
  testing, 26–27, 80*f*
  workers and distribution of, 144*f*
Ability-performance relationship, 311, 311*f*
"Absence frequency", 174
Absences, attitudinal, 174
"Absence severity", 174
Absenteeism, 166, 172–175
Absolute judgment, 178–186
  absolute rating formats in, 186
  behaviorally anchored rating scales in, 180–181, 182*f*

behavioral observation scales in, 181–182
comparative judgment *vs.*, 187
"errors" in absolute ratings in, 183–186
graphic rating scales in, 178–179, 179*f*, 189
mixed standard scales in, 182–183, 184*f*
Absolute ratings. *See also* Ratings
  comparing formats of, 186
  errors associated with, 183–186
Academic setting, performance in, 335
Accidents, 176
Accomplishment record, 421–422
Accuracy and job analysis, 139–140
Achievement (A), 330*f*, 331
Achievement motivation (AM), 80*f*, 335, 455, 456*t*
Activity level, 450
"Actual criterion", 158–159, 159*f*, 223–224, 223*f*

637

SUBJECT INDEX

ADA (Americans with Disabilities Act, 1990), 549, 565, 565t, 572–573
Adaptability
  to jobs, 5
  performance, 213–214, 215b
Adaptive subsystems in organizational functioning, 594t
Advancement, rate of, 177
Adverse impact ratio, 568
Advertising, 279t, 280t
AERC (Association of Executive Search Consultants), 283
Affective commitment, 249
Affectivity, 440
Affirmative action (AA), 569–571
Affirmative action programs (AAPs), 569–571
African Americans, test scores of, 340, 416
Age Discrimination Act, 1967, 565, 565t
Aggregate effectiveness, in staff appraisal, 217–218
Aggregate performance, 162
Aggregate personality, 312
Agreeableness
  in five factor model of personality, 323, 328t, 437t
  realistic interests and, 470
  in task performance, 439
  team performance and, 445
Alcohol and drug use, 176
Alpha level, in criterion-related validity study, 308–309
Alternate ranking, 187
AM (Achievement motivation), 80f, 335, 455, 456t
American Optical Company, 544
American Psychological Association (APA), 578–579, 582
Americans with Disabilities Act (ADA), 1990, 549, 565, 565t, 572–573
Anxiety, 319–320, 506
APA (American Psychological Association), 578–579, 582
Appearance, interviewers and, 506
Applicants
  attraction of, 605–606
  maintaining interest of, 272–273, 274–275b
  outreach to, 270–272
  reactions in assessment centers, 533–534
  reactions to recruiting practices, 284–288
Application blanks, 476–478
  preemployment inquiries and, 477t
  weighted, 478
Application process, 274b

Application reactions
  interviews and, 504–505
  to recruiting practices, 284, 286–288
Appraisal system, 193–194
Appropriateness fit, 94–95
Aptitude batteries, 449
Aptitude testing
  job-related, 407–410, 409t, 410t
  personal inventories vs., 449
  specific, 409, 410t
Aptitude Testing (Hull), 15
Aristotle, 9
Artistic personality, 262
Artistic vocations, 369f, 370t, 469
ASA model (Attraction-selection attrition), 446, 472
Asian cultures, staff appraisal in, 156
Asians, test scores of, 416
Assessee performance, 531
Assessment centers
  applicant reactions in, 533–534
  constructs assessed (dimensions) in, 528–529
  description of, 521t
  development of, 537t
  future directions of, 537–538
  guidelines for, 522–523
  legal issues and, 536
  overview of, 519, 522
  purpose of, 523–524
  exercises of, 524–525, 528
  rating process in, 534–536
  simulation in, 525, 526–527f
  subgroup differences in, 532–533
  technology and, 536
  utility of, 533
  validity in, 529–532
  work samples in, 538–542
Assessors, ratings and, 534–536
Assignment of behaviors to dimensions, 181
Association of Executive Search Consultants (AERC), 283
Assurance, 211
AT&T, 33
"Attitudinal absences", 174
Attitudinal commitment, 249
Attraction-selection-attrition (ASA) model, 446, 472
Attributes
  job analysis and, 134–137
  tasks and, 124–125
  team performance and, 445
"At work" frame of reference, 460
Auditory perception, in cognitive abilities, 406f
Aural/oral methods, 373t, 375

SUBJECT INDEX

Awareness
 in formal posting, 258
 social, 457

B

Bakke case, 574
Banding, 424–426
"Bare bones" analysis, 333–337, 334t
BARS. *See* Behaviorally anchored rating scales (BARS)
Batteries
 aptitude, 449
 multiaptitude, 407–408, 409t
 predictors, 366, 385–386, 417–419
Bausch & Lomb, 544
Behavior
 feelings in, 21
 Lewin's notion about, 19
 past behavior and, 435
 theory of motivation and, 20
 type A, 436
Behavioral commitment, 248–249
Behavioral interview, 498
Behaviorally anchored rating scales (BARS)
 example of, 182f
 flexibility and, 547
 overview of, 180–181
 reasons and steps in, 180–181
 situational interviews and, 499, 600
Behavioral observation scales, 181–182
Behavioral vs. situational interviews, 499, 503
Behavioral withdrawal, 172
Behaviors of dimensions, assignment of, 181
Behavior statements
 retranslation of, 181
 scaling of, 181
Bennett Hand-Tool Dexterity Test, 545
Bias. *See also* Criterion contamination
 in predictions, 416–417, 420
 in staff appraisals, 206–207
BIBs. *See* Biographical information blanks (BIBs)
Binet, Alfred, 13
Binet-Simon scales, 13–15, 403
Biographical data. *See also* Biographical information blanks (BIBs)
 in hiring procedures, 475–485
 and personality measures, 327, 328t
 staffing decisions and, 371t, 380t
Biographical information blanks (BIBs), 478–485
 criterion-related validity of, 481–482
 faking issues in, 482

 scoring key development for, 479t
 steps in, 479–480
Biological approach, in job design, 128
Bisection of a 50-cm line, in Cattell test, 14
Blake vs. City of Los Angeles (1979), 546
Blanks, in personality inventories, 449
Bond strength principle, 8
Borman and Motowidlo performance model, 162
Boundary conditions, 464–465
Branding, 290
Broad auditory perception, in cognitive abilities, 406f
Broad cognitive speediness, in cognitive abilities, 406f
Broad retrieval ability, in cognitive abilities, 406f
Broad visual perception, in cognitive abilities, 406f
Brown, Spearman, 73

C

CA. *See* Cognitive ability (CA)
Calculating weights, 67–69
California Proposition 209, 574
California Psychological Inventory (CPI), 449
Campbell, John, 160
Campbell performance domain models, 160–165
Campbell's Interests and Skills Survey (CISS), 474
Campion, M. A., 128
Campus recruitment, 279t, 280t, 281–282
Candidates selection, requirements for, 258–259
Career choice, 261–269, 263f, 266f
Career conferences, 279t, 280t, 281
Career development/choice, 261–269, 263f, 266f
CARS (Computerized adaptive rating scale), 188
CATs. *See* Computer adaptive tests (CATs)
Cattell, James McKeen, 11, 13–14
Cattell tests, 13–14
Caucasians, test scores of, 340, 416
Central tendency error, 184
Central tendency measures, 43–44, 44t
Certification programs, 426–429
CFA (Confirmatory factor analysis), 85–86, 86f
Changing work roles, 172–173
Checklists, in reports of behavior methods, 190
Choosing a career, 261–269, 263f, 266f

SUBJECT INDEX

CISS (Campbell's Interests and Skills Survey), 474
CIT (Critical incident technique), 118–119, 513
Civil Rights Act of 1964, 25, 340, 382, 573
Civil Rights Act of 1964 and 1991, 564–569, 565t
Civil service agencies, targeted recruiting and, 291
Classical test theory (CTT), 71–72
Classifying, in aptitude test, 410t
Clerical/technical skills, 598t
Closing the deal, in external recruitment, 275–279
Coaching, 423–424, 425
Codes of ethics, 582
Coding, in aptitude test, 410t
Coefficient alpha reliability, 73–75
Coefficient of determination, 347–348
Cognition, subgroups and, 613
Cognitive ability (CA), 397. *See also* Cognitive ability (CA) test
   in model of supervisory ratings, 81–87, 81f
   importance of g and, 410–416, 412t, 414t
   job-related aptitudes and, 407–410, 409t, 410t
   multidimensional concepts of, 404–407
   racial subgroup differences and, 416–426
   staffing decisions and, 368f, 370, 371t, 380t
   theory of structure of, 405, 406f
Cognitive ability (CA) test
   correlation with performance and interview, 60–61, 393
   paper-and-pencil tests, 402–405, 407
   reducing subgroup differences in, 417–424
   use in the United States, 415
   validity of, 412f, 413
Cognitive domain, 371
Cognitive models, SEM techniques and, 185–186
Cognitive predictors, 417, 419
Cognitive speediness, in cognitive abilities, 406f
Cognitive task analysis, 141
Cognitive validity, 412t, 413
Coherence, 465
College students, situational judgment measure for, 520b
Colleges/universities recruitment, 279t, 280t, 281–282

Combining predictors, method of, 386–387
Commitment
   attitudinal, 249
   behavioral, 248–249
   continuance, 249–250
   models of, 250–252, 250f
   normative, 250
   organizational, 247–252
   psychological contract and, 7, 249, 251–252
Commitment-based human resources, 110
Communication skills, 420, 598t, 600
Community agencies, 279t, 280t, 281
Companies, recruit by occupation in, 279t
Comparative judgment *vs.* absolute judgment, 187
Compensation systems, 145
Competencies, heritability of, 11–12
Competency models, 128–130, 130b
Competition, staffing and, 7
Competitive advantage, recruitment as, 292, 293f
Competitiveness, 37–38, 42
Complaints
   employment, 205
   verbal or physical, 176
Componential model, 407
"Compound" personality measures, 460
"Compound" predictors, 388
Comprehension, 367, 368f, 373t, 410t
Computation, in aptitude test, 410t
Computer adaptive tests (CATs), 92
   advantages of, 96–97
   presentation sequence, 95t
Computerized adaptive rating scale (CARS), 188
Computer methods, 373–375, 373t
Concurrent validity study, 303–304, 304f, 306
Conditional reasoning, personality tests and, 454–457, 456t
Conferences, career, 279t, 280t, 281
Confidence intervals, 65
Confidentiality, in search firms, 282
Confirmatory factor analysis (CFA), 85–86, 86f
Conflict in formal posting, 258
Conformity pressures, 140
Connecticut vs. Teal (1982), 566t
Conscientiousness
   in five factor model of personality, 323, 328t, 437, 437t
   leadership and, 442
   in task performance, 439
   team performance and, 445

SUBJECT INDEX 641

Consent decree, 563
Consequence of reliability, 76–77
Consequential validity, 339
Consistency, in process fairness, 286
Consortia for validity studies, 309, 310b
Consortium, small businesses and, 310b
Constructs, 320, 502–503, 528–529
Constructs vs. measurement methods, 370–377
 computer and web-based methods in, 373–375
 interview methods in, 375–376
 measurement methods as predictors in, 376–377
 performance-based methods in, 376
 predictor construct domains in, 371–372, 371t
 predictor measurement methods in, 372
 visual and aural/oral methods in, 375
 written (paper-and-pencil) methods in, 372–373
Construct validity
 definition of, 301–302
 example of, 549–550
 multitrait-mulitmethod matrix and, 329
 of physical ability measures, 327
 rules of, 535–536
Construct variable, 42–43
"Contaminating" sources of variance, 368f, 369
Contamination, in criterion measure, 159–160, 159f
Content dimension, in criteria, 204
Content-oriented validation strategy, 138, 603
Content validation strategy, 138–139
Content validity, 301–302
 response to test content in, 320–322
 SME judgments of, 315–316
 staffing small businesses and, 597
 using, 310b
 of various test formats, 315f
Content validity ratio (CVR), 318–319
Context in personality, 435
Contextual effects, 312
Contextual intelligence, 407
Contextual performance, 162–165, 420
Continuance commitment, 249–250
Contract
 psychological, 7, 249, 251–252
 workers, 6–7
"Control theory" of motivation, 22
Conventional personality, 262
Conventional vocations, 369f, 370t, 469
Convergent validation, 329–332

Convergent validity, 530
Cooperativeness, 598t, 600
Corrections for unreliability, 459–460
Correlation
 definition of, 43
 description of, 59–61, 60f
 predictions based on, 61–66
Correlation coefficient (r)
 about, 12
 description of, 59–61, 60f
 of different sizes, 57f
 limitation of, 58–59
 range of, 55–56
 wonder of, 56–58
Correlation ratio (eta), 58
Cost, in internal recruitment, 253–254
Counter-productive work behaviors (CWBs)
 contextual performance vs., 163–165
 safety and accidents in, 176
 types of, 176–177
 withdrawal in, 172–175
Court cases. See also Legal issues
 Blake vs. City of Los Angeles (1979), 546
 Gratz vs. Bollinger (2003), 566t, 574
 Griggs vs. Duke Power Company (1971), 566t, 574
 Grutter vs. Bollinger (2003), 417, 566t, 575
 individual, 573–576
 staffing process and, 25–28
 United States, 566t
 United States vs. Garland, Texas (2004), 566t, 568, 583, 588
 United States vs. Georgia Power (1973), 566t, 574
 Ward's Cove Packing Company, Inc. vs. Antonio (1989), 566t, 574
 Watson vs. Fort Worth Bank & Trust (1988), 566t
Covariation, 49–54
 in general, 49–50
 multiple scores on the predictor in, 51–54, 51f, 52f, 53f, 54f
 two scores on the predictor in, 50, 50f
C parameter, 91
CPI (California Psychological Inventory), 449
Crawford Small Parts Dexterity Test, 544
Credibility interval, 335–336
Criterion contamination
 in actual and ultimate criterion, 159f
 group characteristic bias in, 159
 Opportunity bias in, 159
 predictors bias in, 159–160
Criterion/criteria. See also Criterion contamination

actual, 158–159, 159f, 223–224, 223f
aspects of
  discriminability in, 166–167
  practicality in, 169
  relevance in, 159, 159f, 165–166
  reliability in, 167–169
deficiency, 159, 159f, 160
definition of, 50, 157–158
distribution of scores, 50f
judgmental measures of, 178–194
  absolute judgments in, 178–186
  appraisal purpose in, 193–194
  employee comparisons in, 186–190
  objective criteria vs., 194–195
  rater training in, 192–193
  reports of behavior in, 190–192
  measurement considerations of, 195–205
objective measures of, 170–178
  absenteeism in, 172–175
  counterproductive work behaviors in, 176–177
  judgmental criteria vs., 194–195
  lateness in, 172–175
  production of output in, 170
  quality in, 171
  rate of advancement in, 177
  safety and accidents in, 176
  tardiness in, 172–175
  training success and work samples in, 171–172
  turnover in, 172–175
relationship with predictors, 311f
relevance, 159, 159f
in validation studies, 337t
weighting of, 420
"Criterion for criteria", 166–167
Criterion-predictor relationship, in staff appraisal, 223–224, 223f
"Criterion problem", 157
Criterion-referenced exams, 428
Criterion-related validity. See also Criterion-related validity studies
  across cultures, 393
  of the assessment centers, 533
  of biographical data, 481–482
  definition of, 300–301
  in evaluating predictors adequacy, 380t, 381
  response to test content in, 320–322
  stability and, 167
Criterion-related validity studies
  conducting, 308–309
  feasibility of, 304–307
  of licensing exams, 428

Criterion scores, distribution of, 50f, 51f, 52f
Critical incident technique (CIT), 118–119, 513
Criticality task ratings, 125t
Critical mass of diversity, 575
Cross-cultural issues, in staff appraisal, 216–217
Cross-validation of multiple regression, 69–70
Crystallized intelligence, in cognitive abilities, 406f
CTT (Classical test theory), 71–72
Cultural adaptability, 5
Cultural influences, predictor development and, 393–394
Cultural stereotypes, test performance and, 422
Customer satisfaction, staffing and, 586–587b
Cutoff scores, 426–428
"Cut scores" for examination performance, 316–319, 317f
CVR (Content validity ratio), 318–319
CWBs. See Counter-productive work behaviors (CWBs)

D

Darwin, Charles, 11–12
Data analytic concepts, 66–77
Data analytic tools
  aims of, 35–37
  background of, 33–35
  covariation in, 49–54
  cross-validation of multiple regression in, 69–70
  definition of measurement, 37
  hierarchical linear models in, 87–90
  human experts vs. linear regression in, 64b
  individual differences measurements in, 43–54
  item response theory in, 90–97, 90f
  key statistical symbols, 36–37b
  manifest and latent variables in, 42–43
  nature of measurement and, 37–42
  normal distribution on tests, 40f
  Pearson product-moment correlation coefficient (r) in, 54–61, 57f
  predictions based on correlation in, 61–66
  structural equation modeling in, 81–82f, 81–87, 87f
  validity in, 77–78

SUBJECT INDEX

Decision making
 in aptitude test, 410t
 immediate hiring and, 604–605
 regarding work environment, 606
 systematic evaluation and, 618–619
Decision speed, in cognitive abilities, 406f
Declarative knowledge, 161
Deductive (subjective) approaches, in criteria, 165–166
Deficiency, in criterion measure, 159, 159t, 160
Demographic attributes, job analysis and, 134–137
Descriptive statements, generation of, 180
Determination of performance dimensions, 180
Development
 of interactive, Internet-based assessment of leadership, 395b, 396f
 staff and tasks for, 125, 126t
Deviance, organization and interpersonal, 177
Dexterity tests, 544–545
Dichotomizing the criterion, 350
Dictionary of Occupational Codes, 474
Dictionary of Occupational Titles (DOT) (U.S. Department of Labor), 113, 119–120, 412
"Differentiation prediction", 341–343, 343f
Difficulty parameter, 91
Dimensions (Constructs assessed), 528–529
Discriminability in criteria, 166–167
Discriminant validation, 329–332
Disparate impact, 566
Disparate treatment, 565
Distal environmental factors, 155f
Distal vs. proximal predictors, 377, 378f, 379
Distress, 441
Diversity, critical mass of, 575
Dominance, 442–443
DOT. See *Dictionary of Occupational Titles* (DOT) (U.S. Department of Labor)
Double cross-validation, 69–70
Dynamometer pressure, in Cattell test, 14

E

EAS (Employee Aptitude Survey), 408, 409t
EC (Emotional control), 330f, 331, 598t, 600

Economic investment, in employee turnover, 240
EEOC (Equal opportunity Employment Commission), 536, 562, 565
Effective interview, 496
Effective staffing, 595–596
Effect size, 59
EI (Emotional intelligence), 457–458
Emotional control (EC), 330f, 331, 598t, 600
Emotional intelligence (EI), 457–458
Emotional stability. See Neuroticism
Empathy, 211–212
Employee Aptitude Survey (EAS), 408, 409t
Employee comparisons, 186–190
 forced distribution method in, 189–190
 method or paired comparisons in, 187–188
 methods of rank order in, 187
"Employee engagement", 246
Employee Relocation Council, 283
Employees
 commitment in internal recruitment, 247–252
 development in staff appraisal, 218–220, 219f
 referrals of, 279t, 280, 280t
 relocation of, 283
 turnover of, 172–175, 242–247
Employment agencies, private, 279t
Employment complaints, 205
Employment legislation, 573–574
 federal, 25–28
Enterprising personality, 262
Enterprising vocations, 369f, 370t, 469
Environment. See also External environment; Larger environment
 decision making regarding, 606
 organizational, 106–113, 262
 work, 16–23
Equality, in outcome fairness, 287
Equal opportunity Employment Commission (EEOC), 536, 562, 565
Equity, in outcome fairness, 287
Error component, in test, 71
"Errors" in absolute ratings, 183–186
Error variance, 71
eta (Correlation ratio), 58
Ethics
 codes of, 582
 staffing practice and, 561f, 581–582
Evaluation
 of predictors adequacy, 379–385, 380t
 of staff, 125, 126t

Examination(s)
 content construction, 314, 315f
 criterion-referenced, 428
 plan, 597–598, 598t
 setting cut scores for, 316–319, 317f
 in validity based on test content, 314–315
Executive Order 11246 and affirmative action, 569–571, 573, 574
Experience, staffing decisions and, 371t, 380t
Experience-based interviews, 498, 503
Experience requirements, 121
Expert witness, staffing specialist as, 587–588
External environment. *See also* Larger environment
 human resources, strategy and, 107f, 111–112
 organizations and, 106–113, 107f
External recruitment, 261–288. *See also* External recruitment research
 applicant reactions in, 284, 286–288
 career development and choice in, 261–269, 263f, 266f
 current challenges for
  meaning of fit in, 288–289
  recruitment as competitive advantage in, 292, 293t
  selection system withdrawal in, 290–291
  targeted recruiting in, 291–292
  theories of recruiting in, 289–290
 methods of, 279–284
 organization decision in, 261
External recruitment research, 269–279
 closing the deal in, 275–279
 maintaining applicant interest in, 272–273, 274–275b
 outreach to potential applicants in, 270–272
Extraversion
 in five factor model of personality, 323, 325t, 328t, 437, 437t, 438
 job satisfaction and, 441
 social and enterprising interests and, 470
 in task performance, 439
 team performance and, 445

F

FACT (Flanagan Aptitude Classification Tests), 408, 409t
Factor analysis, 57, 115, 117, 404
Factors of intellect, 404–405
Fairs, job, 279t, 280t, 281

Fakability, in personality testing, 461–463
Faking in biographical data, 482–483
Fear of failure (FF), 455, 456t
Federal regulations, staffing and, 7
Feedback, in process fairness, 286
Feelings in human behavior, 21
FF (Fear of failure), 455, 456t
FFM. *See* Five factor model (FFM)
Filing names, in aptitude test, 410t
Filing numbers, in aptitude test, 410t
Final rating scale, 181
Fine, Sidney, 101, 113–115
Firms, in validation studies, 337t
Fisherz equivalents, 308
Fit, 277
 appropriateness, 94–95
 meaning of, 288–289
 organizations and, 267–269, 277
 person-environment, 467–468
Fit index for persons, 94–95
Five factor model (FFM)
 background of, 439
 with descriptive terms, 437t
 emotional intelligence and, 458
 O*NET work styles taxonomy and, 448t
 personality and, 443
 personality inventories and, 448–449
 Personality-Related Position Requirements Form and, 134, 436–439, 437t
FJA (Functional job analysis), 113–115, 116f
Flanagan Aptitude Classification Tests (FACT), 408, 409t
Flexibility in working hours, 607
Fluid intelligence, in cognitive abilities, 406f
Forced-choice response formats, 461
Forced choice technique, 190–192
Forced distribution method, 189–190
Formal job posting, drawbacks of, 258
Formal reward system, 103
Formal tasks, characteristics of, 5
Four fifths rule, 27–28, 568
Frame of reference training, 192
Frosting Developmental Test of Visual Perception, 544
Functional job analysis (FJA), 113–115, 116f
Functional utility, 11
Future attractions of the present job, in employee turnover, 239–240
Future-oriented questions, in interviews, 497
Future research, assessment centers and, 537–538

SUBJECT INDEX

G

Gallup Organization, 246
Galton, Sir Francis, 11, 12, 13
GATB (General Aptitude Test Battery), 408, 409t
Gender
 bias in staff appraisals and, 207
 measuring, 39
 subgroup differences and, 504
 test scores and, 340
General Aptitude Test Battery (GATB), 408, 409t
General factor (g), 404, 405, 410–416
General intelligence, in cognitive abilities, 406f
Generalizability theory, 531
Generalization, validity. *See* Validity generalization
General memory, in cognitive abilities, 406f
General *vs.* specific performance, 196–198
Generation of descriptive statements, 180
George and Brief approach, 441
g (General factor), 404, 405, 410–416
Global organizations, staffing and, 614
Glossary of statistical concepts, 36–37b
Gordon Personal Profile (GPP), 449, 614
Graphic rating scales, 178–179, 179f, 189
Gratz vs. Bollinger (2003), 566t, 574
Griggs vs. Duke Power Company (1971), 566t, 574
Group characteristic bias, in criterion contamination, 159
Group level theory, 24–25, 24f
"Growth models", 200–201
Grutter vs. Bollinger (2003), 417, 566t, 575
Guidelines and Ethical Considerations, 519
Guilford-Zimmerman Temperament Survey (G-ZTS), 449

H

Halo error, 184–185
Hanges, P. J., 22
"The Hawthorne Studies," 17, 20
Heritability of competencies, 11–12
Heterotrait-heteromethod blocks, 329, 331
Heterotrait-monomethod triangles, 329, 330f, 331
Hierarchical linear models (HLMs), 34–35, 58, 87–90
High-performance work systems, 103, 144

"High stakes" tests, 422
Hiring procedures. *See also*
 Paper-and-pencil tests;
 Performance-based testing
 biographical data in, 475–476
 definition of predictors, 366
 interests and interest testing in, 468–485
 interviews in, 492–510
 licensure and certification exams in, 426–429
 personality and personality tests in, 436–468
 physical ability testing in, 542–550
 predictor development and, 389–397, 391f, 395b, 396f
 predictor domain in, 370–379, 371t, 373t, 378f
 predictor-response process model in, 367–370, 368f
 predictors adequacy evaluation in, 379–385, 380t
 predictors in staffing practice and, 385–397, 391f
 racial subgroup differences and cognitive ability in, 416–426, 418–419b
 using predictors in staffing practice, 385–389
Hispanics, test scores of, 340
HLMs (Hierarchical linear models), 34–35, 58, 87–90
Holland, John, 469
Holland's theory of vocational choices, 369f, 469–471
Holland's Vocational Preference Inventory (VPI), 474
Holtzman Ink Blot Test, 451
Homegrown job analysis procedures, 123–128
Honesty, in process fairness, 287
HR. *See* Human resources (HR)
HRIS (Human resource information system), 255
Human attributes, tasks and, 124–125
Human capital, 109
Human experts *vs.* linear regression, 64b
Human Performance, 423
Human resource information system (HRIS), 255
Human resources (HR)
 activities of, 100
 commitment-based, 110
 competency models and, 128–130
 external environment and, 107f, 111–112

organizational effectiveness and, 358–359
performance and, 144–145

I

IAR (Individual Achievement Record), 482
Imagery, 451
Immediate hiring, 604–605
Implicit motives, 444
Impression management, 327, 328t, 368f, 506–507
Inaccuracy, job analysis and, 139–140
In-basket exercise, in assessment centers, 524, 525f, 530
Incomplete sentences, in projective tests, 452–454
Increasing chronic lateness, 173–174
Increasing outputs, 172
Index of forecasting efficiency, 347–348
Individual Achievement Record (IAR), 482
Individual attributes
 in general, 2
 staffing challenges and, 3–4, 3f, 4t, 7–8
Individual differences
 about, 9–11
 central tendency measures in, 43–44, 44t
 Charles Darwin on, 11–12
 correlation coefficient and, 12
 measurements of, 44–49
 mental tests and, 13–16
 standard deviations and standard scores in, 46–49
Individual level predictors, 391f
Individual-level staff appraisal factors, 155f
Individual level theory, 24–25, 24f
Individual (person) characteristics, 111
Individual rationality, 20–21
Inductive (objective) procedures, in criteria, 165–166
Industrial Psychology (Viteles), 245
Industrial relations process, 110
Information
 in validation studies, 337t
 Web sites and, 285b
Information processing, 140
Information technology, tasks and, 5
Inquiries, preemployment, 477t
Integration of criterion measurement, 203–205, 304f
Intellect model, 405
Intelligence
 Binet-Simon indicators of, 14

in cognitive abilities, 406f
 componential aspect of, 407
 emotional, 457–458
Intensity dimension, in criteria, 204
Interactionist perspective, 464–465
Interactive voice recording (IVR), 375
Interdisciplinary model of work and job design, 128
Interest inventories, 471, 474–475
Interest measurement, 471–474
Interests
 behavior at work and, 435
 personality and, 468–485
 staffing decisions and, 371t, 380t
 testing, 468–485
 vocational development and choice and, 265
Interest validity, context in, 435
Internal consistency reliability, 167–168
Internal recruitment, 239–261
 benefits of, 252–254
 current challenges for
 meaning of fit in, 288–289
 recruitment as competitive advantage in, 292, 293t
 selection system withdrawal in, 290–291
 targeted recruiting in, 291–292
 theories of recruiting in, 289–290
 employee commitment in, 247–252
 employee turnover in, 242–247
 job satisfaction in, 242–247
 models and methods of, 254–261
International Journal of Selection and Assessment, 394
Internet-based assessment of leadership, 395b
Internet recruiting, 283–284
Interpersonal deviance, 177
Interpersonal effectiveness, in process fairness, 287
Interpersonal skills, 420–421
Interrater agreement, 168–169
Interrater reliability, 75, 168, 499, 535–536
Interrater unreliability, 208–209
Interval measurement, 38–39, 40
Interviewer judgment process, 505
Interviews. *See also* Situational interview
 application reactions and, 504–505
 for assessing constructs, 330f, 331
 in assessment centers, 519–542, 528
 constructs measured by, 502–503
 experience-based, 498, 503
 future directions of, 510
 introduction and purposes of, 492–494
 legal issues with the use of, 507–508
 methods of, 373t, 375–376, 380t

SUBJECT INDEX 647

reliability of, 499, 501
screening, 493–494, 509b
structure of, 494–496
subgroup differences and, 503–504
telephone administered, 501
types of questions in, 496–498
validity of, 499, 509–510
Intrinsic rewards, 276
Inventories
 interest, 471, 474–475
 personality, 448–450
Investigative personality, 262
Investigative vocations, 369f, 370t, 469
Involuntary turnover, 175
IRC (Item response curve), 90–97
IRT (Item response theory), 90–97, 90f
Item characteristics curves, 97f
Item discrimination parameter, 91
Item information curve, 90f
Item response curve (IRC), 90–97
Item response function, 90f, 94f
Item response theory (IRT), 90–97, 90f
Item-total correlation, 323
IVR (Interactive voice recording), 375

J

Jackson Personality Inventory (JPI), 449
JE. *See* Job experience (JE)
J (Judgment), 330f, 331, 598t, 600, 601f
Job analysis. *See also* Larger environment
 contrasting, 112–113
 current challenges for
  accuracy in, 139–140
  cognitive tasks analysis in, 141
  demographic attributes for SME in, 134–137
  dispositional constructs in, 134
  legal necessity in, 138–139
  non existing jobs in, 141–142, 607–608
  organizational perspectives in, 142–147
  social setting for job tasks in, 131–134
 definition of, 100
 effective staffing and, 595
 functions of, 101–104
 historical perspectives on, 113–119
  critical incident technique in, 118–119
  functional job analysis in, 113–115, 116f, 116f
  position analysis questionnaire in, 115, 117–118
 modern perspectives on jobs, 119–131
  Campion in, 128

competency models in, 128–130
homegrown procedures in, 123–128
occupational information network (O*NET) and, 101, 113, 119–122, 604
within the organizational context, 110–111, 111f
in small businesses, 310b
staff appraisal and, 152
staffing and, 104–106, 105f
subject-matter experts (SMEs)and, 104, 105f, 607–609
traditional, 141
validity based on test content in, 314
Job Analysis and Competency Modeling Task Force, 129
Job analysis ratings, 134–137
Job analysts
 definition of tasks, 124
 task statements and, 114–115, 116f
Job characteristics, 3f, 4–5, 4t, 111
Job description
 definition of, 100
 in validation studies, 337t
Job experience (JE)
 in model of supervisory ratings, 81–87, 81f
 job analysis rating and, 136–137
Job performance, 80f
 absolute judgment in, 178–186
 appraisal system in, 193–194
 change over time, 200f
 cognitive ability and, 415
 discriminability and, 166–167
 employee comparisons in, 186–1190
 human resources practices and, 144–145
 models, 160–165
 objectives of, 314
 rater training in, 192–193
 reports of behavior in, 190–192
 specific *vs.* general dimensions of, 196–198
 staff appraisal and, 152, 220–223
 teams and, 444–446
Job-related aptitude testing, 407–410, 409t, 410t
Job relatedness, in process fairness, 286
Job(s). *See also* Job analysis; Job performance; Job satisfaction
 attractions in employee turnover, 239
 autonomy, 137
 characteristics of, 3f, 4–5, 4t, 7–8, 111
 design, 128
 embeddedness, 241–242
 fairs, 279t, 280t, 281

observed and true validity for, 412*t*
offer, 275, 276
from organizational perspectives, 142–147
organization analysis and, 615–616
security in work setting, 242
staffing decisions and, 371*t*, 380*t*
Job satisfaction
confirmatory factor model and, 85*f*
definition of satisfaction, 4
in external recruitment, 261–288
career development and choice in, 261–269, 263*f*, 266*f*
external recruitment research in, 269–279
methods of external recruitment in, 279–284
organization decision in, 261
fit as predictor of, 468
in internal recruitment, 242–247
assessment in, 246–247
multifaceted construct in, 242–243
personal variables in, 244–245
productivity in, 243–244
personality and, 440–441
in staff appraisal, 220–223
work setting and, 3*f*
Job specifications, 105*f*
Campion and, 128
competency models in, 128–130
homegrown job analysis procedures and, 123–128
occupational information network and, 101, 113, 119–122
Job title and code, in validation studies, 337*t*
Journal of Applied Psychology, 451
Journal of Vocational Behavior, 471
JPI (Jackson Personality Inventory), 449
Judgmental criterion measures, 178–194
absolute judgment in, 178–186
appraisal purpose in, 193–194
employee comparisons in, 186–190
objective criteria *vs.*, 194–196
rater training in, 192–193
reports of behavior in, 190–192
Judgment (J), 330*f*, 331, 598*t*, 600, 601*f*
Judgment of 10 seconds time, in Cattell test, 14
"Justice rules", 286–287
Justification mechanisms, 455, 456*t*

K

Kelvin scale, 39
Knowledge, skill, ability, and other (KSAO)

competency modeling and, 129
determinants of performance factors, 161–162
developing measures of constructs, 104, 106
human attributes and, 124–125
job analysis and, 104
of job-based employment, 109
measuring, 595
personnel specialist rating form for, 126*t*
selection, 127
situational judgment test and, 518
supervisors and, 140
tasks in selection, evaluation, and development, 126*t*
tests and, 597
KSAO. *See* Knowledge, skill, ability, and other (KSAO)

L

Labor, staffing and, 7
Language method effects, 368*f*
Language skills, in aptitude test, 410*t*
Larger environment
organizations and, 106–113, 107*f*
staffing and, 3*f*, 4*t*, 7–9
Lateness of workers, 172–175
Latent construct, 158–159, 159*f*, 223–224, 223*f*
Latent growth curve modeling, 201
Latent individual differences, 368, 368*f*
Latent variable, 42–43
Lawful inquiries, in Michigan Department of Civil Rights, 477*t*
Leadership
Internet-based assessment of, 395*b*, 396*f*
measuring, 421
in model of staffing team contexts, 391*f*, 392
personality and, 442–444
Leadership group discussion, in assessment centers, 524
Leadership motive profile, 444
Learning, in cognitive abilities, 406*f*
Least noticeable difference in weight, in Cattell test, 14
Legal issues. *See also* Court cases
assessment centers and, 536
impact on staffing, 596
job analysis and, 138–139
staff appraisal and, 205–206
of staffing practice, 561*f*, 562–576, 563*f*
with the use of interviews, 507–508

SUBJECT INDEX

Legislation, 7, 25–28
Leniency error, 184
"Levels effects," 312
Lewin, Kurt, 19
Licensure, 426–429
"Lie scales", 461
Life outcomes, cognitive ability and, 414–415
Limitations of correlation coefficient r, 58–59
Linear regression $vs.$ human experts, 64b
Linkage research, 217–218
Local high schools, 279t, 280t
Locus of control, 450
Lord, R. G, 22

M

Maintenance subsystems in organizational functioning, 594t
Maintenance $vs.$ transitional performance, 198–202
Majority group selection ratio, 341t
Management professionals, recruitment of, 279t, 280, 280t
Managerial subsystems in organizational functioning, 594t
Managers
  assessment centers and, 534
  attitudes toward personal tests, 483b
  recruitment of, 279t, 280, 280t
  role of, 453–454
Manifest measure, 158–159, 159f, 223–224, 223f
Manifest variable, 42–43
Manipulation tests, 544
Manpower planning structure, 102f
Maximum performance, 202
Mayo, Elton, 17
MBTI (Myers-Briggs Type Indicator), 450
McCormick, Ernest, J., 115
McGregor, D. M., 21–22
Measurement. *See also* Measurement concepts; Measurement methods
  of criterion, 195–205
  of job satisfaction, 246–247
Measurement and structural model, 86–87, 87f
Measurement concepts
  aims of, 35–37
  background of, 33–35
  covariation in, 49–54
  cross-validation of multiple regression in, 69–70
  definition of measurement, 37
  hierarchical linear models in, 87–90

human experts $vs.$ linear regression in, 64b
individual differences measurements, 43–54
item response theory in, 90–97, 90f
key statistical symbols in, 36–37b
manifest and latent variables in, 42–43
nature of measurement, 37–42
normal distribution on tests, 40f
Pearson product-moment correlation coefficient (r) in, 54–61, 57f
predictions based on correlation in, 61–66
reliability in, 71–77
structural equation modeling in, 81–82f, 81–87, 87f
using multiple predictors in, 64b, 66–77
validity in, 77–78
Measurement error, test information and, 93f
Measurement methods
  constructs $vs.$, 370–377
  as predictors, 376–377
Mechanistic approach in job design, 128
Mechanistic organizational forms, 108, 108t
Mediating appraisal factors, 155f
"Medical absences," 174
Meehl, Paul, 64b
Meeting immediate organizational needs, 257
Memory
  in examination plan of, 598t
  in structure of cognitive abilities, 406f
  in tests of ability, 404
Mental ability, 265
Mental tests, 13–16
Meta analysis, staffing and, 612–613
Methods
  of external recruitment, 279–284
  of internal recruitment, 257–261
  of paired comparisons, 187–188
  of rank order, 187
Methods requiring absolute judgment, 178–186
  absolute rating formats and, 186
  behaviorally anchored rating scales in, 180–182, 183f
  behavioral observation scales in, 181–182
  comparative judgment $vs.$, 187
  "errors" with absolute ratings in, 183–186
  graphic rating scales in, 178–179, 179f, 189

mixed standard scales in, 182–183, 184f
Methods requiring employee comparisons, 186–190
    forced distribution and, 189–190
    paired comparisons and, 187–188
    rank order and, 187
Methods requiring reports of behavior, 190–192
    checklists in, 190
    forced choice technique in, 190–192
Michigan Department of Civil Rights, 476, 477t
Miner Sentence Completion Scale (MSCS), 436, 452–454
Minimally competent person, 427–428
Minority group selection ratio, 341t
Minority-hiring rates, 425
Minority representation, banding and, 425
Mixed-purpose interview, 494
Mixed standard scales (MSS), 182–183, 184f
Mock presentations, in assessment centers, 528
Model of predictor-response process, 367–370, 368f
Model of staffing in team context, 391f
Model(s). *See also* Hierarchical linear models (HLMs)
    Borman and Motowidlo performance model, 162
    Campbell performance domain models, 160–165
    of commitment, 250–252, 250f
    competency model, 128–130
    componential, 407
    Five factor model (FFM), 134
    "growth models," 200–201
    hierarchical linear models, 87–90
    of internal recruitment, 254–257
    measurement and structural model, 86–87, 87f
    multilevel staffing, 610–612
Mode of stimulus presentation and required responses, 420–422
Moderated regression, 78–81, 80f
Moderator variables, 78, 336
Monetary reward, in closing the deal, 275, 276
Monitoring exercise, in oral directions test, 598–599
Monster.com, 283
Motivation
    control theory of, 22
    determinants of, 161
    McGregor and, 21–22

staffing decisions and, 371t, 380t
test-taking, 287–288, 422–424
theory of, 20–21, 443
variance and, 368, 368f
Motivational approach, in job design, 128
Motivation-performance relationship, 335–336
Motives, implicit, 444
MSCS (Miner Sentence Completion Scale), 436, 452–454
MSS (Mixed standard scales), 182–183, 184f
MTMM matrix (Multitrait-multimethod), 329, 330f, 331–332
Multiaptitude batteries, 407–408, 409t
Multilevel contexts, predictors in, 390–392, 391f
Multilevel staffing models, 610–612
Multiple-choice response, 368f
Multiple-choice test, in licensing, 428
Multiple hurdle systems, 387–388
Multiple predictors, using, 66–77
Multiple regression, 34, 57
    cross-validation of, 69–70
    for two predictors, 67–77
Multiple scores on the predictor, 51–54, 51f, 52f, 53f, 54f
Multirater feedback performance appraisals, 219–220
Multitrait-multimethod (MTMM) matrix, 329, 330f, 331–332
Myers-Briggs Type Indicator (MBTI), 450

N

nAch (Need for achievement), 436, 443–444, 451–452
nAff (Need for affiliation), 444, 451
Narrowly tailored process, 575
National Academy of Sciences, 26–27
National Association of Scholars, 575
Need for achievement (nAch), 436, 443–444, 451–452
Need for affiliation (nAff), 444, 451
Need for Power (nPow), 436, 444, 451
Needs, in outcome fairness, 287
Neuroticism
    in five factor model of personality, 328t, 437t
    job satisfaction and, 441
    leadership and, 442
    in measuring personality, 323, 325t, 328t
    in task performance, 439
    team performance and, 445
Newspaper advertising, 279t, 280, 280t
Nominal scale, 39–40

SUBJECT INDEX

Noncognitive domain, 371
Noncognitive predictors, using predictor battery of, 417, 419
Nonexisting job, hiring for, 607–609
Nonjob factors, in employee turnover, 240–241
Nonverbal behavior, interviewer judgments and, 506–507
Normal distribution, 45f, 46
Normalization, 49
Normative commitment, 250
nPow (Need for Power), 436, 444, 451
Number of letters remembered on once hearing, in Cattell test, 14
Numbers
 features of, 38–42
 tests of ability on, 404

O

Observed variable, 42
OCB. *See* Organizational citizenship behavior (OCB)
Occupation
 effectiveness of recruiting source by, 280t
 limitations to, 265
 recruit by, 279t
Occupational information network (O*NET)
 overview of, 119–122
 Dictionary of Occupational Titles and, 113
 example of using, 122–123b
 five factor model and, 448t
 for immediate hiring, 604
Occupational requirements, 121
Occupation characteristics, 121
Occupation-specific requirements, 121
OFCCP (Office of Federal Contract Compliance Programs), 562, 571
Office of Federal Contract Compliance Programs (OFCCP), 562, 571
Offlimits, search firms and, 282
O*NET. *See* Occupational Information Network (O*NET)
Openness, in Big Five personality dimensions, 323, 325t, 328t
Openness to experience
 in five factor model of personality, 437t
 investigative and artistic interests and, 470
 job satisfaction and, 440–441
 leadership and, 442
 in task performance, 439–440

Opportunity bias, in criterion contamination, 159
Opportunity to be considered, in process fairness, 286
Opportunity to perform, in process fairness, 286
Oral directions, in aptitude test, 410t
Oral directions/tying test, 598–599, 598t
Ordinal scale, 38–39, 40
Organic organizational forms, 108, 108t
Organization adaptation/withdrawal model, 172–173, 173f
Organizational analysis
 functions of, 101–104
 job analysis, staffing system design, and, 105f
 within the organizational context, 110–111, 111f
Organizational attractiveness, 284
Organizational characteristics, 111
Organizational choice, way of thinking about, 266f
Organizational citizenship behavior (OCB)
 as defined by Organ, 163
 five factor model and, 440
 in hypothetical model of supervisory ratings, 81–87, 81f
 informal tasks and, 5
 job satisfaction and, 243–244
 organizational effectiveness and, 440
 task performance and, 439–440
Organizational climates, role in staffing appraisal, 154
Organizational commitment, 247–252
Organizational consequences, 23–24, 24f
Organizational context, importance of, 153–156
Organizational culture
 personality and, 446–447
 role in staffing appraisal, 154
 team performance and, 445
 workers performance and, 145
Organizational deviance, 177
Organizational effectiveness
 formal reward systems and, 103
 human resources and, 358–359
 organizational citizenship behavior and, 440
 Smith and, 157–158
 staffing and, 23–25
Organizational environment. 106-113, 262
Organizational functioning, managerial subsystems in, 594t
Organizational goals, 6–7, 153–156, 162
Organizational justice theory, 505

SUBJECT INDEX

Organizational needs, meeting immediate, 257
Organizational practices and procedures, staffing and, 3f, 4t, 6–7, 8
Organizational values, 276
Organization development process, 110
Organization-level theory, 24–25, 24f
Organizations. *See also* Organizational culture; Organizational effectiveness
  applying psychology in, 582
  definition of, 1
  and external environment, 106–113, 107f, 262
  global, 614
  interrelationship among, 101, 102f
  psychology in, 582
  rejecting potential job offers in, 274–275b
  relationship between workers and, 6–7
  staff appraisal and, 152–156
  understanding jobs from, 142–147
Orientation programs, 423–424, 425
Other counterproductive work behaviors, 176–177
Otis Quick Scoring Mental Abilities, 403
Outcome fairness, 286
Output
  increasing, 172
  production of, 170
  quality of, 171

P

PAA (Physical Abilities Analysis), 547, 548f
Paired comparisons, method of, 187–188
Paper-and-pencil tests. *See also* Written (paper-and-pencil) methods
  in assessment centers, 528
  cognitive ability in, 404–407, 406f
  of general ability, 402–416
  importance of g in, 410–416, 412t, 414t
  introduction to, 402–404
  job-related aptitudes in, 407–410, 409t, 410t
PAQ (Position analysis questionnaire), 115, 117–118
Parallel forms reliability, 72, 75
Pearson, Karl, 12, 58
Pearson product-moment correlation coefficient (r), 12
  description of, 59–61, 60f
  of different sizes, 57f
  in general, 54–55
  limitations of, 58–59
  range of, 55–56
  wonder of, 56–58
Perceived external alternatives, in employee turnover, 240
Percentiles, 48
Perceptual ability, 412t
Perceptual approach, in job design, 128
Performance. *See also* Job performance
  aggregate, 162
  assessee, 531
  definition of, 157
  general *vs.* specific, 196–198
  human resources and, 144–145
  maximum, 202
  transitional *vs.* maintenance stages of, 198–202
  in work setting, 335
Performance adaptability, in staff appraisal, 213–216, 215b
Performance appraisal, 153, 209, 218–219, 254
Performance-based domain, 372
Performance-based methods, 373t, 376, 380t
Performance-based testing. *See also* Assessment centers; Situational judgment test (SJT)
  work samples and, 538–542
Performance behavior dimensions, 160–161
Performance criterion, reconceptualizing, 419–420
Performance dimensions, determination of, 180
Performance in different setting, 335
Performance tests. *See* Work sample tests
Persistence, 368f
Personal competence, in emotional intelligence, 457
Personal inventories, 449
Personality. *See also* Personality at work; Personality testing
  aggregate, 312
  artistic, 262
  behavior at work and, 435
  constructs, 134
  context in, 435
  conventional, 262
  enterprising, 262
  five factor model of, 436–439, 437t
  in hypothetical model of supervisory ratings, 81–87, 81f
  internal analysis of, 323, 324–325t
  investigative, 262
  job satisfaction and, 440–441
  organizational citizenship behavior and, 440

SUBJECT INDEX

realistic, 262
social, 262
staffing decisions and, 368f, 371t, 380t
team performance and, 444–446
Type A, 450
Personality at work
  conceptual issues in, 463–468
  person-environment fit and, 467–468
  person-situation interaction and, 464–465
  importance of, 439–447
  job satisfaction and, 440–441
  leadership and, 442–444
  organizational culture and, 446–447
  stress and, 441
  task performance and organizational citizenship behavior and, 439–440
  team performance and, 444–446
Personality-related position requirements form (PPRF), 134
Personality testing
  conditional reasoning in, 454–457, 456t
  fakability in, 461–463
  improving the validity of, 459–461
  managers attitudes toward, 483b
  personality inventories in, 448–450
  projective instruments in, 450–454
Personal variables, job satisfaction and, 244–245
Person-environment fit, 467–468
Personnel Management (Scott and Clothier), 15
Personnel Psychology, 415, 451
Person-situation debate, personality in, 466–467
Person-situation interaction, in personality at work, 464–465
Physical abilities. *See also* Physical ability testing
  justifying in selections, 547–549
  staffing decisions and, 368f, 371t, 380t
  types of, 546
Physical Abilities Analysis (PAA), 547, 548f
Physical ability testing, 542–550
  example of construct validation for, 549–550
  overview of, 542–543
  physical ability performance testing in, 545–550
  psychomotor abilities in, 543–545
Physical complaints, 176
Physical fidelity, predictors and, 377

Pictures, in projective tests, 450–452
Plato, 9–10
PMPQ (Professional and Managerial Position Questionnaire), 117–118
Policy capturing, 197–198, 276–277, 289
Political advocate, staffing specialist as, 588–589
Position analysis questionnaire (PAQ), 115, 117–118
Potential bias in staff appraisals, 206–207
Power and precision, website of, 336
PPRF (Personality-related position requirements form), 134
Practical intelligence, 407
Practicality, in criteria, 169
Predictions
  balancing subgroup differences with, 584b
  based on correlation, 61–66
  biased, 416–417, 420
  optimal, 64b
  using multiple predictors, 66–70
Predictive validity study, 303–304, 304f
Predictor batteries, 366, 385–386, 417–419
Predictor-criterion relationships, 67f, 68–69
Predictor measurement methods, 372, 376–377, 380
Predictor response model, 367–369, 368f
Predictor(s)
  of behavior at work, 435
  bias, 159–160
  definition of, 50, 366
  development of, 105f, 389–398
  cultural influences on use and, 393–394
  predictors in multilevel contexts and, 390–392, 391f
  technological advances and, 394, 397
  distal *vs.* proximal, 377, 378f, 379
  effective performance and, 595
  evaluating adequacy of, 379–385, 380t
  model of predictor-response process, 367–370, 368f
  multiple, 66–77
  multiple scores on, 51–54, 51f, 52f, 53f, 54f
  relationship with criteria, 311f
  structure and function of, 370–379
  team-level, 391f
  two scores on, 50, 50f
  use in staffing practice, 385–389
    combining predictors and, 386–387
    compound predictors and, 388–389

multiple hurdle systems and, 387–388
predictor batteries and, 385–386
in validation studies, 337t
validity of, 310–313, 311f
Predictors construct domains, 371–372, 371t, 380t
Preemployment inquires, lawful and unlawful examples of, 477t
Pressure causing pain in Cattell test, 14
Primary validity studies, staffing and, 612–613
Principles for the Validation and Use of Personnel Selection Procedures (Society for Industrial and Organizational Psychology), 579–580
Private employment agencies, 279t
Probability level, in criterion-related validity study, 308–309
Probability of success, in internal recruitment, 253
Problem and setting, in validation studies, 337t
Problem solving, in aptitude test, 410t
Procedural knowledge, 161
Process fairness, 286
Processing speed, in cognitive abilities, 406f
Process variables, 321
Production of output, in criteria, 170
Production subsystem in organizational functioning, 594t
Productivity
  job description and, 243–244
  workers and distribution of, 144f
Productivity-based configuration, 110
Professional and Managerial Position Questionnaire (PMPQ), 117–118
Professional persons, source of recruits for, 280
Professional societies, 279t, 280t
Professional standards, staffing practice and, 576–581
Profiles, in personality inventories, 449
Programs
  Affirmative action programs, 569–571
  certification, 426–429
  orientation, 423–424, 425
Projective personality, 450
Promotion. *See* Rate of advancement
Proximal organizational factors, 155f
Proximal *vs.* distal predictors, 377, 378f, 379
PSI (Psychological Services Inc.), 409, 410t
Psychological construct, 319
Psychological contract, 7, 249, 251–252

Psychological fidelity, predictors and, 377
Psychological investment, in employee turnover, 240
Psychological measures, 40–41
Psychological processes, rating errors and, 185
Psychological Services Inc. (PSI), 409, 410t
Psychological tests, 577
Psychological Tests and Personnel Decisions (Cronbach and Gleser), 353
Psychological withdrawal, 172
Psychology in organizations, 582
Psychometric approach in cognitive ability, 403–407
Psychometric evaluation, 602–603
Psychomotor abilities
  dexterity tests in, 544–545
  observed and true validity for all jobs for, 412t
  sensory tests in, 543–544
p value, 308–309

Q

Qualified applicants, 605–606
Quality measures, in criteria, 171
Quality of work life (QWL), 246
Questionnaires for assessing constructs, 330f, 331
Question propriety, in process fairness, 287
Questions
  in interviews, 496–498
  from structured interviews, 500–501b
QWL (Quality of work life), 246

R

Race, 39, 207, 504
Racial subgroup differences. *See also* Subgroup differences
  banding and, 424–426
  and cognitive ability, 416–426
  cognitive ability tests and, 417–424
Radio-TV advertising, 279t, 280t
Range of r, 55–56, 57f
Rank order, methods of, 187
Rate error training, 192
Rate of advancement, 177
Rate of movement, in Cattell test, 14
Rater accuracy training, 192
Rater training, 192–193
Rating process, in assessment centers, 534–536
Ratings. *See also* Absolute ratings
  assessors and, 534–536
  criticality task rating, 125t

SUBJECT INDEX 655

interviews, potential containments of, 505–507
job analysis, demographic attributes of, 134–137
of personnel specialist KSAOs, 126t
from structured interviews, 500–501b
true, 192
Rating scales
 final, 181
 graphic, 178–179, 179f, 189
 research comparing formats of, 186
Ratio scale of measurement, 38, 39
r (Correlation coefficient), 12, 55–56, 57f, 61
 limitation of, 58–59
Reaction time for sound, in Cattell test
Reading
 comprehension in aptitude test, 410t
 ease formula, 317–318, 318f
Realism of information issue, 271
Realistic job preview (RJP), 271–272
Realistic personality, 262
Realistic vocations, 369f, 370t, 469
Reasoning
 in aptitude test, 410t
 in tests of ability, 404–405
Recognition in work setting, 242
Recruit by occupation, 279t
Recruiters, in perceptions of jobs and organizations, 274b
Recruiting source, effectiveness of, 280t
Recruitment. *See also* External recruitment; Internal recruitment
 applicant reactions to, 284, 286–288
 campus, 279t, 280t, 281–282
 as competitive advantage, 292, 293t
 effective staffing and, 595–596
 interview, 493
 of jobs, organizations, and careers, 616–617
 of management persons, 279t, 280, 280t
 stages of, 270
 targeted, 291–292
 theories of, 289–290
 walk-ins, 280, 280t
 Web sites and effectiveness of, 285b
Redesigning jobs, 606
Reference, frame of, 460
Referrals, employee, 279t, 280, 280t
Regents of the University of California *vs.* Bakke (1978), 566t, 574
Regression
 analysis, 61–66, 62f, 88–89
 equation, 37f, 62f, 63, 78, 88

moderated, 78–81, 80f
multiple, 34, 57
Regulatory action model, 563f
Rehabilitation Act, 1973, 565
Relationship Management, 457
Relevance in criteria, 159, 159f, 165–166
Reliability. *See also* Reliability measurement
 consequence of, 76–77
 in criteria, 167–169
 definition of, 71–72, 212
 interrater, 75, 168, 499, 535–536
 of interviews, 499, 501
Reliability measurement, 35
 coefficient alpha reliability in, 73–75
 consequences of reliability in, 76–77
 interrater reliability in, 75
 parallel forms reliability in, 72
 parallel forms with time interval reliability in, 75
 split-half reliability in, 72–73
 structure and estimation of, 71–77
 test-retest reliability in, 72
Relocation of employees, 283
Reports of behavior
 checklists in, 190
 forced choice technique in, 190–192
Research linkage, 217–218
Research in external recruitment, 269–279
Response, 367, 368f, 373t
Response to test content, in content validity, 320–322
Responsiveness, 211
Restriction of range, 304, 305f, 306
Retail service provider, graphic rating scales for, 179f
Retaking tests, 424
Retention, search firms and, 282
"Retranslation" of behavior statements, 181
Retrieval ability, in cognitive abilities, 406f
Rewards
 job offers and, 605–606
 monetary, 275, 276
 system, 103
RIASEC, 474, 475
RJP (Realistic job preview), 271–272
Role play, in assessment centers, 524–525
Rorschach Ink Blot Test, 451
Rotating jobs, 606

S

Safety, 176
Sample in validation studies, 337t
Satisfaction. *See* Job satisfaction

Satisfactoriness, 3f, 4
Scale(s)
  of behavior statements, 181
  final rating scale, 181
  lie, 461
  of measurement, 39–40
  in personality inventories, 449
  social desirability scales, 461
Schedules, in personality inventories, 449
Scientist, staffing specialist as, 585–586, 586–587b
Scoring mixed standard scale triads, 184f
Screening interview, 493–494
SDs (Standard deviations), 45–49, 45f
SD_y estimates, utility analyses, 354–359
Search firms, 279t, 280, 280t, 282–283
Secretary, in perceptions of jobs and organizations, 274b
Security of job, in work setting, 242
Selection information, in process fairness, 286
Selection interviews, 494, 509b
Selection KSAO, 127
Selection of staff, 125, 126t, 258–259
Selection Process, 617–618
Selection system withdrawal, 290–291
Selection tests, 321–322
Selection utility models, development of, 347–354, 348f, 349f
Select21 project, 142
Self-Awareness, 457
Self-esteem, 450
Self-Management, 457
Self-presentation processes, 140
SEM (Structural equation modeling), 81–82f, 81–87, 87f
Sensation areas, in Cattell test, 14
Sensory tests, 543–544
Sentence completion test, 452
Service contexts, in staff appraisal, 210–213
Service orientation measures, 460–461
Service quality (SERVQUAL)
  performance scale, 211–213
Shrinkage, 176
SHRM (Society for Human Resource Management), 580, 582
Shyness, 450
"Signaling effect," 273
Signs of competitiveness, 37–38, 42
Simon, Theophile, 13
Simple to-p-down hiring, 425
Simulation. *See also* Simulation testing
  for assessing constructs, 330f, 331
  in assessment centers, 525, 526–527f
  phone call, 601

Simulation testing, 510–542. *See also* Assessment centers;
Situational judgment test (SJT)
  about, 510–511, 511f
  work samples in, 538–542
SIOP (Society for Industrial and Organizational Psychology), 129, 313, 516, 579
Site visits, 273
Situational interview
  in examination plan of, 598t
  questions of, 496–497, 497t
  vs. behavioral interviews, 499, 503
Situational judgment test (SJT)
  development of, 514–515t, 520b
  guidelines for writing, 516t
  overview of, 511–513, 516–519
Situationist perspective on behavior, 464
16 Personality Factor Test, 449
SJT. *See* Situational judgment test (SJT)
Skill dimension matrix, examination plan of, 598t
Skills
  clerical/technical, 598t
  communication, 420, 598t, 600
  interpersonal, 420–421
  language, 410t
Sliding band, 425
Small businesses
  internet recruiting and, 284
  qualified applicants and, 605–606
  staffing, 596–604
  validity options for, 310b
SMEs. *See* Subject-matter experts (SMEs)
Snack food delivery industry,
  competency analysis for, 130b
Sociability, 442–443
Social adaptability, 5
Social awareness, 457
Social competence, in emotional intelligence, 457
"Social desirability scales," 461
Social personality, 262
Social vocations, 369f, 370t, 469
Society for Human Resource Management (SHRM), 580, 582
Society for Industrial and Organizational Psychology (SIOP), 129, 313, 560, 579
Software for credibility intervals, 335–336
Sources
  job analysis and, 140
  of variance, 368
Space, in tests of ability, 404
Spearman Brown correction formula, 73–75

SUBJECT INDEX 657

Special publications adverting, 279t, 280, 280t
Specific aptitude tests, 409, 410t
Specific factors, 404
Specific vs. general performance, 196–198
Spelling test, in oral directions test, 598
Split-half reliability, 72–73
SPR (Supervisory performance rating), 81f, 82–87
Stability, 167, 204
Stable periodic lateness, 174
Staff appraisal
 current challenges for, 205–225
  aggregate effectiveness in, 217–218
  criterion-predictor relationship in, 223–224, 223f
  cross-cultural issues in, 216–217
  employee development in, 218–220
  interrater unreliability in, 208–209
  legal context in, 205–206
  performance adaptability in, 213–216, 215b
  potential bias in, 206–207
  satisfaction, well-being, and performance in, 220–223
  service contexts in, 210–213
  team contexts in, 209–210
 definition of, 151
 organizations and, 152–156
Staffing. *See also* Staff appraisal; Staffing practice; Staffing practitioner; Staffing specialist
 competitiveness in, 37–38
 definition of, 1
 developing and measuring criteria, 195–205
 effective, 595–596
 future innovations in, 610–614
 influences on appraisal of, 153–154, 155f, 156
 job analysis and, 104–106, 105f
 legal system impact on, 596
 modern challenges affecting, 4t
 organizational effectiveness and, 23–25
 recommendations for, 596–610
Staffing experts, consensus and, 589–590
Staffing organizations
 definition of, 2
 effective staffing in, 595–596
 subsystems of organizational functioning, 594t
Staffing practice
 current issues in, 590
 ethical considerations and, 561f, 581–582
 legal issues of, 561f, 562–576, 563f
 professional standards and, 576–581
 tripartite view of, 560–562, 561f
 use of predictors in, 385–389
  combining predictors and, 386–387
  compound predictors and, 388–389
  multiple hurdle systems and, 387–388
  predictor batteries and, 385–386
Staffing practitioner
 staffing specialist as, 583, 584b, 585
 in the tripartite model, 561, 561f, 577
Staffing procedures, 105f
Staffing specialist
 as expert witness, 587–588
 organizational goals and, 6
 as political advocate, 588–589
 as practitioner, 583, 584b, 585
 as scientist, 585–586, 586–587b
 in the tripartite model, 561–562
 web-based testing and, 394, 397
Staffing system design, 104, 105f, 106
Standard deviations (SDs), 45–49, 45f
Standard error, 16, 95t
Standard scores, 45f, 46–49
Standards development, 101–102
Standards for Educational and Psychological Tests, 300, 301, 578
Stanford-Binet procedures, 15, 403
Start-up time, in internal recruitment, 252–253
State Employment Services, 408
Statistical symbols, 36–37b
Status image, 276
Stereotype threat, test performance and, 422–423
Stimuli presentation and required responses, 420–421
Strain, 441
Strategic job modeling, 142
Strategy, human resources and, 111–112
Stratum II in Carroll's hierarchy, 405, 406f
Stress
 in formal posting, 258
 job satisfaction and, 441
Stressors, 441
Strong Vocational Interest Blank for Men, 472
Structural equation modeling (SEM), 81–82f, 81–87, 87f
Structured interviews
 advantages of, 503
 components of, 494–496
 sample questions and ratings from, 500–501b

SUBJECT INDEX

Subgroup differences
 in assessment centers, 532–533
 balancing with prediction, 584b
 and cognitive ability, 416–426, 613
 interviews and, 503–504
 minimizing, 418–419b
 predictors evaluation and, 381–383, 383f
Subgroups, 382, 613
Subject-matter experts (SMEs)
 cognitive models and, 185–186
 creation of scoring key, 516–7
 critical incident technique and, 118
 demographic attributes of, 134–137
 job analysis and, 104, 105f, 607–609
 judgments of content validity, 315–316
 situational judgment test and, 513
 staffing procedures and, 136f
 task rating, 124, 125t
Success
 in internal recruitment, 253
 training, 171–172
Supervisors, KSAO and, 140
Supervisory performance rating (SPR), 81f, 82–87
Supervisory style, 276
Supportive subsystem in organizational functioning, 594t
Suppressor effect, 68
Surveys, in personality inventories, 449
Synthetic validity, using, 310b
Systematic evaluation and decision making, 618–619

T

Talent pools, 358–359
Tangibles, 211
Tardiness of workers, 172–175, 173–174
Targeted recruiting, 291–292
Task Behavior (TB), 81f
Task Force on Assessment Center Guidelines, 522
Task performance, 162–163, 420, 439–440
Task(s). *See also* Task statements
 characteristics of, 5
 definition of, 124
 difficulty, 125t
 in job analysis, 124
 for staff development, 125, 126t
 for staff evaluation, 125, 126t
 for staff selection, 125, 126t
 team performance and attributes of, 445
Task statements
 in functional job analysis, 114, 124

and job analysts, 114–115, 116f
 tests and, 597
TAT (Thematic Apperception Test), 436, 443, 451–452
Taxonomy
 of constructs, 502
 of lateness behavior, 173–174
Taylor and Russell approach, 348–349f, 348–350
TB (Task Behavior), 81f
Team contexts, in staff appraisal, 209–210
Team-level predictors, 391f
Team performance, personality and, 444–446
Teams, job performance and, 444–446
Technical adaptability, 5
Technical persons, source of recruits for, 280
Technology
 assessment centers and, 536
 as a challenge, 4t
 and interviews, 508
 measuring psychomotor abilities and, 613–614
 and predictors, 394, 397
Telephone administered interview, 501
Telephone call, in oral directions test, 598
Terman, Lewis, 403
Test battery, use of, 310b
Test-criterion relationships, 303–319
Test information and measurement error, 93f
"Testing the test," 101
Test-retest reliability, 72, 167
Tests
 aptitude, 409, 410t
 Bennett Hand-Tool Dexterity, 545
 Binet-Simon scales, 13–15, 403
 California Psychological Inventory (CPI), 449
 Cattell, 13–14
 definition of, 27
 Employee Aptitude Survey (EAS), 408, 409t
 Flanagan Aptitude Classification Tests (FACT), 408, 409t
 General Aptitude Test Battery (GATB), 408, 409t
 Gordon Personal Profile (GPP), 449
 guidelines, 578
 Guilford-Zimmerman Temperament Survey (G-ZTS), 449
 Holtzman Ink Blot Test, 451
 Jackson Personality Inventory (JPI), 449
 knowledge, skill, ability and other (KSAO) and, 597

SUBJECT INDEX 659

Miner Sentence Completion Scale (MSCS), 436
motivational factors and performance, 422–423
oral directions/typing test, 598–599, 598t
perceptions, 287–288
psychological, 577
retaking, 424
Rorschach Ink Blot Test, 451
sensory, 543–544
of situational judgment, 511–519
16 Personality Factor Test, 449
Strong Vocational Interest Blank for Men, 472
Thematic Apperception Test (TAT), 436, 443, 451–452
Tomkins-Horn Picture Arrangement Test, 452
true score in, 71
utility, 344–359
validity of, 313–339
Web-based, 394, 397
Wechsler Adult Intelligence Scale (WAIS), 403
Test scores, 584b
of African Americans, 340, 416
of Asians, 416
of Hispanics, 340
racial differences and, 340
Test-taking motivation, 287–288, 368f, 422–424
Theft, 176
Thematic Apperception Test (TAT), 436, 443, 451–452
Theoretical meaningfulness of tests, validity of, 319–339
Theories of recruiting, 289–290
Theory of branding, 290
Theory of cognitive ability structure, 405, 406f
Theory of intelligence, 407
Theory of Motivation, 20–21, 443
Theory X philosophy, 21
Theory Y philosophy, 21
360-degree feedback performance appraisals, 219–220
Time delay, in applicant interest, 273
Time for naming colors, in Cattell test, 14
Time in formal posting, 258
Time spent, task and, 124, 125t
Tomkins-Horn Picture Arrangement Test, 452
Total career experiences model, 255, 256t
Trade schools, 279t, 280t
Traditional job analysis, 141
Trainability, definition of, 171

Training
process, 110
rate error, 192
rater, 192–193
Training success and work samples, 171–172
Transitional vs. maintenance stages of performance, 198–202
Tripartite view of staffing practice, 560–562, 561f
"True" ratings, 192
True score, in test, 71
Turmoil in formal posting, 258
Turnover of employees, 172–175, 242–247
Two-way communication, in process fairness, 287
Type A behavior, 436
Type A personality, 450
Typical performance, 202
Typing
in aptitude test, 410t
in oral directions test, 599

U

UGESP (*Uniform Guidelines on Employee Selection Procedures*), 27, 429, 567, 602–603
"Ultimate criterion", 158–159, 159f, 223–224, 223f
Unavoidable lateness, 174
"Unfolding" model of turnover, 175
Uniform Guidelines on Employee Selection Procedures (UGESP), 27, 429, 567, 602–603
Unions, as recruitment source, 279t
United States court cases, 566t
United States vs. Garland, Texas (2004), 566t, 568, 583, 588
United States vs. Georgia Power (1973), 566t, 574
Universities recruitment, 279t, 280t
Unlawful inquires, in Michigan Department of Civil Rights, 477t
Unreliability
interrater, 208–209
using corrections for, 459–460
U.S. Department of Labor, 113, 119, 567
U.S. Employment Service (USES), 113, 279t, 280t, 281
U.S. federal acts and orders, 565t
U.S. workforce, staffing challenges and, 4
User acceptability, 384
USES (U.S. Employment Service), 113, 279t, 280t, 281

660 SUBJECT INDEX

Utility. *See also* Utility analyses
  of assessment centers, 533
  definition of, 384
  functional, 11
  test, 344–359
Utility analyses
  importance of, 345–346
  SD_y estimates key to, 354–359

V

Validating a test, 13
Validation
  convergent, 329–332
  discriminant, 329–332
  information in studies about, 337t
  test-criterion relationships in, 303–313
Validity. *See also* Construct validity; Content validity; Criterion-related validity; Validity generalization
  in assessment centers, 529–532
  based on test content, 313–319
  convergent, 530
  in general, 35, 77–78
  generalization, 332–338, 334t, 337t
  of internal test structure, 322–323, 324–325t
  of interviews, 499, 501, 509–510
  options for small businesses, 310b
  and other variables, 326–327, 328t
  personality testing and, 459–461
  response processes and, 320–322
  testing consequences and, 339–344
  types of, 300–302
Validity coefficient, 307, 351, 352f
Validity generalization
  bare bones study and, 333–338, 334t
  introduction to, 332–333
  options for small businesses, 310b
  relying on results of, 338
Values
  organizational, 276
  staffing decisions and, 371t, 380t
Variability measures, 44–46, 45f
Variables, manifest and latent, 42–43
Verbal ability measure, 419
Verbal complaints, 176
Verbal comprehension, in tests of ability, 404
Visual display, Web sites and, 285b
Visual job families, 544
Visual methods, 373t, 375
Visual perception, 406f
Visual speed and accuracy, in aptitude test, 410t
Vocabulary, in aptitude test, 410t

Vocational choices, Holland's theory of, 369f, 469–471
Vocational development and choice, 265
Vocational interests, 468, 474
Vocational maturity, 267
"Vocational personalities", 468
Vocational Preference Inventory, Holland's (VPI), 474
Vocations, 369f, 370t, 469
Voluntary turnover, 175
VPI (Holland's Vocational Preference Inventory), 474
Vroom, V. H., 20–21

W

WAB (Weight application blank), 478
WAIS (Wechsler Adult Intelligence Scale), 403
Walk-ins recruiting, 280, 280t
"Walk-through performance test", 542
Wal-Mart, 585
Ward's Cove Packing Company, Inc. vs. Antonio (1989), 566t, 574
Watson vs. Fort Worth Bank & Trust (1988), 566t
Web-based methods, 373–375, 373t
Web-based testing, 394, 397
Web sites
  of affirmative action programs, 570
  of American Psychological Association, 578
  of American Psychological Association code of ethics, 582
  of Americans with Disabilities Act, 573
  of Department of Labor, 567
  of Equal opportunity Employment Commission, 565
  and information, 285b
  organizational attractiveness and, 284
  in perceptions of jobs and organizations, 274b
  recruitment effectiveness and, 285b
  of Society for Human Resource Management, 580
  of the Society for Industrial and Organizational Psychology, 579
Wechsler Adult Intelligence Scale (WAIS), 403
Wechsler-Bellevue, 403
Weight application blank (WAB), 478
Weighting, 420
Weights, calculation of, 67–69
Well-being, in staff appraisal, 220–223
Wherry, R. J., 190
Withdrawal

SUBJECT INDEX

behavioral, 172
in counter-productive work behaviors (CWBs), 172–175
in external recruitment, 290–291
psychological, 172
Wonder of correlation coefficient r, 56–58
Word fluency, in tests of ability, 404
Work. *See also* Work and work environment; Workers
behaviors, 176–177
importance of personality at, 439–447
predictors of behaviors at, 435
Work and work environment. *See also* Workers
Kurt Lewin and, 19
McGregor and, 21–22
overview of, 16–19
Vroom and, 20–21
Workers
absenteeism of, 166, 172–175
characteristics of, 120
compensation systems and, 145
contract, 6–7
distributions of ability and productivity of, 144f
functions of, 115, 117t
lateness of, 172–175

McGregor on motivation of, 21–22
organizational culture and, 145
organizations and, 6–7
requirements of, 120
safety and accidents of, 176
tardiness of, 172–175
turnover of, 172–175
Worker satisfaction. *See* Job satisfaction
Working conditions, in work setting, 242
Working framework, 2, 3f
Working hours, flexibility in, 607
Work samples, 538–542
Work sample tests, 171. *See also* Simulation testing
Work setting
performance in, 335
working conditions in, 242
Writing, in situational judgment test, 516t
Written directions, in aptitude test, 410t
Written (paper-and-pencil) methods, 372–373, 373t
Wundt, Wilhelm Laboratory, 11

Z

Zeitgeist, 18

Made in the USA
Monee, IL
24 August 2023